Chemical Synthesis Using Highly Reactive Metals

Chemical Synthesis Using Highly Reactive Metals

Reuben D. Rieke

Published by John Wiley & Sons, Inc., Hoboken, New Jersey
Published simultaneously in Canada

Library of Congress Cataloging-in-Publication Data:

Names: Rieke, Reuben D., 1939– author.
Title: Chemical synthesis using highly reactive metals / Reuben D. Rieke.
Description: Hoboken, New Jersey : John Wiley & Sons, Inc., [2017] |
Includes bibliographical references and index.
Identifiers: LCCN 2016034805 | ISBN 9781118929117 (cloth) |
ISBN 9781118929131 (epub) | ISBN 9781118929148 (Adobe PDF)
Subjects: LCSH: Organometallic compounds–Synthesis. | Reactivity (Chemistry)
Classification: LCC QD411.7.S94 R54 2017 | DDC 547/.050453–dc23
LC record available at https://lccn.loc.gov/2016034805

Cover image courtesy: JacobH/Gettyimages

Set in 10/12pt Warnock by SPi Global, Pondicherry, India

Printed in the United States of America

10 9 8 7 6 5 4 3 2 1

Contents

Preface

It is obvious that such a large body of work as the summary of our active metal research for over 50 years requires the acknowledgement of many people. There is also no doubt that there is one key person without whose lifelong help this book would not be possible. That person is my wife Loretta. From the day we met at the entrance examinations for the chemistry graduate program at the University of Wisconsin–Madison in September 1961 until today, she has been of incredible help. From our early days at the University of North Carolina at Chapel Hill, where she carried out research with my research group, to 1991 when the two of us founded Rieke Metals, Inc., in Lincoln, Nebraska, where she served as Vice President and Business Manager, she has been a cornerstone of my travels through life. Another major force in these efforts is our daughter Elizabeth, who started working part time in Rieke Metals, Inc., and rose to the position of CEO before we sold the company in July 2014. Finally, our son Dennis was a constant supporter of our efforts and an excellent sounding board for our ideas.

Of course, this work would not be possible if I did not have an excellent group of graduate students, postdoctoral students, and undergraduate students. From the initial two students who worked on the active metals, Dr. Phillip Hudnall and Dr. Steven Bales, to my final student, Dr. S. H. Kim, I had an outstanding group of people to work with. This book only covers my research on active metals so my students that worked on radical anion chemistry, electrochemistry, electron paramagnetic chemistry, and quantum mechanical calculations are not mentioned in the book. The active metal students are all referenced in this book in the metal sections that they were involved with. Of special note is my postdoctoral student from Spain, Professor Alberto Guijarro of the University of Alicante, Alicante, Spain, who carried out the beautiful mechanistic studies on the oxidative addition of Rieke zinc with organic halides as well as several synthetic studies. The three years he spent with us were particularly productive. The history of active metals is discussed in the early part of the book. However, special thanks must go to Professor Saul Winstein of UCLA who allowed me to follow my idea of studying through-space interactions by preparing radical anions and determining their EPR spectra. My other mentors, my undergraduate research director at the University of Minnesota–Minneapolis, Professor Wayland E. Noland, and my PhD mentor, Professor Howard E. Zimmerman of the University of Wisconsin–Madison, were also of major help in my early training.

Of final note is the assistance of our cat, Buddy. He always felt that it was his duty to come and sit in the middle of my papers as I was writing this book. When he was banished to the side of the papers, he insisted on placing his head and two front paws on my arm.

Chemical research is a long, hard road but the rewards of discovery are hard to describe. As the old saying goes, the train ride has been long and many times bumpy, but we have not reached the station yet.

1

Genesis of Highly Reactive Metals

Modern life without metals is inconceivable. We find them at every turn in our existence: transportation, buildings and homes, transporting our water, carrying our electricity, modern electronics, cooking utensils, and drinking vessels. Perhaps this is not to be unexpected as 91 of the 118 elements in the periodic table are metals. Accordingly, we can surely expect to find them in all aspects of our lives. The early chemistry of metals or processing of metals is one of the oldest sciences of mankind. Its history can be traced back to 6000 BC. Gold was probably the first metal used by man as it can be found as a relatively pure metal in nature. It is bright and attractive and is easily formed into a variety of objects but has little strength and accordingly was used mainly for jewelry, coins, and adornment of statues and palaces. Copper articles can also be traced to ~6000 BC. The world's oldest crown made of copper was discovered in a remote cave near the Dead Sea in 1961 and dates to around 6000 BC. The smelting of copper ores is more difficult and requires more sophisticated techniques and probably involved a clay firing furnace which could reach temperatures of 1100–1200°C. Silver (~4000 BC), lead (~3500 BC), tin (~1750 BC), smelted iron (~1500 BC), and mercury (~750 BC) constituted the metals known to man in the ancient world. It would not be until the thirteenth century that arsenic would be discovered. The 1700s, 1800s, and 1900s would see the rapid discovery of over 60 new metals. The bulk of these metals were prepared by reducing the corresponding metal salt with some form of carbon or, in a few cases, with hydrogen. A small number of difficult to free metals were eventually prepared by electrochemical methods such as the metals sodium, potassium, and aluminum. Eventually the concept of a metal alloy was understood. It became readily apparent that the presence of one or more different metals dispersed throughout a metal could dramatically change the chemical and physical properties of any metal. The extensive and broad field of metal alloys will not be discussed in this text. The main point to be made is that the presence of a foreign material, whether it be another metal or a nonmetal, can have a significant effect on a metal's chemical and physical properties. Pure metals prepared by different methods have essentially all the same chemical and

Chemical Synthesis Using Highly Reactive Metals, First Edition. Reuben D. Rieke.

physical properties. The one caveat in this statement is particle size or surface area. Whitesides clearly demonstrated the effect of surface area on the rate of Grignard formation at a magnesium surface. Taking this to the extreme, Skell and Klabunde have demonstrated the high chemical reactivity of free metal atoms produced by metal vaporization. These two topics will be discussed in greater depth later in the text. Thus it is clear that preparation of metals which leads to the presence of foreign atoms throughout the metal lattice can have a profound effect on the metal's chemical and physical properties. This will be discussed in greater detail later in the text.

The genesis of highly reactive metals from our laboratories can be traced back to my time spent in a small two-room schoolhouse in a small town of 180 people in southern Minnesota (1947–1949) and then to graduate school at the University of Wisconsin–Madison where I was working on my PhD degree under the direction of Professor Howard E. Zimmerman. My research proposal, which was part of the degree requirements, was the synthesis of the naphthalene-like molecule shown in Figure 1.1. The ultimate goal of the project was to determine if there was through-space interaction between the two 1,3-butadiene units via the bridging ethylene unit (4N + 2 electrons). To verify the through-space interaction, I proposed preparing the radical anion and measuring the electron paramagnetic resonance (EPR) spectrum. EPR became an available experimental technique, thanks to the explosion of solid-state electronics in the 1960s. Simulating the spectrum in conjunction with quantum mechanical calculations should provide a reasonable estimate of the influence of through-space interaction. My postdoctoral mentor, Professor Saul Winstein, at UCLA allowed me to pursue this general idea and we went on to produce the monohomocyclooctatetraene radical anion. The experience gained in this project working with solvated electrons in THF allowed me to write my first proposal as an assistant professor of chemistry at the University of North Carolina at Chapel Hill. The project was the reduction of 1,2-dibromobenzocyclobutene with solvated electrons to generate the radical anion of benzocyclobutadiene as shown in Figure 1.2. The reduction was to be carried out in the mixing chamber of a flow mixing reactor in the sensing region of an EPR spectrometer. However, even at −78°C, the only spectrum we could see was the radical anion of benzocyclobutene. It became clear that the radical anion (II) and/or the dianion was so basic that even at −78°C in extremely dry THF, the anions were protonated to yield benzocyclobutene which was then reduced to the radical anion. Quenching with D_2O verified the presence of **II** and its dianion. In order to trap or stabilize the dianion, we attempted to carry out this chemistry in the presence of $MgCl_2$ and generate the di-Grignard. However, we mistakenly mixed the solvated electrons (we were using potassium naphthalenide) with $MgCl_2$, generating a black slurry of finely divided black

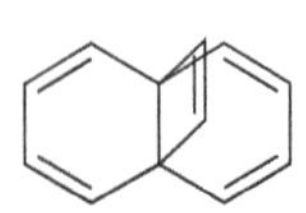

Figure 1.1 Graduate research proposal.

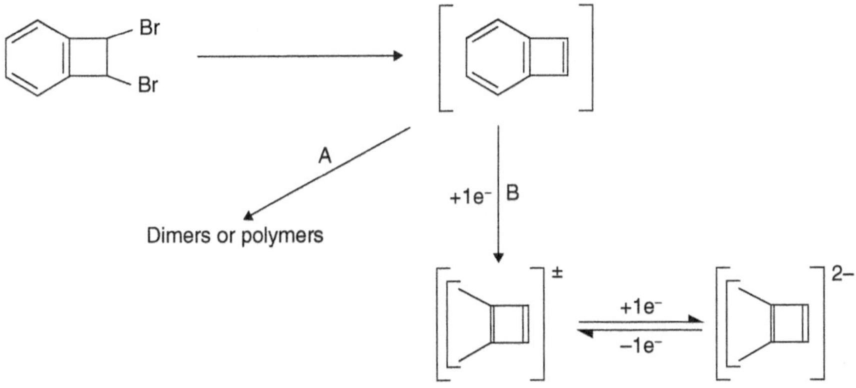

Br
Br
A
+1e⁻ B
Dimers or polymers
±
+1e⁻
−1e⁻
2−

Figure 1.2 First research proposal.

metal. Upon reflection, it became clear that we had generated finely divided magnesium. We quickly determined that this magnesium was extremely reactive with aryl halides and generated the corresponding Grignard reagent. Thus, the field of generating highly reactive metals by reduction of the metal salts in ethereal or hydrocarbon solvents was born.

2

General Methods of Preparation and Properties

2.1 General Methods for Preparation of Highly Reactive Metals

In 1972 we reported a general approach for preparing highly reactive metal powders by reducing metal salts in ethereal or hydrocarbon solvents using alkali metals as reducing agents [1–5]. Several basic approaches are possible, and each has its own particular advantages. For some metals, all approaches lead to metal powders of identical reactivity. However, for other metals one method can lead to far superior reactivity. High reactivity, for the most part, refers to oxidative addition reactions. Since our initial report, several other reduction methods have been reported including metal-graphite compounds, a magnesium-anthracene complex, and dissolved alkalides [6].

Although our initial entry into this area of study involved the reduction of $MgCl_2$ with potassium biphenylide, our early work concentrated on reductions without the use of electron carriers. In this approach, reductions are conveniently carried out with an alkali metal and a solvent whose boiling point exceeds the melting point of the alkali metal. The metal salt to be reduced must also be partially soluble in the solvent, and the reductions are carried out under an argon atmosphere. Equation 2.1 shows the reduction of metal salts using potassium as the reducing agent:

$$MX_n + nK \rightarrow M^* + nKX \quad (2.1)$$

The reductions are exothermic and are generally completed within a few hours. In addition to the metal powder, one or more moles of alkali salt are generated. Convenient systems of reducing agents and solvents include potassium and THF, sodium and 1,2-dimethoxyethane (DME), and sodium or potassium with benzene or toluene. For many metal salts, solubility considerations restrict reductions to ethereal solvents. Also, for some metal salts, reductive cleavage of the ethereal solvents requires reductions in hydrocarbon solvents such as benzene or toluene. This is the case for Al, In, and Cr. When reductions

Chemical Synthesis Using Highly Reactive Metals, First Edition. Reuben D. Rieke.

are carried out in hydrocarbon solvents, solubility of the metal salts may become a serious problem. In the case of Cr [7], this was solved by using $CrCl_3{\cdot}3$ THF.

A second general approach is to use an alkali metal in conjunction with an electron carrier such as naphthalene. The electron carrier is normally used in less than stoichiometric proportions, generally 5–10% by mole based on the metal salt being reduced. This procedure allows reductions to be carried out at ambient temperature or at least at lower temperatures compared with the previous approach, which requires refluxing. A convenient reducing metal is lithium. Not only is the procedure much safer when lithium is used rather than sodium or potassium, but also in many cases the reactivity of the metal powders is greater.

A third approach is to use a stoichiometric amount of preformed lithium naphthalenide. This approach allows for very rapid generation of the metal powders in that the reductions are diffusion controlled. Very low to ambient temperatures can be use for the reduction. In some cases the reductions are slower at low temperatures because of the low solubility of the metal salts. This approach frequently generates the most active metals, as the relatively short reduction times at low temperatures restrict the sintering (or growth) of the metal particles. This approach has been particularly important for preparing active copper. Fujita et al. have shown that lithium naphthalenide in toluene can be prepared by sonicating lithium, naphthalene, and *N,N,N′,N′*-tetramethylethylenediamine (TMEDA) in toluene [8]. This allows reductions of metal salts in hydrocarbon solvents. This proved to be especially beneficial with cadmium [9]. An extension of this approach is to use the solid dilithium salt of the dianion of naphthalene. Use of this reducing agent in a hydrocarbon solvent is essential in the preparation of highly reactive uranium [10].

For many of the metals generated by one of the three general methods in the preceding text, the finely divided black metals will settle after standing for a few hours, leaving a clear, and in most cases colorless, solution. This allows the solvent to be removed via a cannula. Thus the metal powder can be washed to remove the electron carrier as well as the alkali salt, especially if it is a lithium salt. Moreover, a different solvent may be added at this point, providing versatility in solvent choice for subsequent reactions.

Finally, a fourth approach using lithium and an electron carrier such as naphthalene along with $Zn(CN)_2$ yields the most reactive zinc metal of all four approaches [11].

The wide range of reducing agents under a variety of conditions can result in dramatic differences in the reactivity of the metal. For some metals, essentially the same reactivity is found no matter what reducing agent or reduction conditions are used. In addition to the reducing conditions, the anion of the metal salt can have a profound effect on the resulting reactivity. These effects are discussed separately for each metal. However, for the majority of metals, lithium is by far the preferred reducing agent. First, it is much safer to carry out reductions with lithium. Second, for many metals (magnesium, zinc, nickel, etc.), the resulting metal powders are much more reactive if they have been generated by lithium reduction.

An important aspect of the highly reactive metal powders is their convenient preparation. The apparatus required is very inexpensive and simple. The reductions are usually carried out in a two-necked flask equipped with a condenser (if necessary), septum, heating mantle (if necessary), magnetic stirrer, and argon atmosphere. A critical aspect of the procedure is that *anhydrous* metal salts must be used. Alternatively, anhydrous salts can sometimes be easily prepared as, for example, $MgBr_2$ from Mg turnings and 1,2-dibromoethane. In some cases, anhydrous salts can be prepared by drying the hydrated salts at high temperatures in vacuum. This approach must be used with caution as many hydrated salts are very difficult to dry completely by this method or lead to mixtures of metal oxides and hydroxides. This is the most common cause when metal powders of low reactivity are obtained. The introduction of the metal salt and reducing agent into the reaction vessel is best done in a dry box or glove bag; however, very nonhygroscopic salts can be weighed out in the air and then introduced into the reaction vessel. Solvents, freshly distilled from suitable drying agents under argon, are then added to the flask with a syringe. While it varies from metal to metal, the reactivity will diminish with time, and the metals are best reacted within a few days of preparation.

We have never had a fire or explosion caused by the activated metals; however, extreme caution should be exercised when working with these materials. Until one becomes familiar with the characteristics of the metal powder involved, careful consideration should be taken at every step. With the exception of some forms of magnesium, no metal powder we have generated will spontaneously ignite if removed from the reaction vessel while wet with solvent. They do, however, react rapidly with oxygen and with moisture in the air. Accordingly, they should be handled under an argon atmosphere. If the metal powders are dried before being exposed to the air, many will begin to smoke and/or ignite, especially magnesium. Perhaps the most dangerous step in the preparation of the active metals is the handling of sodium or potassium. This can be avoided for most metals by using lithium as the reducing agent. In rare cases, heat generated during the reduction process can cause the solvent to reflux excessively. For example, reductions of $ZnCl_2$ or $FeCl_3$ in THF with potassium are quite exothermic. This is generally only observed when the metal salts are very soluble and the molten alkali metal approach (method one) is used. Sodium–potassium alloy is very reactive and difficult to use as a reducing agent; it is used only as a last resort in special cases.

2.2 Physical Characteristics of Highly Reactive Metal Powders

The reduction generates a finely divided black powder. Particle size analyses indicate a range of sizes varying from 1 to 2 μm to submicron dimensions depending on the metal and, more importantly, on the method of preparation. In cases such as nickel and copper, black colloidal suspensions are obtained that do not settle and cannot be filtered. In some cases even centrifugation is not successful. It should be pointed out that the particle size analysis and surface area studies have been done on samples that have been collected, dried, and sent off for analysis and are thus likely to have experienced considerable sintering. Scanning electron microscopy (SEM) photographs reveal a range from spongelike material to polycrystalline material (Figures 2.1 and 2.2). Results from X-ray powder diffraction studies range from those for metals such as Al and In, which show diffraction lines for both the metal and the alkali salt, to those for Mg and Co, which only show lines for the alkali salt. This result suggests that the metal in this latter case is either amorphous or has a particle size <0.1 μm. In the case of Co, a sample heated to 300°C under argon and then reexamined showed diffraction lines due to Co, suggesting that the small crystallites had sintered upon heating [12].

ESCA (XPS) studies have been carried out on several metals, and in all cases the metal has been shown to be in the zerovalent state. Bulk analysis also clearly shows that the metal powders are complex materials containing in many cases

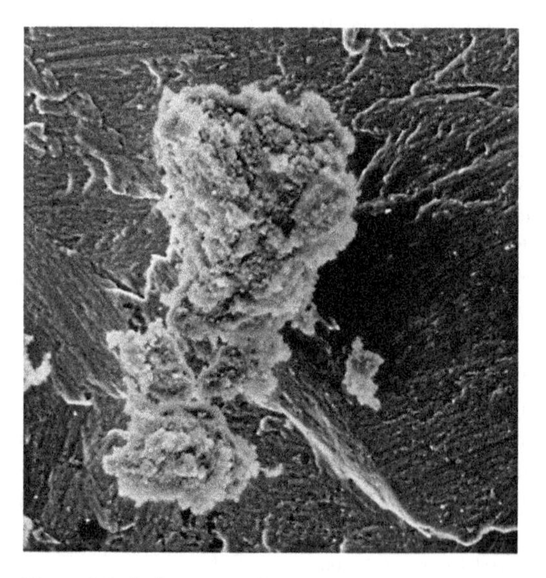

Figure 2.1 Active magnesium.

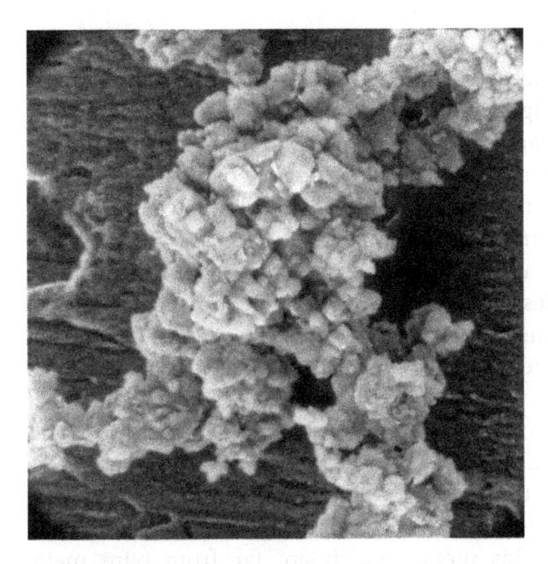

Figure 2.2 Active indium.

significant quantities of carbon, hydrogen, oxygen, halogens, and alkali metal. A BET [13] surface area measurement was carried out on the activated Ni powder showing it to have a specific surface area of 32.7 m^2/g. Thus, it is clear that the highly reactive metals have very high surface areas which, when initially prepared, are probably relatively free of oxide coatings.

2.3 Origin of the Metals' High Reactivity

There are several characteristics of the metal powders prepared by these methods which clearly explain their high reactivity. They all exhibit very high surface areas. Particle sizes of a few microns or in some cases <0.1 μm point to very high surface areas. The BET studies [13] on Ni powder indicated surface areas of over 30 m^2/g. Moreover, the lack of diffraction lines for several metals suggests particle sizes of <0.1 μm. Also the possibility of some metals being amorphous would increase their internal energy and lead to higher reactivity compared to the corresponding highly crystalline counterpart. In addition, the metals are produced under nonequilibrium conditions and exhibit many dislocations and imperfections. This would also be expected to lead to increased chemical reactivity. The metals are also prepared under a pure argon atmosphere which would result in a relatively oxide-free surface being produced. Bulk analysis of the metals is quite varied depending on the metal. However, in all cases, there is a significant amount of other elements generally including carbon, hydrogen, halogens, and alkali metal ions from the alkali metal reducing agent. As will be pointed out in detail later, finely divided metal powders prepared by methods which do not introduce these materials into the metal lattice are all significantly less reactive than Rieke metals. For example, metal powders prepared by metal vaporization methods are far less reactive in oxidative addition reactions compared to the corresponding Rieke metals even though they are of comparable or even smaller particle size [14]. There is also one extremely important difference between the Rieke metals and finely divided metals prepared by other methods, and that is the presence of alkali metal salts. Whitesides' [15] work on magnesium and our studies [16] on zinc clearly show that the rate-determining step in oxidative addition reactions is the electron transfer from the metal surface to the organic halide. As in an electrochemical reduction reaction, the alkali salt can act as an electrolyte and facilitate this electron transfer. In most of the reductions presented in this text, the alkali salt is LiCl or LiBr. We will see later in the text that these alkali salts can also increase the reactivity of the resulting organometallic reagents RMX toward many electrophiles. In summary, the Rieke method of producing metal powders yields metals which are far from pure metal powders. The presence of these foreign materials along with the features mentioned yields metal powders which undergo many new and novel reactions which cannot be achieved by standard metals or their chemically activated counterparts.

References

1 Rieke, R.D.; Hudnall, P.M. *J. Am. Chem. Soc.* 1972, **94**, 7178.
2 Rieke, R.D.; Hudnall, P.M.; Uhm, S. *J. Chem. Soc. Chem. Commun.* 1973, 269.
3 Rieke, R.D.; Bales, S.E. *J. Chem. Soc. Chem. Commun.* 1973, 739.
4 Rieke, R.D.; Bales, S.E. *J. Am. Chem. Soc.* 1974, **96**, 1775.
5 Rieke, R.D.; Chao, L. *Synth. React. Inorg. Met.-Org. Chem.* 1974, **4**, 101.
6 (a) Csuk, R.; Glanzer, B.L.; Furstner, A. *Adv. Organomet. Chem.* 1988, **28**, 85. (b) Savoia, D., Trombini, C., Uamni-Ronchi, A. *Pure Appl. Chem.* 1995, **57**, 1887. (c) Bogdanovic, B. *Acc. Chem. Res.* 1988, **21**, 261. (d) Marceau, P., Gautreau, L., Beguin, F. *J. Organomet. Chem.* 1991, **403**, 21. (e) Tsai, K.L.; Dye, J.L. *Am. Chem. Soc.* 1991, **113**, 1650.
7 Rieke, R.D.; Ofele, K.; Fischer, E.O. *J. Organomet. Chem.* 1974, **76**, C19.
8 Fujita, T.; Watanaba, S.; Suga, K.; Sugahara, K.; Tsuchimoto, K. *Chem. Inad. (London).* 1983, **4**, 167.
9 Burkhardt, E.; Rieke, R.D. *J. Org. Chem.* 1985, **50**, 416.
10 (a) Kahn, B.E.; Rieke, R.D. *Organometallics* 1988, **7**, 463. (b) Kahn, B.E.; Rieke, R.D. *J. Organomet. Chem.* 1988, **346**, C45.
11 Hanson, M.; Rieke, R.D. *Synth. Commum.* 1995, **25**, 101.
12 Rochfort, G.L.; Rieke, R.D. *Inorg. Chem.* 1986, **25**, 348.

13 Kavaliunas, A.V.; Taylor, A.; Rieke, R.D. *Organometallics* 1983, **2**, 377.
14 Klabunde, K.J. *Chemistry of Free Atoms and Particles*; Academic Press: New York, 1980.
15 (a) Rogers, H.R.; Hill, C.L.; Fugiwara, Y.; Rogers, R.J.; Mitchell, H.L.; Whitesides, G.M. *J. Am. Chem. Soc.* 1980, **102**, 217. (b) Rogers, H.R.; Deutch, J.; Whitesides, G.M. *J. Am. Chem. Soc.* 1980, **102**, 226. (c) Rogers, H.R.; Rogers, R.J.; Mitchell, H.L.; Whitesides, G.M. *J. Am. Chem. Soc.* 1980, **102**, 231. (d) Barber, J.J.; Whitesides, G.M. *J. Am. Chem. Soc.* 1980, **102**, 239.
16 Guijarro, A.; Rosenberg, D.M.; Rieke, R.D. *J. Am. Chem. Soc.* 1999, **121**, 4155.

3

Zinc

3.1 General Methods for Preparation of Rieke Zinc

In 1973 we reported the formation of Rieke zinc. For the first time, this zinc was shown to add oxidatively to alkyl and aryl bromides. The Rieke zinc used in those reactions was prepared using anhydrous zinc chloride or zinc bromide and potassium or sodium metal in refluxing tetrahydrofuran (THF) or 1,2-dimethoxyethane (DME) for 4 h. The reaction is very exothermic, and extreme care must be exercised while carrying out this reaction. The reaction should be heated very slowly at first and carried out in a hood. A large oversized flask should be used to allow expansion of a refluxing solution. A deep black zinc powder is generated during the reduction. Particle size analysis indicates that the average size is 17 μm. Powder patterns show both the characteristic lines of KCl and ordinary zinc metals. For preparation details see Method 1 presented later in this chapter.

A wide variety of zinc salts and various reducing agents have been tried, but the aforementioned conditions seem to lead to the most active zinc. The addition of other alkali salts such as KI, NaI, LiI, KBr, LiBr, or LiCl prior to the reduction step does affect the activity of the zinc.

A far superior method of preparing the highly reactive zinc is to use lithium metal as the reducing agent along with an electron carrier such as naphthalene (Method 2). This approach is considerably safer as there is no rapid burst of heat. The dark green lithium naphthalenide also serves as an indicator, signaling when the reduction is over.

Rieke zinc is prepared by placing lithium metal (10 mmol), a catalytic amount of naphthalene (1 mmol), and 12–15 ml of THF in one flask placed in an ice bath. Once this mixture has stirred for about 30–60 s, it will turn dark green, indicating the formation of lithium naphthalenide. Zinc chloride dissolved previously in 12–15 ml of THF is then cannulated dropwise (ca. 3 s per drop) into the lithium naphthalenide, and stirring is continued for 30 min after the transfer is complete. This method is not only safer due to the use of lithium

Chemical Synthesis Using Highly Reactive Metals, First Edition. Reuben D. Rieke.

metal rather than sodium or potassium but also yields a more reactive zinc. A third method sometimes employed to prepare Rieke zinc uses a stoichiometric amount of naphthalene with respect to lithium (Method 3). Both methods yield Rieke zinc with the same reactivity. It should also be noted that the reactivities are similar regardless of the choice of solvent (THF or DME) and that of the halide salt. Further, the electron carrier is not limited to naphthalene. Other carriers such as biphenyl and anthracene have also been used. The zinc settles very rapidly allowing the solvent and electron carrier to be removed if deemed desirable. Washing it a second or third time removes the majority of the electron carrier. Finally, the desired solvent for the following chemistry can be then added. It should be noted that Rieke zinc in THF or other solvents can be purchased commercially (Rieke Metals, LLC). The zinc metal can be transferred readily either by a cannula or by a syringe yielding the highly reactive zinc in a dry solvent ready for further chemistry. Details for the three procedures are given later in the text.

Finally, the most reactive Rieke zinc yet prepared is generated by the reduction of $Zn(CN)_2$ with lithium using an electron carrier such as naphthalene. Some of this zinc's chemistry will be described later in the text. The reaction details are the same as Method 2.

Method 1 Active Zinc Prepared from the Potassium Reduction of Zinc Chloride:

1) Using a dry box or a glove bag with an argon or nitrogen atmosphere, charge an oven-dried, two-necked, round-bottomed flask (250 ml), containing a magnetic stirring bar, with anhydrous zinc chloride (9.54 g, 0.07 mol) and thinly cut potassium metal (5.47 g, 0.14 mol). Fit the flask with a condenser capped with a gas adapter (with stopcock). Close the stopcock and cap the side neck with a rubber septum.
2) Remove the apparatus from the dry box or glove bag, and connect it to a vacuum/argon or nitrogen manifold. Before opening the stopcock to the inert atmosphere from the manifold, evacuate the system (5 min) and refill with argon or nitrogen (1 min) in three cycles.
3) Open the stopcock, and add freshly distilled THF (40 ml) through the septum inlet using a glass syringe.
4) The mixture is heated without stirring until the zinc chloride visibly reduces at the surface of the potassium. The heating is then stopped, and the vigorous exothermic reduction of the zinc chloride proceeds. At this point cooling in a water or ice bath may be required to moderate the progress of the reaction. After the reduction subsides, the mixture is refluxed for 2.5 h with rapid stirring. The active zinc is then ready for use.

Method 2 Active Zinc Prepared from the Lithium Reduction of Zinc Chloride Using Catalytic Naphthalene

1) Using a dry box or glove bag with an argon atmosphere, charge an oven-dried, two-necked flask (50 ml), containing a magnetic stirring bar, with anhydrous zinc chloride (1.09 g, 8 mmol). Secure a gas adapter (with stopcock) to the flask, and cap the side neck with a septum. Charge a second dry, two-necked round-bottomed flask (50 ml) containing a magnetic stir bar, with thinly cut lithium metal (0.11 g, 1.6 mmol) and naphthalene (0.2 g, 1.6 mmol). Fit the flask with a gas adapter (with stopcock), and secure a septum to the side neck. Close the stopcock before removing the flasks from the dry box or glove bag.
2) Remove the flasks from the dry box or glove bag, and connect them to a vacuum/argon manifold. Before opening the stopcocks to the inert atmosphere from the manifold, evacuate the system (5 min) and refill with argon or nitrogen (1 min) in three cycles. Open the stopcocks to positive argon pressure.
3) Add dry, freshly distilled THF (15 ml) through the septum inlet using a glass syringe to the flask containing the lithium metal and naphthalene. The stirring mixture will turn green in <30 s. Add freshly distilled THF (20 ml) through the septum inlet using a dried glass syringe to the flask containing the zinc chloride. This addition should be performed with rapid stirring.
4) Transfer the stirring zinc chloride solution to the flask containing the stirring green mixture of lithium and naphthalene in THF, using a cannula, dropwise, over a period of 1.5 h. Perform the addition slowly enough so that the green color of lithium naphthalenide persists. If the mixture becomes clear during the addition of the zinc chloride, stop the addition, and allow the mixture to stir until the green color returns before resuming the addition of the zinc chloride solution. When addition of the zinc chloride is complete, stir the mixture until all the residual lithium is consumed. The resulting black slurry of active zinc is then ready for use. The rate of addition of the zinc chloride solution is crucial. If the addition of zinc chloride is performed over a period of four or more hours, the active zinc formed may not settle completely from the THF solution. The reduction of the zinc chloride in approximately 1.5–2 h produces a mossy form of active zinc which rapidly settles.
5) If the presence of naphthalene or lithium chloride (from the reduction) is not desired in the active zinc, they can be removed at this point by repeated washing with dry THF. After the reduction is complete, turn the stir plate off and allow the active zinc to settle (1–2 h). Monitor the progress of the settling by shining a strong light through the slurry by use of a flashlight. Remove the THF solution, by cannula, down to the surface of the settled zinc. Tip the flask slightly to facilitate the removal of the last portion of THF. Add freshly distilled THF (25 ml), and stir for several minutes. Turn off the stirring, allow the zinc to settle for a few minutes, and remove the supernatant by cannula. Repeat the washing cycle two additional times.

Method 3 Active Zinc Prepared from the Stoichiometric Lithium Naphthalenide Reduction of Zinc Chloride

1) Using a dry box or glove bag with an argon atmosphere, charge an oven-dried, two-necked, round-bottomed flask (50 ml), containing a magnetic stirring bar, with anhydrous zinc chloride (2.09 g, 15.4 mmol). Secure a gas adapter (with stopcock) to the flask, and cap the side neck with a septum. Charge a second dry, two-necked, round-bottomed flask (50 ml) containing a magnetic stirring bar with thinly cut lithium metal (0.213 g, 30.6 mmol) and naphthalene (3.99 g, 31.2 mmol). Fit the flask with a gas adapter (with stopcock), and secure a septum to the side neck. Close the stopcocks before removing the flasks from the dry box or glove bag.
2) Remove the flasks from the dry box or glove bag, and connect them to a vacuum/argon manifold. Before opening the stopcocks to the inert atmosphere from the manifold, evacuate the system (5 min) and refill with argon or nitrogen (1 min) in three cycles. Open the stopcocks to positive argon pressure.
3) Add dry, freshly distilled THF (15 ml) through the septum inlet using a glass syringe to the flask containing the lithium metal and naphthalene. The stirring mixture will turn green within 30 s, and the stirring should be continued for 2 h at room temperature until the lithium metal is totally consumed. Add freshly distilled THF (20 ml) through the septum inlet using a dry glass syringe to the flask containing the zinc chloride. Perform this addition with rapid stirring.
4) Transfer the zinc chloride solution to the flask containing the stirring lithium naphthalenide solution, using a cannula, dropwise, over a period of 15 min. The resulting black slurry of active zinc is ready for use.

3.2 Direct Oxidative Addition of Reactive Zinc to Functionalized Alkyl, Aryl, and Vinyl Halides

The first report of the oxidative addition of zinc metal to organic halides dates back to the work of Frankland [1–4] around 1850. He discovered that dialkylzinc compounds could be prepared from zinc metal and methyl iodide or ethyl iodide. However, the reaction did not proceed with alkyl bromides or chlorides. Also, no aryl halides were found to undergo the oxidative addition reaction. Several approaches have been reported since that time to increase the reactivity of the zinc metal. The majority of these modifications have employed zinc–copper couples [5–8] or zinc–silver couples. However, all of these procedures still only worked with alkyl iodides. Noller used a mixture of alkyl iodides and bromides but found that the mixture must contain a large percent of alkyl iodide [9].

Methods to activate the metal itself include grinding or milling the metal to yield a finely divided metal powder. However, this zinc powder was of limited reactivity and only worked with alkyl iodides. Shriner's approach for activation employed successively washing zinc dust with 20% HCl, water until neutral, then acetone, and finally anhydrous ether [10]. The resulting zinc dust was air dried and used immediately. Cornforth et al. [11] have described a modification of this procedure which involves the washing of zinc dust with 2% HCl, then ethanol, acetone, and finally ether. The resulting zinc dust is dried in vacuo with a crystal of iodine. They reported that the use of this activated zinc in the Reformatsky reaction gave improved yields.

In 1962, Gaudeman used THF as the solvent for the oxidative reactions and found that the reaction could be extended to allylic and benzylic bromides. Also, alkyl iodides were easily reacted. Knochel has since made considerable advances in activating the metal by using 1,2-dibromoethane and chlorotrimethylsilane [12, 13]. Recently, Knochel has reported that adding alkali salts such as LiCl can be used to activate zinc metals. The importance of the alkali salts generated in the Rieke method was pointed out in our first reports in the 1970s.

The oxidation addition of Rieke zinc has been exceptionally successful. The reaction proceeds rapidly and in quantitative yields with alkyl iodides, bromides, and even chlorides. They react with cyclic and multicyclic alkyl halides in quantitative yields. Significantly, aryl iodides and aryl bromides can be readily converted to the corresponding R–Zn–X reagents. In some cases, even aryl chlorides can be reacted with Rieke zinc. Demonstrating the unusual reactivity of the zinc is the fact that even vinyl iodides and bromides can be converted into the corresponding organozinc halides. In all of the aforementioned molecules, it is to be noted that most any functional group will tolerate the reaction conditions. Among these groups include esters, amides, nitriles, ketones, imines, and aldehydes. Table 3.1 contains a few examples of the many thousands of halides converted into the corresponding RZnX reagents using Rieke zinc [14].

Table 3.1 demonstrates the three general classes of organic halides and the general conditions used to carry out the oxidative addition reactions. It must be emphasized, however, that the organic halide's reactivity can vary greatly depending on the exact chemical structure. However, in general, all benzylic type halides are done in the same manner. The active zinc (prepared by one of the three methods presented or purchased commercially from Rieke Metals, LLC) in dry THF is placed in a three-necked round-bottomed flask in an ice bath. A thermometer is placed in the zinc slurry so the temperature can be monitored throughout the reaction. The temperature should always be maintained below 5°C throughout the oxidative addition to minimize homocoupling of the benzylic halide. The benzylic halide is dissolved in dry THF in a separate flask. Both flasks are kept under an atmosphere of good quality argon

Table 3.1 Preparation of organozinc compounds.

$$RX + Zn \rightarrow RZnX$$

Entry	Organic halide	Zn:RX ratio	Temperature (°C)	Time (h)	Yield (%)
1	$Br(CH_2)_6Cl$	1.2:1	23	4	100
2	$Br(CH_2)_7CH_3$	1.2:1	23	6	100
3	$Br(CH_2)_3CO_2Et$	1:1	23	3	100
4	*p*-IC_6H_4Cl	2.1	23	3	100
5	*p*-BrC_6H_4CN	2.1	Reflux	3	90
6	*p*-BrC_6H_4CN	3.1	Reflux	3	100
7	*p*-$BrC_6H_4CO_2Et$	2.1	Reflux	2	100
8	*o*-$BrC_6H_4CO_2Et$	2.1	Reflux	2	100
9	*m*-$BrC_6H_4CO_2Et$	3:1	Reflux	4	100
10	$Cl(CH_2)_3CO_2Et$	3:1	Reflux	4	100
11	2-Bromopyridine	1.3:1	Reflux	1	100
12	2-Bromothiophene	1.3:1	23	3	100
13	2-Bromothiazole	1.3:1	23	3	100
14	1-Bromoadamantane	1.3:1	Reflux	3	95
15	Cyclohexyl bromide	1.2:1	23	2	98
16	Benzyl bromide	1.3:1	0	3	95
17	α-Bromostyrene	1.4:1	Reflux	3	96
18	4-Fluoro-1-bromobenzene	2:1	Reflux	4	97

or nitrogen. The benzylic halide is then added dropwise to the active zinc. The drop rate generally can be between 1 and 3 drops per second. The drop rate should be controlled to keep the temperature below 5°C. Generally, the addition takes between 1 and 4 h. Stirring for an additional hour generally leads to complete conversion. In some less reactive halides, one might have to stir the reaction for 1–2 h at room temperature to complete the reaction. The mixture is then allowed to settle overnight and the RZnX/solvent can be removed by a cannula or syringe. If one prefers to speed up the process, the RZnX/solvent can be centrifuged as soon as the reaction is complete to obtain the RZnX solution free of excess zinc powder. A detailed procedure for the preparation of a benzylic zinc halide is presented later in the text.

The reaction of alkyl iodides and bromides is basically done as described previously except the reaction is done at room temperature. The reaction temperature is closely followed as the halide is slowly added to the reactive zinc.

The heat of reaction will normally increase the temperature of the reaction to 50°C or 60°C. In some cases, the temperature may increase to reflux. Accordingly, the reaction flask should always be equipped with a reflux condenser. In many cases shortly after all the halide has been added, the reaction will be complete as indicated by gas chromatography. In some cases, if the halide is slow to react, the mixture may be heated to reflux for 1–2 h. After the reaction is complete, it can be allowed to settle overnight, and the RZnX solution removed by syringe or cannula for further reaction. Also, once the completed reaction reaches room temperature, it can be centrifuged to facilitate the process. A typical detailed procedure for an alkyl bromide is presented later in the text. Also, it should be pointed out that some aryl iodides are so reactive that they can be prepared by this general approach. In general, aryl bromides, chlorides, and even many iodides require refluxing to bring the reaction to completion. The reactions are carried out as described for alkyl bromides. Normally, 15–20% of the aryl halide is added slowly at room temperature to see how exothermic the reaction is. In most cases, the temperature will only increase a few degrees whereupon the reaction can be slowly heated to reflux. Once at reflux, the remaining aryl halide can be added slowly over the next 2–4 h. When all the halide has been added, the solution can be refluxed until the reaction is complete. The simplest approach to follow the reaction is to remove samples and check the progress by GC. A detailed procedure for a typical aryl bromide is presented later in the text. Reaction of vinyl iodides and bromides is essentially identical to the procedure for aryl bromides. Reaction of alkyl chlorides is also similar except in some cases an alkali iodide such as LiI, KI, or NaI may have to be added to bring the reaction to completion. While most halides can be reacted to completion with 1.3–1.5 equiv of active zinc, in some very difficult reactions, more reactive zinc may have to be added. Finally, the most reactive zinc yet prepared is made by reduction of $Zn(CN)_2$. While this zinc has limited use because of the cyanide present, it can be used in special cases of a highly unreactive organic halide. The resulting organozinc halides undergo the usual cross-coupling reactions such as with acid chlorides as will be presented later in the text.

Typical Preparation of 3-Fluorobenzylzinc Bromide

In an oven-dried 2 l 2-necked round-bottomed flask charged with argon was added Zn* (45 g, 0.688 mol, 450 ml of Zn* solution at 0.1 g/ml). The halide, 3-fluorobenzyl chloride (0.5 mol), was diluted with 150 ml THF in a 500 ml round-bottomed flask under argon. The Zn* solution was cooled in an ice bath to 0°C, and the halide solution was added dropwise, with stirring, over 30 min. The reaction temperature was maintained around 0°C throughout the addition. After 2 h the reaction was determined by GC to be complete and allowed to settle overnight. The supernatant was cannulated to a bottle under argon and diluted to 1 l.

Typical Preparation of 4-Cyanobutylzinc Bromide

In an oven-dried 1 l 2-necked round-bottomed flask equipped with a condenser and charged with argon was added Zn* (50 g, 0.765 mol, 500 ml of Zn* solution at 0.1 g/ml). 4-Bromovaleronitrile (0.6 mol) was weighed into a 100 ml round-bottomed flask and charged with argon. The halide was added neat via cannula to the Zn* solution dropwise. The reaction was allowed to reflux gently during the addition. After 2 h the reaction was determined by GC to be complete and allowed to settle overnight. The supernatant was cannulated to a bottle under argon and diluted to 1 l.

Typical Preparation of 4-Bromophenylzinc Iodide

In an oven-dried 2 l 2-necked round-bottomed flask equipped with a condenser and charged with argon was added Zn* (54.9 g, 0.841 mol, 550 ml of Zn* solution at 0.1 g/ml). The halide, 1-bromo-4-iodobenzene (0.6 mol), was dissolved with 250 ml THF in a 500 ml round-bottomed flask under argon. The halide solution was added dropwise, via cannula, with stirring over 20 min. The solution was refluxing gently by the end of the addition. The reaction was allowed to stir for 2 h and determined by GC to be complete. Allowed excess Zn* to settle overnight and the supernatant was cannulated into a bottle under argon and diluted to 1 l.

Typical Preparation of 3-Methyl-2-Pyridlyzinc Bromide

In an oven-dried 1 l 2-necked round-bottomed flask equipped with a condenser and charged with argon was added Zn* (60 g, 0.918 mol, 600 ml of Zn* solution at 0.1 g/ml). 2-Bromo-3-methylpyridine (0.6 mol) was weighed into a 100 ml round-bottomed flask and charged with argon. The halide was added neat via cannula to the Zn* solution dropwise. The reaction was then refluxed with a heating mantle for 3 h. The reaction was determined to be complete by GC and allowed to settle overnight. The supernatant was cannulated to a bottle under argon and diluted to 1 l.

3.3 Reactions of Organozinc Reagents with Acid Chlorides

It is clear from the preceding chapter that the oxidative addition of Rieke zinc is general and can be used with most any organic halide. The reaction proceeds with alkyl iodides, bromides and even most chlorides. Significantly, the reaction proceeds with aryl iodides, bromides, and even some chlorides. Most unusual of all, the reaction proceeds with vinyl iodides and bromides. Moreover, the yields are always in the range of 90–95%. In all cases, the molecules can contain a wide range of functionality. However, a major shortcoming of these organozinc halides, RZnX, is their limited reactivity with electrophiles. In fact their nucleophilicity is so low that they will not even add to the carbonyl group of aldehydes. In order to enhance this reactivity, a number of transmetallations have been developed. Knochel et al. [15] have reported that organozinc reagents react with CuCN·LiBr at −35°C, giving RCu(CN)ZnCl which will react with a wide range of functionalized acid chlorides to yield a wide range of ketones. We explored the scope of this reaction using the wide range of organozinc halides we could prepare using Rieke zinc [14]. The reaction is very general and the yields are good to excellent. Both aryl and alkyl acid chlorides work equally well. Representative examples of this reaction are shown in Table 3.2. Diorganozinc bromides and iodides were also prepared and reacted with 2 equiv of acid chloride, affording the corresponding diketones in moderate to excellent yields. Once again, this reaction will tolerate most any functional group. Details of a typical reaction are presented later in the text. Negishi [16] demonstrated that a number of palladium catalysts can also be used for this coupling reaction. Table 3.3 contains ten examples of cross-coupling a range of organozinc reagents with a wide range of acid chlorides to yield highly functionalized ketones. As an alternative to using CuCN·LiBr or the Negishi approach, we developed the use of CuI as an alternative catalysis. The reaction is not as general as the use of CuCN·LiBr or the use of palladium. However, it proceeds well in a large number of cases and has several advantages such as not using cyanide ions and is very inexpensive.

Table 3.4 contains several representative examples. In the succeeding text you will find detailed procedures for the three general methods of preparing highly functionalized ketones.

Typical Generation of Organozinc Halides from Organic Halides and Active Zinc and Their Copper-Mediated Coupling with Acid Chlorides

Ethyl 4-iodobenzoate (1.934 g, 7.00 mmol) was added neat, via a syringe, to the active zinc (15.40 mmol) at room temperature. The reaction mixture was stirred for 3 h at room temperature. The solution was allowed to stand for about 3 h to allow the excess zinc to settle from the dark brown organozinc iodide solution. The top solution was then transferred carefully via a cannula to another two-necked flask under an argon atmosphere and cooled to −20°C. A solution prepared by mixing CuCN (0.651 g, 7.27 mmol) and anhydrous LiBr (1.273 g, 14.66 mmol) in THF (10 ml) was added at −20°C. The reaction mixture was gradually warmed to 0°C and stirred at 0°C for about 15 min. The solution was then cooled to −25°C, and valeryl chloride (0.851 g, 7.02 mmol) was added neat via a syringe. The mixture was then worked up by pouring into a saturated NH_4Cl aqueous solution (20 ml) and extracting with diethyl ether (3 × 20 ml). The combined organic layers were dried over anhydrous $CaCl_2$.

Table 3.2 Reactions of organozinc halides mediated by copper with acid chlorides.

$$RX + Zn^* \rightarrow [RZnX] \xrightarrow{CuCN\cdot 2LiBr} [RCu(CN)ZnX] \xrightarrow{R'COCl} RCOR'$$

Entry	RX	R′COCl	Zn^* : RX : R′COCl	Product	Yield (%)
1	$Br(CH_2)_7CH_3$	PhCOCl	1.5 : 1.0 : 0.9	$PhCO(CH_2)_7CH_3$	92
2	$Br(CH_2)_6CN$	PhCOCl	1.1 : 1.0 : 0.8	$PhCO(CH_2)_6CN$	94
3	$Br(CH_2)_6Cl$	PhCOCl	1.0 : 1.0 : 1.0	$PhCO(CH_2)_6Cl$	85
4	$BrCH_2CH_2Ph$	PhCOCl	1.2 : 1.0 : 0.9	$PhCOCH_2CH_2Ph$	97
5	$Br(CH_2)_3CO_2Et$	$CH_3(CH_2)_3COCl$	1.0 : 1.0 : 0.9	$CH_3(CH_2)_3CO(CH_2)_3CO_2Et$	91
6	$Br(CH_2)_3CO_2Et$	PhCOCl	1.0 : 1.0 : 0.9	$PhCO(CH_2)_3CO_2Et$	95
6′	$Cl(CH_2)_3CO_2Et$	PhCOCl	1.0 : 1.0 : 0.9	$PhCO(CH_2)_3CO_2Et$	91
7	*p*-BrC_6H_4Me	$CH_3(CH_2)_3COCl$	3.5 : 1.0 : 0.9	$CH_3(CH_2)_3COC_6H_4$-*p*-Me	86
8	*p*-$BrC_6H_4CO_2Et$	$CH_3(CH_2)_3COCl$	2.0 : 1.0 : 1.0	$CH_3(CH_2)_3COC_6H_4$-*p*-(CO_2Et)	83
9	*p*-IC_6H_4Cl	$CH_3(CH_2)_3COCl$	1.5 : 1.0 : 1.0	$CH_3(CH_2)_3COC_6H_4$-*p*-Cl	90
10	*p*-$IC_6H_4CO_2Et$	PhCOCl	2.0 : 1.0 : 1.0	$PhCOC_6H_4$-*p*-(CO_2Et)	88
11	*m*-$BrC_6H_4CO_2Et$	PhCOCl	4.0 : 1.0 : 0.9	$PhCOC_6H_4$-*m*-(CO_2Et)	83
12	*o*-$BrC_6H_4CO_2Et$	PhCOCl	2.0 : 1.0 : 0.9	$PhCOC_6H_4$-*o*-(CO_2Et)	92
13	*o*-$BrC_6H_4CO_2Et$	$CH_3(CH_2)_3COCl$	2.0 : 1.0 : 0.9	$CH_3(CH_2)_3COC_6H_4$-*o*-(CO_2Et)	94
14	*p*-BrC_6H_4CN	$CH_3(CH_2)_3COCl$	2.5 : 1.0 : 1.0	$CH_3(CH_2)_3COC_6H_4$-*p*-CN	71
15	*p*-BrC_6H_4CN	PhCOCl	3.0 : 1.0 : 0.9	$PhCOC_6H_4$-*p*-CN	73
16	*o*-BrC_6H_4CN	PhCOCl	3.0 : 1.0 : 0.9	$PhCOC_6H_4$-*o*-CN	98

17	*o*-BrC_6H_4CN	$CH_3(CH_2)_3COCl$	3.0:1.0:0.9	$CH_3(CH_2)_3COC_6H_4$-*o*-CN	97
18	*p*-$BrC_6H_4COCH_3$	$CH_3(CH_2)_3COCl$	2.0:1.0:0.9	$CH_3(CH_2)_3COC_6H_4$-*p*-($COCH_3$)	80
19	Br	$CH_3(CH_2)_3COCl$	3.0:1.0:0.9	O $CH_3(CH_2)_3$	82
20	$PhCH_2Cl$	PhCOCl	1.5:1.0:0.9	$PhCOCH_2Ph$	81
21	$Br(CH_2)_4Br$	PhCOCl	3.0:1.0:2.0	$PhCO(CH_2)_4COPh$	78
22	*p*-IC_6H_4I	CH_3COCl	3.0:1.0:2.0	$CH_3COC_6H_4$-*p*-($COCH_3$)	76

Table 3.3 Reactions of organozinc halides mediated by $Pd(PPh_3)_4$ and with acid chlorides.

$$RznX + E^+ \xrightarrow[\text{room temperature}]{Pd(PPh_3)_4} \text{product}$$

	[RZnX]	E^+	[RznX] : E^+	Product	% yield (GC%)
1	*m*-$H_3COC_6H_4CH_2ZnBr$	C_6H_5COCl	1:0.95	*m*-$H_3COC_6H_4CH_2COC_6H_5$	86
2	*p*-$H_3COC_6H_4CH_2ZnBr$	C_6H_5COCl	1:0.95	*p*-$H_3COC_6H_4CH_2COC_6H_5$	94
3	$EtO_2C(CH_2)_6ZnBr$	*t*-BuCOCl	1:0.95	$EtO_2C(CH_2)_6COC(CH_3)_3$	80
4	$EtO_2C(CH_2)_5ZnBr$	*p*-$H_3COC_6H_4COCl$	1:0.9	*p*-$H_3COC_6H_4CO(CH_2)_5CO_2Et$	71
5	*m*-ClC_6H_4ZnBr	$Cl(CH_2)_3COCl$	1:1	*m*-$ClC_6H_4CO(CH_2)_3Cl$	83
6	*p*-NCC_6H_4ZnBr	*m*-$EtO_2CC_6H_4I$	1:0.95	*p*-$NCC_6H_4C_6H_4CO_2Et$-*m*	54
7	*p*-$EtO_2CC_6H_4ZnI$	*p*-$EtO_2CC_6H_4I$	1:1.1	*p*-$EtO_2CC_6H_4C_6H_4CO_2Et$-*p*	67
8	*o*-$H_3SC_6H_4ZnI$	*p*-$H_3COC_6H_4COCl$	1:0.9	*o*-$H_3SC_6H_4COC_6H_4COCH_3$-*p*	83
9	*p*-$EtO_2CC_6H_4ZnI$	*m*-$O_2NC_6H_4COCl$	1:1	*p*-$EtO_2CC_6H_4COC_6H_4NO_2$-*m*	71
10	*o*-$H_3CCO_2C_6H_4ZnI$	*m*-BrC_6H_4COCl	1:0.8	*o*-$H_3CCO_2C_6H_4COC_6H_4Br$-*m*	60

Table 3.4 Reactions of organozinc halides mediated by copper iodide and with acid chlorides.

$$RZnX + RCOCl \xrightarrow[CuI \cdot 2LiBr]{0°C} product$$

	[RZnX]	Acid chloride	[RznX] : RCOCl	Product	% yield (GC%)
1	$EtO_2C(CH_2)_3ZnBr$	*o*-$HC_3C_6H_4COCl$	1 : 0.8	*o*-$HC_3C_6H_4CO(CH_2)_3CO_2Et$	78
2	$EtO_2C(CH_2)_3ZnBr$	*p*-IC_6H_4COCl	1 : 1	*p*-$IC_6H_4CO(CH_2)_3CO_2Et$	83 (95)
3	$EtO_2C(CH_2)_4ZnBr$	*p*-$H_3COC_6H_4COCl$	1 : 1	*p*-$H_3COC_6H_4CO(CH_2)_4CO_2Et$	58 (84)
4	$EtO_2C(CH_2)_6ZnBr$	F_3C-C_6H_4COCl	1 : 0.9	*p*-F_3C-$C_6H_4CO(CH_2)_5CO_2Et$	71 (85)
5	$EtO_2C(CH_2)_5ZnBr$	$H_3C(CH_2)_4COCl$	1 : 1	$H_3C(CH_2)_4CO(CH_2)_5CO_2Et$	61 (82)
6	$NC(CH_2)_5ZnBr$	*m*-FC_6H_4COCl	1 : 0.9	*m*-$FC_6H_4CO(CH_2)_5CN$	72 (91)
7	$NC(CH_2)_3ZnBr$	*o*-ClC_6H_4COCl	1 : 1.1	*o*-$ClC_6H_4CO(CH_2)_3CN$	79 (95)
8	$Cl(CH_2)_4ZnBr$	*p*-ClC_6H_4COCl	1 : 1	*p*-$ClC_6H_4CO(CH_2)_4Cl$	96
9	*p*-$BrC_6H_4CH_2ZnBr$	*p*-BrC_6H_4COCl	1 : 0.9	*p*-$BrC_6H_4CH_2COC_6H_4Br$-*p*	86
10	*p*-NCC_6H_4ZnBr	*p*-BrC_6H_4COCl	1 : 1	*p*-$BrC_6H_4CH_2COC_6H_4CN$-*p*	60 (86)
11	*o*-NCC_6H_4ZnBr	*p*-BrC_6H_4COCl	1 : 1	*p*-$BrC_6H_4CH_2COC_6H_4CN$-*o*	75
12	*o*-$H_3COC_6H_4ZnI$	*o*-BrC_6H_4COCl	1 : 0.9	*o*-$H_3COC_6H_4COC_6H_4Br$-*o*	72

The resultant crude product was chromatographed on flash silica gel using gradient elution (hexanes to remove naphthalene first, then hexanes/ethyl acetate) to give ethyl-4-(1-oxopentyl)benzoate (1.360 g, 5.81 mmol) as a white crystalline solid in 83% isolated yield: mp 47.5–48.0°C.

Synthesis of 4-Methoxy-2′-Thiomethylbenzophenone Using Tetrakis(triphenylphosphine)palladium(0) as Catalyst

In an argon dry box, an oven-dried 500 ml 2-necked round-bottomed flask was charged with $Pd(PPh_3)_4$ (4.47 g, 3.87 mmol). The catalyst was dissolved with 50 ml THF, and 2-thiomethylphenylzinc iodide (150 ml at 0.5 M, 75 mmol) was added. Anisoyl chloride (11.9 g, 69.7 mmol) was weighed into a 50 ml round-bottomed flask and charged with argon. The acid chloride was added via cannula over 30 min as the temperature peaked at 38°C. The reaction was determined to be complete by GC after stirring overnight. The reaction was quenched with 3 M HCl and extracted two times with ether. The combined organics were washed with saturated sodium bicarbonate and brine. The organics were dried and concentrated to a red oil. Purification was done by column chromatography using ethyl acetate and hexanes to yield 15 g of an orange oil.

Synthesis of Ethyl 7-(3,4-Difluorophenyl)-7-Oxoheptanoate Using Copper Iodide as Catalyst

In an argon dry box, an oven-dried 1 l 2-necked round-bottomed flask was charged with CuI (4.8 g, 25 mmol) and LiBr (4.3 g, 50 mmol). The catalyst was dissolved with 100 ml THF, and 6-ethoxy-6-oxohexylzinc bromide (500 ml at 0.5 M, 0.250 mol) was added. The reaction was cooled to 0°C. 3,4-Difluorobenzoyl chloride was weighed into a 100 ml round-bottomed flask and charged with argon. The acid chloride was added neat via cannula over 6 min. After stirring for 30 min, the reaction was determined to be complete by GC. The reaction was quenched with 3 M HCl and extracted with ether two times. The combined organics were washed with 7.5% NH_4OH solution, saturated sodium bicarbonate, and brine. The organics were dried and concentrated to a yellow oil and purified by vacuum distilling, yielding 50.8 g of white solid. The typical copper iodide coupling used 10% CuI based off of the organometallic. Organometallics can be either zinc or magnesium. The organometallic is typically added to the CuI solution and then the electrophile. It can be done in reverse, but if the electrophile is an acid chloride, there will be more THF esters. The coupling can be done at low temperature (about −35°C) or at room temperature where the reaction is allowed to get hot during the addition. Many times, the reactions that get hot progress better and are more complete. Reactions can be cooled to reduce impurities that may be making purification difficult.

The typical palladium coupling is done using about 0.5% based off of the organometallic. We use Pd(0) but Pd(+2) can also be used. Again, the organometallics can be either zinc or magnesium. The organometallic is typically added to the palladium solution and then the electrophile. If Pd(+2) is used, the organometallic is added first to generate the Pd(0). The reactions are typically done at room temperature and allowed to get hot. Usually the heat of the reaction is enough to drive the reaction to completion, but occasionally the reaction is heated to get it to completion.

Cyanide-Based Rieke Zinc

The lithium reduction of zinc(II) cyanide using naphthalene or biphenyl as a catalytic electron carrier yields a more reactive form of Rieke zinc. This new form of Rieke zinc is able to undergo direct oxidative addition to alkyl chlorides under mild conditions and tolerates the presence of nitriles and bulky tertiary amides [17]. Table 3.5 shows representative reactions of alkyl zinc chloride reagents with benzoyl chloride. The activation of the zinc surface could originate by the adsorption of the Lewis base cyanide ion on the metal surface. The adsorbed cyanide ion can affect the metal's reactivity in two possible ways. One possible mode of activation would be the reduction of the metal's work function in the vicinity of the adsorbed cyanide ion, and the second could be that the cyanide ion is acting as a conduction path for the transfer of the metal's electrons to the alkyl chloride. One or both processes could account for the observed enhanced chemical reactivity.

3.4 Reactions of Organozinc Reagents with α,β-Unsaturated Ketones

When we started our studies in 1985 on the addition of RZnX reagents to α,β-unsaturated ketones, no general set of reaction conditions were available [14]. Luche [18] has reported the efficient conjugate addition of alkyl halides to α,β-unsaturated ketones mediated by a Zn/Cu couple in aqueous media and suggested that this process occurs via a radical mechanism rather than by the free organometallic species. Lithium [19] and magnesium [20] triorganozincates undergo 1,4-addition with α,β-unsaturated ketones. However, only one of the three organic moieties is transferred in the process. This problem has been solved to some degree by utilizing only 1 equiv of the alkyllithium and 2 equiv of methyllithium in forming these reagents [21]. The methyl group appears to be a good, nontransferable "dummy" ligand for the lithium trialkylzincates. However, since trialkylzincate reagents are derived from the Grignard or lithium precursors, they offer no distinct advantage over the various types of cuprate reagents. Copper(I) salts are widely used to mediate the 1,4-addition of organozinc species to α,β-unsaturated ketones [22]. As seen in the work described previously, convenient methods for forming the organozinc species

Table 3.5 Reactions of alkylzinc chloride reagents with benzoyl chloride.

Entry	Organohalide	Zn*:R–Cl	PhCOCl	Product	Yield (%)
1	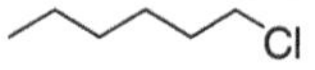 Cl	3:1	0.8	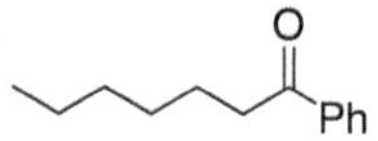 O, Ph	74
2	 Cl	4:1	0.8	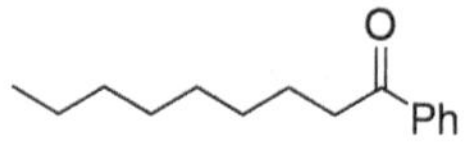 O, Ph	65
3	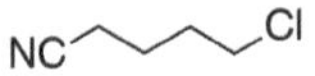 NC, Cl	2:1	0.7	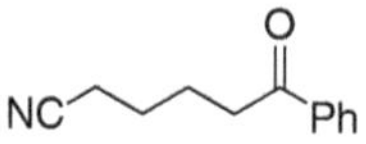 NC, O, Ph	41
4	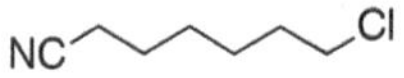 NC, Cl	2.5:1	0.9	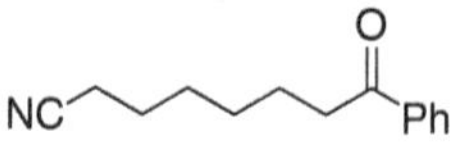 NC, O, Ph	68
5	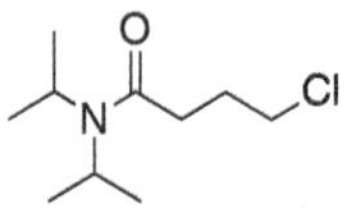 O, N, Cl	3:1	0.9	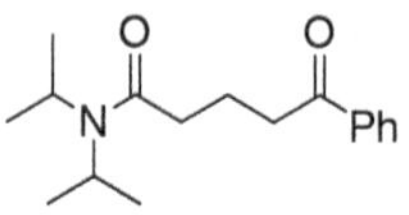 O, N, O, Ph	61

are very important to the utility of these reagents in the synthetic transformation. Active zinc allows a straightforward preparation of these reagents [14].

Several variations were attempted using (3-carbethoxypropyl)zinc bromide as a target in order to optimize the 1,4-addition process. The best approach involved forming the zinc cuprate from the soluble CuCN/LiBr complex according to Knochel's procedure [22]. A 92% yield (GC) of 1,4-adduct was obtained (entry 1, Table 3.6). Another approach taken in an effort to optimize the conjugate addition process was the formation of the so-called "higher-order" cuprate species as developed by Lipshutz [23]. Based on this chemistry, we have attempted to prepare a similar "higher-order" cuprate species from lithium 2-thienylcyanocuprate in conjunction with the organozinc compounds (Equation 3.1):

$$\mathrm{LiCu(CN)_2\text{-}th + RZnX \rightarrow ``R(2\text{-}th)Cu(CN)ZnX"} \tag{3.1}$$

Substituting lithium 2-thienylcyanocuprate for the CuCN/LiBr complex and omitting boron trifluoride etherate from the reaction of the organozinc halide species with 2-cyclohexenone indeed gave a reasonable yield (76% GC) of the 1,4-adduct (entry 3, Table 3.6). The reaction rate, temperature, and yield were similar to the reactions employing CuCN/LiBr. However, contrary to all previous reports by Lipshutz, a noticeable amount (ca. 9% GC yield) of the product resulting from the 1,4-addition of the 2-thienyl group was observed. This tendency was even more pronounced when the reaction was carried out in the presence of boron trifluoride etherate. In this reaction, 1,4-addition of the 2-thienyl group was the major product (59% GC yield) along with the usual 1,4-adduct (ca. 10% GC yield) (entries 3 and 4, Table 3.6).

Table 3.6 Copper-mediated conjugate additions of organozinc halides with α,β-unsaturated ketones.

Entry	Organic halide	Enone	Cu(I)[a]	Additives[b]	Products	Yield (%)[c]
1	$Br(CH_2)_3CO_2Et$	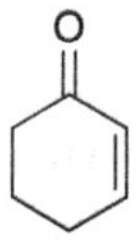 I	A	$BF_3{\cdot}OEt_2$	II ($(CH_2)_3CO_2Et$)	92 (74)
		I		TMSCl	II	
2	$Br(CH_2)_3CO_2Et$	I	A	TMSCl	II	75
3	$Br(CH_2)_3CO_2Et$	I	B	TMSCl	II	76 (55)
					III (S)	9 (8)
4	$Br(CH_2)_3CO_2Et$	I	B	$BF_3{\cdot}OEt_2$	III	59 (39)
					II	10
5	$Br(CH_2)_6Cl$	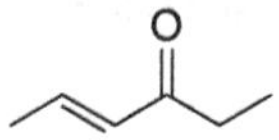 IV	A	$BF_3{\cdot}OEt_2$ TMSCl	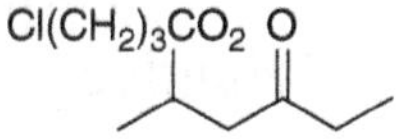 $Cl(CH_2)_3CO_2$ O	72 (77)
6	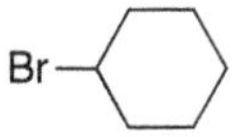 Br	I	A	$BF_3{\cdot}OEt_2$ TMSCl	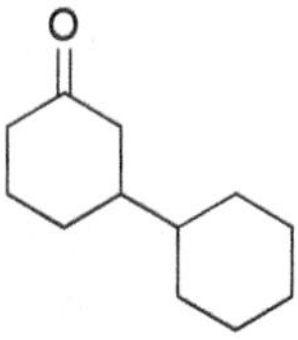 O	58 (66)
7	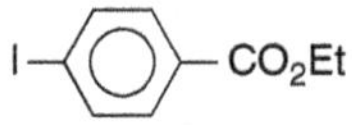 I–C$_6$H$_4$–CO_2Et	I	A	$BF_3{\cdot}OEt_2$ TMSCl	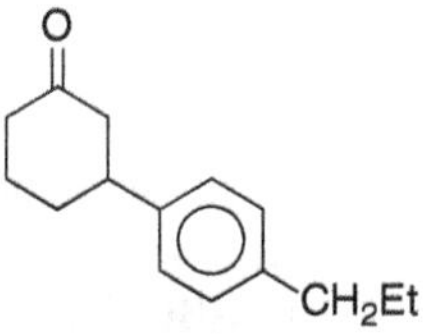 O, CH_2Et	68

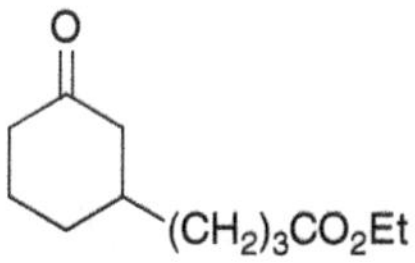

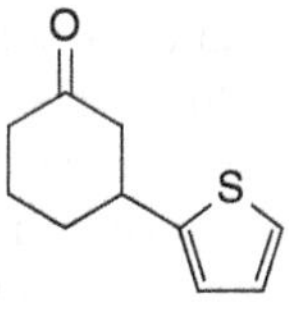

[a] Reactions were performed in the presence of 0.9 equiv of the following copper salts: (A) CuCN.2LiBr in THF or (B) lithium 2-thienyl-cyanocuprate (0.25M solution in THF).
[b] Added at -78°C just prior to the addition of enone.
[c] GC yield (isolated yield).

A detailed experimental procedure for the copper-mediated conjugate addition of organozinc reagents to α,β-unsaturated ketones is given later in the text.

Typical Copper-Mediated Conjugate Addition Reaction of Organozinc Halides to α,β-Unsaturated Ketones

Ethyl 4-bromobutanoate (0.705 g, 3.62 mmol) was added neat, via syringe, to the active zinc (4.00 mmol) at room temperature. The reaction mixture was stirred at room temperature for 18 h, giving a dark solution of the alkylzinc bromide species. A solution prepared by mixing CuCN (0.327 g, 3.65 mmol) and anhydrous LiBr (0.636 g, 7.32 mmol) in THF (10 ml) was added at 0°C. The reaction mixture was stirred at 0°C for 15 min and then cooled to −78°C. $(CH_3)_3SiCl$ (0.719 g, 6.62 mmol) and $BF_3{\cdot}Et_2O$ (0.800 g, 5.64 mmol) were added neat via a syringe, and the solution was stirred for 10–15 min. A solution of 2-cyclohexenone (0.269 g, 2.80 mmol) in THF (10 ml) was added dropwise over 20 min to the reaction mixture at −78°C. The reaction mixture was stirred at −78°C for 3 h and slowly warmed to 0°C. After being stirred at 0°C for 1–2 h, the reaction mixture was then worked up and purified to give 3-(3-carboethoxypropyl)-cyclohexanone (0.442 g) as an oil in 74% yield (92% GC yield).

3.5 Reactions of Organozinc Reagents with Allylic and Alkynyl Halides

Of high synthetic value, the allyl moiety is an integral feature of many natural products and biosynthetic intermediates. Substitution reactions of allylic halides with organometallic reagents have provided an important route for the synthesis of these allylic compounds. Many factors affect the regiochemistry of these reactions, including the nature of the leaving group [24], the degree of substitution at the two ends of the allylic system [25], the solvent system [26], and the nature of the nucleophile and the catalyst [26]. In general, substitution of allylic substrates with or without complete allylic rearrangement is still an unpredictable process. While considerable research has been done on the coupling reactions of allylic halides with a variety of organometallic reagents, in only a few instances were the desired results obtained. Miyaura et al. [27] have found that $[(C_3H_7)_3BCH_3]Cu$ reacted with cinnamyl chloride in THF to give a 96% yield of exclusively the γ-product (S_N2'). However, this is the only example reported with 100% allylic rearrangement. Mourik and Pabon [28] indicated that when >1.5 equiv of *n*-butyl- or vinyllithium cuprate/dimethyl sulfide complex was treated with 1-chloro-2-butene at −76°C in diethyl ether, almost negligible allylic rearrangement was seen. Yoshida et al. [29] synthesized the unsaturated ester via the CuCN-catalyzed allylation of zinc esters in THF/DMA at 60°C. Although the yields were high, the regioselectivity was poor

[S_N2 : S_N2' = (15–28) : (85–72)]. Nakamura et al. [30] stated that allylation of organozinc reagents in the presence of 5 mol% $CuBr·Me_2S$ catalyst took place in a highly S_N2'-selective manner, whereas in the presence of a nickel catalyst, excellent S_N2 selectivity was obtained. However, in the S_N2'-selective allylation reactions, only simple alkylzinc compounds (methyl to butyl) were used, and in the S_N2 selective allylation reactions, only primary allylic chlorides were used and surprisingly no homocoupling occurred. Probably, the most successful regioselective allylation of γ-attack of allylic halides was reported by Yamamoto [31]. He found that $RCu·BF_3$ attacked the γ-position of the allylic substrate irrespective of the degree of substitution at the two ends of the allylic system (84–99.7%). We [14] have been able to carry out the highly regioselective γ-alkylation of allylic halides using highly functionalized organozinc compounds mediated by CuCN·2LiBr at 0°C. Not only are the additions highly regioselective but the organozinc reagents can also be highly functionalized. Both alkylzinc halides and arylzinc halides react with 3-chloro-1-butene to give 100% γ-attack (S_N2') (entries 5 and 7, Table 3.7). Alkylzinc halides react with crotyl chloride and cinnamyl chloride to give about 97% of the S_N2' products (entries 1–3 and 8–10, Table 3.7). Even 1-bromo-3-methyl-2-butene, with the highly hindered γ-position, yields up to 95% of S_N2' products (entries 11–13, Table 3.7). The arylzinc halide reacts with crotyl chloride to give up to 80% of the S_N2' product. Temperature did not seem to affect the regioselectivity of attack of the alkyl or aryl zinc halides. Reactions that run at −78°C for about 10 min and then gradually warmed to room temperature had almost the same regioselectivity as the reaction run at 0°C. A detailed experimental procedure for the copper-mediated reaction of RZnX with allylic halides is presented later in the text.

Typical Reaction of Organozinc Halides with Allylic Halides

A solution prepared by mixing CuCN (0.495 g, 5.53 mmol) and anhydrous LiBr (0.965 g, 11.11 mmol) in THF (10 ml) under an argon atmosphere was precooled to −20°C and added to 6-cyanohexyl zinc bromide (5.46 mmol, in about 10 ml of THF) at −20°C, and the mixture was then slowly warmed to 0°C. 1-Chloro-2-butene (0.544 g, 6.01 mmol, 1.1 equiv) was added neat, via a syringe. The reaction mixture was stirred at this temperature for about 30 min. The reaction mixture was worked up by pouring into a saturated NH_4Cl aqueous solution (20 ml) and extracting with diethyl ether (3 × 20 ml). The combined organic layers were dried over anhydrous $CaCl_2$. The resultant crude product was chromatographed on flash silica gel using gradient elution (hexanes to remove naphthalene first and then hexanes/ethyl acetate) to give a 91% yield of S_N2' and S_N2 mixture. The ratio determined by the ^1H NMR spectroscopy was S_N2' : S_N2 = 97 : 3.

We also found it possible to make 2,3-disubstituted-1,3-butadienes using CuCN·LiBr and cross-coupling with 1,4-dichloro-2-butyne. This reaction was

Table 3.7 Reactions of RZnX with allylic halides mediated by CuCN·2LiBr.

$$RX + Zn^* \longrightarrow RZnx \xrightarrow{CuCN \cdot 2LiBr} [RCu(CN)ZnX]$$

[RCu(CN)ZnX] + $R^1(R^2)C{=}CH{-}CH_2Y$ → S_N2 + S_N2'

Entry	RX	Allylic halide	Products S_N2':S_N2	Yield (%)
1	$Br(CH_2)_3CO_2Et$	$H_3C{-}CH{=}CH{-}CH_2Cl$	96:4	83
2	$Br(CH_2)_6CN$		97:3	91
3	$Br(CH_2)_6Cl$		98:2	94
4	*p*-$BrC_6H_4CO_2Et$		80:20	86
5	$Br(CH_2)_3CO_2Et$	$CH_3CH(Cl)CH{=}CH_2$	100:0	87
6	$Br(CH_2)_6CN$		97:3	87
7	*p*-$IC_6H_4CO_2Et$		100:0	93
8	$Br(CH_2)_3CO_2Et$		97:3	86
9	$Br(CH_2)_6CN$	$Ph{-}CH{=}CH{-}CH_2Cl$	98:2	88
10	$Br(CH_2)_2Ph$		95:5	94
11	$Br(CH_2)_3CO_2Et$	$(H_3C)(CH_3)C{=}CH{-}CH_2Br$	98:2	88
12	$Br(CH_2)_6CN$		95:5	90
13	$Br(CH_2)_2Ph$		95:5	89

very regiospecific, giving S_N2' products. Unfortunately, arylzinc compounds did not yield the S_N2' product when reacted with 1,4-dichloro-2-butyne. However, when reacted with 1,4-ditosyloxy-2-butyne, aryl and alkyl organozincs gave excellent yields of the S_N2' products. Table 3.8 shows examples of this type of transformation. The use of Rieke zinc allows a wide variety of functional groups to be included [32]. This same catalyst can be used to make 2-halo-olefins incorporating a high degree of functionality by reacting with 2,3-dibromopropene [33]. Examples of this type of reaction are shown in Table 3.9.

One reaction starting with *p*-cyanobenzyl bromide is shown later in the text along with the experimental details.

Table 3.8 Reaction of organozinc halides with Y—CH_2—C≡C—CH_2—Y.

$$2RZnBr + Y{-}CH_2{-}C{\equiv}C{-}CH_2{-}Y \xrightarrow[THF]{CuCN \cdot LiBr}$$ (Y=Cl or TsO)

No.	Y	R	Yield (%)	No.	Y	R	Yield (%)
1.	Cl	$-(CH_2)_7CH_3$	95	8.	Cl	$-CH_2C_6H_4$-*p*-CN	93
2.	Cl	$-(CH_2)_6Cl$	92	9.	TsO	$-(CH_2)_7CH_3$	(98)
3.	Cl	$-(CH_2)_3CO_2Et$	95	10.	TsO	$-(CH_2)_3CO_2Et$	(82)
4.	Cl	$-(CH_2)_2CN$	84	11.	TsO	$-C_6H_5$	88
5.	Cl	$-CH_2$-cyclohexyl	87	12.	TsO	$-CH_2C_6H_5$	(95)
6.	Cl	cyclohexyl	82	13.	TsO	$-C_6H_4$-*p*-COMe	93
7.	Cl	$-CH_2C_6H_5$	91	14.	TsO	$-C_6H_4$-*p*-CN	97

Table 3.9 Preparation of 2-bromo-1-alkene.

$$RX + Zn^* \xrightarrow[Temp.]{Time} RZnX \xrightarrow[CuCN\ 2LiBr]{CH_2{=}C(Y)CH_2Y}$$

No.	RX	Y	Rx:Zn*	Temperature (°C)	Time (h)	Yields (%)
1.	*p*-BrC_6H_4CN	Br	1:2.5	Reflux	1	85
2.	*p*-$BrC_6H_4COCH_3$	Br	1:2.0	Reflux	1	81
3.	*p*-$IC_6H_4CO_2Et$	Br	1:1.5	Reflux	0.3	90
4.	*p*-BrC_6H_4Cl	Br	1:1.7	Reflux	1.5	82
5.	$Br(CH_2)_3CO_2Et$	Br	1:1.2	rt	2	96
6.	$Br(CH_2)_3CN$	Br	1:1.5	Reflux	0.5	94
7.	$Br(CH_2)_6Cl$	Br	1:1.2	rt	2	88
8.	$Br(CH_2)_3CO_2Et$	Cl	1:1.2	rt	2	95
9.	$Br(CH_2)_3CN$	Cl	1:1.5	rt	1	78
10.	*m*-$IC_6H_4CO_2Et$	Cl	1:1.5	Reflux	1	83

Preparation of 2,3-Di(*p*-Cyanobenzyl)-1,3-Butadiene Reaction

1) Prepare active zinc (50 mmol) in a two-necked, round-bottomed flask (50 ml) fitted with a gas adapter (with stopcock) according to the procedures of Method 3 in THF (25 ml).
2) Charge an oven-dried, round-bottomed flask (25 ml), containing a stirring bar, with 4-cyanobenzyl bromide (4.3 g, 35 mmol). Seal the flask with a rubber septum, and subject it to three pump/inert atmosphere refill cycles. While maintaining a blanket of argon, add THF (10 ml) and stir until the 4-cyanobenzyl bromide is completely dissolved. Transfer the resulting solution by cannula to the active zinc. Stir the reaction mixture at room temperature for 2 h. Stop the stirring and allow the active zinc to settle overnight.
3) Charge an oven-dried, two-necked, round-bottomed flask (100 ml), containing a stirring bar, with CuCN (3.1 g, 35 mmol) and LiBr (3 g, 35 mmol). Fit the flask with a gas adapter (with stopcock) and a septum. Connect the flask to a manifold system and subject it to three evacuation/inert gas refill cycles. Add dry, distilled THF (15 ml) and stir the mixture until the salts dissolve. Cool the solution to −20°C.
4) Transfer the supernatant, from step 2, by cannula, to the CuCN/LiBr solution at −20°C. Stir the solution for 15 min and then warm to 0°C. Add 1,4-dichloro-2-butyne (1.84 g, 15 mmol). Stir the solution for 30 min and then at room temperature for 1 h.
5) Quench the reaction mixture by pouring into a saturated ammonium chloride solution (20 ml). Extract the aqueous layer with diethyl ether (3 × 20 ml), then dry the combined organic layers over $MgSO_4$, filter, and concentrate under reduced pressure. Isolate the product from the crude reaction mixture by flash chromatography (ethyl acetate : hexanes, 1 : 5, v/v); 2,3-di(*p*-cyanobenzyl)-1,3-butadiene is obtained in 93% yield.

3.6 Negishi Cross-Coupling of Vinyl and Aryl Organozinc Halides

As we already pointed out in Section 3.3, the cross-coupling of highly functionalized RZnX reagents with acid chlorides is readily carried out using Negishi's approach with $Pd(PPh_3)_4$ as a catalyst. The Negishi method also is very effective in the cross-coupling of aryl and vinyl halides. The reaction proceeds in excellent yields, producing highly functionalized symmetrical and unsymmetrical biaryls, and symmetrical and unsymmetrical butadienes (Table 3.10). Attempts to cross-couple alkyl or aryl zinc reagents with 1,2-dibromobenzene failed. Similarly, the 1,4-dibromozinc butane reagent prepared from 1,4-dibromobutane also failed to undergo a cyclizational reaction with 1,2-dibromobenzene.

Table 3.10 Coupling reactions of RZnX with aryl and vinyl halides catalyzed by $Pd(PPh_3)_4$.

$$RZnZ + R'Y \xrightarrow[\text{THF}]{\text{5mol\% Pd(PPh}_3)_4} RR'$$

Entry	RZnX	R′Y	Product	Yield (%)[a]
1	$EtO_2C(CH_2)_3ZnBr$	*p*-$BrC_6H_4COCH_3$	$EtO_2C(CH_2)_3$–C_6H_4–$COCH_3$	86
2	$EtO_2C(CH_2)_3ZnBr$	*p*-BrC_6H_4CN	$EtO_2C(CH_2)_3$–C_6H_4–CN	93
3	$EtO_2C(CH_2)_3ZnBr$	*p*-$BrC_6H_4NO_2$	$EtO_2C(CH_2)_3$–C_6H_4–NO_2	90
4	EtO_2C–C_6H_4–ZnI	*p*-BrC_6H_4CN	EtO_2C–C_6H_4–C_6H_4–CN	80
5	EtO_2C–C_6H_4–ZnI	*p*-$IC_6H_4CO_2Et$	EtO_2C–C_6H_4–C_6H_4–CO_2Et	94
6	EtO_2C–C_6H_4–ZnBr (meta)	*p*-BrC_6H_4CN	EtO_2C–C_6H_4–C_6H_4–CN (meta-EtO_2C)	82
7	NC–C_6H_4–ZnBr	*p*-BrC_6H_4CN	NC–C_6H_4–C_6H_4–CN	95
8	NC–C_6H_4–ZnBr	*p*-$IC_6H_4CO_2Et$	NC–C_6H_4–C_6H_4–CO_2Et	82
9	*o*-NC–C_6H_4–ZnBr	*m*-$BrC_6H_4CO_2Et$	*o*-CN–C_6H_4–C_6H_4–*m*-CO_2Et	93
10 11	$C_6H_5C(=CH_2)ZnBr$	$CH_3C(=CH_2)Br$	$C_6H_5C(=CH_2)C(CH_3)=CH_2$	85 (91) 86
12 13	*o*-EtO_2C–C_6H_4–CH=CH–ZnBr	$H_2C=CHBr$	*o*-EtO_2C–C_6H_4–CH=CH–CH=CH_2	95 93

[a] Isolated yields. GC yields are shown in parentheses.

Experimental details for a typical cross-coupling of an arylzinc halide with an aryl bromide are presented later in the text.

Typical Procedure for the Reaction of RZnX with Aryl and Vinyl Halides

(4-Carbethoxyphenyl)zinc iodide (2.16 mmol, in about 10 ml of THF) was transferred via a cannula to a THF solution of 5 mol% $Pd(PPh_3)_4$ (0.127 g, 0.11 mmol) and 4-bromobenzylnitrile (0.400 g, 2.19 mmol) at room temperature under an argon atmosphere. The solution was then stirred for 3 h. The mixture was then worked up by pouring into a saturated NH_4Cl aqueous solution (20 ml) and extracting with diethyl ether (3 × 20 ml). The combined organic layers were dried over anhydrous $CaCl_2$. The resultant crude product was chromatographed on flash silica gel using gradient elution (hexanes to remove naphthalene first and then hexanes/ethyl acetate) to give ethyl 4-(4-cyanophenyl)-benzoate (0.433 g, 1.73 mmol) in 80% as a crystalline solid: mp 114–115°C.

Preparation of Aryl Ketones via Ni-Catalyzed Negishi Coupling Reactions

The transition metal-catalyzed cross-coupling reactions of organometallics with acid chlorides have been the most widely used approach for the preparation of ketones [34]. To this end, numerous efficient synthetic methodologies using mainly organomagnesium [35], -copper [36], -manganese [37], -lithium [38], -boron [39], -stannane [40], and -zinc [41] have been developed. Among those, the most considerable effort has been devoted to utilizing organozinc reagents simply due to the highly efficient functional group tolerance. Consequently, transition metal-catalyzed coupling reactions of organozinc with acid chlorides have been considered as one of the most effective routes for the ketone synthesis. In most cases of coupling reactions, copper and palladium were the most frequently used catalysts. Unfortunately, this route contains some limitations such as the inconvenience of using a copper catalyst and the high price of the palladium catalyst. Therefore, there is still a need to develop an economically and environmentally more efficient method for the synthesis of ketones, albeit Cu- or Pd-catalyzed coupling reactions of organozincs are useful tools.

To date, relatively few outstanding studies on the coupling reaction of organozinc with acid halides have been reported to avoid these restrictions. For example, Gosmini and coworkers [42] reported a method for the preparation of aromatic ketones using the cobalt-catalyzed coupling reaction of organozinc prepared in situ with acid chlorides or carboxylic anhydrides. The preparation of ketones using the coupling reaction of mixed organozincs with acid chlorides in the presence of tri-*n*-butylphosphine has been reported [43, 44]. It was performed by the palladium–phosphinous acid-catalyzed cross-coupling reaction of organozinc with acid halides. Compared to other organometallics, however, there has been considerably less progress in developing effective catalytic systems for producing ketone compounds using organozinc reagents. One of the more interesting routes in the Negishi coupling with acid halides has been accomplished by Rovis [45]. This approach employed organozinc reagents along with a Ni precatalyst in the presence of an appropriate ligand. In spite of good to excellent isolated yields, a limited number of organozinc reagents were used in this study.

We have developed several efficient routes for the synthesis of organozinc reagents and their subsequent coupling reactions [46]. However, a significant improvement in this methodology would be the development of a nickel catalyst for constructing carbonyl compounds. We reported a versatile application of readily available organozinc reagents for ketone synthesis in the presence of a nickel catalyst [47].

As mentioned before, most of the coupling reactions of organozincs for the preparation of ketones were performed with a copper or palladium catalyst. In search of a more efficient catalyst in terms of easy availability, we first attempted the coupling reaction of 2-(ethoxycarbonyl)phenylzinc bromide (**A**) prepared by the direct insertion of active zinc [46] with benzoyl chloride.

To investigate the proper conditions, several nickel catalysts were employed, and the results were summarized in Table 3.11. Even though each catalyst required a little different reaction time (20 min to 24 h) at room temperature,

Table 3.11 Catalyst screening.

CO_2Et, ZnBr (**A**, 1.2 eq) + COCl (1.0 eq) → THF, rt → O, CO_2Et (**1a**)

Entry	Catalyst[a]	Time[b]	Result[c] (%)
1	$Ni(acac)_2$	20 min	89
2	$Ni(dppe)Cl_2$	24 h	80
3	$Ni(PMe_3)_2Cl_2$	1 h	82
4	$Ni(PPh_3)_2Cl_2$	30 min	85
5	$Co(acac)_2$	1 h	(20)[d]

[a] 2 mol% used.
[b] No acid chloride observed.
[c] Isolated yield (based on benzoyl chloride).
[d] Determined (conversion) by GC.

the coupling product (**1a**) was obtained in good to excellent isolated yields in the presence of a nickel catalyst. It was of interest that the coupling reaction was completed in the absence of any extra ligand under the given conditions with 2 mol% of a nickel catalyst [48]. As described in Table 3.11, among the nickel catalysts used in this study, $Ni(acac)_2$ is shown to be a very efficient catalyst for the preparation of ketones. Generally, the coupling reaction was conducted at room temperature in THF. The subsequent cross-coupling reaction using 2 mol% of $Ni(acac)_2$ as a catalyst was completed in 20 min at room temperature and resulted in the formation of the product (**1a**) in 89% isolated yield (Table 3.11, entry 1). All the rest of the coupling reactions in this study were carried out with $Ni(acac)_2$ catalyst at room temperature.

Table 3.12 shows the general applications of this strategy for the preparation of highly functionalized benzophenones. As aforementioned, a catalytic amount of $Ni(acac)_2$ was employed, and the reaction was carried out at room temperature in THF. To compare the results depending upon the benzoyl chlorides, the same reaction conditions were maintained during the coupling reaction in progress (1 h instead of 20 min). A bromine atom on the benzoyl chloride tolerated the reaction conditions (Table 3.12, entries 1–3), and the corresponding coupling products (**2a–2c**) were achieved in good yields. Interestingly, the presence of iodine on the benzoyl chloride was also not

Table 3.12 Coupling reactions with benzoyl chlorides.

CO_2Et / ZnBr (1.2 eq, **A**) + COCl / FG (1.0 eq) → 2% $Ni(acac)_2$, THF/rt/1.0 h → O / CO_2Et / FG **2a–2h**

Entry	FG	Product	Yield (%)[a]
1	2-Br	**2a**	75
2	3-Br	**2b**	82
3	4-Br	**2c**	83
4	2-I	**2d**	70
5	3-I	**2e**	75
6	4-I	**2f**	79
7	2-NO_2	**2g**	60
8	4-NO_2	**2h**	65

[a] Isolated yield (based on benzoyl chloride).

affected resulting in the formation of iodine-substituted benzophenones (**2d–2f**, Table 3.12) in excellent yields. 2-Nitro- and 4-nitrobenzoyl chlorides also effectively reacted with organozinc **A** to give rise to the ketones (**2g** and **2h**, Table 3.12) in 60% and 65% isolated yield, respectively, under the same conditions.

In order to examine the generality of this catalytic system, we expanded our study to include a wide variety of acid chlorides to be completed with **A** [47]. The results are summarized in Table 3.13. Under the same conditions (2 mol% of $Ni(acac)_2$ in THF) used before, the coupling reaction with a simple alkyl acid chloride as well as chlorine-substituted acid chloride was completed in 20 min at room temperature, affording the corresponding product (**3a**, **3b**) in 79% and 81% isolated yields (Table 3.13, entries 1 and 2), respectively. A bulky acid chloride was also reacted with **A** to give rise to the ketone (**3c**) in excellent yield (Table 3.13, entry 3). Significantly, ketone (**3d**) containing two ester functionalities was successfully prepared from the coupling reaction with ethyl 2-chloro-2-oxoacetate in good yield (Table 3.13, entry 4). Not only alkyl acid chlorides but also other heteroaryl acid chlorides were good coupling partners to prepare heteroaryl ketones. As depicted in Table 3.13, thiophene, furan, and pyridine carbonyl chlorides were coupled with **A** under mild conditions to afford the corresponding ketones (Table 3.13, **3e**, **3f**, and **3g**) in good yields, respectively. Even though a slightly longer reaction time (24 h at refluxing temperature) was required in the case of pyridine carbonyl chloride, the coupling reaction occurred selectively on the acid chloride (Table 3.13, entry 7). Another significant result was formed in the reaction with 4-(chloromethyl) benzoyl chloride (Table 3.13, entry 8). As noted in Table 3.13, a chlorine atom in the benzylic position was intact during the coupling reaction under the conditions used in this study and was not attacked in the cross-coupling reaction (**3h**, Table 3.13).

The utility of this methodology in the preparation of a number of ketones was extended to the coupling reactions with a variety of organozinc reagents. Again, all of the organozinc reagents used in this study were prepared by the direct insertion of active zinc into the corresponding halides, and the results obtained from this study are described in Table 3.14. For most of the cases, the coupling reactions were conducted in the presence of 2 mol% $Ni(acac)_2$ at room temperature in THF and completed within 30 min except in the case of 3-bromothiophene-2-ylzinc bromide (Table 3.14, entry 6). Halogen-substituted phenylzinc iodide underwent the coupling reaction with aryl acid chlorides under mild conditions and successfully gave the aryl ketones (**4a** and **4b**, Table 3.14). No significant effect on the coupling reaction was observed when an electron-donating group was present (Table 3.14, entry 3). Again, the aforementioned mild conditions worked well for the coupling reactions with heteroarylzinc reagents. 5-(Ethoxycarbonyl)thiophene-2-ylzinc bromide reacted with 2-thiophenecarbonyl chloride to give rise to the formation

Table 3.13 Coupling reactions with acid chlorides.

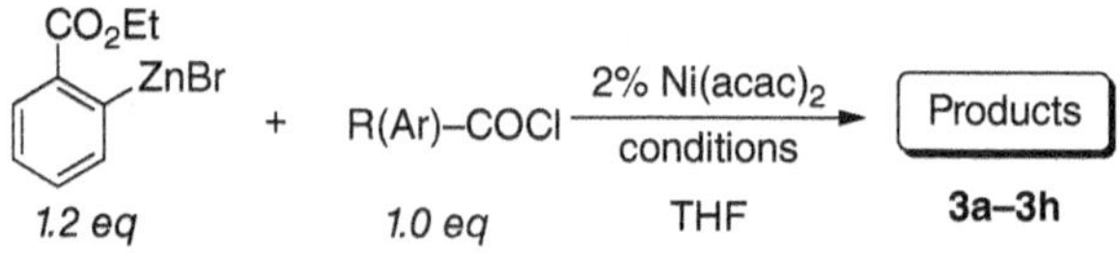

(2-(Ethoxycarbonyl)phenyl)zinc bromide (1.2 eq) + R(Ar)–COCl (1.0 eq) → [2% $Ni(acac)_2$, conditions, THF] → Products **3a–3h**

Entry	Acid chloride	Conditions	Product	Yield (%)[a]
1	$CH_3(CH_2)_3COCl$	rt/20 min	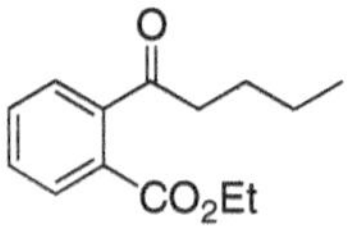 **3a**	79
2	$ClCH_2(CH_2)_3COCl$	rt/20 min	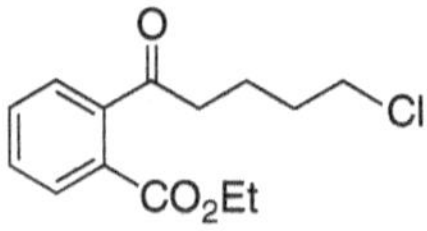 **3b**	81
3	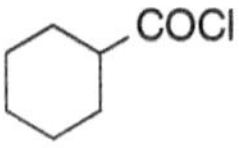 cyclohexyl–COCl	rt/60 min	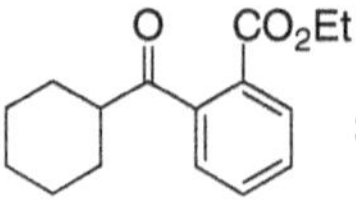 **3c**	89
4	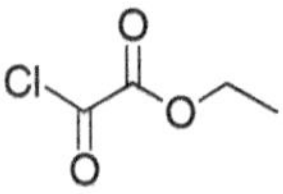 Cl–CO–CO–OEt	rt/10 min	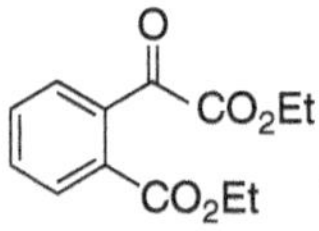 **3d**	75
5	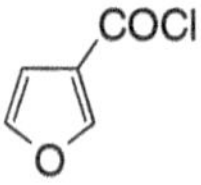 3-furyl–COCl	rt/30 min	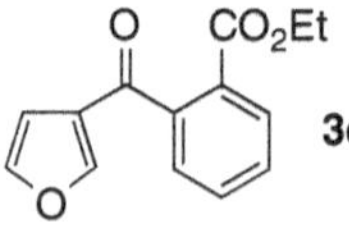 **3e**	86
6	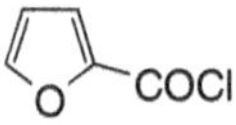 2-furyl–COCl	rt/30 min	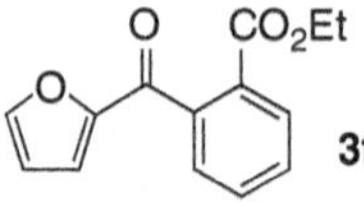 **3f**	79
7	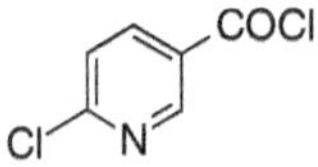 6-chloropyridin-3-yl–COCl	Reflux/24 h	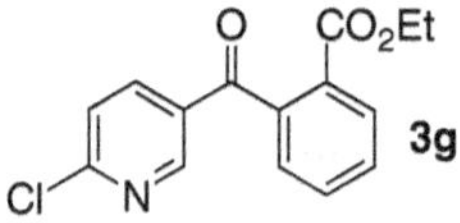 **3g**	68
8	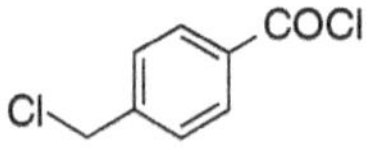 4-($ClCH_2$)C_6H_4–COCl	rt/30 min	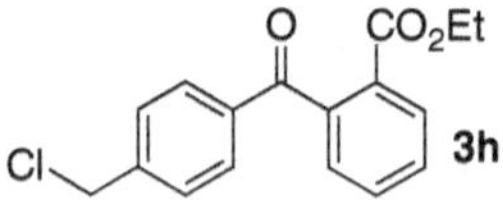 **3h**	66

[a] Isolated yield (based on acid chloride).

Table 3.14 Preparation of functionalized ketones.

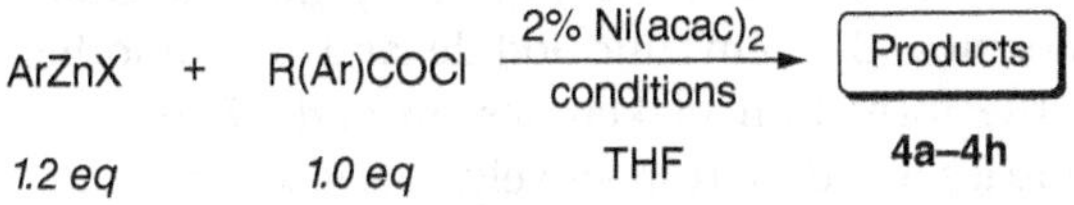

$$\underset{1.2\ eq}{ArZnX} + \underset{1.0\ eq}{R(Ar)COCl} \xrightarrow[\text{conditions}]{2\%\ Ni(acac)_2} \underset{4a\text{–}4h}{\boxed{\text{Products}}}$$

THF

Entry	Organozinc	Acid chloride	Conditions	Product	Yield[a] (%)
1	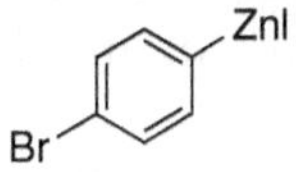 ZnI, Br	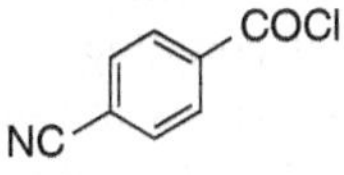 COCl, NC	rt/30 min	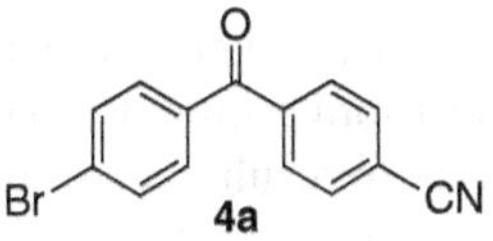 Br, CN, **4a**	84
2	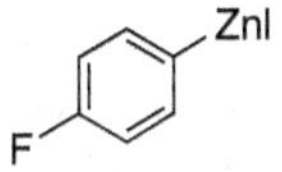 ZnI, F	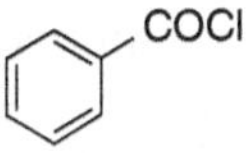 COCl	rt/30 min	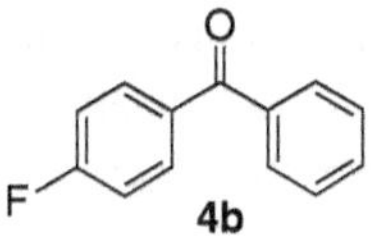 F, **4b**	86
3	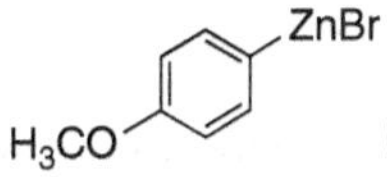 ZnBr, H_3CO	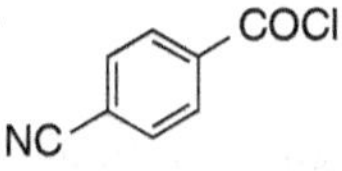 COCl, NC	rt/30 min	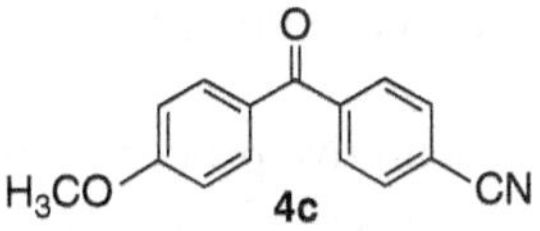 H_3CO, CN, **4c**	89
4	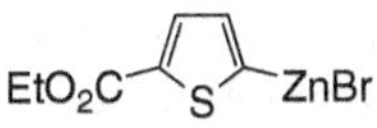 EtO_2C, S, ZnBr	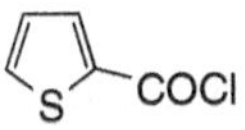 S, COCl	rt/30 min	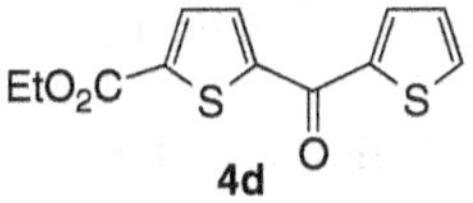 EtO_2C, S, S, **4d**	81
5	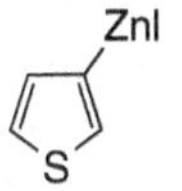 ZnI, S	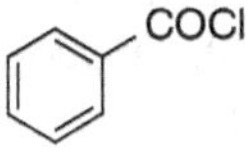 COCl	rt/30 min	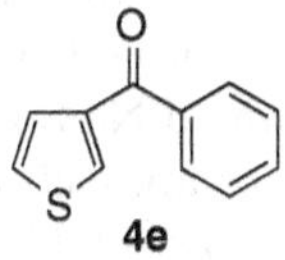 S, **4e**	83
6	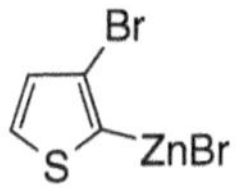 Br, S, ZnBr	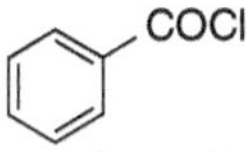 COCl	rt/60 min	 S, Br, **4f**	77
7	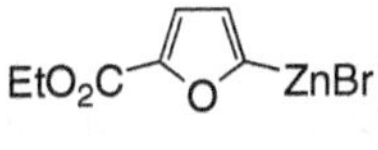 EtO_2C, O, ZnBr	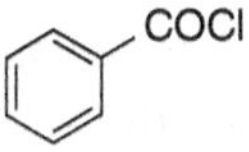 COCl	rt/30 min	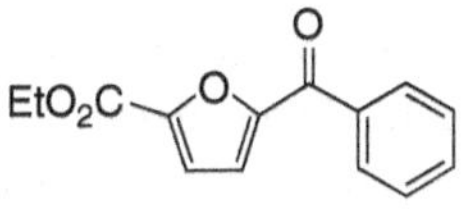 EtO_2C, O, **4g**	78
8	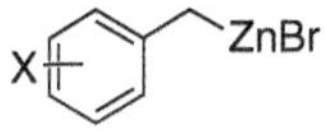 X, ZnBr	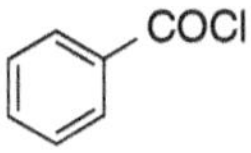 COCl	rt/10 min	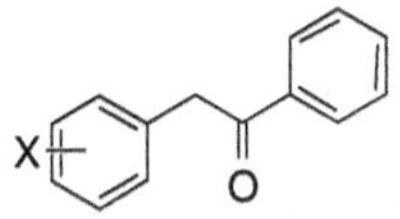 X	87 (**4h**)
9	X = 3-F, 4-Cl		rt/10 min	X = 3-F (**4h**), 4-Cl (**4i**)	72 (**4i**)

[a] Isolated yield (based on acid chloride).

of heteroaryl ketone (**4d**, Table 3.14) in 81% yield. The coupling reactions of similar organozincs, 3-thienylzinc iodide and 3-bromothien-2-ylzinc bromide, led to the formation of ketones **4e** and **4f** in 83% and 77% yields (Table 3.14, entries 5 and 6), respectively. It is also significant that 5-(ethoxycarbonyl)furan-2-ylzinc bromide was coupled with benzoyl chloride to yield ethyl 5-benzoylfuran-2-carboxylate **4g** in 78% yield (Table 3.14, entry 7). Additionally, it was found that treatment of benzylzinc reagents under the same conditions generated functionalized coupling products **4h** and **4i** in 10 min at room temperature (Table 3.14, entries 8 and 9).

From these results, it can be concluded that the mild conditions and nickel catalyst have resulted in a wider tolerance of functional groups and a greater scope of the reactions. Significantly, this strategy affords a more economically and environmentally useful and valuable synthetic procedure for the preparation of ketones utilizing organozinc reagents, which are readily available.

Typical Reaction Procedure

A representative procedure of coupling reaction: In a 25 ml round-bottomed flask, $Ni(acac)_2$, (0.06 g, 2 mol%), and 10 ml (5 mmol) of 0.5 M solution of 2-(ethoxycarbonyl)phenylzinc bromide in THF were added into the flask at room temperature. Next, 6-chloronicotinoyl chloride (0.70 g, 4 mmol) dissolved in 5.0 ml of THF was added. The resulting mixture was refluxed overnight, then cooled down to room temperature, quenched with saturated NH_4Cl solution, and then extracted with ethyl acetate (30 ml× 3). The combined organics were washed with saturated $Na_2S_2O_3$ solution and brine and dried over anhydrous $MgSO_4$. A flash column chromatography (50% EtOAc/50% heptane) gave 0.78 g of **3g** as a yellow solid in 68% isolated yield. mp = 48–51°C.

3.7 Intramolecular Cyclizations and Conjugate Additions Mediated by Rieke Zinc

Active zinc is an effective mediator in intramolecular conjugate additions [49]. For example, a spirodecanone was formed from the 1,4-addition of the organozinc reagent, which was readily available from the primary iodide (Equation 3.2) [50]. Other types of ring closures also occur (Equations 3.3 and 3.4),

Zn*
rt, 24 h

67–74%

(3.2)

Zn*
rt, 1 h
65–70%

(3.3)

Zn*
rt, 50 min
52–56%

(3.4)

and are thought to proceed by a mechanism that does not involve a free radical pathway.

The direct insertion of active zinc into the carbon-iodide bond of 6-iodo-3-functionalized-1-hexenes forms the corresponding primary alkylzinc iodides, which then undergo an intramolecular insertion of the olefinic π-bond into the zinc–carbon bond to form methyl cyclopentanes (Equation 3.5 and 3.6) [51]. When R is methyl, the product diastereomeric ratio is high. This is a significant finding in that this step is a regiospecific 5-*exo-trig* cyclization, which can occur in the presence of functional groups. The cyclization produces an intermediate that can be elaborated further with various electrophiles. This methodology has been extended to ω-*sec*- alkylzinc reagents (Equation 3.7) [52].

RO, I, Zn*, [RO, ZnI], RO, ZnI

R = CH_3CO, *tert*-BuCO

(3.5)

Me, I, Zn*, [Me, ZnI], 1. rt, 2. I_2, Me, I

d.r. = 93/7
80%

(3.6)

Zn* ZnI 1. CuCN•2 LiCl 2. Ethyl propiolate 3. H+ ZnI Bu CO2Et Bu

(3.7)

3.8 The Formation and Chemistry of Secondary and Tertiary Alkylzinc Halides

Highly functionalized secondary and tertiary alkylzinc halides offer special synthetic value to the synthetic chemist if they can be transformed into a wide range of final structures. In this section, we will demonstrate how these alkylzinc halides can be readily generated from Rieke zinc and a wide range of secondary and tertiary halides. Moreover, they can tolerate most any functionality. The tertiary alkylzinc halides are of special interest as the construction of quaternary centers has always been a serious synthetic challenge and are frequently found in a wide spectrum of bioactive molecules. As will be demonstrated, the secondary and tertiary zinc halides are readily converted into a wide range of structures in high yields. Rieke zinc readily yields alkylzinc bromides under mild conditions from the direct oxidative addition of Rieke zinc to secondary and tertiary alkyl bromides. This procedure eliminates the need for the proximal functional group activation of the carbon–bromine bond and zinc insertion proceeds readily under mild conditions. 2-Butylzinc bromide can be generated from 2-bromobutane and Rieke zinc at ambient temperature, but the reaction time is conveniently reduced in refluxing THF to 2.5 h. This zinc reagent coupled with benzoyl chloride in high yield (95%) using the soluble copper cyanide/lithium bromide complexes catalytically (Table 3.15, entry 1). 2-Bromobutane was sufficiently reactive so that Rieke zinc could be used stoichiometrically. For most cases, better success was achieved if 1.1 equiv of the highly reactive zinc to the alkyl halide were used. After complete consumption of the alkyl halide, the zinc was allowed to settle (ca. 4–6 h) and the cross-coupling was then conducted. 3-Bromopentane (0.9 equiv) readily underwent oxidative insertion in refluxing THF to form the organozinc bromide reagent and likewise coupled with benzoyl chloride (0.9 equiv) and 5-chlorovalery chloride (0.9 equiv) to give good yields of the product ketones (Table 3.15, entries 2–3). Cycloalkyl bromides reacted similarly

Table 3.15 Formation and coupling reactions of *sec*- and *t*-alkylzinc bromides.

Entry	Alkyl bromide	Electrophile	Product	Yield (%)
1	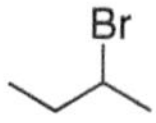 Br	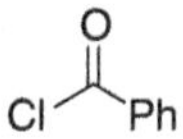 O, Cl, Ph	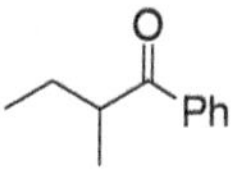 O, Ph	95
2	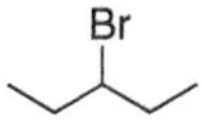 Br	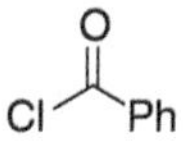 O, Cl, Ph	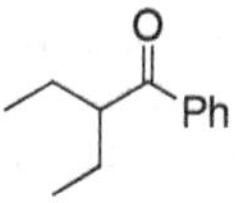 O, Ph	87
3	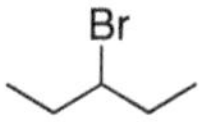 Br	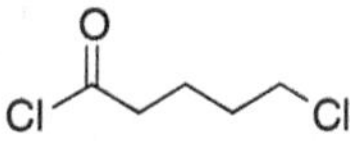 O, Cl, Cl	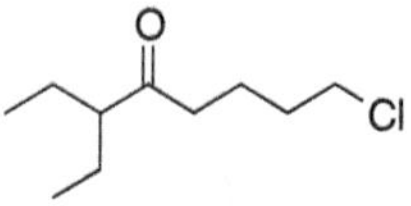 O, Cl	62
4	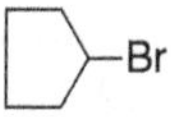 Br	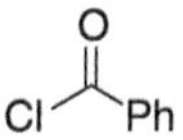 O, Cl, Ph	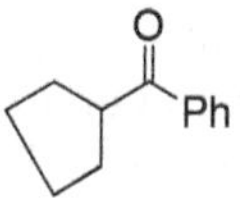 O, Ph	84
5	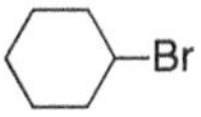 Br	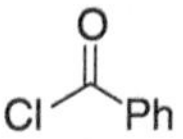 O, Cl, Ph	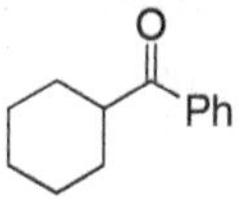 O, Ph	99
6	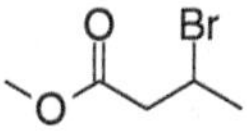 O, O, Br	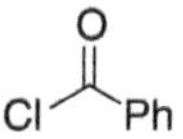 O, Cl, Ph	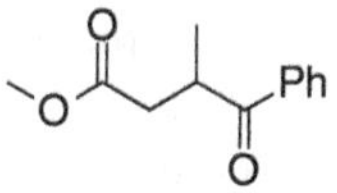 O, O, Ph, O	75
7	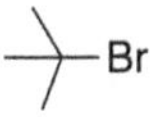 Br	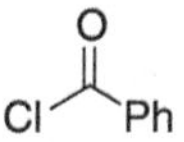 O, Cl, Ph	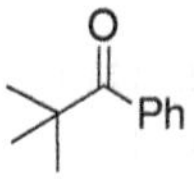 O, Ph	75
8	 Br	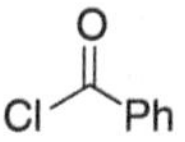 O, Cl, Ph	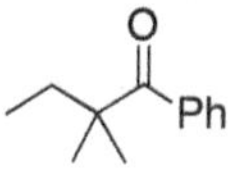 O, Ph	86
9	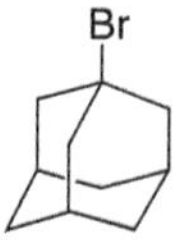 Br	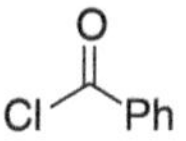 O, Cl, Ph	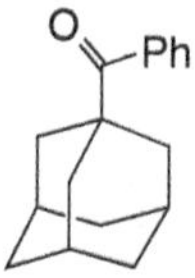 O, Ph	72

(Table 3.15, entries 4–5, 0.9 and 0.7 equiv benzoyl chloride, respectively). *t*-Butyl bromide easily reacted at rt in 1 h to form the zinc reagent. The tertiary alkylzinc bromides (Table 3.15, entries 7–9) couple slowly with benzoyl chloride (0.7 equiv), taking 4–8 h for completion. These results are significant, in that tertiary alkylzinc bromides can be formed readily from active zinc insertion in high yield, which would be difficult or impossible by other methodologies which do not tolerate functionality.

A representative procedure: to a slurry of Rieke zinc (7.98 mmol) in THF (25 ml) under argon was added 2-bromobutane (7.95 mmol), and the mixture was refluxed for 2.5 h. The resulting light brown solution was cooled to rt and was transferred via cannula to a THF (10 ml) solution of CuCN (1.5 mmol) and LiBr (1.5 mmol) at −45°C. Benzoyl chloride (7.95 mmol) was added neat, and the mixture was warmed slowly to rt over 4 h. The reaction mixture was quenched with 3 M HCl (20 ml) and extracted with ether (3 × 20 ml), and the combined organic layers were washed with water (20 ml), dried over $MgSO_4$, and concentrated. 2-Methyl-1-phenylbutanone (7.52 mmol, 95%) was isolated from the crude reaction mixture by flash chromatography (hexanes/ethyl acetate).

We have already reported in Section 3.4 that alkylzinc halides can be added conjugatively to α,β-unsaturated ketones. However, the 1,4-additions require the presence of CuCN·2LiCl or CuCN·2LiBr. Conditions have been developed for the 1,4-addition of secondary and tertiary alkylzinc bromides which do not require the presence of a copper cyanide catalyst [53]. Before our work was reported, the direct use of organozinc halides in synthetically useful conjugate additions to enones has been limited to alkynylzinc halides using TMSI as a Lewis acid [54]. Organozinc reagents are capable of forming the 1,4-addition products without a copper catalyst if a lithium or magnesium triorganozincate approach is used [55]. Direct use of primary alkylzinc halide reagents has been demonstrated not to be effectual in the formation of Michael products [56–58]. We reported an efficient procedure to generate secondary and tertiary alkylzinc bromides [59] in high yield, and that 1,4-addition of these reagents to enones proceeds directly, in good yield, without the presence of a copper(I) catalyst (Scheme 3.1).

tert-Butylzinc bromide (1) reacted with cyclohexenone (2) to form the 1,4-addition product (Table 3.16, entry 3) in good yield in THF, using BF_3Et_2O (1.5 equiv) and TMSCl (2 equiv) as a Lewis acid combination at −78°C. Unfortunately, under these reaction conditions, the use of other secondary and tertiary alkylzinc bromides gave only low yields of the desired products (Table 3.16, entries 4, 7, and 11). Performing the reactions in a mixed

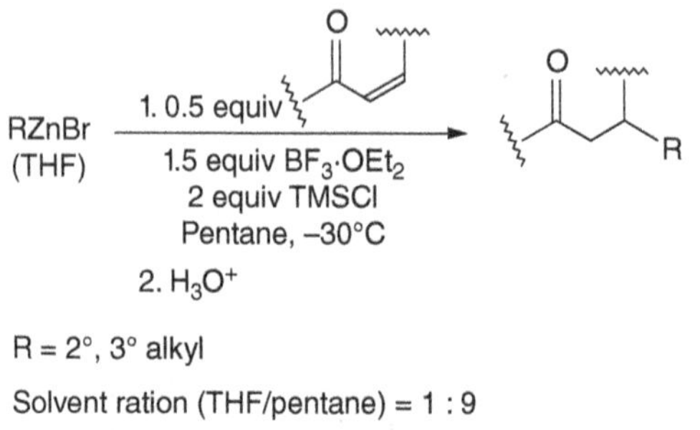

RZnBr (THF)
1. 0.5 equiv (enone)
1.5 equiv $BF_3·OEt_2$
2 equiv TMSCl
Pentane, −30°C
2. H_3O^+

R = 2°, 3° alkyl

Solvent ration (THF/pentane) = 1 : 9

Scheme 3.1 1,4-Addition of 2°, 3° alkylzinc bromides.

Table 3.16 Conjugate addition of secondary and tertiary alkylzinc bromides to enones.[a]

Entry	RZnBr	Enone[b]	Product[c]	Yield (%)[d]
1	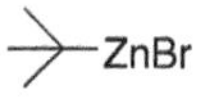 ZnBr	A	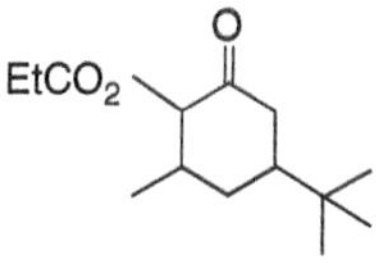 O $EtCO_2$	54[e,f]
2	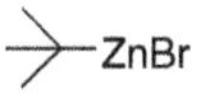 ZnBr	B	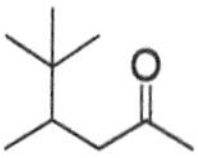 O	71
3	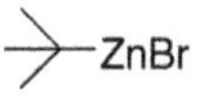 ZnBr	C	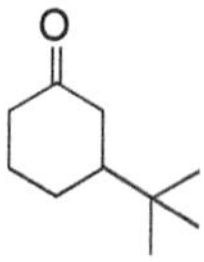 O	89[e]
4	 ZnBr	C	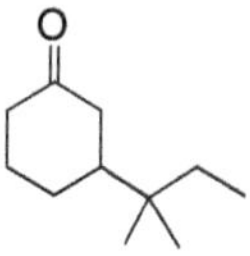 O	35[e]
5	 ZnBr	C	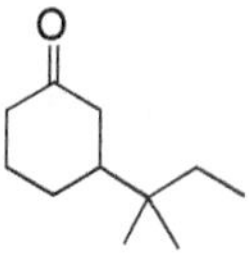 O	74
6	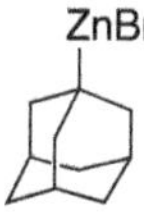 ZnBr	C	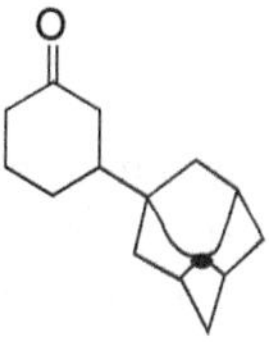 O	54
7	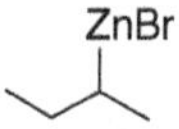 ZnBr	C	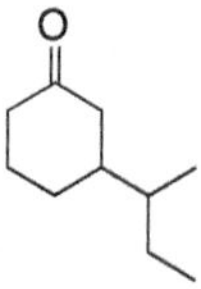 O	34[e,f]
8	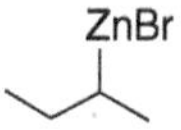 ZnBr	C	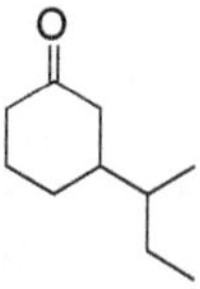 O	65[f]
9	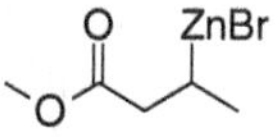 O ZnBr O	C	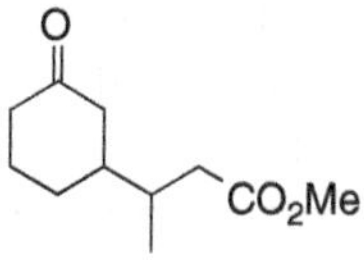 O CO_2Me	51[f]

(Continued)

Table 3.16 (Continued)

Entry	RZnBr	Enone[b]	Product[c]	Yield (%)[d]
10	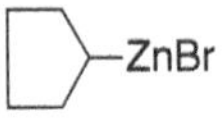 –ZnBr	B	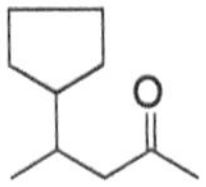 O	40
11	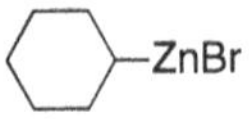 –ZnBr	C	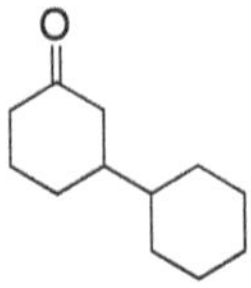 O	36[e]
12	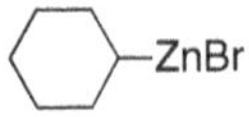 –ZnBr	C	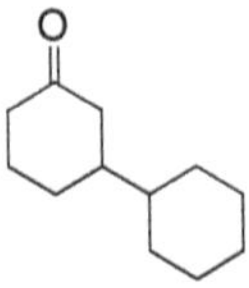 O	73

[a] Reaction conditions: RZnBr (1 M, 10 ml THF), pentane (90 ml), 2 equiv of TMSCl, 1.5 equiv of $BF_3 \cdot OEt_2$, 0.5 equiv of enone, −30°C.
[b] Enones: A, ethyl 6-methyl-2-oxo-3-cyclohexene-l-carboxylate; B, 3-penten-2-one; C, 2-cyclohexen-l-one.
[c] NMR, IR, HRMS, and elemental analyses were consistent with structure or in agreement with the literature.
[d] Isolated yield.
[e] Reaction performed in THF.
[f] Ratio of diastereomers not determined.

polar/nonpolar solvent system proved to be advantageous in that when the alkylzinc bromide reagents (prepared in THF) were transferred to solutions of the enones in pentane, forming heterogeneous reaction mixtures, the yields of the resulting ketones increased dramatically (entries 5, 8, 12). The 1,4-addition reactions were most efficiently performed at −30°C, as the reactions were sluggish at −78°C, except for the case of **1**, which readily transferred the *tert*-butyl group at the lower temperature. A slight modification in the structure from the *tert*-butyl group, as in the case of 2-methyl-2-butylzinc bromide (entry 4), resulted in a sluggish reaction in THF with **2** and performed well only in a mixed solvent system (entry 5). The reaction of **1** with **2** in pentane alone gave a low yield of the desired product (17%). Primary alkyl- and arylzinc bromides did not react with enones under any of the aforementioned reaction conditions.

Though the use of a mixed solvent system in the 1,4-addition enabled an effective transfer of alkyl groups to enones, the product yields suffered when the active zinc was not washed with THF prior to the reaction with the starting alkyl bromides. Washing removed most of the lithium salts and naphthalene. Accordingly, it appeared that the reaction was inhibited by the presence of lithium salts. To confirm this observation, the reaction of **1**, in THF, prepared from washed active zinc, with enone **2** in the presence of added lithium chloride (1 equiv), led to a 54% yield of product (cf. entry 3). The possibility of performing the 1,4-addition catalytically in $BF_3{\cdot}OEt_2$ was investigated. The reaction of 2-butylzinc bromide with enone **2** using $BF_3{\cdot}OEt_2$ catalytically (10 mol%) and 2 equiv of TMSCl resulted in a 50% yield of product, as opposed to a yield of 65% (entry 8) with 1.5 equiv of $BF_3{\cdot}OEt_2$. Reactions conducted without the use of TMSCl resulted in the appearance of side products.

The reactions run in the mixed solvent system proceeded well at −30°C, with typical reaction times of 2–12 h. The transfer of bulky groups in the organozinc was efficient as in the cases of adamantly- and cyclohexylzinc bromides (entries 6 and 12). The tolerance of the methyl ester under the reaction conditions was apparent in the methyl-3-butyrylzinc bromide reaction with **1**, as a 51% yield (entry 9) was realized. Additionally, the nature of the enone was important in the reaction rate. The reactions of cyclic enones proceeded more readily than those of acyclic enones, the latter requiring longer reaction times.

In summary, methods have been developed which allow for the first time the 1,4-addition of secondary and tertiary alkylzinc bromides to α,β-unsaturated ketones without the need for a copper catalyst or without the presence of cyanide ion. The method is general and will tolerate functional groups. The ability to carry out conjugate additions without copper or cyanide ions represents an economical and exceedingly "environmentally friendly" approach. Moreover, the reaction temperature need only be −30°C, rather than −78°C as most procedures require.

A representative procedure: to a flask charged with finely cut lithium (24.8 mmol), naphthalene (2.5 mmol) and THF (10 ml) under argon were transferred by cannula to a THF (20 ml) solution of zinc chloride (11.8 mmol) dropwise such that addition was complete in ≈1.5 h with moderate stirring. The stirring was stopped when the lithium was totally consumed, and the active zinc was allowed to settle. The supernatant was then removed. Fresh THF (25 ml) was added, the mixture briefly stirred and then allowed to settle, and the supernatant subsequently removed. The active zinc was then washed an additional time. THF (10 ml) was added to the flask, and the active zinc was ready for use. To the stirring mixture of active zinc was added methyl 3-bromobutyrate (7.59 mmol), and the mixture was refluxed for 2 h. The reaction mixture was allowed to cool to room temperature, and the zinc settled in 4 h. The alkylzinc bromide reagent was added dropwise to a solution of 2-cyclohexene-1-one (4.14 mmol), $BF_3{\cdot}OEt_2$ (11 mmol), and TMSCl (15 mmol) in pentane (90 ml) at −30°C, such that addition was complete in 20 min. The heterogeneous reaction mixture was stirred for 3.5 h at −30°C. The reaction was then quenched by addition to saturated NH_4Cl (30 ml) and was then taken up in ether (20 ml). The aqueous layer was extracted with ether (2 × 20 ml). The combined organics were washed sequentially with water (20 ml) and brine (20 ml), dried over $MgSO_4$, and concentrated.

Methyl 3-(3′-oxocyclohexyl)butanoate (2.10 mmol, 51%) was isolated from the crude mixture by flash chromatography (hexanes/ethyl acetate).

3.9 Electrophilic Amination of Organozinc Halides

This study provides the first facile approach for the electrophilic amination of organozinc halides [60]. The reaction of several functionalized primary, secondary, and tertiary organozinc bromides, benzylzinc bromide, functionalized arylzinc halides, and heteroarylzinc bromide with di *tert*-butyl azodicarboxylate is described. The reaction products, *N,N′*-di-*tert*-butoxycarbonylhydrazino derivatives, are obtained in excellent yields for most aliphatic substrates and good yields for aromatic substrates. These compounds are direct precursors of hydrazine and amino derivatives by deprotection. The process constitutes the first synthetically useful electrophilic amination of organozinc derivatives and takes advantage of the broad functional group tolerance of the organozinc chemistry.

The only examples of electrophilic amination of organozinc compounds described in the literature correspond to the reaction of an excess of diethyl/diisopropyl zinc with several chloroamines [61]. The reaction afforded moderate to low yields of substitution products, often as a mixture of amines [62, 63]. On the other hand, electrophilic amination using dialkyl azodicarboxylates has been reported for other nucleophilic species, including lithium enolates and silyl enol ethers [64–70].

We explored the scope of this reaction with a variety of organozinc halides, easily prepared by reaction of the corresponding organic bromide or iodide with highly reactive zinc (Scheme 3.2) [60].

Thus, in a representative example, 2-bromothiophene (1 g, 1 mmol) was added to a 50 ml centrifuge tube containing active zinc (Zn*, 1.5 equiv) in dry THF under an Ar atmosphere. After 30 min stirring at rt, the reaction mixture was centrifuged and cannulated to another flask at 0°C. Di-*tert*-butyl azodicarboxylate (1 mmol in THF) was then added dropwise over 5 min. After stirring (1 h), the reaction was quenched with sat. $NaHCO_3$, extracted (Et_2O), concentrated, and purified by flash chromatography (silica gel, hexanes : ethyl acetate) to afford pure **2g** as a white solid (Table 3.17, **2g**, 80%). In the case of aromatic bromides, reflux conditions were required (Table 3.17, **h** and **i**). An excess of active zinc was used in all cases to assure that oxidative addition went to completion within reasonable times. Di-*tert*-butyl hydrazinodicarboxylates

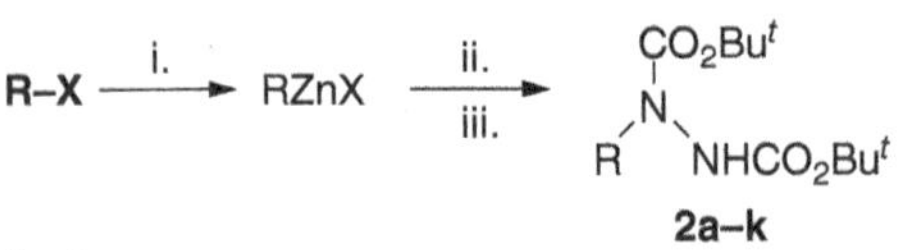

R–X —(i.)→ RZnX —(ii., iii.)→ R–N(CO_2Bu^t)–$NHCO_2Bu^t$

1a–k **2a–k**

Scheme 3.2 (i) Zn*(1.5–3 equiv), THF. (ii) DTBAD (1 equiv), 0°C. (iii) Sat. $NaHCO_3$. X=Br(**1a–1i**, **1k**), I(**1j**).

Table 3.17 Preparation of compounds **2**.[a]

	Starting halide (RX, 1)	Conditions[b] Zn*:1; *T* (°C); time (h)	Product 2, NMR/isolated yield[c]
1a	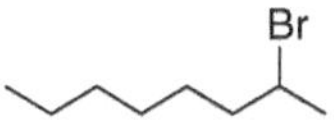	2:1; 25°; 3	**2a**, 94/94
1b	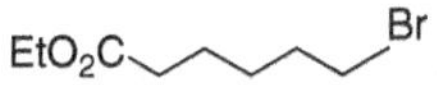	2:1; 25°; 3	**2b**, 95/90
1c	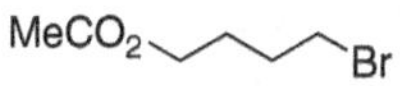	2:1; 25°; 3	**2c**, 90/90
1d	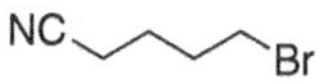	2:1; 25°; 1	**2d**, 94/90
1e		2:1; 25°; 3	**2e**, 90/81
1f	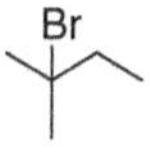	2:1; 25°; 0.5	**2f**, 73/75
1g	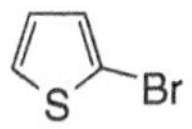	1.5:1; 25°; 0.5	**2g**, 80/80
1h	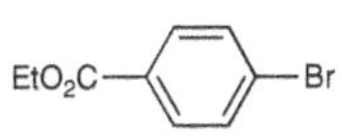	3:1; 70°; 3	**2h**, –/40
1i	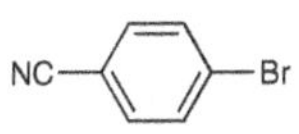	3:1; 70°; 3	**2i**, 65/66
1j	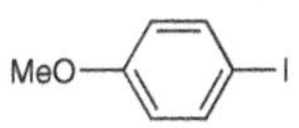	3:1; 25°; 3	**2j**, 55/–
1k	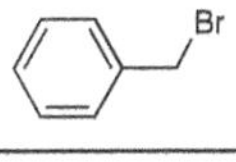	1.5:1; 25°; 0.5	**2k**, 90/90

[a] All products were ≥95% pure (300 and/or 200 MHz ^{1}H-NMR) and were fully characterized by spectroscopic means (IR, ^{1}H-, and ^{13}C-NMR, MS).
[b] Corresponding to the reaction of **1** with Zn*.
[c] ^{1}H-NMR yield using bibenzyl as a standard added to the crude mixture. Isolated yield after flash chromatography (silica gel, hexanes:ethyl acetate). Both based on starting material **1** or DTBAD.

(i.e., **2a–2k**) have been reported to have largely uninterpretable ^{1}H-NMR spectra at room temperature [71]. We also found that NMR procurement at either 300 MHz, 25°C in $CDCl_3$, or 500 MHz, 100°C in DMSO-d_6, proved to be unsuccessful. Line-broadening and coalescence phenomena were observed throughout, as a consequence of the restricted rotation of the hydrazine and amide bonds in these biplanar hydrazine derivatives [72]. Low field NMR (200 MHz, 100°C in DMSO-d_6), however, gave satisfactory results [73]. Complete spectroscopic characterization of the products **2** was done under these conditions.

The reaction products are direct precursors of the corresponding hydrazine derivatives by removal of the *tert*-butoxycarbonyl groups with CF_3CO_2H [64, 65, 70, 74]. Reductive cleavage of the N—N bond can be accomplished using H_2/Ni Raney to afford the amines [65, 66, 74]. Other deprotective protocols are also widely described [64, 67, 70, 71, 75]. In summary, the new reported reaction expands the range of uncatalyzed processes of organozinc reagents. It is the first synthetically applicable electrophilic amination of organozinc reagents. It proceeds with excellent yields for most alkyl organozinc bromides, as well as benzylic ones. Good to medium yields are obtained in the reaction with aromatic and heteroaromatic organozinc bromides or iodides. Excellent functional group tolerance is also exhibited in the process, similar to other reactions mediated by organozinc reagents. Finally, the reaction can be considered as environmentally friendly as it does not require any transition metal catalyst.

3.10 Reformatsky and Reformatsky-Like Reagents and Their Chemistry

One of the oldest organozinc reagents known to the synthetic chemist is the Reformatsky reagent. The Reformatsky reaction (Equation 3.8) is one of the standard methods used for the preparation of β-hydroxy esters [10]. The standard conditions generally employ refluxing

$$R_2C{=}O\ (\mathbf{1}) + Br{-}CH_2{-}COOC_2H_5 + Zn \longrightarrow R{-}C(R)(CH_2{-}COOC_2H_5){-}O{-}Zn{-}Br\ (\mathbf{2}) \xrightarrow{H_3O^{\oplus}} R{-}C(R)(OH){-}CH_2{-}COOC_2H_5\ (\mathbf{3}) \quad (3.8)$$

benzene or benzene–ether solvents, and yields are generally modest at best. One of the major difficulties in carrying out the Reformatsky reaction is the initiation of the reaction and then the control of the very exothermic reaction. Several improvements have been put forth [76], and two publications provide major improvements on the standard reaction [77, 78]. Rathke and Lindert [77] have shown that the reaction can be carried out at room temperature in trimethyl-borate-tetrahydrofuran (TMB-THF), producing high yields. Ruppert and White [78] have used a continuous-flow procedure which also gave very good yields. However, the use of TMB somewhat complicates the workup procedure, and the continuous-flow procedure still requires refluxing benzene.

We have found that Rieke zinc is especially attractive for the generation of the Reformatsky reagent [79–81]. The unusual reactivity of Rieke zinc is further demonstrated by its reaction with ethyl α-bromoacetate, which proceeds rapidly in THF or diethyl ether at −5°C. The reaction was completed in 0.5 h, which was confirmed by GPC analysis. Furthermore, the active zinc reacted with ethyl α-chloroacetate rapidly in ether solvents at room temperature.

When a 1:1 mixture of ethyl α-bromoacetate and ketone was reacted with the active zinc at −5°C, followed by refluxing for various times, yields of the β-hydroxy ester on the order of 70–85% were obtained. It was then discovered that a superior solvent system was diethyl ether. In this case, the highly reactive zinc was generated in THF by the usual procedure, then the THF was removed, and freshly distilled dry diethyl ether is added to the black powder. A one-to-one mixture of ketone or aldehydes and α-bromoester is then added dropwise at ice-bath temperatures. Finally, the reaction mixture is stirred at room temperature for 1 h followed by normal workup procedures. The ability to use diethyl ether at ice bath or room temperatures would appear to make the "activated zinc" procedure highly desirable.

The "activated zinc" will also react rapidly with α-chloroesters in THF. However, the yields are not as high and generally are in the 70–80% range. The Reformatsky reagent can be prepared separately and then added dropwise to the carbonyl compound at room temperature. In addition, the Reformatsky reagent in THF or diethyl ether can be purchased from Rieke Metals, LLC. The shelf life of the Reformatsky reagent is limited and slowly decomposes over a period of 10–20 days even if it is stored in a refrigerator. The best results are obtained if it is used as soon as possible.

Synthesis of Reformatsky Reagent in THF

A 250 ml round-bottomed flask, under Ar, was charged with 50 ml of Rieke Zn* solution (5 g at 0.1 g/ml; 76 mmol). The flask was cooled between 5 and 10°C using a water bath with a little ice. The *tert*-butyl chloroacetate was added dropwise over 10 min. The temperature was maintained between 5 and 20°C. After stirring for 2 h, the reaction was determined to be complete by GC. The reaction was centrifuged to remove excess Rieke Zn*, bottled under Ar, and diluted to 100 ml with fresh THF (0.5 M).

Synthesis of Reformatsky Reagent in Diethyl Ether

The THF supernatant was removed from a bottle of Rieke Zn* and washed 3 times with anhydrous diethyl ether. A 250 ml round-bottomed flask, under Ar, was charged with 50 ml of this Rieke Zn* solution (5 g at 0.1 g/ml; 76.5 mmol) in diethyl ether. The flask was cooled to 0°C using an ice water bath. The *tert*-butyl-chloroacetate was added dropwise over 10 min. The temperature was allowed to reach 24°C during addition. After stirring for 2 h at room temperature, the reaction was determined to be complete by GC. The reaction was centrifuged to remove excess Rieke Zn*, bottled under Ar, and diluted to 100 ml with fresh anhydrous diethyl ether (0.5 M).

3.11 Configurationally Stable Organozinc Reagents and Intramolecular Insertion Reactions

The active zinc may also lend itself toward the possibility of forming configurationally stable organometallics. For example, *cis*-4-*tert*-butylcyclohexyl iodide inserted zinc and was quenched with D_2O at low temperature to give the trans-monodeuterated product (Equation 3.9) [82].

Active zinc is an effective mediator in intramolecular conjugate additions. For example, a spirodecanone was formed from the 1,4-addition of the organozinc reagent, which was readily available from the primary iodide (Equation 3.10) [50]. Other types of ring closures also occur (Equations 3.11 and 3.12) and are thought to proceed by a mechanism that does not involve a free radical pathway.

Zn*, THF; IZn; D_2O, −75°C; D; *cis* : *trans* 1 : 99 (3.9)

Zn*, rt, 24 h; 67–74% (3.10)

Zn* (3.11)

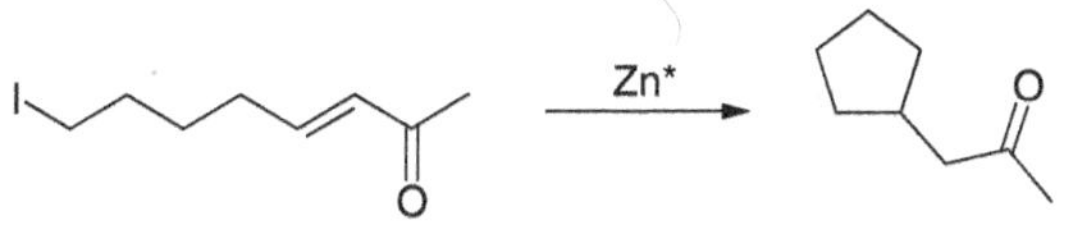

(3.12)

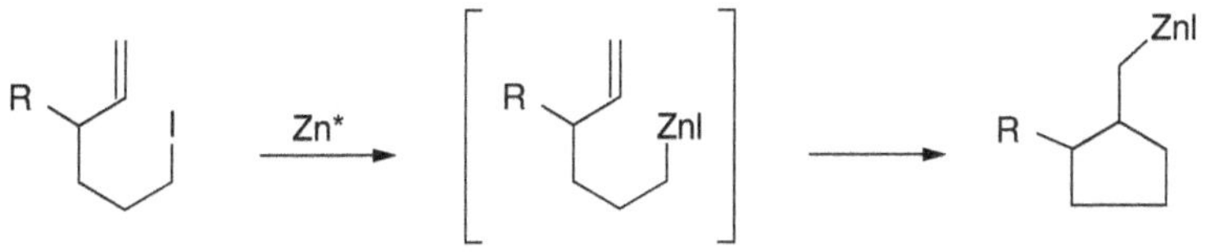

R = CH_3CO, *tert*-BuCO, CH_3CO_2, Me, PivO

(3.13)

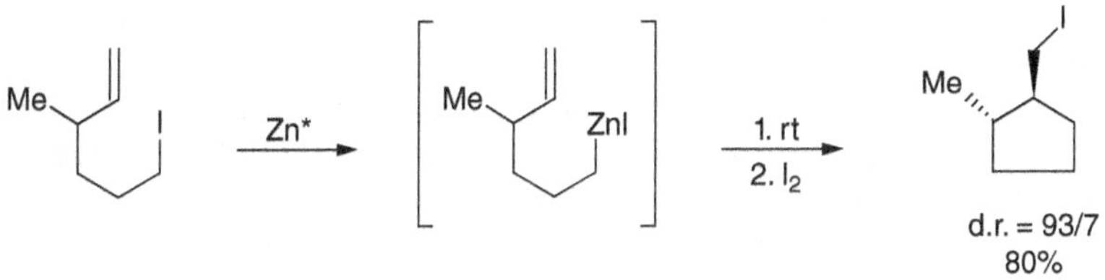

d.r. = 93/7
80%

(3.14)

The direct insertion of active zinc into the carbon–iodine bond of 6-iodo-3-functionalized-1-hexenes forms the corresponding primary alkylzinc iodides, which then undergo an intramolecular insertion of the olefinic π-bond into the zinc–carbon bond to form methyl cyclopentanes (Equations 3.13 and 3.14) [51]. When R is methyl, the product diastereomeric ratio is high. This is a significant finding in that this step is regiospecific 5-*exo*-*trig* cyclization, which can occur in the presence of functional groups. The cyclization produces an intermediate that can be elaborated further with various electrophiles. This methodology has been extended to ω-alkenyl-*sec*-alkylzinc reagents (Scheme 3.3) [52].

3.12 Preparation of Tertiary Amides via Aryl, Heteroaryl, and Benzyl Organozinc Reagents

Amides are of special interest in synthetic organic chemistry because of their wide spectrum in natural compounds. To introduce this moiety, two general approaches have been widely utilized, the addition of an amine to the corresponding carboxylic acid derivative and the coupling reaction of an organometallic with a carbamoyl chloride. Among those, the latter has been more intensively investigated due to the wide variety of organometallic reagents that are available.

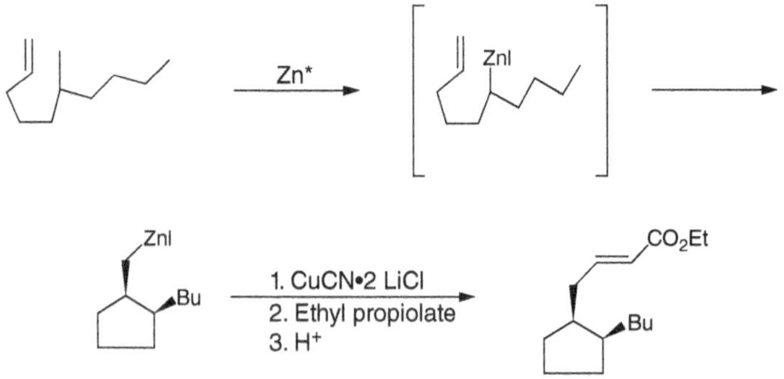

Scheme 3.3 Olefinic π-bond insertion into the carbon—zinc bond.

Grignard reagents have been coupled with carbamoyl chlorides to yield the corresponding amides in the presence of a Ni catalyst [83]. Rouden and coworkers described the preparation of tertiary amides via the coupling reactions of organocuprates with carbamoyl chlorides [84]. In this study, organolithium, organomagnesium, and a limited number of organozinc reagents were used as precursors for making the organocuprates. The palladium-catalyzed cross-coupling reactions between organotin and carbamoyl chlorides also provided the corresponding amides [85]. Of the protocols using organometallic reagents, Suzuki-type reactions with a palladium catalyst and an appropriate base have been the most widely employed for the coupling reaction with carbamoyl chlorides [86]. Direct use of benzylic halides with a carbamoylsilane is yet another approach for the preparation of tertiary amides [87]. In addition to these general synthetic routes, a three-component coupling reaction, aminocarbonylation, for the preparation of a variety of amides was recently reported. It has been accomplished by employing halides (or triflates) and a Weinreb amide at atmospheric CO pressure in the presence of a palladium catalyst along with a ligand such as Xantphos [88].

In spite of the aforementioned methods, a new general approach generating highly functionalized amides would be highly desirable. As is well known, organozinc reagents have been widely utilized in organic synthesis mainly due to their general applicability [89]. We reported a facile protocol for the preparation of tertiary amides utilizing a wide range of organozinc halides [90].

Our initial studies to determine the appropriate reaction conditions and the best catalyst involved the coupling reaction of 2-ethoxycarbonylphenylzinc bromide **2** with diethylcarbamoyl chloride (**A**). The coupling reactions were carried out in THF at refluxing temperature in the presence of 5 mol% of the following catalysts: $Pd(PPh_3)_4$, $Pd(PPh_3)_2Cl_2$, $Ni(dppe)Cl_2$, and $Ni(PPh_3)_2Cl_2$, respectively. In each case, some of the desired product (**1b**) was formed and confirmed by GC-MS analysis of the reaction mixture. Among these, it was found that the use of $Pd(PPh_3)_2Cl_2$ gave the highest conversion to the corresponding coupling product (**1b**). Therefore, this catalytic system was employed in this study yielding the corresponding amides in moderate to good yields. The results are summarized in Table 3.18.

A substituted arylzinc reagent, 4-ethylphenylzinc bromide (**1**) [91], was coupled with carbamoyl chloride **A** in THF at refluxing temperature in the presence of 5 mol% $Pd(PPh_3)_2Cl_2$, resulting in the formation of the corresponding amide (**1a**) in 50% isolated yield (Table 3.18, entry 1). As depicted previously, 2-ethoxycarbonylphenylzinc bromide (**2**) was also reacted with **A** under the same conditions, and the title compound (**1b**) was obtained in 52%

Table 3.18 Coupling reactions of arylzinc halides.

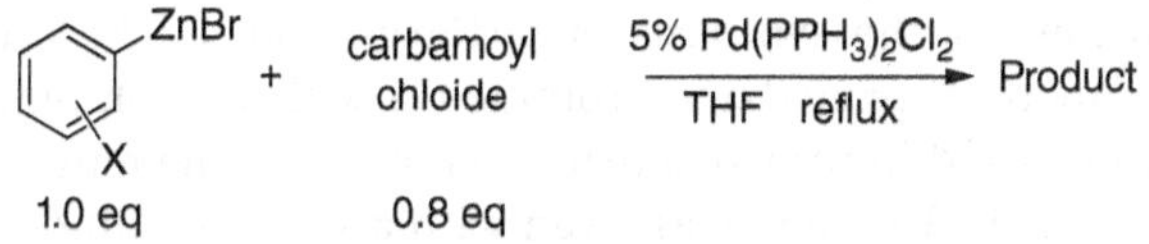

Entry	RZnX	Carbamoyl chloride[a]	Product		Yield (%)[b]
	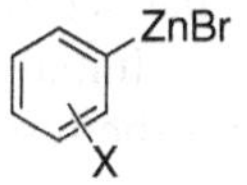		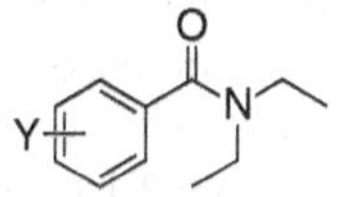		
1	X: 4-CH_2CH_3 (**1**)	A	Y: 4-CH_2CH_3	**1a**	50
2	2-CO_2Et (**2**)		2-CO_2Et	**1b**	52
3	4-CO_2Et (**3**)		4-CO_2Et	**1c**	70
			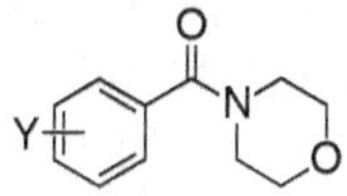		
4	X: 4-CO_2Et (**3**)	B	Y: 4-CO_2Et	**1d**	68
5	4-CH_3 (**4**)		4-CH_3	**1e**	62
6	4-Bromophenylzinciodide (**5**)		4-Br	**1f**	48
7	2-CH_2CH_3 (**6**)		2-CH_2CH_3	**1g**	52
			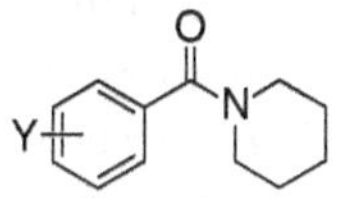		
8	X: 2-OMe (**7**)		Y: 2-OMe	**1h**	66
9	2,6-Me_2 (**8**)	C	2,6-Me_2	**1i**	42

[a] **A**: Diethylcarbamoyl chloride; **B**: morpholinecarbonyl chloride; **C**: piperidinecarbonyl chloride.
[b] Isolated yield (based on carbonyl chloride).

isolated yield (Table 3.18, entry 2). A higher yield was observed from the coupling reaction of **A** with an organozinc reagent (**3**), producing 70% isolated yield (Table 3.18, entry 3). Interestingly, a similar result was achieved from the coupling reaction with carbamoyl chloride **B**. As shown in entry 4 (Table 3.18), the coupling reaction of organozinc **3** with morpholinecarbonyl chloride (**B**) led to the desired product (**1d**) in the highest yield. Again, alkyl-substituted organozinc reagents (**4** and **6**) were readily coupled with **B**, affording the amides (**1e** and **1g**) in 62% and 52% isolated yields (Table 3.18, entries 5 and 7), respectively. Significantly, a bromine atom remained intact under the conditions used in this study. As described in Table 3.18, the coupling reaction of 4-bromophenylzinc iodide (**5**) with **B** proceeded smoothly, producing 4-(bromophenylmorpholino)methanone (**1f**) in moderate yield (Table 3.18, entry 6). Piperidinecarbonylchloride (**C**) can also be used as a coupling partner for the synthesis of tertiary amides. The cross-coupling reaction of 2-methoxyphenylzinc bromide (7) with **C** took place under the same conditions used in previous reactions, yielding the amide (**1h**) in 66% isolated yield (Table 3.18, entry 8). Even with a sterically hindered organozinc, 2,6-dimethylphenylzinc bromide (**8**), the desired coupling product, 2,6-(dimethylphenylpiperidin-1-yl)methanone (**1i**), was obtained in moderate yield (Table 3.18, entry 9).

With these promising results in hand, our studies were expanded to see if this protocol could be used for a much broader range of amides. Initial studies included several heteroaromatic organometallic reagents, and the results are summarized in Table 3.19. The first attempts were conducted with the readily available organozincs 3-bromo-2-thienylzinc bromide (**9**) and 3-thienylzinc iodide (**10**) under the same reaction conditions used in previous coupling reactions. The expected coupling products (**2a** and **2b**) were successfully obtained in 61% and 51% isolated yields (Table 3.19, entries 1 and 2), respectively. Diethylcarbamoyl chloride (**A**) was also effectively coupled with the organozinc **10** to give rise to the amide **2c** in 68% yield (Table 3.19, entry 3). Unfortunately, other heteroarylzinc reagents, pyridyl- and coumarinylzinc halides, failed to yield useful results [92]. No reasonable amounts of the desired product were observed by switching the metal catalyst to $Pd(PPh_3)_4$, Ni(dppe)Cl_2, and $Ni(PPh_3)_2Cl_2$. Instead, by-products from the ring opening of THF by carbamoyl chloride were the major components in the reaction mixture (confirmed by GC-MS) [93].

Cunico and Pandey reported the palladium-catalyzed preparation of α-aryl tertiary amides using benzylic halides and carbamoylsilane [87], and, in this report, a disadvantage of using aminocarbonylation of benzyl halides was also described [90]. Accordingly, to expand the scope of this chemistry, we next focused on the use of benzylic organometallic reagents to develop a facile route for the synthesis of α-aryl tertiary amides. Since the appropriate benzylzinc halides were easily prepared via the direct insertion of highly active zinc, they were employed in the

Table 3.19 Coupling reactions of heteroarylzinc halides.

$$\underset{1.0\ \text{equiv}}{\text{HetAr}-\text{ZnX}} + \underset{0.8\ \text{equiv}}{\text{carbamoyl chloride}} \xrightarrow[\text{THF reflux}]{5\%\ \text{Pd(PPh}_3)_2\text{Cl}_2} \underset{\mathbf{2a-2c}}{\text{amides}}$$

Entry	RZnX	Carbamoyl chloride[a]	Product	Yield (%)[b]
1	(9)	B	2a	61
2	(10)	B	2b	51
3	(10)	A	2c	68
4		B		NR[c]
5		B		NR[c]
6		B		NR[c]
7		B		NR[c]

[a] **A**: Diethylcarbamoyl chloride, **B**: morpholinecarbonyl chloride.
[b] Isolated yield (based on carbonyl chloride), otherwise mentioned.
[c] No desired product, instead, by-products only (confirmed by GC–MS).

coupling reaction. As depicted later in the text, the coupling reactions were carried out under the conditions (5 mol% $Pd(PPh_3)_2Cl_2$/THF/reflux) used in the previous reactions, and the results are summarized in Table 3.20.

As we expected, benzylzinc halides were readily coupled with carbamoyl chlorides to give rise to the desired amides in moderate yields. Coupling

Table 3.20 Coupling reaction of benzylzinc halides.

Benzylzinc (1.0 equiv) + carbamoyl chloride (0.8 equiv) $\xrightarrow[\text{THF reflux}]{5\%\ Pd(PPh_3)_2Cl_2}$ product **3a–3g**

Entry	RZnX	Carbamoyl chloride	Product	Yield[a] (%)
1	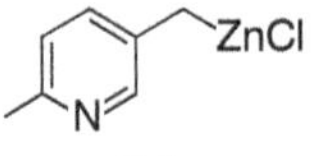 (11)	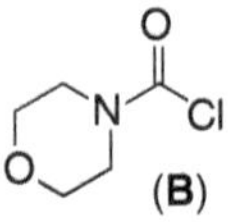 (B)	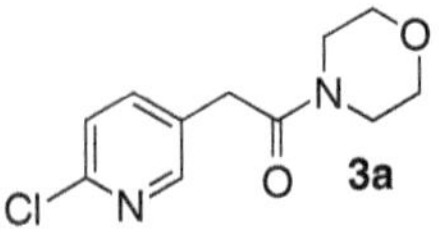 3a	54
2	(11)	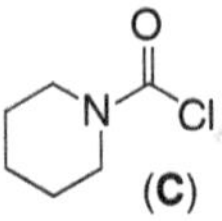 (C)	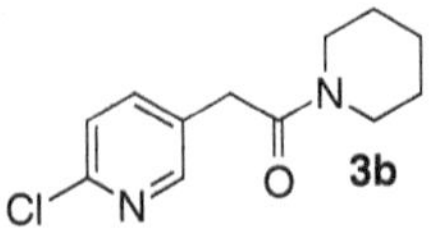 3b	46
3	(11)	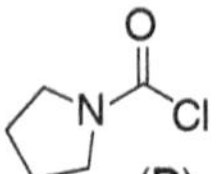 (D)	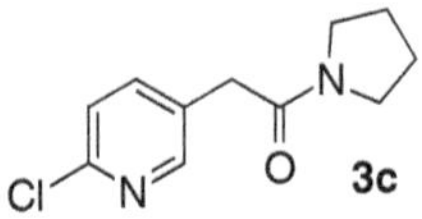 3c	51
4	(11)	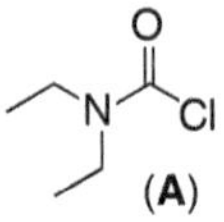 (A)	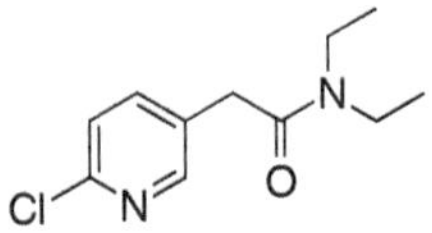	NR[b]
5	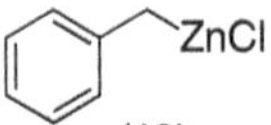 (12)	(A)	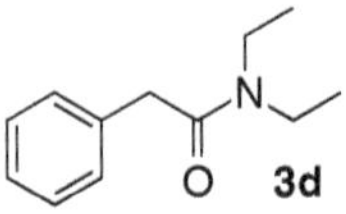 3d	68
6	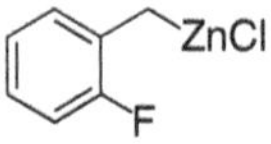 (13)	(B)	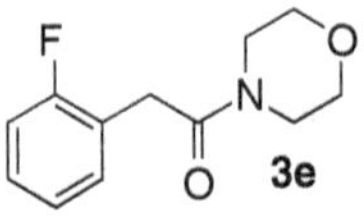 3e	50
7	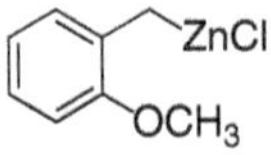 (14)	(B)	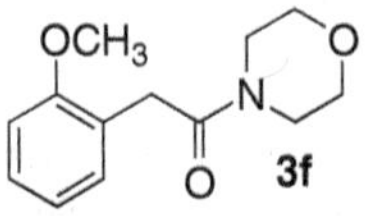 3f	52
8	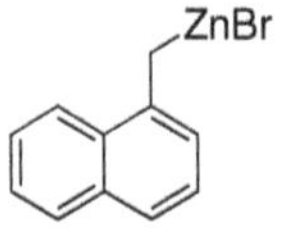 (15)	(B)	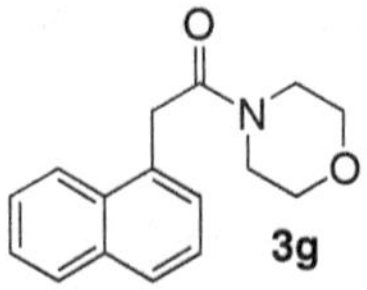 3g	55

[a] Isolated yield (based on carbonyl chloride).
[b] No desired product, instead, by-products only (confirmed by GC–MS).

reactions of 2-chloropyridin-5-yl-methylzinc chloride (**11**) with several carbamoyl chlorides (**B**, **C**, and **D**) were converted to the corresponding amides (**3a**, **3b**, and **3c**) in moderate yields (Table 3.20, entries 1–3). However, the expected product was not observed from the coupling with diethylcarbamoyl chloride (**A**) (Table 3.20, entry 4). Again, the ring-opening byproduct of THF was the only product formed in this reaction. It was of interest that the coupling reaction of **A** with benzylzinc chloride (**12**) proceeded smoothly, providing an amide (**3d**) in 68% isolated yield (Table 3.20, entry 5). Coupling reactions of 2-fluorobenzylzinc chloride (**13**) and 2-methoxybenzylzinc chloride (**14**) with **B** took place successfully and resulted in the formation of the corresponding amides (**3e** and **3f**) in moderate yields (50% and 52%, Table 3.20, entries 6 and 7), respectively. A bulky benzylzinc reagent (**15**) was also efficiently coupled with **B**, leading to the amide (**3g**) in 55% yield (Table 3.20, entry 8).

In summary, we have described the development of a facile and general protocol for the synthesis of tertiary amides. It has been accomplished by a transition metal-catalyzed cross-coupling reaction of organozinc reagents with several carbamoyl chlorides under mild conditions [90]. The use of $Pd(PPh_3)_2Cl_2$ as a catalyst was important for the success. Both functionalized and nonfunctionalized aryl and benzyl organozinc reagents have shown good reactivity, providing the corresponding amides in good to moderate yields. The method has been extended to utilize heteroaromatic zinc halides. However, some limitations have also been found in this study. Nevertheless, this method provides an alternative protocol and reaction conditions that avoid the use of carbon monoxide as a carbonylating reagent for the preparation of tertiary amides.

3.13 Preparation of 5-Substituted-2-Furaldehydes

The majority of the work presented in this chapter was published in a special Howard E. Zimmerman Memorial Issue [94]. This work was published to help celebrate the life and many contributions of Howard E. Zimmerman.

Widespread applications of furan-containing derivatives can be found in a wide range of chemical industries such as pharmaceuticals, textiles, dyes, fossil fuels, photosensitizers, cosmetics, and even in agrochemicals [95]. This rich variety of applicability is attributed to the preparation of numerous furan derivatives. In general, the preparation methods for these derivatives consist of two approaches, ring construction and derivatization of preformed furan rings. Among these, the latter case is a more readily accessible route since many of these derivatives are commercially available.

It is significant that the furan moiety is an interesting structural unit playing a prominent role especially in natural products that show a wide range of biological activity [96]. Of special interest are aryl-substituted furan intermediates, which have been intensively used in the preparation of furan-containing pharmaceuticals [96a]. Due to the particular interest in this area, considerable time and effort have been spent in the development of a versatile synthetic methodology for the preparation of 2,5-disubstituted furan derivatives.

Among the better known procedures as depicted in Scheme 3.4, transition-metal-catalyzed cross-coupling reactions of the corresponding organometallic reagents are one of the predominant approaches [97]. Of these organometallics, unfortunately, little work has been performed using arylmetallic reagents. O'Doherty described the preparation of 5-aryl-2-furaldehydes using palladium-catalyzed cross-coupling reaction of protected furylstannes and/or furylzincs (routes **A** and **B**, Scheme 3.4) [98]. A similar approach using a one-pot, four-step sequence palladium-catalyzed cross-coupling reaction of triorganozincates was reported by Gauthier et al. (route **B**, Scheme 3.4) [99]. The organometallic reagents used in these studies were prepared by the lithiation of the protected furans followed by transmetallation. Generally, cryogenic conditions are required for the lithiation of organic compounds. McClure et al. has reported a one-pot synthesis of 5-aryl-2-furaldehydes via the Suzuki coupling reaction prepared using protected furan moiety route **B** (Scheme 3.4) and also the regioselective palladium-catalyzed direct arylation of 2-furaldehyde (route **C**, Scheme 3.4) [100]. More recently, simple 2-substituted furan derivatives were prepared by iron- and palladium-catalyzed coupling reactions using the Grignard and Suzuki coupling reagents [101]. Knochel has also reported the regio- and chemoselective synthesis of highly substituted furans using Grignard reagents [102].

Despite the present methodologies, there is still a need to introduce a convenient route for the preparation of a variety of 5-substituted furaldehydes. We reported an alternative synthetic route providing a unique way of preparing highly functionalized 2-furaldehydes under mild conditions.

Our chemistry was focused on the utilization of organozinc reagents that were readily accessible. It is well known that the use of organozinc reagents is

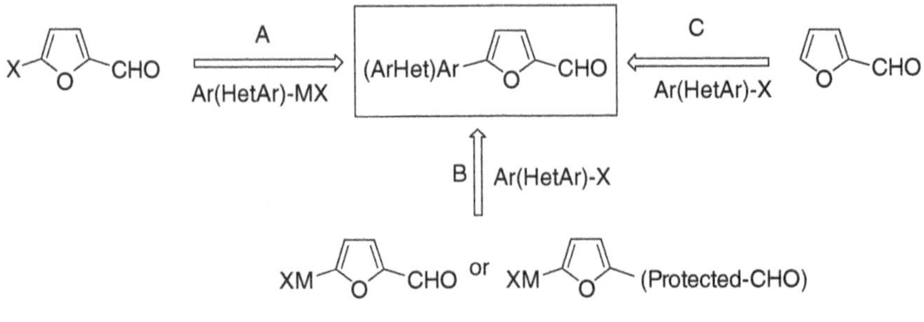

A: X-furan-CHO → (Ar(HetAr)-MX) → (ArHet)Ar-furan-CHO
C: furan-CHO → (Ar(HetAr)-X) → (ArHet)Ar-furan-CHO
B: XM-furan-CHO or XM-furan-(Protected-CHO) → (Ar(HetAr)-X) → (ArHet)Ar-furan-CHO

Scheme 3.4 Representative synthetic routes for 5-substituted 2-furaldehydes.

advantageous over the other organometallics such as the Grignard, Suzuki, and Stille coupling reactions mainly because of the functional group tolerance of the organozincs. The arylzinc reagents used in the coupling reactions with 5-bromo-2-furaldehyde [103] in this study (Tables 3.21, 3.22, and 3.23) were easily prepared by the direct insertion of highly active zinc (Rieke zinc) into the corresponding aryl halides [104]. It was also of interest that all of the subsequent cross-coupling reactions of the resulting organozincs were efficiently carried out in the presence of a catalytic amount of Pd(0) catalyst under very mild conditions affording the cross-coupling products in good to excellent yield. A wide range of 5-substituted 2-furaldehydes were provided through this methodology, and the results are summarized in Tables 3.21, 3.22, and 3.23.

Results and Discussion

Our first attempt was conducted with arylzinc halides bearing various functionalities. As described in Table 3.21, many functionalized arylzinc halides underwent the coupling reaction with 5-bromo-2-furaldehyde under the mild conditions (1 mol% $Pd[P(Ph)_3]_4$, room temperature) and successfully gave 5-aryl-substituted 2-furaldehydes. In the case of electron-withdrawing groups (Table 3.21, entries 1–4), excellent yields (82–93%) were achieved with the exception of 3-cyanophenylzinc iodide (42%, **1d**, entry 4, Table 3.21). For most of the cases, the reactions were completed in 1 h at room temperature. However, when 4-methylphenylzinc bromide bearing an electron-donating group was treated with 5-bromo-2-furaldehyde under the same conditions (room temperature, 1 h) used previously (Table 3.21, entry 7), the corresponding cross-coupling product (**1g**, Table 3.21) was obtained in slightly reduced yields (55%). GC-MS analysis showed that the major impurity was the unreacted 5-bromo-2-furaldehyde. Thus, in the following coupling reaction containing another electron-donating group, 4-methoxyphenylzinc bromide, a longer reaction time was employed to lead the coupling reaction to completion, yielding **1h** (Table 3.21, entry 8). With this coupling product (**1h**), a more interesting result was observed. The color of the isolated product was immediately changed from light yellow to greenish black upon storage in air. We assumed that this was caused by air oxidation, but no further study of this observation was carried out.

In addition to the results previously, the Pd(0)-catalyzed coupling reaction is also applicable to the synthesis of several different types of 5-heteroaryl-2-furaldehydes. Again, the aforementioned mild reaction conditions worked well for the following coupling reactions. The results are summarized in Table 3.22. Expansion of this strategy was first conducted by the coupling reaction of thienylzinc bromides containing a halogen atom, 1,3-dioxane, and ester functionalities. As described in entries 1–3 in Table 3.22, the corresponding products (**2a**, **2b**, and **2c**, respectively) were obtained in good to excellent isolated yields.

Table 3.21 Coupling reactions with arylzinc halides.

ArZnX (1.2 equiv) + 5-bromo-2-furaldehyde (1.0 equiv) → (1% Pd[P(Ph)$_3$]$_4$, THF, r.t.) 5-aryl-2-furaldehydes **1a–1h**

Entry	Arylzinc	Time (h)	Product	Yield (%)[a]
1	4-Cl-C$_6$H$_4$ZnI	1.0	**1a**	93
2	4-EtO$_2$C-C$_6$H$_4$ZnI	0.5	**1b**	82
3	4-NC-C$_6$H$_4$ZnBr	1.0	**1c**	91
4	3-NC-C$_6$H$_4$ZnI	0.5	**1d**	42
5	3-OCOCH$_3$-C$_6$H$_4$ZnI	0.5	**1e**	92
6	3-Cl-4-F-C$_6$H$_3$ZnI	1.0	**1f**	90
7	4-H$_3$C-C$_6$H$_4$ZnBr	1.0	**1g**	55
8	4-H$_3$CO-C$_6$H$_4$ZnBr	24	**1h**	(96)[b]

[a] Isolated yield (based on furaldehyde).
[b] Conversion by GC. No isolated product.

Significantly, these functionalities derived from the corresponding organozinc reagents could be used for the further modification along with the aldehydes at the 2-position of the furan ring. Unfortunately, product **2b** appears to be an unstable compound and cannot be stored for long periods of time. It is of interest that an unsymmetrical furan–furan linkage (**2d**, Table 3.22) has been constructed in 83% isolated yield by treatment with 5-ethoxycarbonyl-2-furylzinc bromide (Table 3.22, entry 4). We also examined the use of heteroarylzinc bromides possessing two hetero atoms in a ring compound such as 2-thiazoylzinc

Table 3.22 Coupling reactions with heteroarylzinc halides.

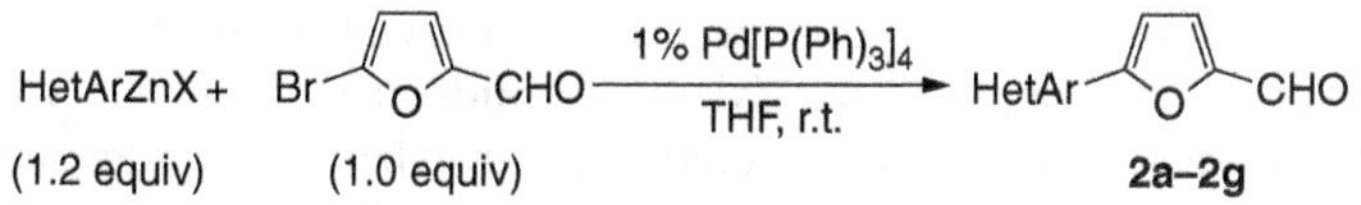

HetArZnX (1.2 equiv) + Br–(furan)–CHO (1.0 equiv) → [1% $Pd[P(Ph)_3]_4$, THF, r.t.] → HetAr–(furan)–CHO **2a–2g**

Entry	Organozinc	Time (h)	Product	Yield (%)[a]
1	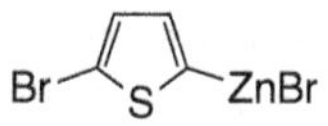 Br, S, ZnBr	1.0	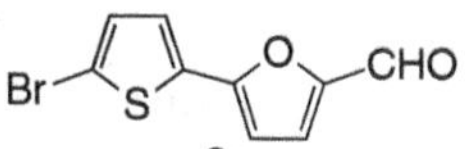 Br, S, O, CHO **2a**	75
2	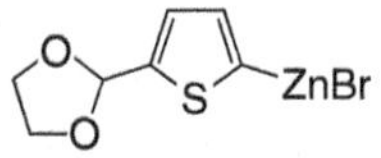 O, O, S, ZnBr	0.5	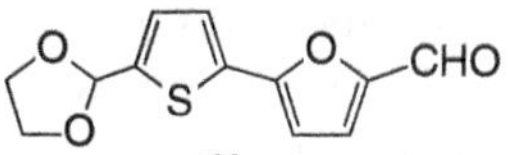 O, O, S, O, CHO **2b**	95
3	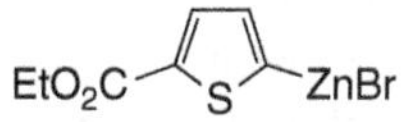 EtO_2C, S, ZnBr	0.5	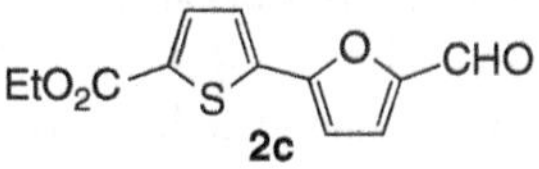 EtO_2C, S, O, CHO **2c**	92
4	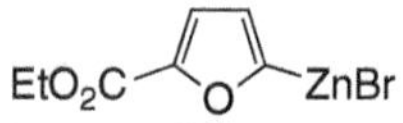 EtO_2C, O, ZnBr	1.0	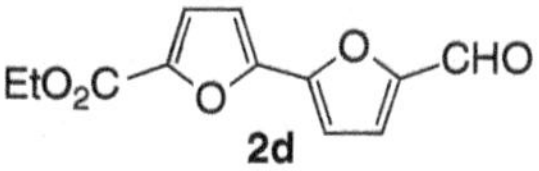 EtO_2C, O, O, CHO **2d**	83
5	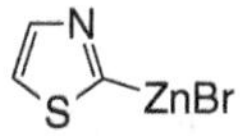 N, S, ZnBr	24	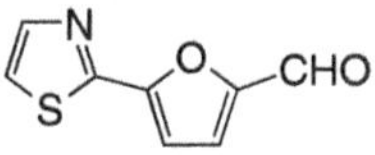 N, S, O, CHO **2e**	43
6	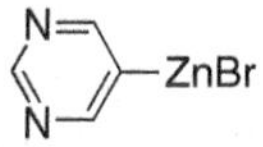 N, N, ZnBr	24	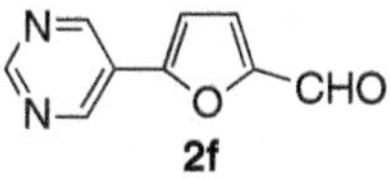 N, N, O, CHO **2f**	33
7	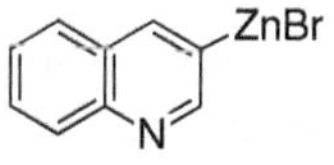 N, ZnBr	24	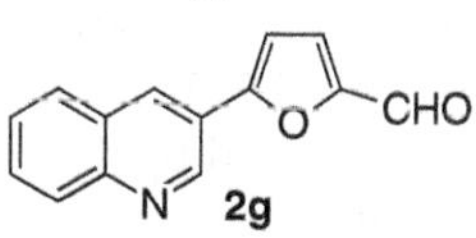 N, O, CHO **2g**	41

[a] Isolated yield (based on furaldehyde).

and 5-pyrimidylzinc bromides. As noted in entries 5 and 6 in Table 3.22, yields (43 and 33%) were somewhat lower than other heteroarylzincs. 3-Quinolinylzinc bromide appeared to be a good coupling partner for 5-bromo-2-furaldehyde, resulting in the formation of **2g** in moderate yield (Table 3.22, entry 7).

To expand the scope of this chemistry, several pyridylzinc halides were chosen and employed in the coupling reaction with 5-bromo-2-furaldehyde. It is worth noting that 5-pyridyl-2-furaldehydes have been frequently used for synthetic intermediates in the preparation of pharmaceuticals [99]. The pyridylzinc halides used in this study were also easily prepared by the direct oxidative addition of active zinc to the corresponding halides [105]. As summarized in Table 3.23, the coupling reactions were carried out under the same reaction conditions as used in the previous study (a catalytic amount of $Pd[P(Ph)_3]_4$ at room temperature in THF). An excellent yield was obtained from the reaction of simple 2-pyridylzinc bromide in 1 h (Table 3.23, entry 1). Sterically hindered organozinc, 3-methyl-2-pyridylzinc bromide, required a prolonged reaction time affording **3b** in lower yield (Table 3.23, entry 2). The coupling reaction of 5-methyl-2-pyridylzinc bromide, however, was completed in 6 h at room temperature to produce **3c** in good yield (Table 3.23, entry 3). An extended time (24 h) was required for the methoxy-substituted pyridylzinc bromide at room temperature in THF (Table 3.23, entry 4). 2-Pyridylzinc bromide containing a fluorine atom was successfully employed in the coupling reaction with 5-bromo-2-furaldehyde to give the coupled product **3e** in 60% yield (Table 3.23, entry 5). Along with the 2-pyridylzinc bromides, 3-pyridylzinc bromides (Table 3.23, entries 6–8) were also easily reacted with 5-bromo-2-furaldehyde at room temperature in THF to afford the corresponding products (**3f**, **3g**, and **3h**, Table 3.23) in good to excellent yields. Moreover, in the reaction with 2-chloro-4-pyridylzinc bromide, the coupling reaction proceeded smoothly to give **3i** in 66% yield (Table 3.23, entry 9).

Even though those approaches used previously provided a variety of furan derivatives, we developed a somewhat different approach for the preparation of 5-substituted 2-furaldehydes containing a unique functionality in the 5-position. One of the reasons is that some of the organozinc reagents bearing especially hydroxyl- and amino-functionalities are not readily available using the direct insertion method. More interestingly, introducing a carbonyl group in the 5-position is not possible using the methodology used in previous approaches. In contrast to this fact, it was successfully accomplished by the direct preparation of the furylzinc reagent and its subsequent use in the coupling reactions in our study (route **B** in Scheme 3.4).

As shown in Scheme 3.5, 5-(1,3-dioxolan-2-yl)-2-furanylzinc bromide **1** was easily prepared as expected by the direct insertion of active zinc to 2-(5-bromofuran-2-yl)-1,3-dioxolane under mild conditions. To confirm the formation of the corresponding organozinc reagent, an aliquot of the reaction mixture was quenched with iodine and analyzed by GC and GC-MS. Both analyses

Table 3.23 Pd-catalyzed coupling reactions with pyridylzincs.

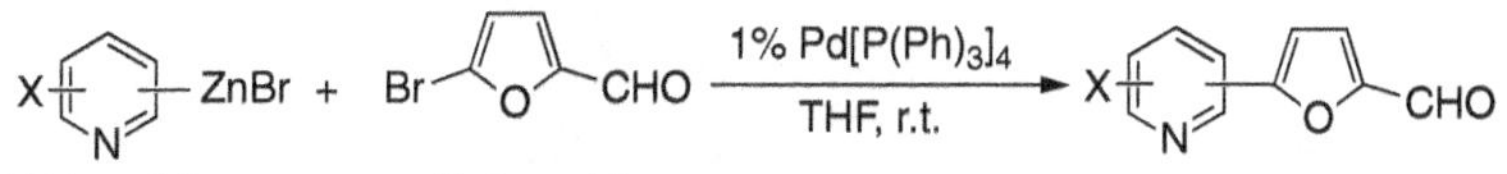

X–pyridyl–ZnBr (1.2 equiv) + Br–furyl–CHO (1.0 equiv) → 1% $Pd[P(Ph)_3]_4$, THF, r.t. → X–pyridyl–furyl–CHO **3a–3i**

Entry	RZnX	Time (h)	Product	Yield (%)[a]
1	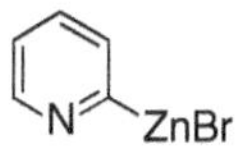 N, ZnBr	1	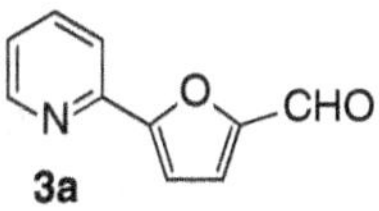 N, O, CHO **3a**	92
2	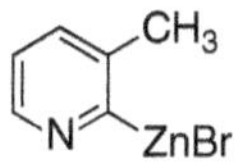 CH_3, N, ZnBr	24	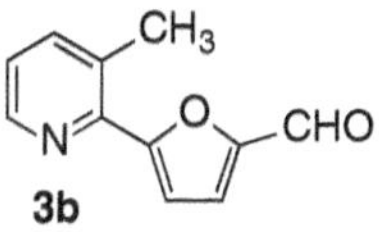 CH_3, N, O, CHO **3b**	56
3	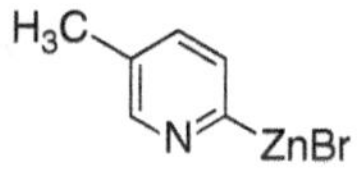 H_3C, N, ZnBr	6	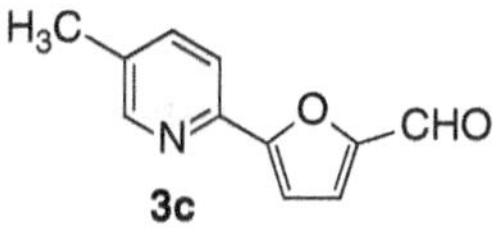 H_3C, N, O, CHO **3c**	80
4	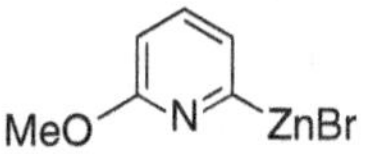 MeO, N, ZnBr	24	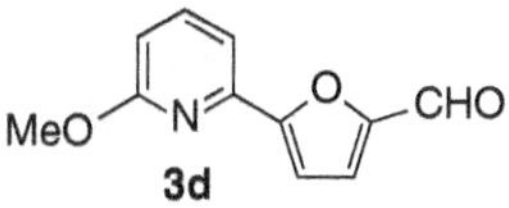 MeO, N, O, CHO **3d**	50
5	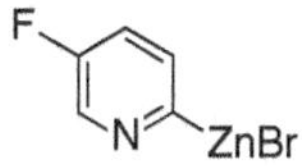 F, N, ZnBr	6	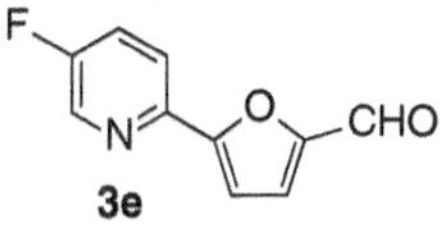 F, N, O, CHO **3e**	60
6	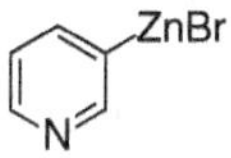 ZnBr, N	6	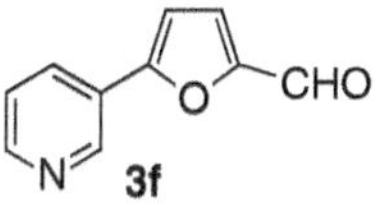 O, CHO, N **3f**	85
7	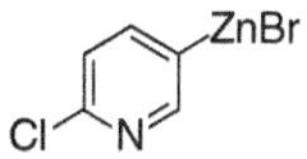 ZnBr, Cl, N	1	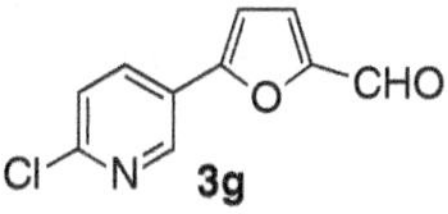 O, CHO, Cl, N **3g**	95
8	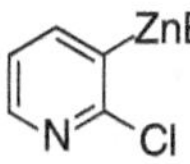 ZnBr, N, Cl	6	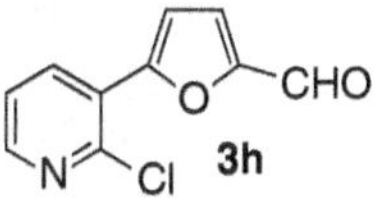 O, CHO, N, Cl **3h**	72
9	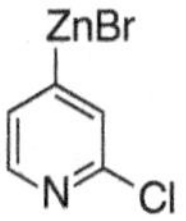 ZnBr, N, Cl	6	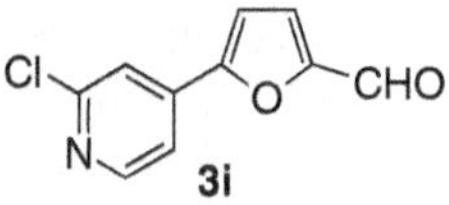 Cl, O, CHO, N **3i**	66

[a] Isolated yield (based on furaldehyde).

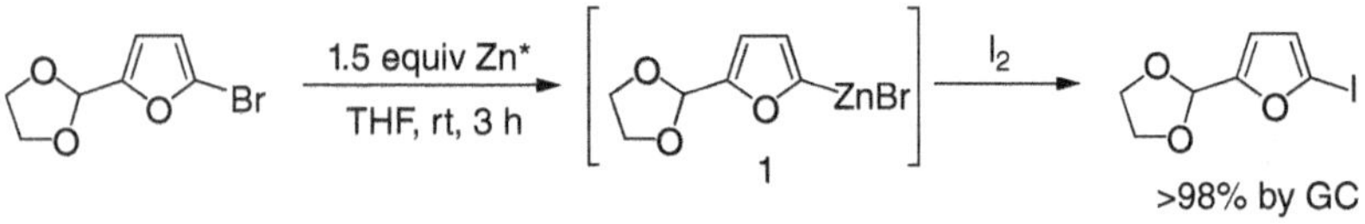

Scheme 3.5 Preparation of 5(1,3-dioxolan-2-yl)-2-furanylzinc bromide (**1**).

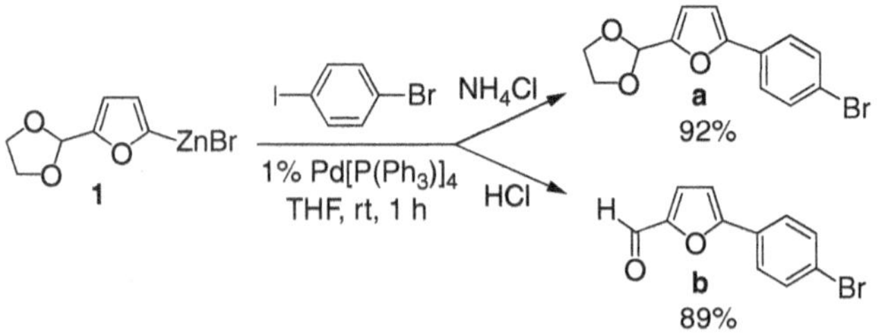

Scheme 3.6 Preparation of 5-aryl-substituted furans.

clearly showed the formation of 2-(5-iodofuran-2-yl)-1,3-dioxolane. From this result, it could be inferred that the corresponding reagent (**1**) was successfully formed. To find out the usefulness of the resulting organozinc reagent, subsequent coupling reactions were performed with a variety of different types of electrophiles such as aromatic halides, haloamines, haloalcohols, and carboxylic acid chlorides.

Prior to the general applications of 5-(1,3-dioxolan-2-yl)-2-furanylzinc bromide **1**, a typical Pd-catalyzed C—C bond forming reaction with 1-bromo-4-iodobenzene was carried out. As described in Scheme 3.6, two derivatives were achieved depending upon the workup procedure.

The coupling reaction was completed in 1 h at room temperature in THF in the presence of 1 mol% Pd[P(Ph$_3$)]$_4$. As is typical, an acidic workup procedure gave rise to the 5-(4-bromophenyl)-2-furaldehyde (**b**) in 89% isolated yield. Meanwhile, a protected furaldehyde (**a**) was obtained from the workup procedure using ammonium chloride in excellent yield.

As described in the aforementioned report [99], these types of molecules are easily accessible via the cross-coupling reaction of 2-bromo-5-furaldehyde with the corresponding organozinc reagents. In order to produce more complex molecules, several new reaction conditions were tried. Table 3.24 shows the results observed from Pd-catalyzed cross-coupling reactions with a variety of aryl bromides. In an effort to evaluate the overall feasibility of the organozinc **1**, coupling reactions with 2-bromo-5-furaldehyde were carried out first to obtain a pseudosymmetrical bifuraldehyde under the conditions depicted in Table 3.24.

The reaction proceeded smoothly at room temperature and was completed in 30 min. Even though the formation of the cross-coupling product was

Table 3.24 Pd-catalyzed synthesis of 2-(5-arylfuran-2-yl)1,3-dioxolanes.

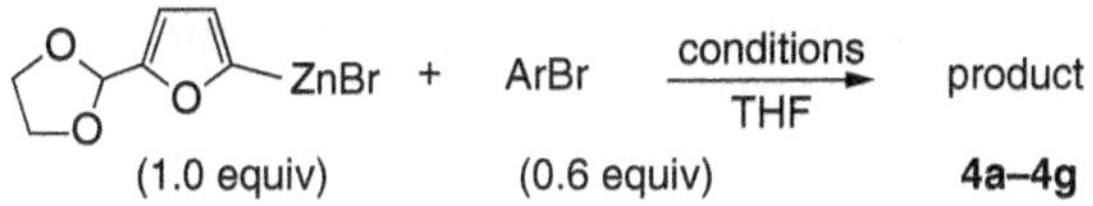

2-(1,3-dioxolan-2-yl)furan-5-ZnBr (1.0 equiv) + ArBr (0.6 equiv) —conditions, THF→ product **4a–4g**

Entry	Halide	Conditions	Product	Yield (%)[a]
1	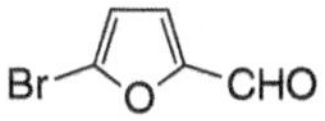 Br-furan-CHO	1% $Pd(PPh_3)_4$ rt 30 min	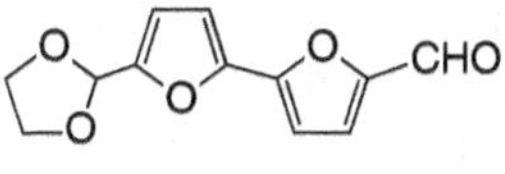 **4a**	(98%)[b]
2	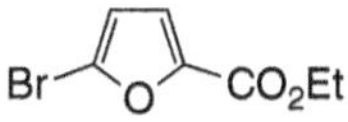 Br-furan-CO_2Et	1% $Pd(PPh_3)_4$ rt 24 h	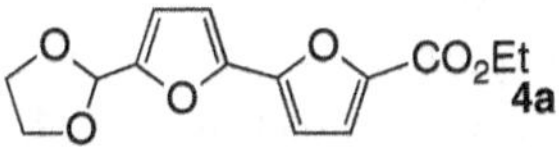 **4a**	63
3	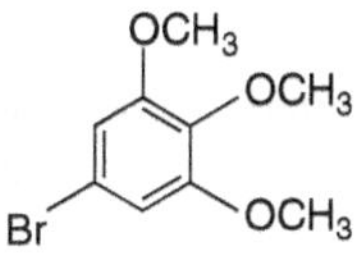 Br-trimethoxybenzene (OCH_3, OCH_3, OCH_3)	2% $Pd(OAc)_2$ 4% SPhos rt overnight	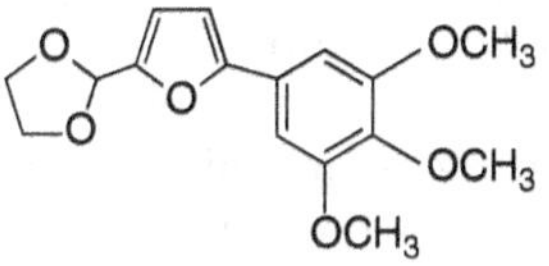 **4c**	73
4	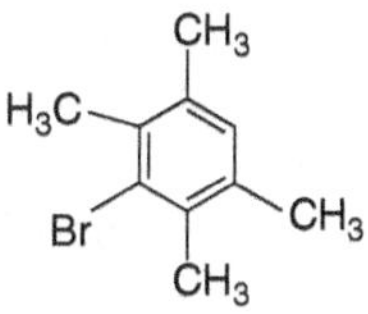 Br-tetramethylbenzene (CH_3, H_3C, CH_3, CH_3)	5% $Pd(PPh_3)_2Cl_2$ rt 24 h	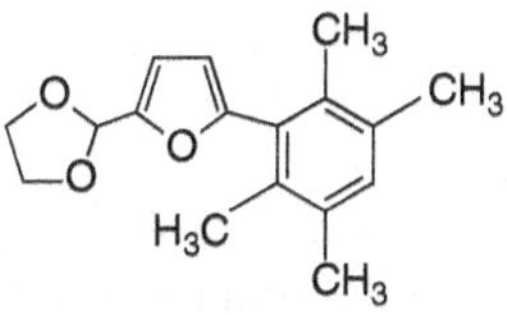 **4d**	64
5	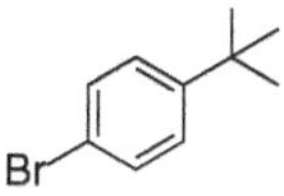 4-*tert*-butylbromobenzene	5% $Pd(PPh_3)_2Cl_2$ rt 2 h	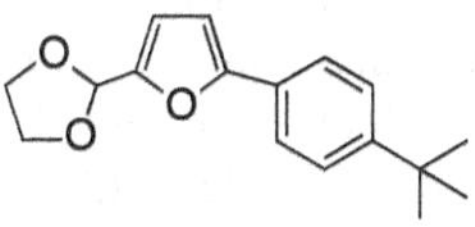 **4e**	82
6	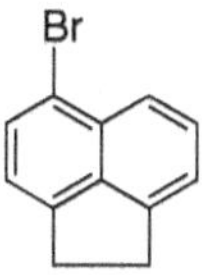 5-bromoacenaphthene	5% $Pd(PPh_3)_2Cl_2$ rt 2 h	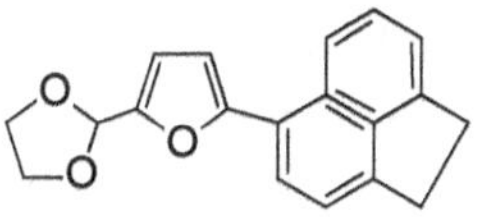 **4f**	93
7	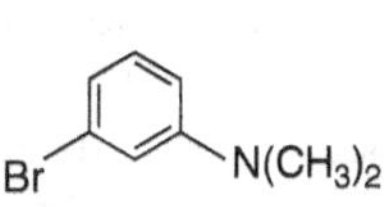 Br-C6H4-$N(CH_3)_2$	5% $Pd(PPh_3)_2Cl_2$ rt 5 h	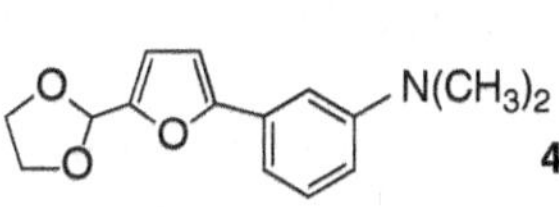 **4g**	92

[a] Isolated yield (based on aryl halide), otherwise mentioned.
[b] Conversion by GC. No isolated product.

confirmed by GC and GC-MS, the separation of coupling product was not possible due to the instability of the product in the atmosphere (Table 3.24, entry 1). However, the similar compound **4b** that has an ester functionality was successfully produced in 63% isolated yield (Table 3.24, entry 2). The next attempt was to couple some aromatic bromides for which the corresponding organozinc reagents were not readily available from the direct insertion method using the active zinc route [106]. Use of the $Pd(OAc)_2$/SPhos catalytic system successfully afforded the coupling product **4c** in 73% isolated yield (Table 3.24, entry 3). In the following several reactions (Table 3.24, entries 4–7), it should be emphasized that no extra ligand was necessary for the coupling reaction in the presence of $Pd[P(Ph_3)]_2Cl_2$. It worked effectively in the coupling reaction at room temperature with tetramethylphenyl, *tert*-butylphenyl and acenaphthyl bromides, leading to the corresponding products **4d**, **4e**, and **4f** in excellent yields (Table 3.24, entries 4–6), respectively. This condition was also very effective with an electron-rich *N,N*-dimethylaminophenyl bromide (Table 3.24, entry 7).

We then attempted the coupling reaction with haloaromatic compounds containing a hydroxyl or an amino functional group. There are very limited examples of coupling reactions of organozinc compounds with haloaromatic alcohols and amines [105] and no reported coupling reaction with furanylzinc bromide. In our study, the coupling reaction was easily accomplished using 2 mol% $Pd(OAc)_2$ and 4 mol% SPhos in THF at room temperature. As described in Table 3.25, it was found that some of the coupling products were not stable enough to be obtained as an isolated product in the atmosphere. It was of interest that the stability of the coupling product is dependent upon the position of the functional group of the aromatic ring. For instance, 4-iodophenyl was coupled well with **1** under mild conditions affording the corresponding product, **5a**, in 92% isolated yield (Table 3.25, entry 1). In contrast, even though the formation of the expected coupling products (**5b** and **5c**) was confirmed by GC-MS analysis of the reaction aliquot from the reaction using 3-iodophenol and 2-iodophenol, respectively, the isolated products (**5b** and **5c**) immediately decomposed upon solvent removal after column chromatography (Table 3.25, entries 2 and 3). In the case of aniline, a similar result was also observed. Again, we were not able to isolate the coupling product **5d** using 4-iodoaniline (Table 3.25, entry 4). Meanwhile, 3-iodoaniline was coupled with **1** giving rise to a stable coupling product **5e** in 80% isolated yield as an orange, oily product (Table 3.25, entry 5).

Subsequent investigation of this chemistry was focused on introducing a carbonyl group in the 5-position of furan. To this end, copper-catalyzed coupling reactions with an acid chloride were applied since this methodology has been one of the most widely used strategies in Negishi coupling. The first attempt was carried out with benzoyl chloride in a standard fashion (10 mol% CuI and 20 mol% LiCl). The coupling product **6a** was achieved in excellent isolated yield (93%, Table 3.26, entry 1). Alkyl acid chlorides (Table 3.26, entries

Table 3.25 Coupling with haloamines and alcohols.

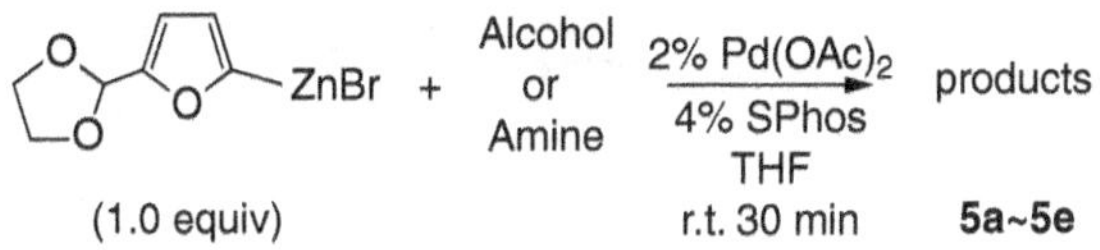

ZnBr (1.0 equiv) + Alcohol or Amine → 2% $Pd(OAc)_2$, 4% SPhos, THF, r.t. 30 min → products **5a~5e**

Entry	Halide	Product	Yield (%)[a]
1	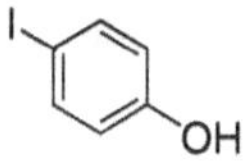 I, OH; 0.4 equiv	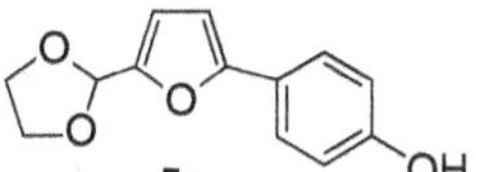 OH; **5a**	92
2	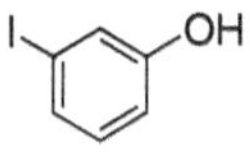 I, OH; 0.4 equiv	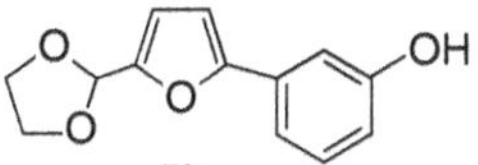 OH; **5b**	(>98%)[b]
3	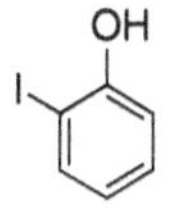 OH, I; 0.4 equiv	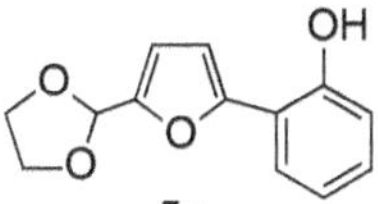 OH; **5c**	(>98%)[b]
4	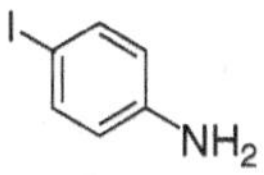 I, NH_2; 0.8 equiv	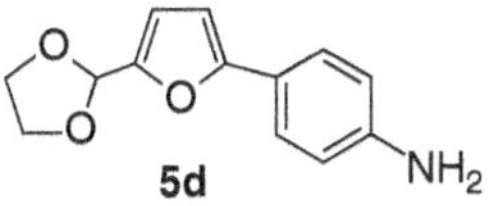 NH_2; **5d**	(>98%)[b]
5	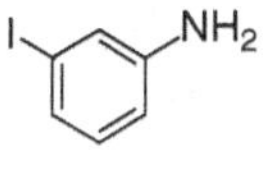 I, NH_2; 0.8 equiv	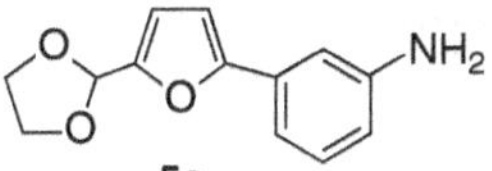 NH_2; **5e**	80

[a] Isolated (based on halide), otherwise mentioned.
[b] Conversion by GC. No isolated product.

3 and 4) were also coupled with **1** to generate ketones **6c** and **6d** in good yields. It should be mentioned that the heterocyclic acid chlorides were also successfully employed in the coupling reaction with 1, providing unsymmetrical heterocyclic ketones **6e** and **6f** in moderate to good yields (Table 3.26, entries 5 and 6). An interesting result was that trifluoroacetic anhydride was also a good coupling partner, and the coupling reaction with **1** gave ketone **6g** in moderate yield (Table 3.26, entry 7). Finally, a S_N2'-type reaction was performed with allyl bromide resulting in the formation of 2-(5-allylfuran-2-yl)-1,3-dioxolane **6h** in 70% yield (Table 3.26, entry 8).

Table 3.26 Cu-catalyzed coupling reaction.

ZnBr + Electrophile $\xrightarrow[\text{20\% LiCl, THF}]{\text{10\% CuI}}$ E

(1.2 equiv) (1.0 equiv) **6a~6h**

Entry	Electrophile	Condition	Product	Yield (%)[a]
1	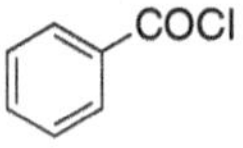 COCl	0°C ~ rt/1 h	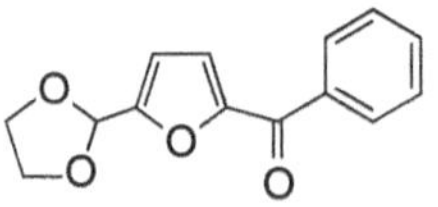 **6a**	93
2	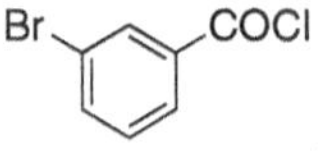 Br, COCl	0°C ~ rt/1 h	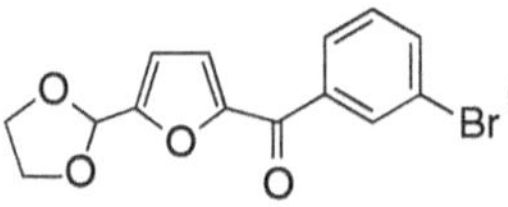 Br **6b**	83
3	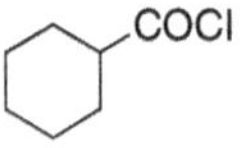 COCl	0°C ~ rt/1 h	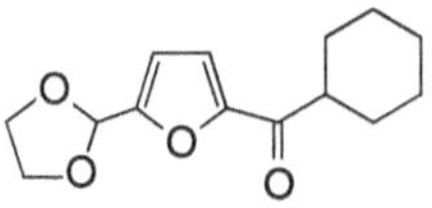 **6c**	90
4	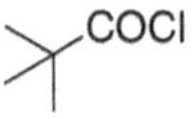 COCl	0°C/1 h	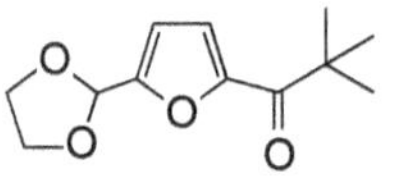 **6d**	89
5	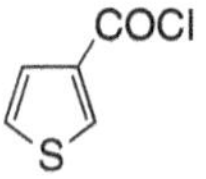 COCl, S	0°C ~ rt/1 h	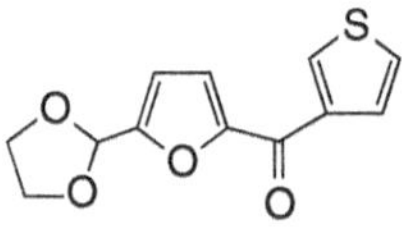 S **6e**	75
6	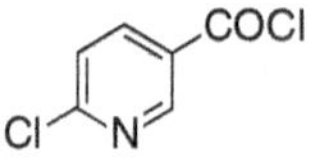 COCl, Cl, N	rt 1 h	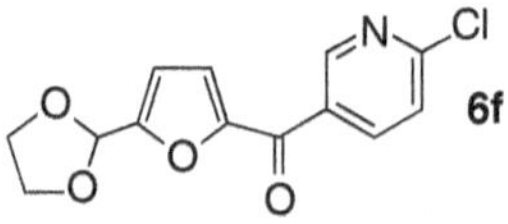 N, Cl **6f**	83
7	$(CF_3CO)_2O$	0°C/1 h	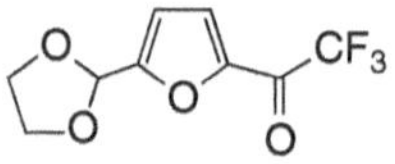 CF_3 **6g**	65
8	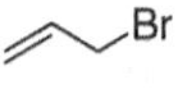 Br	rt 1 h	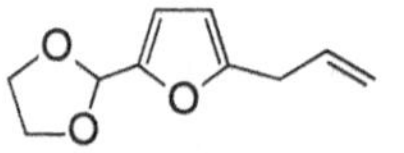 **6h**	70

[a] Isolated yield (based on electrophile).

In conclusion, facile synthetic routes for the preparation of a wide range of 5-substituted 2-furaldehydes are possible. They were accomplished through either Pd-catalyzed cross-coupling reaction of various aryl- and heteroaryl-zinc halides with 5-bromo-2-furaldehyde (route **A**) or utilization of a new organozinc reagent, 5-(1,3-dioxolan-2-yl)-2-furanylzinc bromide **1**, which was easily prepared by the direct insertion of highly active zinc to 2-(5-bromofuran-2-yl)-1,3-dioxolane (route **B**). Of special note is the uniqueness of the route **B**, representing a first example of the direct synthesis of the corresponding organozinc halide. The subsequent coupling reactions of **1** in various types of reaction conditions led to the formation of somewhat different furan derivatives, such as a furan possessing a hydroxyl or aminophenyl substituent and a furan bearing a carbonyl group directly attached at the 5-position. It is also of significance that all of the cross-coupling reactions were carried out under mild conditions.

General Procedure for Pd-Catalyzed Cross-Coupling Reactions

5-(4-Chlorophenyl)furan-2-carbaldehyde **1a**. A 50 ml round-bottomed flask equipped with a stirring bar, a thermometer, and a septum was charged with 0.1 g of $Pd[P(Ph)_3]_4$, and then 20 ml of a 0.5 M solution of 4-chlorophenylzinc bromide (10 mmol) in THF was added into the flask via a syringe. Next, 1.40 g (8 mmol) of 5-bromofuran-2-carbaldehyde was cannulated while being stirred at room temperature. After 1.0 h of stirring at room temperature, the reaction mixture was quenched with saturated 3 M HCl solution and then extracted with ether, which was washed with saturated $Na_2S_2O_3$ and brine and then dried over $MgSO_4$. The mixture was purified by a flash column chromatography on a silica gel column (10% EtOAc/90% heptane) to afford 0.77 g of **1a** as a white solid in 93% isolated yield: mp 127–128°C.

3.14 Preparation and Chemistry of 4-Coumarylzinc Bromide

Due to the particular interest in the coumarin moiety in natural products [107] as well as in material chemistry [108], numerous studies have focused on the development of a versatile synthetic methodology for the preparation of coumarin derivatives. Among the well-known procedures, palladium-catalyzed cross-coupling reactions of organometallic reagents with coumarin derivatives are one of the predominant approaches, especially for the preparation of 4-substituted coumarins (route **A** in Scheme 3.7). Scheme 3.7 illustrates the schematic diagram of the convenient synthetic routes utilizing organometallic reagents for the 4-substituted coumarin derivatives.

Although various methods were reported for the synthesis of 4-substituted coumarin, inconvenient reaction conditions were utilized in most of the previously reported reactions [109]. As mentioned before, route **A** in Scheme 3.7 is the most widely used in the synthesis of the title compound. For example, Yang described the preparation of 4-substituted coumarin using palladium-catalyzed cross-coupling reaction of 4-tosylcoumarins [110a] and nickel-catalyzed

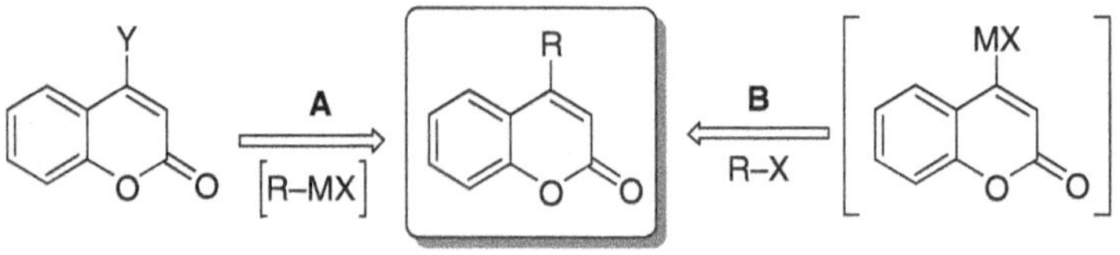

Scheme 3.7 Synthetic routes for 4-substituted coumarin.

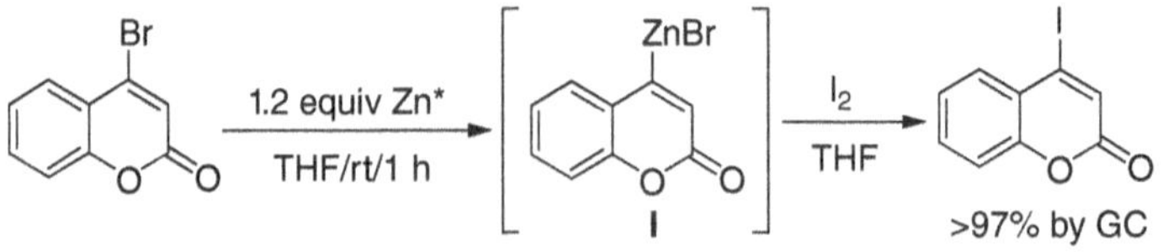

Scheme 3.8 Preparation of 4-coumarinylzinc bromide (I).

cross-coupling reaction of 4-diethylphosphonooxycoumarins with organozinc [110b]. A similar approach using palladium-catalyzed cross-coupling reaction of 4-trifluoromethylsulfonyloxycoumarins or 4-toluenesulfonyloxycoumarins with organostannes was also reported by Wattanasin and Schio, respectively [111]. The Suzuki-type coupling reaction with 4-halocoumarin was employed to generate the 4-substituted coumarins [112].

Even though the route **A** could provide a variety of 4-substituted coumarins, a drawback to this procedure is that some of the organometallic reagents are not readily available. Therefore, in spite of the present methodologies, there is still a need to explore a versatile synthetic methodology for the construction of a chemical library of 4-substituted coumarin derivatives. We considered the alternative route to the synthesis of 4-substituted coumarins, depicted by route **B** in Scheme 3.7. We assumed that this route could provide a versatile way of introducing a variety of different substituents at the 4-position of coumarin. The direct preparation and application of 4-coumarinylzinc bromide represents a novel and versatile approach to many new substituted coumarins [113].

The 4-coumarinylzinc bromide (**1**) was easily prepared by the direct insertion of active zinc to 4-bromocoumarin under mild conditions [114]. The 4-coumarinylzinc bromide was found to undergo cross-coupling with a wide range of electrophiles (Scheme 3.8).

Prior to the study of this reagent in the coupling reactions, an aliquot of the zinc solution was treated with iodine and analyzed by GC and GC-MS to confirm the formation of the corresponding organozinc reagent. Both analyses clearly showed the formation of 4-iodocoumarin. In addition, ^{1}H NMR spectroscopic investigation was completed with the isolated product of the reaction aliquot after quenching [115]. From these results, it could be concluded that the corresponding organozinc reagent (**1**) was successfully obtained.

As described in earlier reports [110–112], introducing aryl and alkyl substituents directly to the C4-position of coumarin has been primarily carried out via palladium-catalyzed cross-coupling reactions with the corresponding organometallic reagents (route **A** in Scheme 3.7). We have taken a different approach which allows the introduction of a carbonyl group in the C4-position. Table 3.27 shows the results observed from Pd-catalyzed cross-coupling reactions with a variety of aryl acid chlorides. These reactions yield a carbonyl group directly bonded to the C4-position of coumarin. All of the coupling reactions were completed in 30 min at room temperature in THF in the presence of 1 mol% $Pd(PPh_3)_4$. Not only a simple acid chloride but also halogenated benzoyl chlorides were successfully coupled with **1** under very mild conditions, affording the corresponding products (Table 3.27, **1a**, **1b**, and **1c**) in excellent yields. Both coupling reactions with benzoyl chlorides possessing an electron-withdrawing group (CF_3) and an electron-donating group (OCH_3) gave rise to the products (Table 3.27, **1d** and **1e**) in 91% and 92% isolated yield, respectively. A benzoyl chloride possessing a benzylic halide reacted under the conditions used in this study to give the product **1f** in good yields (Table 3.27, entry 6). Unfortunately, the desired coupling product was not obtained from the

Table 3.27 Coupling reaction with benzoyl chloride.

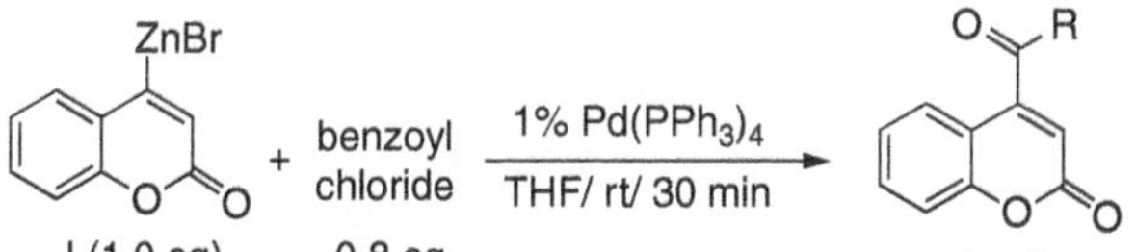

ZnBr; I (1.0 eq) + benzoyl chloride 0.8 eq; 1% $Pd(PPh_3)_4$, THF/ rt/ 30 min → 1a–1f (O, R)

Entry	Acid chloride	Product[a] (%)
	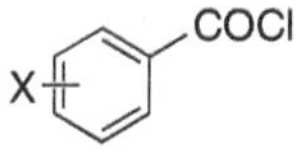 X–C6H4–COCl	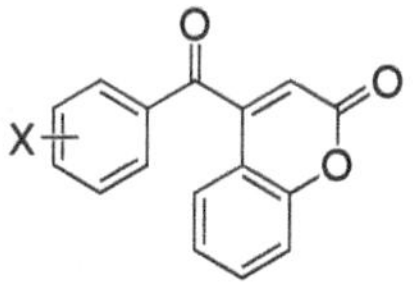
1	X = H (**1a**)	(90)
2	X = 3-Br (**1b**)	(86)
3	X = 4-F (**1c**)	(88)
4	X = 4-CF_3 (**1d**)	(91)
5	X = 4-OMe (**1e**)	(92)
6	X = 3-CH_2Cl (**1f**)	(79)

[a] Isolated yield (based on acid chloride).

coupling reactions with alkyl carbonyl chlorides under the same reaction conditions.

Since the heterocyclic moiety found in many natural compounds plays a critical role for biological activity, the next attempt was to couple **1** with several heteroaryl acid chlorides. Table 3.28 illustrates the reaction conditions and the results. Once again, it should be emphasized that all the coupling reactions were carried out under mild conditions. 5-Bromonicotinoyl chloride and 6-chloronicotinoyl chloride were efficiently employed for the coupling reaction

Table 3.28 Coupling reaction with heteroaryl acid chloride.

$$\underset{1.0\ \text{equiv}}{\mathbf{I}} + \underset{0.8\ \text{equiv}}{\text{acid chloride}} \xrightarrow[\text{THF/rt/30 min}]{1\%\ \text{Pd(PPh}_3)_4} \underset{\mathbf{2a-2e}}{\text{product}}$$

Entry	Acid chloride	Product[a] (%)	Yield[a] (%)
1	Br, COCl, N	O, O, N, O, **2a**	75
2	COCl, Cl, N	O, O, Cl, N, O, **2b**	70
3	COCl, S	O, O, S, O, **2c**	85
4	O, COCl	O, O, O, O, **2d**	79
5	COCl	O, O, O, **2e**	81

[a] Isolated yield (based on acid chloride).

giving rise to the products (**2a** and **2b**) in moderate yields (Table 3.28, entries 1 and 2). 3-Thiophenecarbonyl chloride and 2-furoyl chloride were also coupled with **1** under the same conditions, and the coupling products (Table 3.28, **2c** and **2d**) were obtained in 85% and 79% yields, respectively.

In an effort to evaluate the overall feasibility of the organozinc **1**, coupling reactions with arylhalides were performed. The aforementioned route A can also provide the 4-aryl-substituted coumarins. However, as illustrated in Scheme 3.7, the strategy (route **B**) was to use a totally new organometallic reagent instead of known organometallic reagents. Palladium-catalyzed coupling reaction of **1** with iodobenzene (Table 3.29, entry 1) and bromobenzenes (Table 3.29, entries 2, 3, and 4) proceeded smoothly, resulting in the formation of 4-aryl-substituted coumarins (Table 3.29, **3a**, **3b**, **3c**, and **3d**) in good yields. Heteroaryl halides also proceeded smoothly to yield 4-heteroaryl-substituted coumarins in good yields. Interestingly, the reaction conditions used before worked well for the coupling reaction with heteroaryl compounds. 2-Bromothiophene reacted with **1** to afford **3e** in 85% yield (Table 3.29, entry 1). Good results were obtained from the coupling reactions using bromothiophene and bromofuran bearing a functional group (Table 3.29, entries 6 and 7). The coupling reaction with 2-bromopyridine also proceeded well to generate **3h** in good yield (Table 3.29, entry 8).

In order to demonstrate the exceptional versatility of this reaction, two other very different types of halides were reacted. A S_N2'-type reaction was performed with allyl bromide resulting in the formation of 4-allylcoumarine (**s1**) in 83% yield (Scheme 3.9). Additionally, the coupling reaction was investigated with a haloaromatic amine. As illustrated in Scheme 3.9, the coupling reaction was easily accomplished with 4-iodoaniline using 2 mol% $Pd(OAc)_2$ and 4 mol% SPhos in THF at refluxing temperature affording 4-(4′-aminophenyl)coumarin (**s2**) in moderate yield [116].

In conclusion, a novel synthetic route for the preparation of 4-substituted coumarin derivatives has been demonstrated. It has been accomplished by utilizing a simple coupling reaction of a readily available 4-coumarinylzinc bromide (**I**), which was prepared via the direct insertion of active zinc to 4-bromocoumarin. The subsequent coupling reactions with a variety of different electrophiles have been carried out under mild conditions, providing a new class of 4-substituted coumarins.

3.15 Preparation and Cross-Coupling of 2-Pyridyl and 3-Pyridylzinc Bromides

Heterocyclic compounds which contain a pyridine ring are frequently found in natural products. Accordingly, they are of special interest in pharmaceutical, agrochemical, and medicinal chemistry [117]. Also, a number of pyridine derivatives have been used in material chemistry [118]. Bipyridine groups are

Table 3.29 Coupling reaction with arylhalide.

$$\underset{1.0\ \text{equiv}}{\mathbf{I}} + \underset{0.7\ \text{equiv}}{\text{aryl halide}} \xrightarrow[\text{THF/rt/6 h}]{1\%\ Pd(PPh_3)_4} \underset{\mathbf{3a-3h}}{\text{product}}$$

Entry	Halide	Product	Yield[a] (%)	Entry	Halide	Product	Yield[a] (%)
1	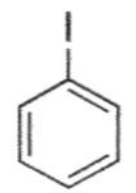 I	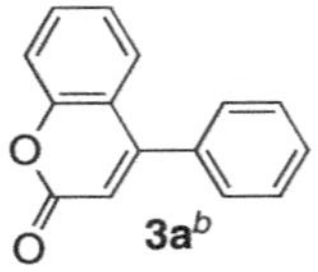 3a[b]	83	5	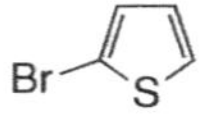 Br, S	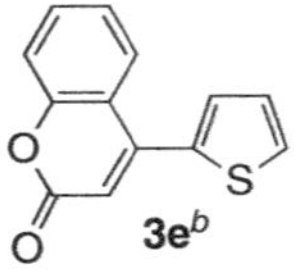 3e[b]	81
2	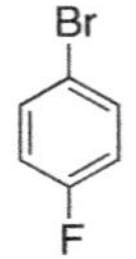 Br, F	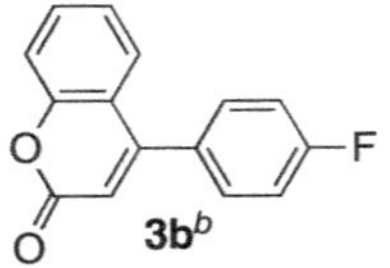 F, 3b[b]	75	6	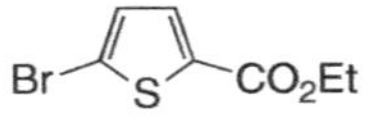 Br, S, CO_2Et	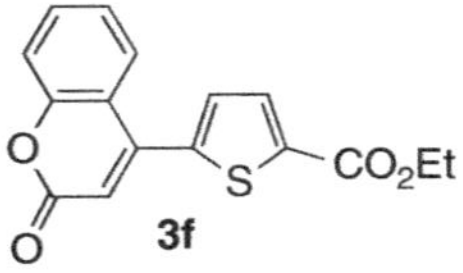 S, CO_2Et, 3f	84
3	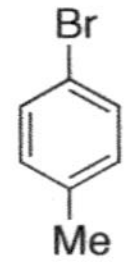 Br, Me	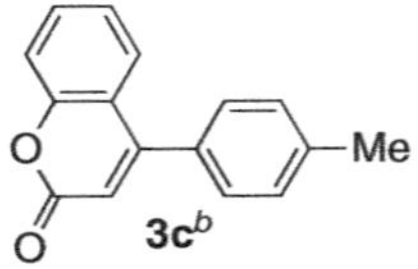 Me, 3c[b]	77	7	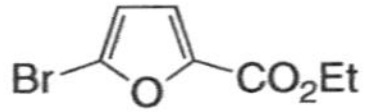 Br, O, CO_2Et	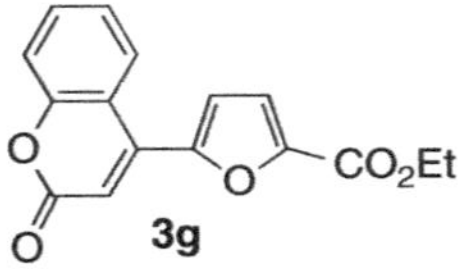 O, CO_2Et, 3g	88
4	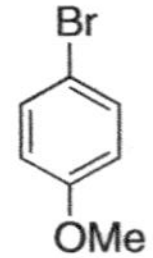 Br, OMe	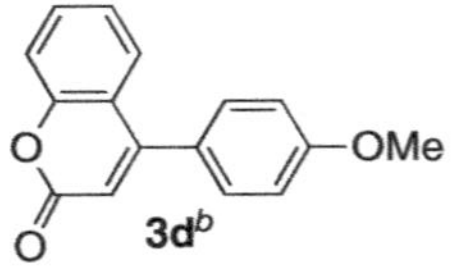 OMe, 3d[b]	76	8	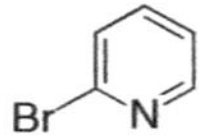 Br, N	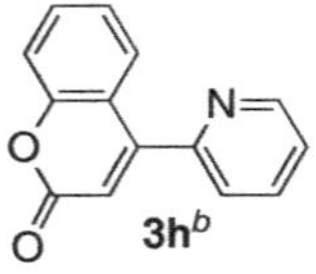 N, 3h[b]	79

[a] Isolated yield (based on aryl halide).
[b] All the spectroscopic data are consistent with literature value (Ref. [4]).

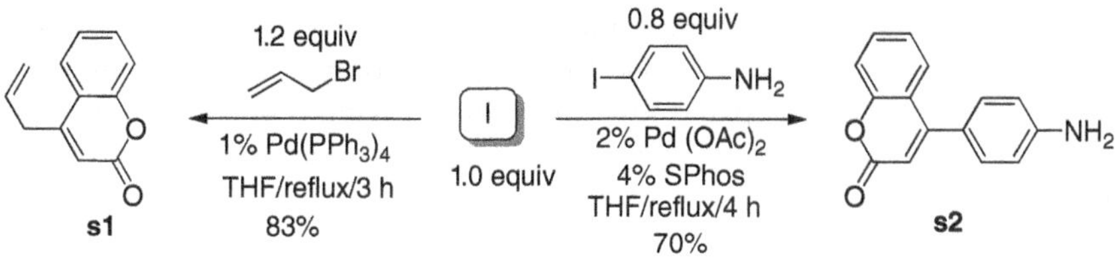

Scheme 3.9 Expansion of coupling reaction.

found to be a key element in antibiotics such as caerulomycins and collismycins. Another example is pyridylprimidines, which are used as fungicides as well as tyrosine kinase inhibitors [119]. In addition, pyridine-containing oligomers are frequently found in liquid crystals [120].

As described previously, the pyridine moiety has played a very significant role in a wide range of organic compounds. Consequently, new practical synthetic approaches for introducing a pyridine ring into complex organic molecules are of high value. To this end, preparation of pyridyl derivatives is mostly performed by transition metal-catalyzed cross-coupling reactions of pyridylmetallic reagents. However, the preparation of electron-deficient pyridinyl organometallic reagents has been a challenging subject mainly because of some difficulties such as instability and formation of by-products.

Most of the 2-pyridyl derivatives have been prepared using the Suzuki [121], Stille [122], Grignard [123], and Negishi [124] coupling reactions in the presence of a transition metal catalyst. Among these, the Suzuki coupling reaction is the most intensively studied, and a very extensive body of work has been developed [125]. Several outstanding studies on the direct arylation of pyridine have been reported to avoid inevitable difficulties. For examples, Rh(I)- [126] and Au(I)- [127] catalyzed arylation of pyridines, Pd-catalyzed arylation of pyridine *N*-oxide with unactivated arenes [128], and haloarenes [129] have been developed. Also, the direct arylation of pyridine *N*-oxide by Grignard reagents was reported [130].

Even though there are many examples of the preparation of 2-pyridylmetallic halides from the reaction of halopyridines, a limited number of studies have been reported on the preparation of 3-pyridylmetallic halides. 3-Pyridylmagnesium [131], 3-pyridylzinc [132], 3-pyridylindium [133] halides, and Suzuki reagents [134] are the most widely used reagents for the preparation of pyridine-containing compounds. Lithiation of 3-halopyridine followed by transmetallation with appropriate metals (Mg, Zn, In) afforded the corresponding 3-pyridylmetallic halides. However, this route has limitations such as cryogenic conditions, several side reactions, and limited functional group tolerance [135]. Very few studies have been reported on the direct synthesis of 3-pyridylmetallic halide reagents. Most of these reports included the treatment of 3-iodo or 3-bromopyridine with highly active metals [136]. Also, the subsequent coupling reactions were carried out with limited electrophiles.

Interestingly, in our continuing study on the preparation and application of organozinc reagents, we found that 2-pyridylzinc bromide and 3-pyridylzinc bromide were easily prepared by treatment of 2-bromopyridine and 3-bromopyridine with active zinc under mild conditions, respectively. Significantly, the resulting organozinc reagents were found to react with a variety of different electrophiles with/without transition metal catalysts, affording the coupling products in good yields [137].

Results and Discussion

In general, the preparation of 2-pyridyl organometallics is mostly performed by lithiation of 2-halopyridine at cryogenic conditions followed by transmetalation with an appropriate metal halide. As mentioned previously, this procedure causes some limitations on the use of the 2-pyridyl organometallics. In our study, readily available 2-bromopyridine was treated at rt with active zinc prepared by the Rieke method [138]. The oxidative addition of the active zinc to carbon–bromine bond was completed in an hour at refluxing temperature to give rise to the corresponding 2-pyridylzinc bromide (**P1**).

In order to investigate the reactivity of the 2-pyridylzinc bromide, it was treated with benzoyl chlorides. As summarized in Table 3.30, the coupling ketone products were obtained in moderate yields. It should be emphasized that the coupling reaction with acid chlorides described in Table 3.30 was carried out in the absence of any transition metal catalyst under mild conditions. Generally, a copper catalyst is widely used for the coupling reactions of organozinc reagents [139]. Halobenzoyl chlorides were easily coupled with 2-pyridylzinc bromide (**P1**) at rt to give the corresponding ketones (Table 3.30, **1a**, **1b**, **1c**, **1d**, and **1e**) in moderate yields. Both benzoyl chlorides containing an electron-withdrawing group (CN and NO_2) and an electron-donating group (Me and MeO) also successfully afforded the corresponding ketones (Table 3.30, **1f**, **1g**, **1h**, and **1i**). Even with nitrobenzoyl chloride, ketone **1j** (Table 3.30) was obtained in moderate yield. According to GC-MS analysis of the reaction mixture, a major by-product was the coupling product obtained from the reaction of the acid chloride with THF.

More results obtained from the catalyst-free coupling reactions are shown in Table 3.31. Treatment of **P1** with chloronicotinoyl chlorides (Table 3.31, entries 1 and 2) at rt for 3 h provided the corresponding ketones (**2a** and **2b**) in 62% and 53%, respectively. Alkyl carbonyl chlorides were also coupled with 2-pyridylzinc bromide resulting in the formation of the ketones (Table 3.31, **2c** and **2d**) in 42% and 63% yields.

We also explored the Pd-catalyzed C—C bond forming reaction of **P1**. Even though 2-pyridylaryl derivatives were successfully prepared via the aforementioned direct arylation methods, relatively harsh conditions (excess amount of

Table 3.30 Coupling reaction of 2-pyridylzinc bromide (**P1**) with benzoyl chlorides.[a]

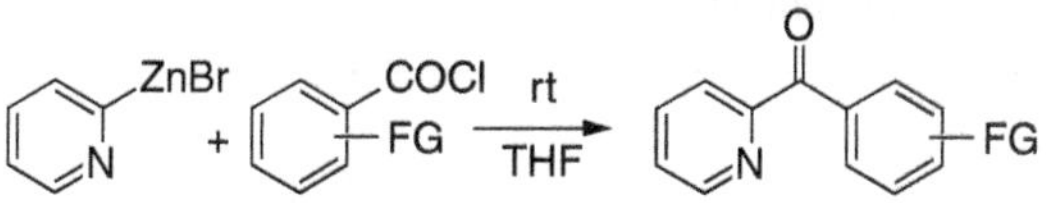

P1 (1.0 eq) + (0.8 eq) → (rt, THF) → **1a–1j**

Entry	FG	Product	Yield (%)[b]
1	2-F	2-F (**1a**)	65
2	3-F	3-F (**1b**)	52
3	2-Br	2-Br (**1c**)	45
4	4-Br	4-Br (**1d**)	47
5	4-I	4-I (**1e**)	36
6	3-CN	3-CN (**1f**)	54
7	4-CN	4-CN (**1g**)	50
8	4-Me	4-Me (**1h**)	64
9	3,4-$(OMe)_2$	3,4-$(OMe)_2$ (**1i**)	40
10	4-NO_2	4-NO_2 (**1j**)	47

[a] No catalyst was used.
[b] Isolated yield (based on electrophile).

Table 3.31 Coupling reaction of **P1** with acid chlorides.[a]

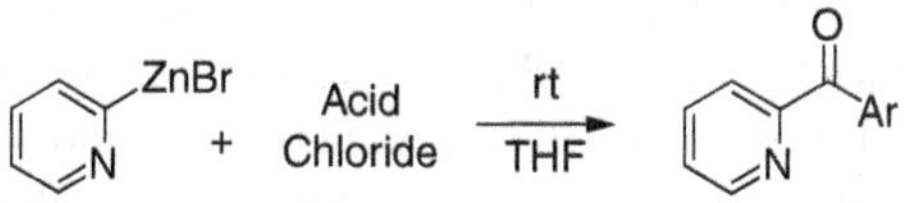

P1 (1.0 eq) + Acid Chloride (0.8 eq) → (rt, THF) → **2a–2d**

Entry	Acid chloride	Product	Yield (%)[b]
1	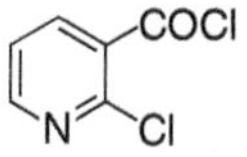	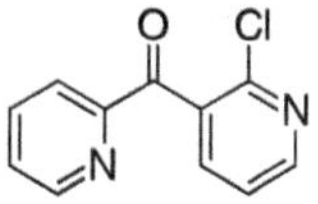 **2a**	62
2	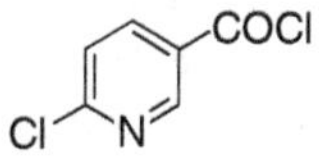	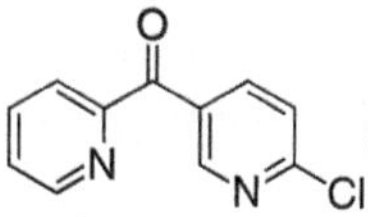 **2b**	53
3	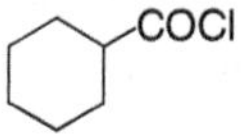	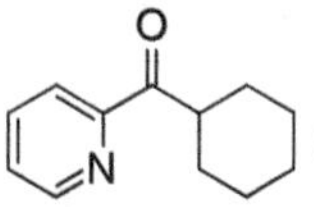 **2c**	42
4	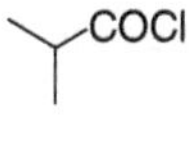	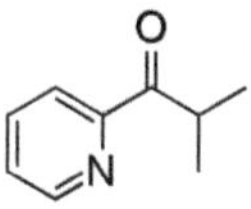 **2d**	63

[a] No catalyst used.
[b] Isolated yield (based on electrophile).

Table 3.32 Study of substituent effect.

Entry	X	Product, X	Yield (%)[a]
1	H (**P1**)	H (**3a**)	85
2	3-CH_3 (**P2**)	3-CH_3 (**3b**)	58
3	4-CH_3 (**P3**)	4-CH_3 (**3c**)	77
4	5-CH_3 (**P4**)	5-CH_3 (**3d**)	79
5	6-CH_3 (**P5**)	6-CH_3 (**3e**)	57
6	6-OCH_3 (**P6**)	6-OCH_3 (**3f**)	54

[a] Isolated yield (based on 3-iodothiophene).

reactant, high temperature, protection/deprotection step, and addition of additives) were required.

Prior to the Pd-catalyzed coupling reaction with a variety of aryl halides, a preliminary test was performed using a Pd(0) catalyst to find any effect of substituents in the C—C bond forming reactions. Several different types of 2-pyridylzinc bromides (Table 3.32, **P1–P6**) were coupled with 3-iodothiophene in the presence of 1 mol% of $Pd(PPh_3)_4$ in THF at rt, and the results are summarized in Table 3.32. In general, good yields (Table 3.32, entries 1, 3, and 4) were obtained from using 2-pyridylzinc bromide (**P1**), 4-methyl-2-pyridylzinc bromide (**P3**), and 5-methyl-2-pyridylzinc bromide (**P4**). Reactions with 3-methyl-2-pyridylzinc bromide (**P2**), 6-methyl-2-pyridylzinc bromide (**P5**), and 6-methoxy-2-pyridylzinc bromide (**P6**) resulted in moderate yields (Table 3.32, entries 2, 5, and 6).

An additional study was carried out to investigate the steric effect on cross-coupling reaction using 2-pyridylzinc bromides (**P2** and **P4**). As shown in Table 3.33, the steric hindrance (76% vs. 89%, 51% vs. 84% isolated yield) was clearly observed from the coupling reactions with 5-bromofuran-2-carboxylic acid ethyl ester and 5-bromothiophen-2-carboxylic acid ethyl ester (Table 3.33, entries 1 vs. 2 and 3 vs. 4), respectively. The results clearly demonstrate that the steric bulk around the reaction site reduces the coupling ability of the corresponding organozinc reagents.

With the preliminary results, we expanded this methodology to the coupling reactions with a variety of haloaromatic compounds. The results are described

Table 3.33 Steric effect on cross-coupling reaction.

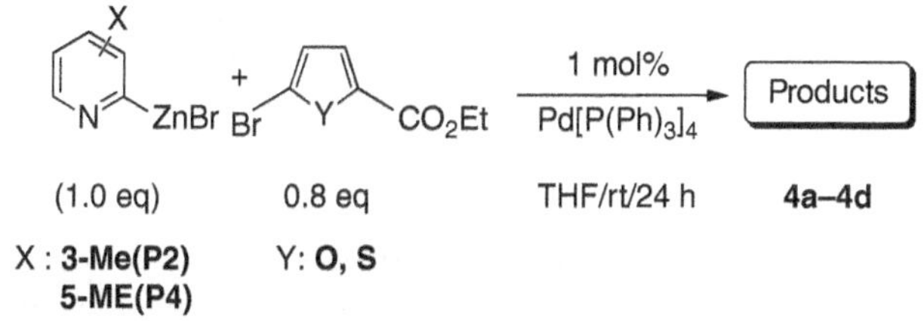

+ Br-(Y)-CO_2Et —1 mol% Pd[P(Ph)$_3$]$_4$, THF/rt/24 h→ Products

(1.0 eq) 0.8 eq **4a–4d**

X : **3-Me(P2)** Y: **O, S**
5-ME(P4)

Entry	RZnBr	Y	Product	Yield (%)[a]	
1	**P2**	O	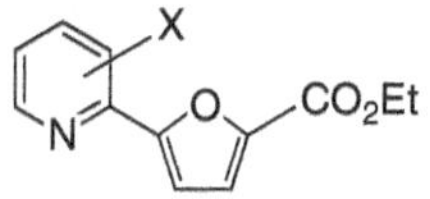	X; 3-Me(**4a**)	76
2	**P4**			X; 5-Me(**4b**)	89
3	**P2**	S	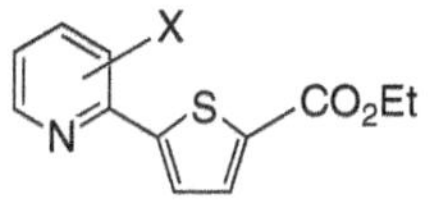	X; 3-Me(**4c**)	51
4	**P4**			X; 5-Me(**4d**)	84

[a] Isolated (based on electrophile).

in Table 3.34. Interestingly, the mild conditions worked well to complete the coupling reaction of 2-pyridylzinc bromide (**P1**). As shown in Table 3.34, several different types of functionalized aryl halides and heteroaryl halides were coupled with **P1** in the presence of 1 mol% of Pd(PPh_3)$_4$ at rt in THF. Functionalized iodobenzenes were first treated with 2-pyridylzinc bromide, and the coupling products (**5a–5d**) were obtained in good to excellent yields (Table 3.34, entries 1–4). Monosubstituted thiophene was also easily coupled with 2-pyridylzinc bromide to give rise to 2-(2′-pyridyl)thiophene (**5e**) in 68%. Di- and trisubstituted thiophenes (Table 3.34, entries 6 and 7) were also good coupling partners to give interesting thiophene derivatives (**5f** and **5g**) in high yields. The coupling reaction with a furan derivative resulted in the formation of **5h** in an excellent yield (Table 3.34, entry 8).

More interesting materials were prepared by the coupling reaction of various 2-pyridylzinc bromides with halo heterocyclic derivatives, and the results are summarized in Table 3.35. A selective C—C bond forming reaction occurred in the reactions with 2-bromo-3-hexyl-5-iodothiophene and 2-bromo-5-chlorothiophene, giving **6a** and **6d** in 41% and 64% isolated yield, respectively (Table 3.35, entries 1 and 4). A slightly longer reaction time was required to complete the coupling reaction with 2-bromothiazole and 2-bromoquinoline with 4-methyl-2-pyridylzinc bromide (**P3**) (Table 3.35, entries 2 and 3). Moderate yields (60% and 40%) were obtained from these reactions. Significantly, another selective C—C bond forming reaction was achieved from

Table 3.34 Pd-catalyzed coupling of **P1** with arylhalide.

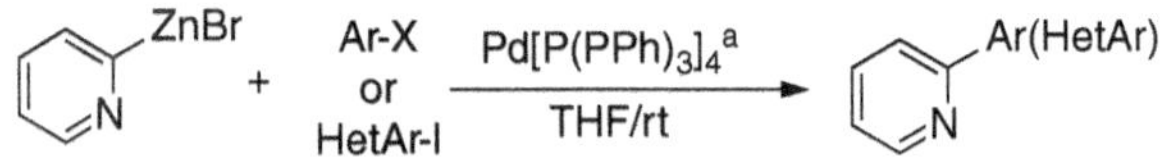

P1(1.0 eq) + Ar-X or HetAr-I (0.8 eq) → $Pd[P(PPh)_3]_4$[a], THF/rt → **5a–5h**

Entry	Electrophile	Time (h)	Product	Yield (%)[b]
1	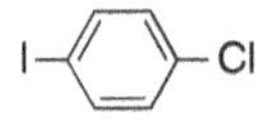 I–C6H4–Cl	24	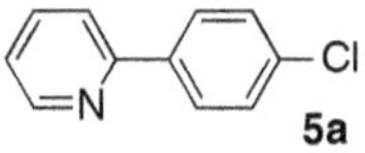 **5a**	81
2	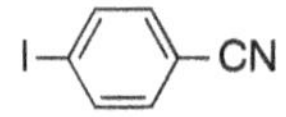 I–C6H4–CN	24	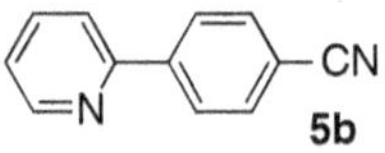 **5b**	88
3	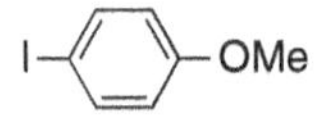 I–C6H4–OMe	24	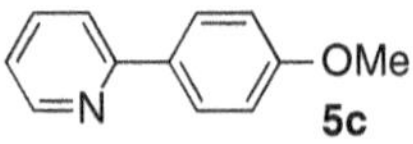 **5c**	68
4	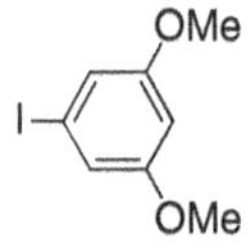 I–C6H3(OMe)2	4	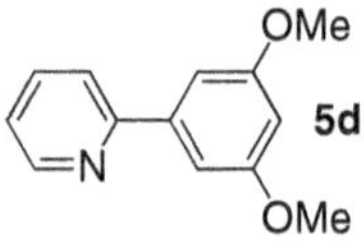 **5d**	90
5	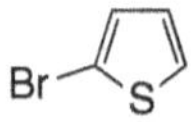 2-Br-thiophene	24	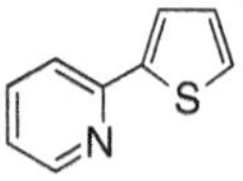 **5e**	68
6	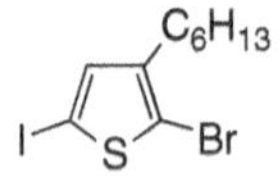 I-thiophene(C6H13)Br	24	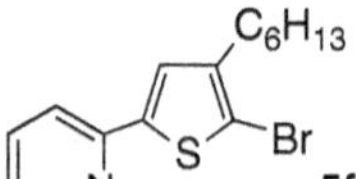 **5f**	89
7	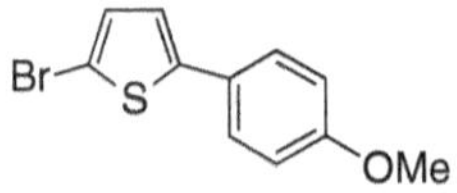 Br-thiophene-C6H4-OMe	24	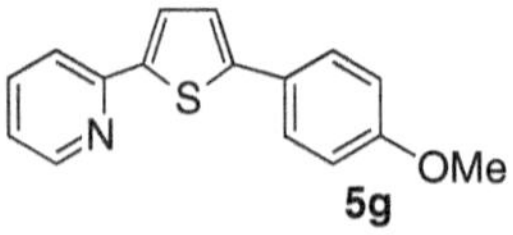 **5g**	80
8	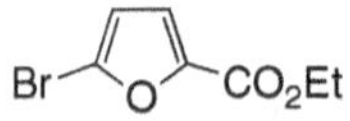 Br-furan-CO2Et	3	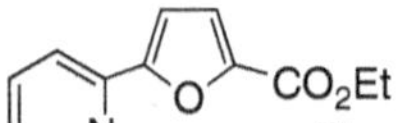 **5h**	91

[a] Performed with 1 mol%.
[b] Isolated (based on electrophile).

the coupling reaction with symmetrically disubstituted thiophene, 2,5-dibromothiophene (Table 3.35, entry 5), affording **6e**, which could be used for further application. Even though slightly different reaction conditions (Pd-II catalyst and refluxing temperature) were applied to carry out the coupling

Table 3.35 Coupling reactions of **P2 ~ P6** with heteroaryl halides.

Entry	RZnBr	Electrophile[a]	Conditions[b]	Product	Yield (%)[c]
1	P2	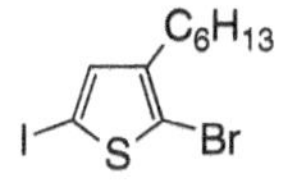	A	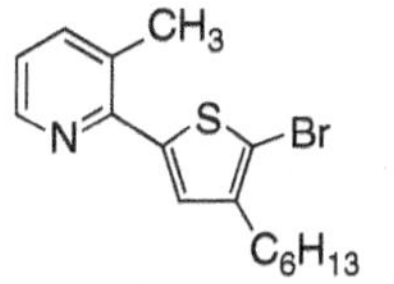 6a	41
2	P3	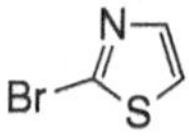	B	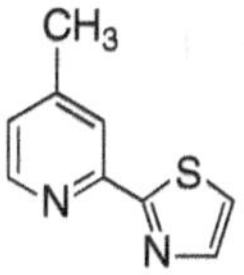 6b	60
3	P3	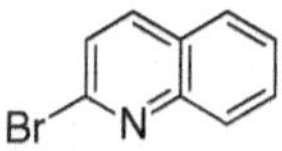	B	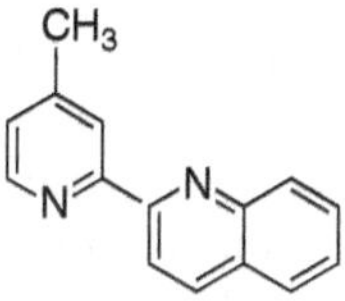 6c	40
4	P4	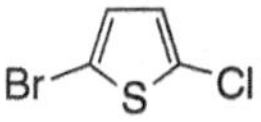	A	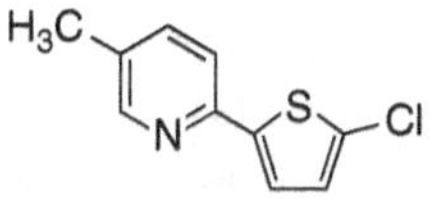 6d	64
5	P4	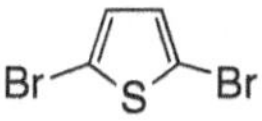	A	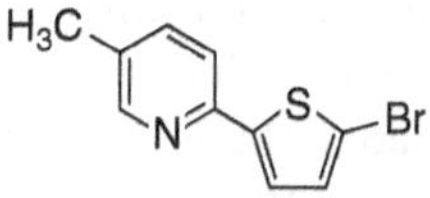 6e	51
6	P6	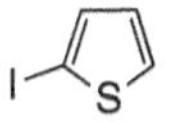	A	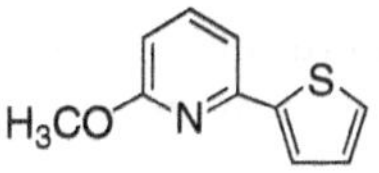 6f	47
7[d]	P4	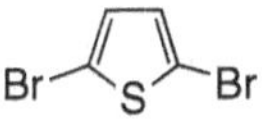	C	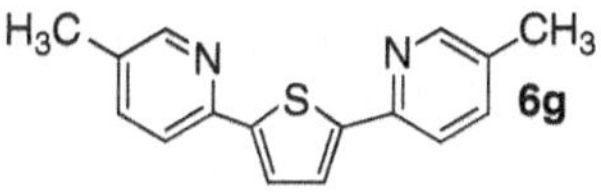 6g	68
8[d]	P5	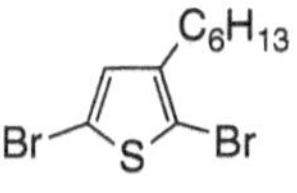	C	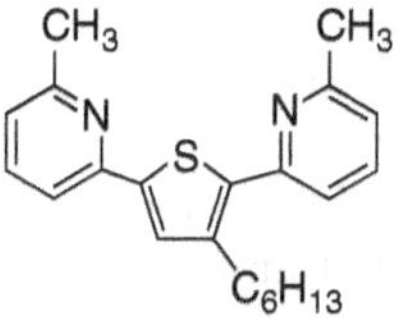 6h	23

[a] 0.8 equiv of electrophile used otherwise mentioned.
[b] **A**: $Pd[P(Ph)_3]_4$/rt/24 h; **B**: $Pd[P(Ph)_3]_4$/rt/72 h; **C**: $Pd[P(Ph)_3]_2Cl_2$/reflux/24 h.
[c] Isolated yield (based on electrophile).
[d] 2.2 equiv of organozinc used.

reactions with dibromothiophenes, symmetrically disubstituted derivatives (**6g** and **6h**) were easily prepared by its twofold reaction (Table 3.35, entries 7 and 8). These types of linear oligomers are important materials for optoelectronic device applications [140].

As described in many previous reports, bipyridine units are very important for many natural products as well as many molecules used in material chemistry [141]. Significantly, this structural moiety can be readily prepared utilizing 2-pyridylzinc bromides. As described in Table 3.36, not only symmetrical 2,2′-bipyridine (**7a**) but also several different types of unsymmetrical 2,2′-bipyridines (**7b–7h**) were prepared in moderate yields. Again, the coupling reaction was completed in the presence of 1 mol% $Pd(PPh_3)_4$ in THF at rt, and, in general, 3-methyl-2-pyridylzinc bromide (**P2**) and 6-methyl-2-pyridylzinc bromide (**P6**) produced 2,2′-bipyridines, **7e** and **7h**, in low yields (Table 3.36, entries 5 and 8). It should be emphasized that the preparation of bipyridines using readily available 2-pyridylzinc bromides (**P1–P6**) could be a very practical approach because considerable effort has been directed toward the preparation of unsymmetrical 2,2′-bipyridines.

As aforementioned, some natural products have the bipyridine unit in their structure. Therefore, we also tried to make an intermediate which could be utilized for the preparation of the natural products. Scheme 3.10 shows two examples. 2,3-Bipyridine (**s1a**) was prepared by the coupling reaction of **P6** with 5-bromo-nicotinic acid methyl ester in the presence of $Pd(PPh_3)_4$ in THF at refluxing temperature affording the coupling product in 64% isolated yield (route **A**, Scheme 3.10). Under similar conditions, 2,2′-bipyridine (**s1b**) was achieved in moderate yield (65%) by Pd(0)-catalyzed cross-coupling reaction of **P1** (route **B**, Scheme 3.10). As depicted in Scheme 3.10, further manipulation of **s1a** and **s1b** would result in the formation of many natural products.

Including our study, most of the electrophiles used in the previously mentioned metal-catalyzed cross-coupling reactions of 2-pyridylmetallics contain relatively nonreactive functional groups toward organometallics, such as ester, ketone, nitrile, halogen, and ether. For the preparation of a variety of 2-pyridyl derivatives, highly functionalized electrophiles are necessary as the coupling partner in the reactions. Therefore, we have performed the cross-coupling reactions of 2-pyridylzinc bromides with haloaromatic compounds containing relatively acidic protons. To this end, haloaromatic amines, phenols, and alcohols are reasonable candidates as coupling reactants. By utilizing this strategy, 2-substituted aminophenyl and hydroyphenyl pyridines have been successfully prepared under mild conditions.

Since Pd(II) catalysts along with an appropriate ligand have been used in the coupling reactions of organozinc reagents with haloaromatic amines and alcohols [142], it seemed reasonable to try these conditions. The coupling reactions worked well with 2-pyridylzinc bromide (**1a**), and the results are summarized in Table 3.37. The reaction of **P1** with 4-iodoaniline in the presence

Table 3.36 Preparation of 2,2′-bipyridines.[a]

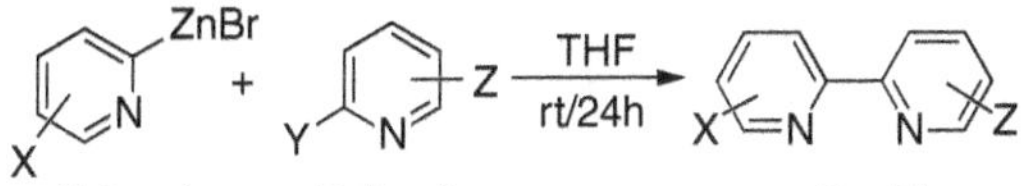

ZnBr-pyridine (X) (1.0 eq) + Y-pyridine-Z (0.8 eq) → (THF, rt/24h) 7a–7h

Entry	X	Y	Z	Product	Yield (%)[b]
1	H(**P1**)	I	H	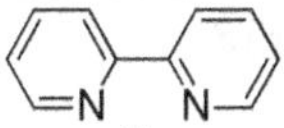 7a	60
2	H(**P1**)	Br	6-Me	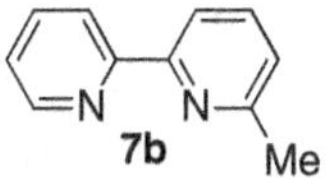 7b	65
3	H(**P1**)	I	5-Br	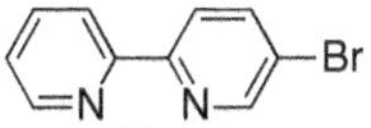 7c	72
4	H(**P1**)	Br	6-OMe	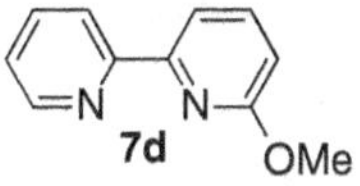 7d	53
5	3-Me(**P2**)	I	5-Br	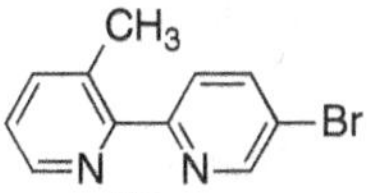 7e	30
6[c]	4-Me(**P3**)	Br	5-Me	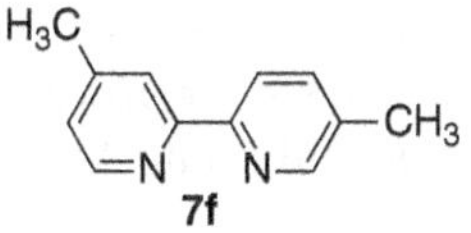 7f	75
7	5-Me(**P4**)	I	5-Br	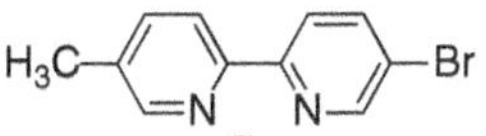 7g	63
8[c]	6-OMe(**P6**)	Br	6-Me	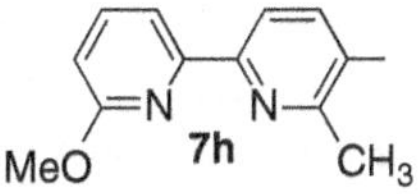 7h	26

[a] Performed in the presence of 1 mol% of $Pd[P(Ph)_3]_4$.
[b] Isolated yield (based on electrophile).
[c] Carried out for 72 h at rt.

of 1 mol% $Pd(OAc)_2$ and 2 mol% SPhos gave rise to the cross-coupling product, **8a**, in 90% yield (Table 3.37, entry 1). Two more reactions, methyl substituted 2-pyridylzinc bromide (**P3**) with 4-iodoaniline and **P1** with 4-bromoaniline resulted in relatively low yields (Table 3.37, entries 2 and 3). However, a significantly

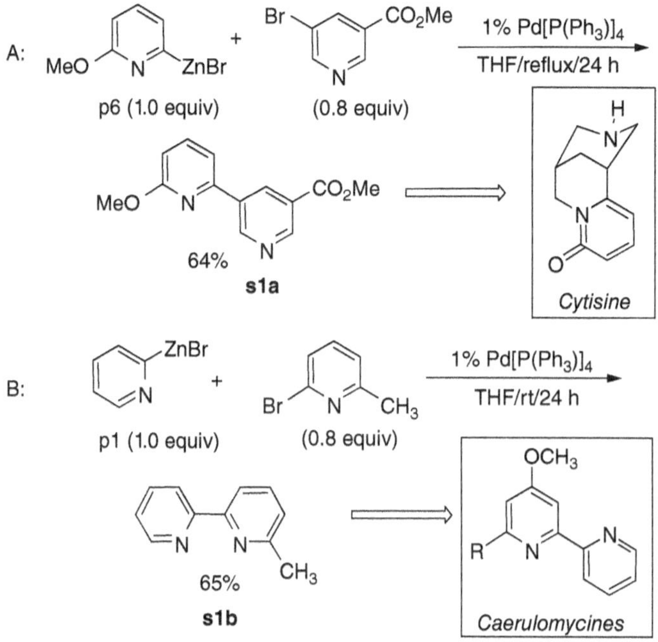

Scheme 3.10 Preparation of intermediates.

improved yield was obtained by the simple change of reaction temperature (Table 3.37, entry 4). An elevated reaction temperature also worked well for the reaction of **P3** with 3-iodoaniline, leading to **8c** in 85% isolated yield (Table 3.37, entry 5). As described in the previous report [142a], we also found that the presence of an extra ligand (SPhos) was critical for the completion of the coupling reaction.

At this point, even though similar conditions with the previous work [142a] were used, it should be emphasized that a more practical procedure, especially for the large-scale synthesis, has been demonstrated in our study. For example, the organozinc solution was added into the flask containing Pd(II) catalyst, ligand (SPhos), and electrophile at a steady-stream rate at rt. However, in the previous report, a very slow addition of organozinc reagent into the reaction flask was crucial in order to obtain high yields [143].

As mentioned previously, the extra ligand (SPhos) was necessary when using the Pd(II) catalyst for the coupling reactions in our study and others. From an economic point of view as well as ease of workup, a ligand-free reaction condition would be highly beneficial. Thus, with the preliminary results (Table 3.37, entries 1–5) in hand, we have investigated the SPhos-free Pd-catalyzed coupling reactions of 2-pyridylzinc bromides with haloanilines. The reactions were performed by employing a Pd(0) catalyst, and the results are summarized in

Table 3.37 Coupling reaction with haloaromatic amines.

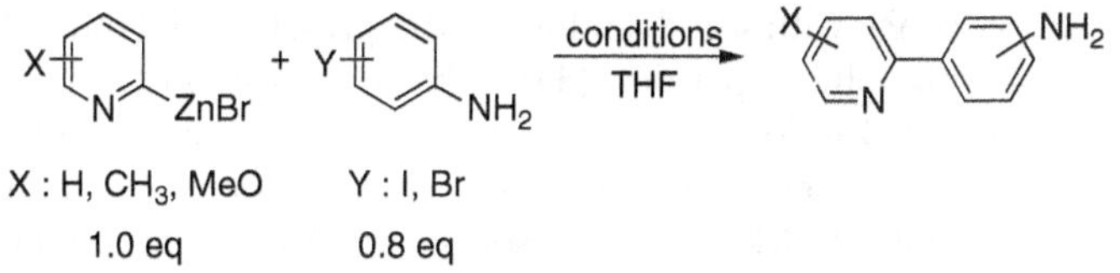

X: H, CH_3, MeO (1.0 eq) + Y: I, Br (0.8 eq) → conditions, THF

Entry	X	Y	Conditions[a]	Product	Yield (%)[b]
1	H(**P1**)	4-I	**A**	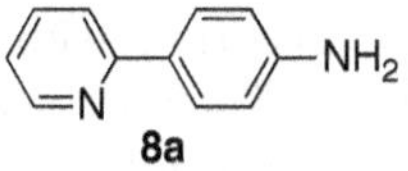 **8a**	90
2	4-CH_3(**P3**)	4-I	**A**	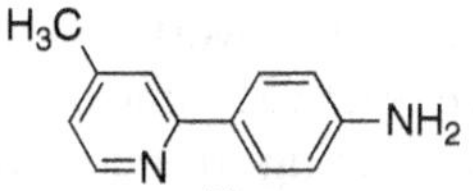 **8b**	50
3	H(**P1**)	4-Br	**A**	**8a**	50
4	H(**P1**)	4-Br	**B**	**8a**	89
5	4-CH_3(**P3**)	3-I	**B**	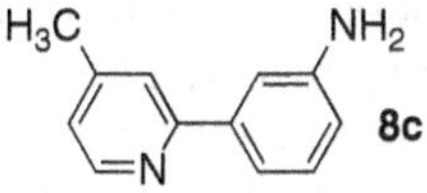 **8c**	85
6	H(**P1**)	4-I	**C**	**8a**	89
7	H(**P1**)	3-I	**C**	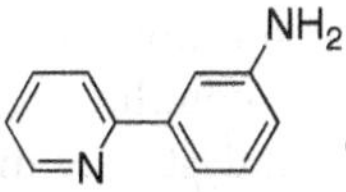 **8d**	64
8	H(**P1**)	2-I	**C**	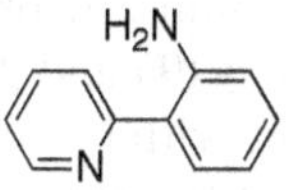 **8e**	74
9	3-CH_3(**P2**)	4-I	**D**	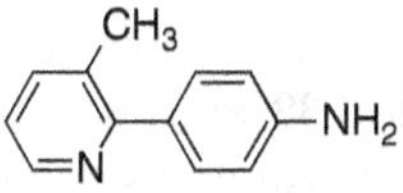 **8f**	68
10	6-OMe(**P6**)	4-Br	**D**	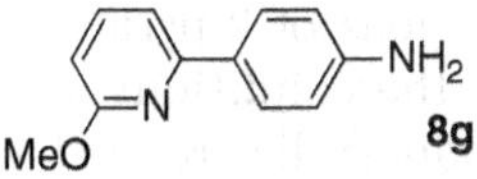 **8g**	Trace[c]

[a] **A**: 1% $Pd(OAc)_2$/2% SPhos/rt/24 h; **B**: 1% $Pd(OAc)_2$/2% SPhos/reflux/24 h; **C**: 1% $Pd[P(Ph)_3]_4$/rt/24 h; **D**: 1% $Pd[P(Ph)_3]_4$/reflux/24 h.

[b] Isolated yield (based on aniline).

[c] By GC–MS.

Table 3.37 (entries 6–10). Significantly, the Pd(0)-catalyzed coupling reactions were not affected by the presence of acidic protons (NH_2) [144].

The reaction of **P1** with 4-iodoaniline in the presence of 1 mol% $Pd(PPh_3)_4$ provided 2-(4-aminophenyl)pyridine (**8a**) with similar results (89% isolated yield, Table 3.37, entry 6). 3-Iodoaniline and 2-iodoaniline were also coupled with **P1** under the same conditions (Table 3.37, condition **C**), affording the aminophenyl pyridines (**8d** and **8e**) in 64% and 74% isolated yields, respectively (Table 3.37, entries 7 and 8). Another successful coupling reaction (Table 3.37, entry 9) was achieved from a sterically hindered 3-methyl-2-pyridylzinc bromide (**P2**), resulting in 68% isolated yield with the formation of **8f**. Unfortunately, no satisfactory coupling reaction occurred with 4-bromoaniline using the Pd(0) catalyst (Table 3.37, entry 10). With the results obtained from the coupling reactions with haloaromatic amines, it can be concluded that the Pd(0)-catalyzed reaction of 2-pyridylzinc bromides works effectively with iodoaromatic amines and also the relatively more reactive bromoaromatic amines.

Another interesting reaction of 2-pyridylzinc bromides would be the coupling reaction with phenols or alcohols, which also have an acidic proton. Encouraged by the results described previously, the coupling reactions with iodophenols were carried out in the presence of Pd(0) catalyst. As shown in Table 3.38, 4-iodophenol and 3-iodophenol were coupled with **P1**, affording the corresponding hydroxyphenyl pyridine products (**9a** and **9b**) in excellent yields (Table 3.38, entries 1 and 2). A slightly disappointing result (25%) was obtained from 2-iodophenol (Table 3.38, entry 3). The reason is not clear, but it is presumably because the coupling was next to the hydroxyl group. A similar outcome has also been reported in another study [143]. In the case of bromophenolic alcohols, no coupling reaction took place with the Pd(0) catalyst. Instead, the Pd(II) catalyst was more efficient for the coupling reaction. 4-Bromophenol and 6-bromo-2-naphthol were nicely coupled with **P1**, resulting in the coupling products (**9a** and **9e**) in 86 and 92% (Table 3.38, entries 5 and 6). Unlike the reactions with bromophenols, it is of interest that the coupling products (**9f** and **9g**) of **P5** and **P1** were efficiently achieved from the Pd(0)-catalyzed reactions with 4-bromobenzyl alcohol and 3-bromo-5-methoxybenzyl alcohol (Table 3.38, entries 7 and 8), respectively.

Interestingly, unsymmetrical amino-bipyridines were produced from the coupling reactions of 2-pyridylzinc bromides with halogenated aminopyridines under the conditions used previously. As shown in Scheme 3.11, 2-amino-5-iodopyridine reacted with **P1** to afford 2,3-bipyridine (**s2a**) in 59% isolated yield in the presence of 1 mol% of $Pd(PPh_3)_4$ catalyst (route **A**, Scheme 3.11).

However, in the case of 2-amino-5-bromopyridine, the Pd(II) catalyst was more efficient for the coupling reaction and the reaction proceeded smoothly to give 2,3-bipyridine (**s2b**) in 37% yield (route **B**, Scheme 3.11). It is of interest that the bipyridyl amines can be used as intermediates for the synthesis of

Table 3.38 Coupling reaction with haloaromatic alcohols.

X-pyridyl-ZnBr (1.0 eq) + Haloalcohol (0.5 eq) —conditions, THF/reflux/24 h→ X-pyridyl-Ar—OH **9a–9g**

Entry	X	Alcohol	Conditions[a]	Product	Yield (%)[b]
1	H(**P1**)	4-iodophenol	**A**	**9a**	95
2	H(**P1**)	3-iodophenol	**A**	**9b**	80
3	H(**P1**)	2-iodophenol	**A**	**9c**	25
4	4-CH_3(**P3**)	4-iodophenol	**A**	**9d**	54
5	H(**P1**)	4-bromophenol	**A**	**9a**	0[c]
			B		86
6	H(**P1**)	6-bromo-2-naphthol	**B**	**9e**	92
7	6-CH_3(**P5**)	4-bromobenzyl alcohol	**A**	**9f**	60
8	H(**P1**)	5-bromo-2-methoxybenzyl alcohol	**A**	**9g**	65

[a] **A**: 1% $Pd[P(Ph)_3]_4$; **B**: 1% $Pd(OAc)_2$/2% SPhos.
[b] Isolated yield (based on alcohol).
[c] No coupling observed by GC.

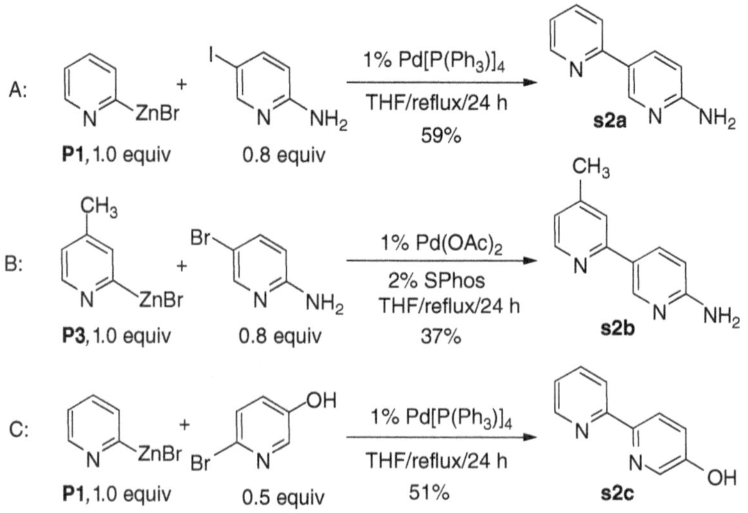

Scheme 3.11 Preparation of amino and hydroxyl bipyridines.

highly functionalized molecules after transformation of the amino group to a halogen [145].

Treatment of 2-pyridylzinc bromide (**P1**) with a halopyridine bearing a hydroxyl group provided another functionalized bipyridine. Interestingly, the relatively reactive bromopyridyl alcohol, 2-bromo-5-hydroxypyridine, was coupled with **P1** using the Pd(0) catalyst. As a result, the corresponding hydroxyl 2,2′-bipyridine (**s2c**) was obtained in 51% isolated yield (route **C**, Scheme 3.11). The hydroxyl group on 2,2′-bipyridine can also be converted to a halogen to make halobipyridines by using several different methods [146].

As mentioned earlier, direct preparation of 3-pyridylzinc reagents has been another challenging subject in organometallic chemistry. In our continuing study of heterocyclic organic reagents [147], it has been found that Rieke zinc in the presence of certain additives exhibits a very high reactivity to 3-bromopyridine. The corresponding 3-pyridylzinc bromide was easily prepared by the direct insertion of active zinc to 3-bromopyridine and the resulting 3-pyridylzinc bromide was successfully applied to the cross-coupling reaction with a variety of electrophiles under mild conditions.

The first attempt to synthesize 3-pyriylzinc bromide from the direct reaction of active zinc and 3-bromopyridine in THF at rt and refluxing temperature resulted in low conversion (70%) to the organozinc reagent. Almost the same result was obtained from an extended reaction time (reflux/24 h). However, a dramatic improvement in the oxidative addition of active zinc has been achieved by adding 10–20 mol% of lithium chloride to the reaction mixture. More than 99% conversion of 3-bromopyridine to 3-pyridylzinc bromide was obtained in 2 h at refluxing temperature in THF. As we pointed out in 1989 [138], the rate-limiting step in the oxidative addition is electron transfer (ET). Accordingly, this process will be accelerated by the presence of alkali salts, which are generated in the reduction process of forming the active metals or additional salts can be added to the reaction mixture [139b].

In order to confirm the formation of 3-pyridylzinc bromide, the resulting organozinc reagent was first treated with iodine, affording 90% 3-iodopyridine and 3% pyridine. Next, the resulting 3-pyridylzinc bromide (**P7**) was added to a variety of different electrophiles to give the corresponding coupling products in moderate to good yields. The results are summarized in Table 3.39. Palladium-catalyzed cross-coupling reactions with aryl iodides (Table 3.39, **a–c**) were completed in 1 h at rt to give 3-pyridylbenzene derivatives (**10a**, **10b**, and **10c**) in good yields (Table 3.39, entries 1–3). A longer reaction time was required with aryl iodides (**d** and **e**) bearing a substituent in the 2-position (Table 3.39, entries 4 and 5). This is probably due to steric hindrance. With this result in hand, heteroaryl iodides (**f–h**) were also coupled with 3-pyridylzinc bromide, affording the heteroaryls (**10f–10h**) in good yields. Coupling reactions with heteroaryl bromides (**i** and **j**) needed also a longer reaction time and afforded the coupling product (Table 3.39, **10i** and **10j**) in 86% and 29% isolated yields, respectively. As shown in entries 4 and 7 in Table 3.39, the carbon—iodine bond was selectively reacted in coupling reactions with the organozinc reagent (**P7**) for carbon—carbon bond formation under the conditions used here. Even though a low yield was obtained from 2,6-dibromopyridine (**j**), the coupling product (**10j**) bearing a bromine atom can serve as a valuable intermediate for the preparation of a variety of materials. Interesting, it was also possible to obtain an aromatic ketone (**10k**) in moderate yield from the reaction of **P7** with benzoic acid anhydride in the presence of a palladium(0) catalyst.

To expand the utility of 3-pyridylzinc bromide, several other copper-catalyzed coupling reactions were also investigated and the results are summarized in Table 3.40. S_N2'-type reactions have been tried with allyl halides, affording the resulting products (Table 3.40, **11a** and **11b**) in moderate to good yields. In the presence of TMSCl, the silyl enol ether (Table 3.40, **11c**) was obtained from the conjugate addition intermediate. Like other general organozinc reagents, 3-pyridylzinc bromide (**P7**) was successfully used for the copper-catalyzed synthesis of ketone compounds. As shown in Table 3.40, 10 mol% of CuI promoted the coupling reaction with **P7** to give the ketones (Table 3.40, **10k**, **11d–11f**) in moderate yields under the conditions described in Table 3.40.

This study was expanded to several analogs of 3-bromopyridine. As described in Table 3.41, 3-bromoquinoline and 3-bromoisoquinoline were treated with active zinc along with 20 mol% of lithium chloride. It was found that the oxidative addition of active zinc was completed in 2 h at refluxing temperature to give the corresponding organozinc reagents (**q1** and **q2**). The subsequent coupling reactions of **q1** were performed with aryl iodide (Table 3.41, entry 1) and heteroaryl

Table 3.39 Pd-catalyzed coupling reactions of 3-pyridylzinc bromide (**P7**).

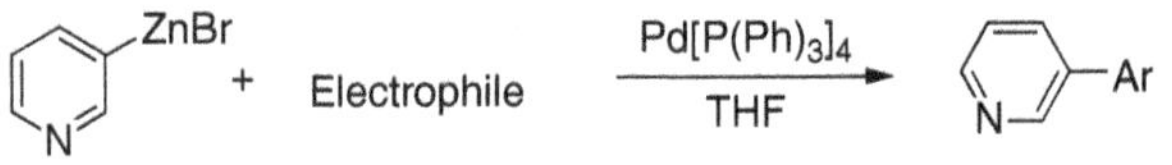

ZnBr + Electrophile $\xrightarrow[\text{THF}]{\text{Pd[P(Ph)}_3\text{]}_4}$ Ar

Entry	Electrophile[a]	Conditions[b]	Product	Yield (%)[c]
1	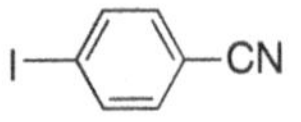 I, CN **a**	rt/1 h	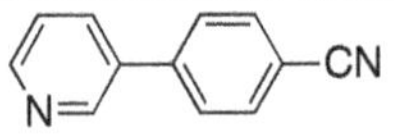 N, CN **10a**	65
2	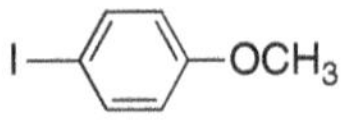 I, OCH_3 **b**	rt/1 h	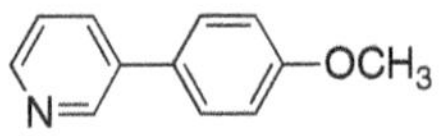 N, OCH_3 **10b**	81
3	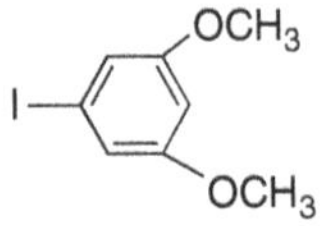 I, OCH_3, OCH_3 **c**	rt/1 h	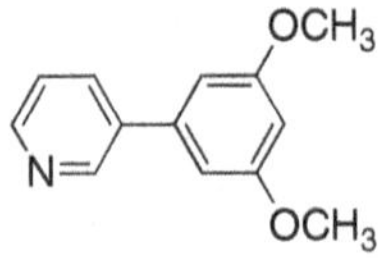 N, OCH_3, OCH_3 **10c**	63
4	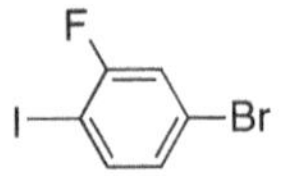 F, I, Br **d**	rt/24 h	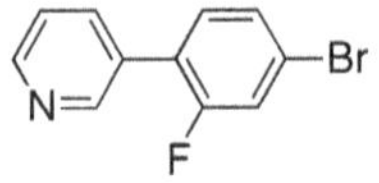 N, Br, F **10d**	63
5	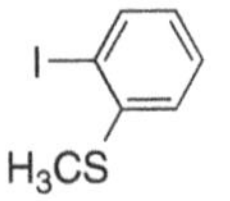 I, H_3CS **e**	rt/24 h	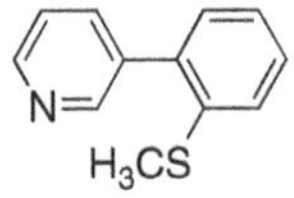 N, H_3CS **10e**	32
6	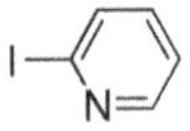 I, N **f**	rt/1 h	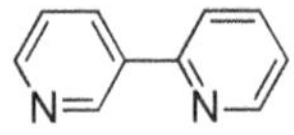 N, N **10f**	71
7	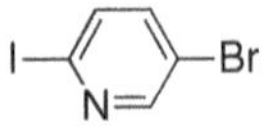 I, N, Br **g**	rt/1 h	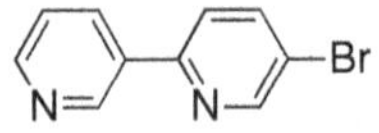 N, N, Br **10g**	62
8	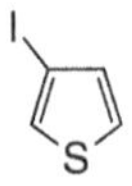 I, S **h**	rt/1 h	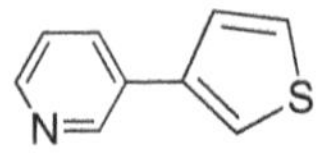 N, S **10h**	71
9	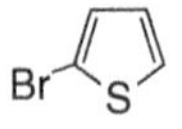 Br, S **i**	rt/48 h	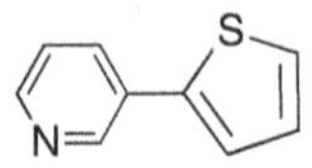 N, S **10i**	86
10	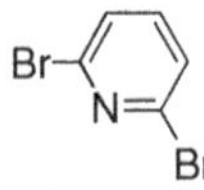 Br, N, Br **j**	rt/48 h	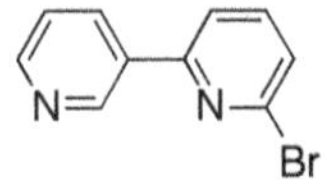 N, N, Br **10j**	29
11	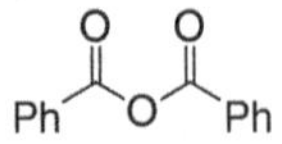 O, O, Ph, O, Ph **k**	rt/12 h	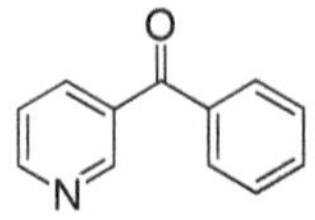 O, N **10k**	38

[a] 0.8 equiv of electrophile used.
[b] 1 mol% of $Pd[P(Ph)_3]_4$ used.
[c] Isolated yield (based on electrophile).

Table 3.40 Copper-catalyzed coupling reaction of **P7**.

3-Pyridylzinc bromide (ZnBr) + Electrophile —CuI, THF→ Product **11a–11f**

Entry	Electrophile[a]	Conditions[b]	Product	Yield (%)[c]
1	3-Bromocyclohexene (Br)	0°C/10 min	**11a**	71
2	2,3-Dichloropropene (Cl, Cl)	0°C/10 min	**11b** (Cl)	50
3	Cyclohexenone (=O)	TMSCl/0°C ~ rt	**11c** ($OSiMe_3$)	48
4	Benzoyl chloride (COCl)	0°C ~ rt/12 h	**10k**	50
5	Cyclohexanecarbonyl chloride (COCl)	0°C ~ rt/12 h	**11d**	69
6	Cyclopropanecarbonyl chloride (COCl)	0°C ~ rt/12 h	**11e**	38
7	Thiophene-3-carbonyl chloride (COCl)	0°C ~ rt/12 h	**11f**	50

[a] 0.8 equiv of electrophile used.
[b] 10 mol% CuI used.
[c] Isolated yield (based on electrophile).

iodides (Table 3.41, entries 2 and 3) in the presence of a palladium catalyst, affording the corresponding products (**12a–12c**) in moderate to good yields. Entries 4 and 5 in Table 3.41 showed the results of the coupling reactions of **q2** with heteroaryl iodide and allyl chloride. From these reactions, more new heteroaryl compounds (Table 3.41, **12d** and **12e**) were obtained in moderate yields.

Table 3.41 Preparation of quinoline and isoquinoline derivatives via heteroarylzinc reagent.

Entry	Organozinc	Electrophile[a]	Conditions[b]	Product	Yield (%)[c]
1	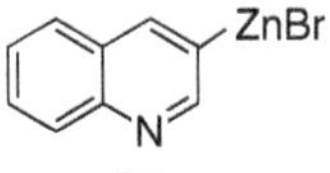 ZnBr, N **Q1**	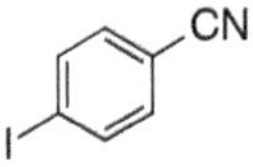 CN, I	$Pd[P(Ph)_3]_4$ rt/1 h	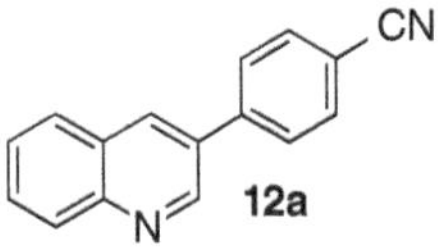 CN, N **12a**	70
2	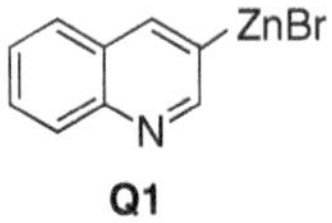 ZnBr, N **Q1**	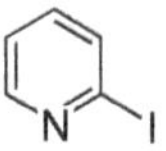 N, I	$Pd[P(Ph)_3]_4$ rt/1 h	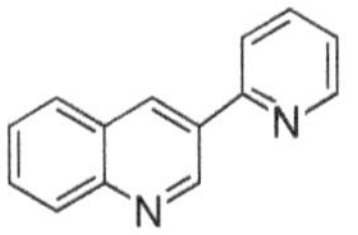 N, N **12b**	65
3	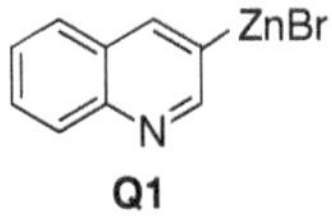 ZnBr, N **Q1**	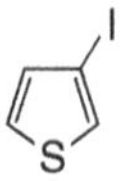 I, S	$Pd[P(Ph)_3]_4$ rt/1 h	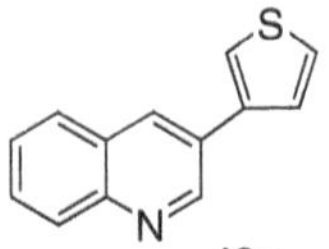 S, N **12c**	53
4	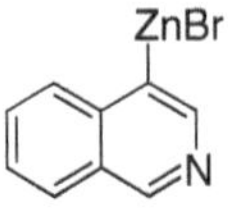 ZnBr, N **Q2**	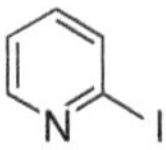 N, I	$Pd[P(Ph)_3]_4$ rt/1 h	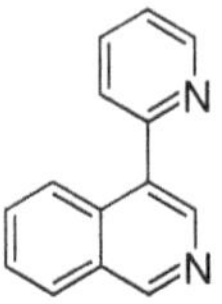 N, N **12d**	37
5	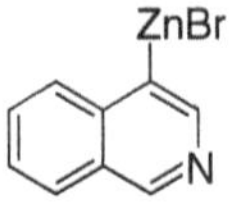 ZnBr, N **Q2**	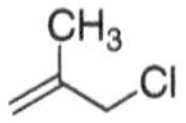 CH_3, Cl	CuI[d] 0°C ~ rt/1 h	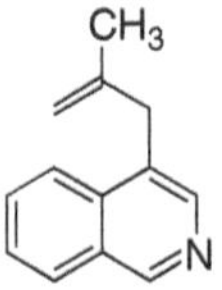 CH_3, N **12e**	35

[a] 0.8 equiv of electrophile used.
[b] 1 mol% Pd-catalyst used.
[c] Isolated yield (based on electrophile).
[d] 10 mol% use.

Since we obtained successful results from the coupling reaction of 2-pyridylzinc bromides with aromatic haloamines and alcohols, we also demonstrated the same strategy for the coupling reaction of the 3-pyridylzinc bromides. It was found that the readily available 3-pyridylzinc bromide was easily coupled with haloaromatic compounds bearing relatively acidic protons under mild conditions, affording the corresponding cross-coupling products in moderate to excellent yields. Of primary interest, we report the general procedure for the

Table 3.42 Preliminary test for the coupling reaction of **P7**.

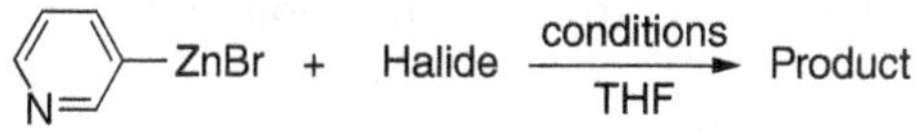

P7 ZnBr + Halide —conditions / THF→ Product

Entry	Halide[a]	Conditions	Product	Yield (%)[b]
1	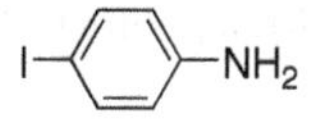 I–C6H4–NH2	1% $Pd(OAc)_2$ 2% SPhos rt/30 min	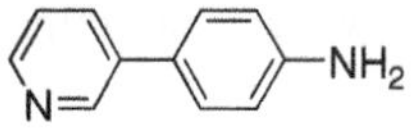 **13a**	80
2	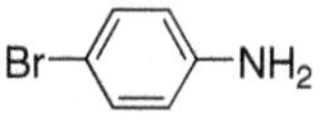 Br–C6H4–NH2	1% $Pd(OAc)_2$ 2% SPhos rt/3 days	**13a**	80
3	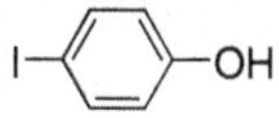 I–C6H4–OH	1% $Pd(OAc)_2$ 2% SPhos rt/24 h	**13a**	Trace[c]
4	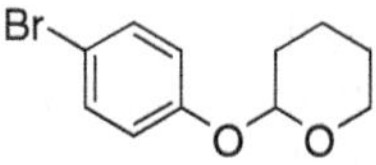 Br–C6H4–O–THP	1% $Pd(OAc)_2$ 2% SPhos rt/24 h	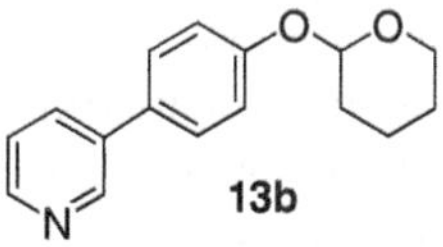 **13b**	44
5	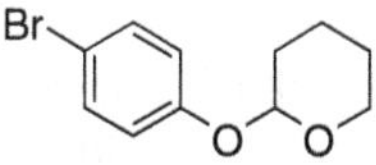 Br–C6H4–O–THP	1% $Pd(OAc)_2$ rt/3 days	**13b**	Trace[c]
6	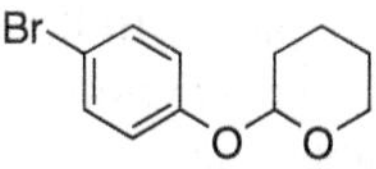 Br–C6H4–O–THP	$Pd(PPh_3)_4$ rt/24 h	**13b**	0[c]

[a] 0.8 equiv of amine, 0.5 equiv of alcohol used.
[b] Isolated (based on halide).
[c] Monitored by GC.

transition metal-catalyzed cross-coupling reactions of 3-pyridylzinc bromides with haloanilines and halophenols providing 3-(aminophenyl)pyridines and 3-(hydroxyphenyl)pyridines. The results also include the preparation of quinoline and isoquinoline derivatives as well as other pyridine derivatives.

The first approach included the reaction of 3-pyridylzinc bromide (**P7**) with iodoaniline in the presence of 1% of $Pd(OAc)_2$ along with 2% of SPhos in THF (Table 3.42, entry 1). The coupling reaction was completed at rt in 30 min, affording 4-pyridin-3-yl-phenylamine (**13a**) in 80% isolated yield. Even though a little longer reaction time was required, the coupling product (**13a**) was also obtained in good yield from the reaction with 4-bromoaniline under the same conditions (Table 3.42, entry 2). Interestingly, trace amounts of the coupling product was detected by GC from the reaction with 4-iodophenol using the same conditions (Table 3.42, entry 3). A protected bromophenyl gave rise to the coupling product (**13b**) in moderate yield under the same conditions (Table 3.42, entry 4). The critical role of an extra ligand (SPhos) for the completion of the coupling reaction was also noticed. Only trace amounts of product was detected in the absence of SPhos (Table 3.42, entry 5). It was also found that a Pd(0) catalyst was not effective for the coupling with a protected bromophenol (Table 3.42, entry 6).

Even though the conditions used in the preliminary tests worked well, an extra ligand (SPhos)-free reaction condition would be more interesting. Thus, once again, the (SPhos)-free Pd-catalyzed coupling reactions of 3-pyridylzinc bromides with haloanilines were carried out. This was accomplished by using 1 mol% of $Pd(PPh_3)_4$ without any extra additives, and the corresponding coupling products were obtained in moderate to excellent yields. The results are summarized in Table 3.43. As observed in the coupling reaction of 2-pyridylzinc bromide (**P1**) with haloaromatic amines, no significant effect of the presence of acidic protons was observed from the study of **P7**.

The coupling reactions were easily accomplished by the addition of 3-pypridylzinc bromide into the mixture of haloaniline and Pd(0) catalyst in THF. The organozinc solution was added into the reaction flask in one portion via a syringe at rt. The lack of a significant heat of reaction should be a useful feature for large-scale synthesis.

The reaction of **P7** with 4-iodoaniline in the presence of 1 mol% $Pd(PPh_3)_4$ provided 4-pyridin-3-yl-phenylamine (**13a**) in 86% isolated yield (Table 3.43, entry 1). 3-Iodoaniline and 2-iodoaniline were also coupled with **P7** under the same conditions (at rt for 1.0 h), affording the aminophenyl pyridines (**14a** and **14b**) in 85% and 61% isolated yield, respectively (Table 3.43, entries 2 and 3). However, even at an elevated temperature, a low yield of **13a** was obtained from the coupling reaction using 4-bromoaniline (Table 3.43, entry 4). With the results obtained from the simple 3-pyridylzinc bromide, we also expanded these reaction conditions to the wide range of functionalized 3-pyridylzinc reagents. All of these 3-pyridylzinc bromides were easily prepared under the same conditions used for the preparation of **P7**. For the sterically hindered 4-methyl-3-pyridylzinc bromide (**P8**), a slightly severe condition (refluxing for 6 h) was required to complete the coupling, resulting in the formation of **14c** in 48% isolated yield (Table 3.43, entry 5). However, the reaction with 2-methyl-5-pyridylzinc bromide (**P10**) provided the coupling product (**14d**) in a good yield (Table 3.43, entry 6). Interestingly, excellent results were achieved from the coupling reactions using chlorine-substituted 3-pyridylzinc reagents. As shown in Table 3.43, a 92% isolated yield of **14e** and **14f** was obtained from both reactions of 2-chloro-3-pyridylzinc bromide (**P11**) and 2-chloro-5-pyridylzinc

Table 3.43 Preparation of aminophenylpyridines.

X-pyridyl-ZnBr + Aniline → (1% $Pd(PPh_3)_4$, THF) → Product

X : H, CH_3, Cl, OMe; 1.0 eq (RZnX); 0.8 eq (aniline)

Entry	RZnX	Aniline	Conditions	Product	Yield (%)[a]
1	3-pyridyl-ZnBr **P7**	4-iodoaniline	rt/1 h	4-(3-pyridyl)aniline **13a**	86
2	**P7**	3-iodoaniline	rt/1 h	3-(3-pyridyl)aniline **14a**	65
3	**P7**	2-iodoaniline	rt/1 h	2-(3-pyridyl)aniline **14b**	61
4	**P7**	4-bromoaniline	Reflux/24 h	**13a**	15
5	4-CH_3-pyridin-3-yl-ZnBr **P8**	4-iodoaniline	Reflux/6 h	**14c**	48
6	H_3C-pyridyl-ZnBr **P10**	4-iodoaniline	rt/1 h	**14d**	85
7	2-Cl-pyridin-3-yl-ZnBr **P11**	4-iodoaniline	rt/3 h	**14e**	92
8	Cl-pyridyl-ZnBr **P12**	4-iodoaniline	rt/3 h	**14f**	92
9	H_3CO-pyridyl-ZnBr **P13**	4-iodoaniline	Reflux/24 h	H_3CO-pyridyl-C_6H_4-NH_2	0[b]

[a] Isolated yield (based on amine).

[b] No reaction. Starting materials were recovered (confirmed by GC and GCMS).

bromide (**P12**) after 3 h stirring at rt (Table 3.43, entries 7 and 8). Unfortunately, no satisfactory coupling reaction occurred with 2-methoxy-5-pyridylzinc bromide (**P13**) using the Pd(0) catalyst (Table 3.43, entry 9). Similar results were also obtained from the reaction with iodophenols (Table 3.37, entry 10). From the results described previously, it can be concluded that Pd(0)-catalyzed coupling reactions of 3-pyridylzinc bromides work effectively with iodoanilines under mild conditions.

Another interesting reaction of 3-pyridylzinc bromide would be the coupling reaction with phenols, which also have an even more acidic proton [144]. Encouraged by the results achieved from the reaction with haloanilines, the coupling reactions with iodophenols were carried out using Pd(0) catalyst. As shown in Table 3.44, 2.0 equiv of organozinc reagent were reacted with halophenols at refluxing temperature in the presence of 1 mol% of $Pd(PPh_3)_4$ in THF. In the case of **P7**, even though the coupling reaction with iodophenols worked fairly at rt, increasing the reaction temperature worked more effectively to complete the coupling reaction. Therefore, all of the coupling reactions with halophenols in this study were conducted at refluxing temperature for 12 h. 4-Iodophenol was coupled with **P7**, affording the corresponding 3-(4-hydroxyphenyl) pyridine (**15a**) in excellent yield (Table 3.44, entry 1). A slightly disappointing result (49%) was obtained from 3-iodophenyl (Table 3.44, entry 2). Unlike the reaction with 4-bromoaniline, treatment of **P7** with 4-bromophenol gave rise to the product **15a** in moderate yield (Table 3.44, entry 3). More coupling reactions of methyl and chlorine-substituted 3-pyridylzinc bromides (**P10**, **P11**, and **P12**) were also performed with 4-iodophenol. The reactions occurred successfully to result in good yields (81%, Table 3.44, entry 4) to moderate yields (61% and 60%, Table 3.44, entries 5 and 6, respectively). Again, no coupling product was obtained from 2-methoxy-5-pyridylzinc bromide (**P13**).

Since the coupling reactions of 3-pyridylzinc bromides provided very positive results of preparing highly functionalized pyridine derivatives, this study was applied to several analogs of 3-bromopyridine. The results are described in Table 3.45. Quinolinylzinc bromide (**q1**) and isoquinolinylzinc bromide (**q2**) were prepared easily by using the previously mentioned procedure. Subsequent coupling reactions of 3-quinolinylzinc bromide with 4-iodoaniline and 4-iodophenol afforded the corresponding products (Table 3.45, **16a** and **16b**) in 89% and 64% isolated yield, respectively. The coupling reaction of 4-isoquinolinylzinc bromide (**q2**) with 4-iodophenol was carried out at rt for 3 h to give the product (**16c**) in excellent yield (Table 3.45, entry **3**). Unfortunately, coupling of 5-pyrimidinylzinc bromide showed no product at all (Table 3.45, entry 4).

Further application of this practical synthetic approach has been performed by the coupling reaction with different types of alcohols. 2-Methoxy-5-bromobenzyl alcohol and 6-bromo-2-naphthol were nicely coupled with **P7** under the reaction conditions given in Scheme 3.12, resulting in the coupling products

Table 3.44 Preparation of hydroxyphenylpyridines.

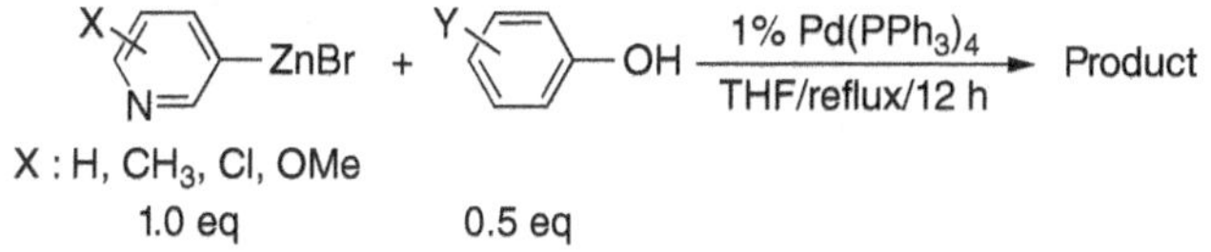

Reaction scheme: X-substituted 3-pyridyl–ZnBr (X : H, CH_3, Cl, OMe; 1.0 eq) + Y-substituted phenol (0.5 eq) → Product [1% $Pd(PPh_3)_4$, THF/reflux/12 h]

Entry	RZnX	Y	Product	Yield (%)[a]
1	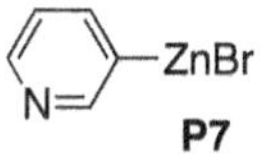 3-pyridyl-ZnBr **P7**	4-I	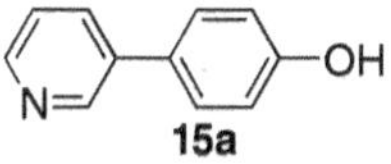 4-(pyridin-3-yl)phenol **15a**	90
2	**P7**	3-I	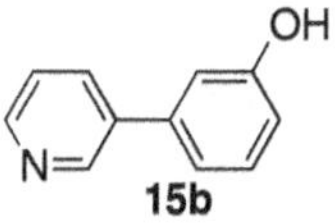 3-(pyridin-3-yl)phenol **15b**	49
3	**P7**	4-Br	**15a**	40
4	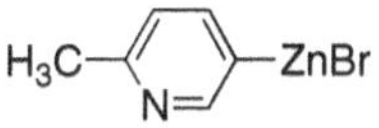 H_3C-pyridyl-ZnBr **P10**	4-I	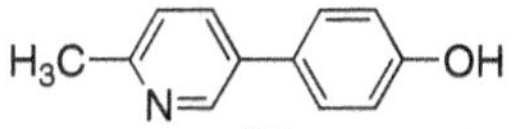 H_3C-pyridyl-phenyl-OH **15c**	81
5	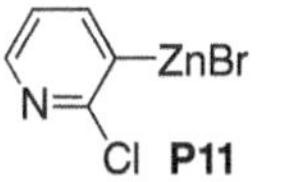 2-Cl-3-pyridyl-ZnBr **P11**	4-I	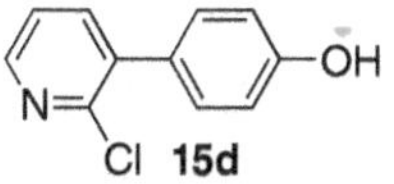 2-Cl-pyridyl-phenyl-OH **15d**	61
6	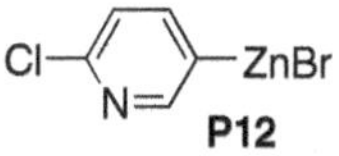 Cl-pyridyl-ZnBr **P12**	4-I	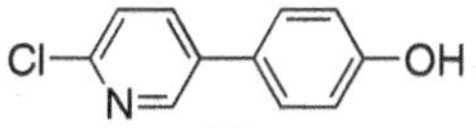 Cl-pyridyl-phenyl-OH **15e**	60
7	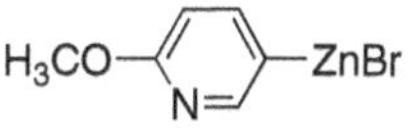 H_3CO-pyridyl-ZnBr **P13**	4-I	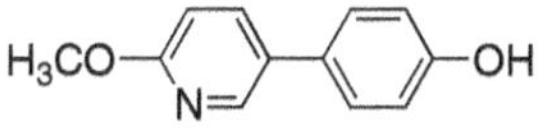 H_3CO-pyridyl-phenyl-OH	0[b]

[a] Isolated yield (based on alcohol).
[b] Carried out at refluxing temperature for 24 h. No reaction. Starting materials were recovered (confirmed by GC and GCMS).

(**s3a** and **s3b**) in 48% and 81% yield, respectively (routes **A** and **B**, Scheme 3.12). Interestingly, unsymmetrical amino-bipyridines were produced from the coupling reactions of **P7** with 2-amino-6-bromopyridine (route **C**, Scheme 3.12). It was accomplished by the reaction of **P7** with 2-amino-6-bromopuridine in the presence of 1 mol% of $Pd(PPh_3)_4$ in THF at refluxing temperature for 12 h, affording 2,3-bipyridine (**s3c**) in 65% isolated yield. It is of interest that the

Table 3.45 Preparation of quinoline and isoquinoline derivatives.

Entry	RZnX	Halide	Conditions	Product	Yield (%)[a]
1	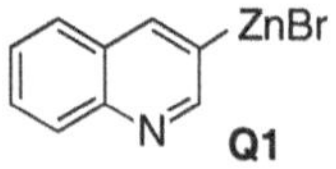 ZnBr, N, Q1, 1.0 eq	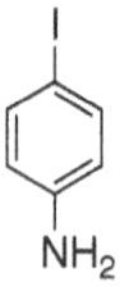 I, NH_2, 0.8 eq	1% $Pd(PPh_3)_4$ THF/rt/3.0 h	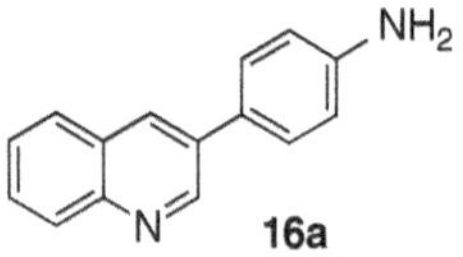 NH_2, N, 16a	89
2	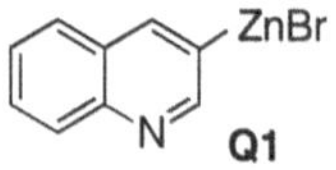 ZnBr, N, Q1, 1.0eq	 I, OH, 0.5eq	1% $Pd(PPh_3)_4$ THF/rt/12 h	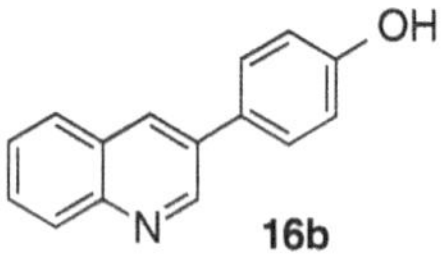 OH, N, 16b	64
3	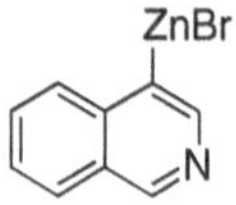 ZnBr, N, Q2, 1.0 eq	 I, OH, 0.5 eq	1% $Pd(PPh_3)_4$ THF/rt/3 h	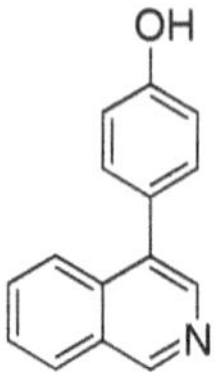 OH, 16c, N	90
4	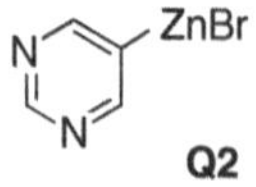 N, ZnBr, N, Q2, 1.0 eq	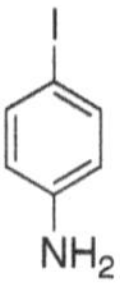 I, NH_2, 0.8 eq	1% $Pd(PPh_3)_4$ THF/reflux/24 h		0[b]

[a] Isolated yield (based on halide).
[b] No reaction. Starting materials were recovered (confirmed by GC and GCMS).

resulting product (**s3c**) can be utilized for further applications after transformation of the amino group to a halogen [145].

Conclusions

We have demonstrated a practical synthetic route for the preparation of 2-pyridyl and 3-pyridyl derivatives. It has been accomplished by utilizing a simple coupling reaction of readily available 2-pyridylzinc bromides and 3-pyridylzinc bromides, which were prepared via the direct insertion of active zinc to the corresponding bromopyridines, or they can be purchased (Rieke

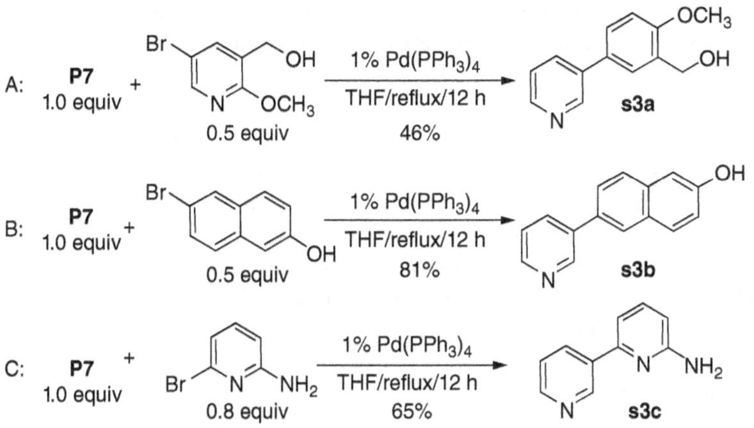

A: **P7** 1.0 equiv + 5-bromo-2-methoxy-3-pyridinemethanol (0.5 equiv) → 1% $Pd(PPh_3)_4$, THF/reflux/12 h, 46% → **s3a**

B: **P7** 1.0 equiv + 6-bromo-2-naphthol (0.5 equiv) → 1% $Pd(PPh_3)_4$, THF/reflux/12 h, 81% → **s3b**

C: **P7** 1.0 equiv + 6-bromo-2-aminopyridine (0.8 equiv) → 1% $Pd(PPh_3)_4$, THF/reflux/12 h, 65% → **s3c**

Scheme 3.12 Coupling of alcohols and amines.

Metals, LLC). The subsequent coupling reactions with a variety of different electrophiles have been performed under mild conditions affording the coupling products [147].

Experimental

General

All reactions were carried out under an argon atmosphere with dry solvent, and a vacuum line was employed for all manipulations of air-sensitive compounds. Commercially available reagents were used without further purification. Active zinc was prepared by the literature procedure [138]. Column chromatography was carried out on silica gel.

Preparation of 2-Pyridylzinc Bromide (P1)

A 250 ml RBF was charged with active zinc (9.81 g, 150 mmol) in 100 ml of THF. 2-Bromopyridine (15.8 g, 100 mmol) was then added to the flask via a cannula while being stirred at rt. After the addition was completed, the resulting mixture was allowed to stir at refluxing temperature for 2 h. The oxidative addition was monitored by GC analysis of the reaction mixture. After settling overnight, the supernatant was transferred into a 500 ml bottle and then diluted with fresh THF to 200 ml.

Preparation of 3-Pyridylzinc Bromide (P7)

An oven-dried 100 ml round-bottomed flask was charged with 3.3 g of active zinc (50 mmol) in 30 ml of THF and 0.2 g of lithium chloride (20 mol%) under the positive pressure of argon gas. 3.9 g of 3-bromopyridine (25 mmol) was added into the solution of active zinc at rt. The resulting mixture was then stirred at refluxing temperature for 2 h, cooled down to rt, and allowed to settle overnight. The supernatant was used for the subsequent coupling reactions.

General Procedure for Copper-Free Coupling Reactions

In a 50 ml round-bottomed flask, isobutyryl chloride (1.27 g, 12 mmol) and 10 ml of THF were placed. Next, 20 ml of 0.5 M solution of 2-pyridylzinc bromide (**P1**) in THF (10 mmol) was added into the reaction flask via a syringe. The resulting mixture was stirred at rt for 4 h. The reaction was monitored by GC analysis, quenched with saturated NH_4Cl solution, and then extracted with ether (30 ml × 3). Combined organics were washed with saturated $NaHCO_3$ solution and brine and then dried over anhydrous $MgSO_4$. A vacuum distillation gave 0.94 g of 2-methyl-1-pyridin-2-yl-propan-1-one (**2d**) as a colorless oil in 63% isolated yield.

Pd-Catalyzed Coupling Reaction with 4-Iodoanisole (10b)

In a 50 ml round-bottomed flask, $Pd[P(Ph)_3]_4$ (0.10 g, 1 mol%) was placed. Next, 20 ml of 0.5 M solution of 3-pyridylzinc bromide (**P7**) in THF was added into the flask at rt. 4-Iodoanisole (1.80 g, 8 mmol) dissolved in 10 ml of THF was added via a syringe. The resulting mixture was stirred at rt for 1.0 h, quenched with saturated NH_4Cl solution, and then extracted with ethyl acetate (30 ml × 3). Combined organics were washed with saturated $NaHCO_3$ solution and brine and then dried over anhydrous $MgSO_4$. A flash column chromatography (20% EtOAc/80% heptane) gave 1.20 g of **10b** as a light yellow solid in 81% isolated yield. mp = 56–58°C.

Preparation of Bipyridines

$Pd[P(Ph)_3]_4$ (0.10 g, 1 mol%) and 5-bromo-2-iodoaniline (2.26 g, 8 mmol) were placed in a 50 ml round-bottomed flask. Next, 20 ml of 3-pyridylzinc bromide (**P7**) (0.5 M in THF, 10 mmol) was added via a syringe. The resulting mixture was allowed to stir at rt for 1 h, quenched with saturated NH_4Cl solution, and then extracted with ethyl acetate (30 ml × 3). Combined organics were washed with saturated $Na_2S_2O_3$ solution and brine and then dried over anhydrous $MgSO_4$. Flash chromatography on silica gel (20% EtOAc/80% heptane) gave 1.17 g of 5-bromo-2,3-bipyridine (**10g**) as a beige solid in 62% isolated yield mp = 75–76°C.

Pd-Catalyzed Coupling Reaction with Haloanilines

$Pd[P(Ph)_3]_4$ (0.10 g, 1 mol%) and 4-iodoaniline (1.75 g, 8 mmol) were placed in a 50 ml round-bottomed flask. Next, 20 ml of 3-pyridylzinc bromide (**P7**) (0.5 M in THF, 10 mmol) was added via a syringe. The resulting mixture was stirred at rt for 1 h, quenched with saturated NH_4Cl solution, and then extracted with ethyl acetate (30 ml × 3). Combined organics were washed with saturated $Na_2S_2O_3$ solution and brine and then dried over anhydrous $MgSO_4$. Flash chromatography on silica gel (30% EtOAc/70% heptane) gave 1.17 g of 3-(4-aminophenyl)pyridine (**13a**) as a beige solid in 86% isolated yield. mp = 114–116°C.

Pd-Catalyzed Coupling Reactions with Halophenols

$Pd[P(Ph)_3]_4$ (0.10 g, 1 mol%) and 4-iodophenol (1.10 g, 5 mmol) were placed in a 50 ml round-bottomed flask. Next, 20 ml of 2-pyridylzinc bromide (**P1**) (0.5 M in THF, 10 mmol) was added via a syringe. The resulting mixture was heated to reflux for 24 h while being stirred, cooled down to rt and quenched with saturated NH_4Cl solution, and then extracted with ethyl acetate (30 ml × 3). Combined organics were washed with saturated $Na_2S_2O_3$ solution and brine and then dried over anhydrous $MgSO_4$. Flash chromatography on silica gel (10% EtOAc/90% heptane) gave 0.80 g of 4-pyridin-3-yl-phenol (**9a**) as a white solid in 95% isolated yield. mp = 159–160°C.

Copper-Catalyzed S_N2 Addition Reactions

CuI (0.50 g, 10 mol%) and LiCl (0.20 g, 20 mol%) were placed in a 100 ml round-bottomed flask. Next, 50 ml of 3-pyridylzinc bromide (**P7**) (0.5 M in THF, 25 mmol) was added via a syringe. The mixture was cooled down to 0°C using an ice bath; 4.0 g (25 mmol) of 3-bromocyclohexene was added via a syringe while being stirred in the ice bath. After being stirred at 0°C, the solution was quenched with saturated NH_4Cl solution and then extracted with ethyl ether (30 ml × 3). The combined organics were washed with 7% NH_4OH solution and brine and then dried over anhydrous $MgSO_4$. Flash chromatography on silica gel (50% ether/50% pentane) gave 2.5 g of 3-cyclohex-2-enyl-pyridine (**11a**) as a yellow oil in 71% isolated yield.

Pd-Catalyzed Bimolecular Coupling Reactions

Bis-(triphenylphosphine palladium (II) dichloride), $Pd[P(Ph)_3]_2Cl_2$ (0.50 g) was placed in a 50 ml round-bottomed flask. Next, 25 ml of 5-methyl-2-pyridylzinc bromide (**P4**) (0.5 M in THF, 12.5 mmol) was added via a syringe. 1.21 g (5.0 mmol) of 2,5-dibromothiophene was added into the flask. The resulting mixture was heated to reflux for 24 h while being stirred. The reaction was cooled down to rt and quenched with saturated NH_4Cl solution and then extracted with ethyl acetate (30 ml × 3). The combined organics were washed with saturated $Na_2S_2O_3$ solution and brine and then dried over anhydrous Na_2SO_4. Flash chromatography on silica gel (10% EtOAc/90% heptane) gave 0.90 g of 2,5-di(5-methylpyridin-2-yl) thiophene (**6g**) as a yellow solid in 68% isolated yield. mp = dec at 170°C.

Preparation of Quinolinylzinc Reagents and Subsequent Coupling Reactions

Preparation of quinolinylzinc bromide: A 250 ml round-bottomed flask was charged with active zinc (2.70 g, 41 mmol) in 30 ml of THF and lithium chloride

(0.23 g, 20 mol%) under an argon atmosphere. 5.70 g (27.5 mmol) of 3-bromoquinoline (or 3-bromoisoquinoline) was added into the solution of active zinc at rt. The resulting mixture was then stirred at refluxing temperature for 2 h, cooled down to rt, and settled overnight. Then the supernatant was used for the subsequent coupling reactions.

Coupling reaction: In a 50 ml round-bottomed flask, CuI (0.19 g, 10 mol%) and LiCl (0.08 g, 20 mol%) were placed. Next, 20 ml of 3-isoquionolinylzinc bromide (**q2**) (0.5 M in THF, 10 mmol) was added via a syringe. The mixture was cooled down to 0°C using an ice bath. 0.80 g (9 mmol) of 3-chloro-2-methylpropene was added via a syringe while being stirred in the ice bath. After being stirred at ambient temperature for 1.0 h, the solution was quenched with saturated NH_4Cl solution. It was extracted with ethyl acetate (30 ml × 3). Combined organics were washed with 7% NH_4OH solution and brine and then dried over anhydrous $MgSO_4$. Flash chromatography on silica gel (20% EtOAc/80% heptane) gave 0.57 g of 4-(2-methyl-allyl)-isoquinoline (**12e**) as a beige solid in 35% isolated yield. mp = 59–61°C.

3.16 Preparation of Functionalized α-Chloromethyl Ketones

α-Chloromethyl ketones are useful synthetic reagents in organic synthesis. However, it is difficult to prepare α-chloromethyl ketones via direct chlorination of methyl ketones. Thus, much work has been done to develop alternate methods of synthesis. However, many of these procedures require the use of organometallic reagents that preclude the presence of secondary functional groups [148]. Other methods require oxidizing agents [149] or conditions which can promote rearrangements [150]. We have developed a simple straightforward method for the preparation of α-chloromethyl ketones under extremely mild conditions [151]. Using Rieke zinc prepared by the reduction of $ZnCl_2$ in THF with lithium naphthalenide, we have found that alkyl and aryl bromides readily undergo oxidative addition, yielding the corresponding organozinc bromides. These organozinc reagents smoothly cross-couple with chloroacetyl chloride mediated by the soluble copper cyanide/lithium bromide complex [152] as shown in Table 3.46.

In Scheme 3.13, 1-bromohexane reacts cleanly with Rieke zinc to give the organozinc, which when catalyzed by copper cross-couples with chloroacetyl chloride to give the corresponding α-chloromethyl ketone in 75% isolated yield. Significantly, a number of functional groups will tolerate these reaction conditions. Respectable yields were obtained with alkyl as well as aryl nitriles and esters. The reaction appears to be totally general and will work with any organozinc halide including alkyl, aryl, and vinyl halides. Significantly, the reaction tolerates all functional groups. A typical experimental procedure is presented later in the text.

Table 3.46 Reaction of organozinc reagent mediated by copper with chloroacetyl chloride.

Entry	RX	Products[a]	Yield (%)[b]
1	$CH_3(CH_2)_5Br$	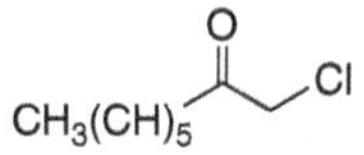 $CH_3(CH)_5$ O Cl	75
2	$CH_3(CH_2)_7Br$	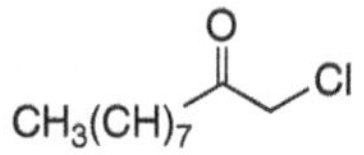 $CH_3(CH)_7$ O Cl	74
3	$CH_3(CH_2)_9Br$	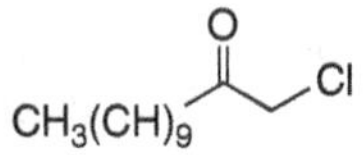 $CH_3(CH)_9$ O Cl	72
4	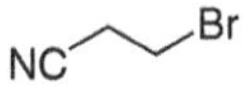 NC Br	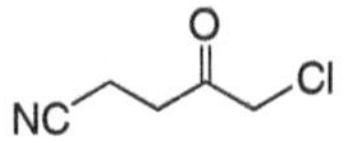 NC O Cl	52
5	 NC Br	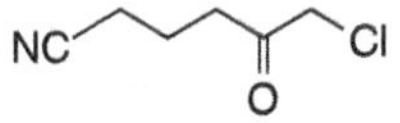 NC O Cl	86
6	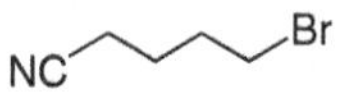 NC Br	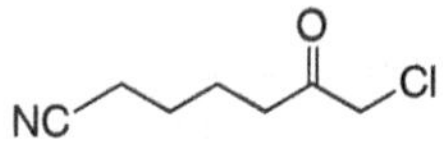 NC O Cl	81
7	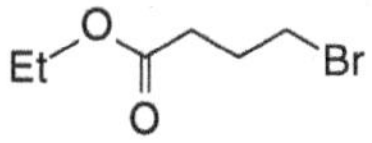 Et O O Br	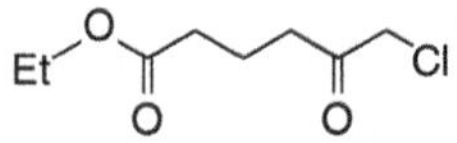 Et O O O Cl	87
8	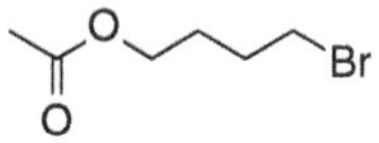 O O Br	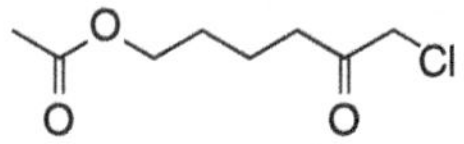 O O O Cl	60
9	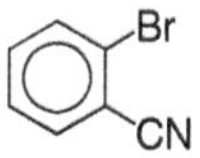 Br CN	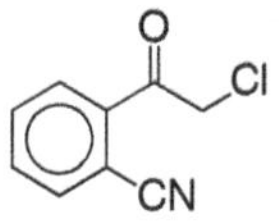 O Cl CN	85
10	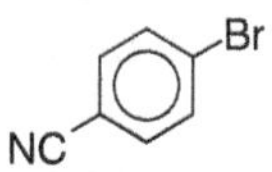 Br NC	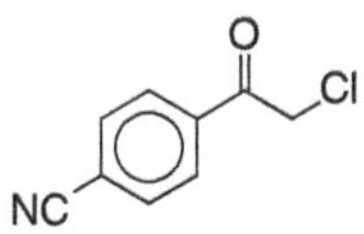 O Cl NC	79
11	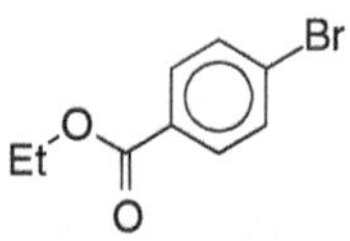 Br Et O O	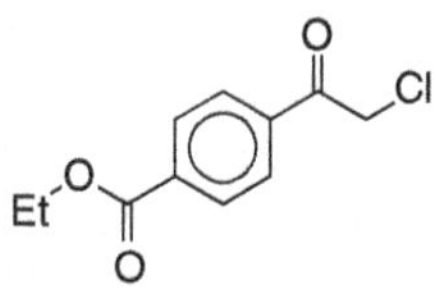 O Cl Et O O	72

[a] All products are characterized by ^{1}H NMR, ^{13}C NMR, IR, and HRMS.
[b] Isolated yield.

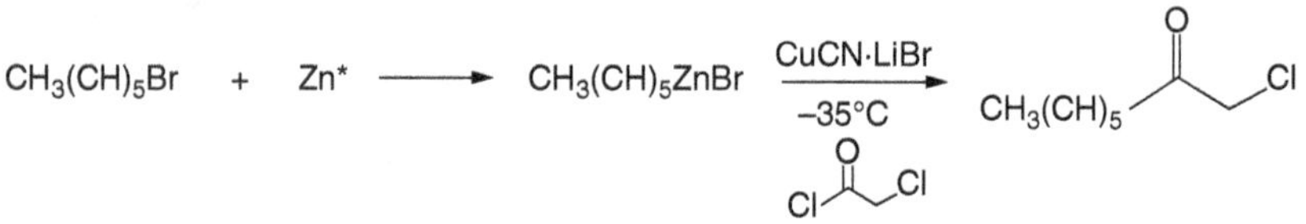

$CH_3(CH)_5Br$ + Zn* ⟶ $CH_3(CH)_5ZnBr$ $\xrightarrow[-35°C]{CuCN\cdot LiBr}$ $CH_3(CH)_5C(O)CH_2Cl$

(reagent: $ClC(O)CH_2Cl$)

Scheme 3.13 Preparation of α-chloroketones.

1-Chloro-6-cyano-2-hexanone: (Table 3.46, entry 5). Lithium (29.12 mmol) and naphthalene (2.92 mmol) in freshly distilled THF (13 ml) were stirred under argon until the green color indication of the formation of lithium naphthalenide appeared (ca. 30 s). $ZnCl_2$ (14.11 mmol) in 15 ml THF was cannulated dropwise into the lithium naphthalenide, taking care to maintain the green color. After the transfer was complete the Rieke zinc was stirred for 30 min and allowed to settle (ca. 2 h). The supernatant was cannulated off and replaced with fresh THF. The active zinc may also be purchased from Rieke Metals, LLC. 4-Bromobenzonitrile (10.91 mmol) was added neat via syringe and stirred for 30 min, after which the excess was allowed to settle. A solution of CuCN (2.83 mmol) and LiBr (3.82 mmol) was dissolved in 10 ml of THF and cooled to −40°C. The RZnBr solution was cannulated into the CuCN solution, taking care not to transfer any active zinc, and the chloroacetyl chloride (9.40 mmol) was added neat via syringe. The reaction was warmed to room temperature with stirring and quenched with NH_4Cl. The aqueous layer was extracted with diethyl ether (3 × 30 ml), and the combined organic layers dried over $MgSO_4$. Removal of solvent and flash chromatography on silica gel afforded 1-chloro-6-cyano-2-hexane (1.179 g, 86% yield).

3.17 Rieke Zinc as a Reducing Agent for Common Organic Functional Groups

Rieke zinc was developed in 1972 and produced a highly reactive form of zinc. The vast amount of studies on this metal was to produce organozinc reagents. It was not until 2006 that Don Kaufman et al. reported on the use of Rieke zinc as a reducing agent for several functional groups under very mild conditions [153, 154]. They found that Rieke zinc is able to reduce conjugated aldehydes, alkynes, esters, and nitro compounds under very mild reaction conditions in high yields. These reductions are performed in a mixture of THF, the solvent in which Rieke zinc is prepared, and a proton donor such as water or methanol or both. The compounds in Table 3.47 were reduced by a procedure patterned after that used by White [155]. The compounds were added as methanol solutions to a slurry of Rieke zinc in THF. Water was then slowly added, and the resulting mixture heated at reflux for the indicated times. Products were isolated by extraction with ether from the reaction mixture.

Table 3.47 Reduction of compounds with Rieke zinc.

Entry	Reactant	Product	Yield (%)[a]	Reaction time (h)
1	$C_6H_5NO_2$	$C_6H_5NH_2$	90	3
2	C_6H_5CHO	$C_6H_5CH_2OH$	100	3
3	$C_6H_5CH{=}CHCHO$	$C_6H_5CH{=}CHCH_2OH$	100	3
4	cyclopentenyl–CHO	cyclopentenyl–CH_2OH	nd	3
5	$C_6H_5COCH_3$	$C_6H_5CHOHCH_3$	nd	12
6	C_6H_5CN	$C_6H_5CN_2NH_2$	nd	24
7	$C_6H_5C(CH_3){=}CH_2$	$C_6H_5CH(CH_3)_2$	nd	24
8	cyclooctene	cyclooctane	nd	24
9	$HC{\equiv}C(CH_2)_5CH_3$	$H_2C{=}CH(CH_2)_5CH_3$	nd	24
10	$C_6H_5C{\equiv}CH$	$C_6H_5CH{=}CH_2$	100	4
11[b]	Fec–$C{\equiv}CH$	Fec–$CH{=}CH_2$	100	0.5
12	H_3C–C_6H_4–$C{\equiv}CH$	H_3C–C_6H_4–$CH{=}CH_2$	100	4
13[c]	$C_6H_5C{\equiv}C{-}CO_2Et$	$C_6H_5C(H){=}C(H){-}CO_2Me$	100	3
14	$EtO_2C{-}C{\equiv}C{-}CO_2Et$	$EtO_2C{-}CH_2CH_2{-}CO_2Et$	100	3
15	$C_6H_5C{\equiv}C{-}CH_3$	$C_6H_5CH{=}CH{\cdot}CH_3$	nd	24

nd, not detected spectroscopically. All starting material recovered.

[a] Yields of the crude reaction product were determined by proton NMR (300 MHz).

[b] Fec represents the ferrocenyl group.

[c] 100% *trans* stereochemistry observed. Also note that transesterification occurred.

The mechanism involved in zinc reductions, including Rieke zinc, of alkynes and other functional groups is assumed to be a dissolving metal reduction [156–158] much the same as that proposed for the reductions of alkynes with sodium metal. This mechanism, as studied by Olah [156], involves both the formation of a radical anion and a radical; it accounts for the trans product predominating unless the reaction is conducted at high temperatures. Olah also notes that, though studied less thoroughly, other dissolving metal reductions are believed to occur via a similar mechanism. He further states that "interestingly, activated zinc in the presence of a proton donor (alcohol), although a dissolving metal reagent, reduces disubstituted alkynes to *cis*-alkenes" [156].

The results are consistent with the dissolving metal model as they were able to reduce only conjugated functional groups as shown in Table 3.47. Whether it is the anion radical or the radical that is formed in the rate-determining step, both would be more stabilized in a conjugated system. Evidently, in the absence of that conjugation, reduction with zinc is not possible. One of the most intriguing Rieke reduction results was found in the reduction of alkynes. The only previous attempts to use Rieke zinc to reduce alkynes were propargylic alcohols, 1,3-diynes and 1,3-enynes; *cis*-alkenes were the major products of this work done by White [155]. It is of interest to note that they obtained the *trans*-alkene in contrast to the *cis*-isomer suggested by Olah and obtained by White. For example, ethyl phenylpropiolate was reduced by Rieke zinc in quantitative yield to *trans*-ethyl cinnamate. Of additional interest, they found that both phenylacetylene and *p*-methylphenylacetylene were readily and quantitatively reduced to their respective alkenes (Table 3.47, entries 10 and 12), while 1-phenylpropyne showed no reaction with Rieke zinc (Table 3.47, entry 15). In all three of these alkynes, the triple bond is conjugated with a phenyl group and so might be expected to react similarly. It may be that the small electron-releasing tendency of the methyl group directly attached to the triple bond in 1-phenylpropyne destabilizes the radical or radical ion just enough to prevent this molecule from being reduced. However, it might also be argued that the *para*-methylphenylacetylene should have a similar effect.

While it is true that "regular" powdered zinc can be used to achieve many of the same reductions as Rieke zinc [159], it should be noted that much milder reaction conditions can be employed with Rieke zinc [158]. For example, proton donors as weak as water or methanol can be used with Rieke zinc, whereas stronger acids such as hydrochloric acid, phosphoric acid, or acetic acid are needed for regular powdered zinc. For instance, ethyl phenylpropiolate was quantitatively reduced to its alkene with Rieke zinc using water/methanol, while powdered zinc gave no reaction with this alkyne under similar conditions. Thus, an attractive feature of using Rieke zinc for organic reductions is that it can be used under conditions that would less likely alter other acid sensitive functional groups in compounds being reduced.

The General Procedure for Dissolving Zinc Metal Reduction

A three-necked flask, fitted with a reflux condenser and septa, was purged with Ar for 15 min. The Rieke zinc/THF slurry (purchased from Rieke Metals, LLC, 1001 Kingbird Road, Lincoln, NE 68521) was transferred to the flask via syringe. After the Zn suspension was heated to reflux, a methanol solution of the organic compound to be reduced was added, followed by addition of water. The ratio of THF : methanol : water used was 7 : 5 : 1. A representative reduction: to 8.0 ml of a stirred suspension of 5% Rieke zinc, 6.12×10^{-3} mol, a solution of 4.08×10^{-3} mol of phenylacetylene dissolved in 5.60 ml of methanol was slowly added, followed by dropwise addition of 1.10 ml of water. After refluxing for 2.5 h under argon, the solution was cooled and 150 ml of ether were added. The resulting mixture was filtered through Celite and extracted successively with 10% ammonium chloride (25 ml), 10% sodium bicarbonate (25 ml), and saturated sodium chloride (25 ml). After drying over anhydrous magnesium sulfate, the solvents were stripped by rotary evaporation. The yield and composition of the resulting crude product were determined by proton NMR.

3.18 Detailed Studies on the Mechanism of Organic Halide Oxidative Addition at a Zinc Metal Surface

Discovered over a century ago, the reaction of organic halides with magnesium [160], and later with lithium [161], is still the most direct and commonly used method to prepare the corresponding organometallic compounds [162]. These processes have been extensively studied, yet there still remains some degree of uncertainty in several aspects concerning mechanistic details of the reactions [163]. Within some limits, the mechanisms of the formation of organomagnesium and organolithium compounds show great similarities [164]. Another common feature is that neither of these reactions shows much selectivity for the organic structure. As had been stated, "the reaction of organic bromides with magnesium is among the least selective of organic reactions" [165]. This lack of selectivity is sometimes attributed to the participation of radicals as key intermediates in the reaction. However, radical-mediated reactions are not necessarily poorly behaved with respect to selectivity, as can be inferred from several recent elegant studies on the topic [166].

In addition to these metals, zinc has become an important tool for the synthetic organic chemist. Over 150 years ago, zinc metal was found to react with organic iodides to yield organozinc compounds [167]. Recently, considerable attention has been focused on zinc chemistry and its synthetic applications [168]. Organozinc reagents tolerate a broad range of functionalities [169], and, since the introduction of highly reactive Rieke zinc [170], virtually any organozinc reagent can be prepared from the corresponding organic bromide. It is generally accepted that the mechanism of these reactions is similar to those of the analogous reactions of magnesium and lithium [171]. However, few mechanistic studies regarding the formation of organozinc halides are available.

We carried out several studies on the reaction of highly reactive zinc with organic bromides. The studies are designed to gain a better understanding of the process and its differences with other related processes. These studies consist of (i) kinetic–structural studies, (ii) stereochemical studies of the course of the reaction, (iii) indirect detection of intermediates, and (iv) linear-free energy relationships. The oxidative addition shows a pronounced structure-reactivity dependence, in contrast to that shown by other metals. The stereochemical outcome of the reaction and radical probes however suggests that radicals are taking an active part in the reaction. The reaction contemplated here can therefore be described as another example of a radical-mediated selective reaction, and it has straightforward synthetic applications. Some synthetic work was done in this direction to demonstrate how this structure-reactivity dependence can be used to obtain selective organozinc formation in unsymmetrical dibromides [172].

Results and Discussion

Competitive Kinetics

The study of the relationship between the structure of a substrate and reactivity exhibited in a given reaction yields extremely valuable information about the nature of the process [173]. It permits, if not to assign a mechanism, to relate a process to others with similar structure-reactivity profiles, so mechanistic inferences can be recognized. When the difference in reaction rates between two different substrates is not too large (this will depend on the limits dictated by the detection/quantification technique utilized), competitive kinetics allow one to obtain reliable rate constant ratios of the reaction substrates [165]. In our case, competitive kinetic methods adopted very well to our purpose, that is, the study of the rates of reaction of organic halides with highly reactive zinc. The rate of reaction of organic bromides depends, in principle, on the concentration of organic halide, $[R_1Br]$ or $[R_2Br]$, and on some undefined physical characteristics of the metal surface, $f(Zn)$, according to Equations 3.15 and 3.16. The simultaneous

Reaction of alkyl bromides with Zn*
Kinetic and mechanistic considerations

$$R_1Br + R_2Br \xrightarrow[\text{ii. } H_3O^+]{\text{i. } Zn^*} \begin{cases} \xrightarrow{k_1} R_1Br + R_1H \\ \xrightarrow{k_2} R_2Br + R_2H \end{cases}$$

$$-d\,[R_1Br]/dt = k_1\,[R_1Br]^x\,f(Zn) \quad (3.15)$$

$$-d\,[R_2Br]/dt = k_1\,[R_2Br]^x\,f(Zn) \quad (3.16)$$

determination of the concentration of two different organic bromides in solution reacting with the same metallic surface permits Equations 3.15 and 3.16 to be simplified. Dividing both, a third expression (Equation 3.17) arises in which two variables, t and $f(\text{Zn})$, are canceled. By assuming that $x = 1$

$$-d[\text{R}_1\text{Br}]/d[\text{R}_2\text{Br}] = k_1/k_2[\text{R}_1\text{Br}]/[\text{R}_2\text{Br}]^x \tag{3.17}$$

$$\ln([\text{R}_1\text{Br}]_t/[\text{R}_1\text{Br}]_0) = k_1/k_2 \ln([\text{R}_2\text{Br}]_t/[\text{R}_2\text{Br}]_0) \tag{3.18}$$

and integrating Equation 3.17, Equation 3.18 is obtained. The experimental kinetic data fit Equation 3.18 very well (Figure 3.1). Plots of $\ln([\text{R}_1\text{Br}]/[\text{R}_1\text{Br}]_0)$ versus $\ln([\text{R}_2\text{Br}]/[\text{R}_2\text{Br}]_0)$ were linear to elevated conversions (up to >95%).

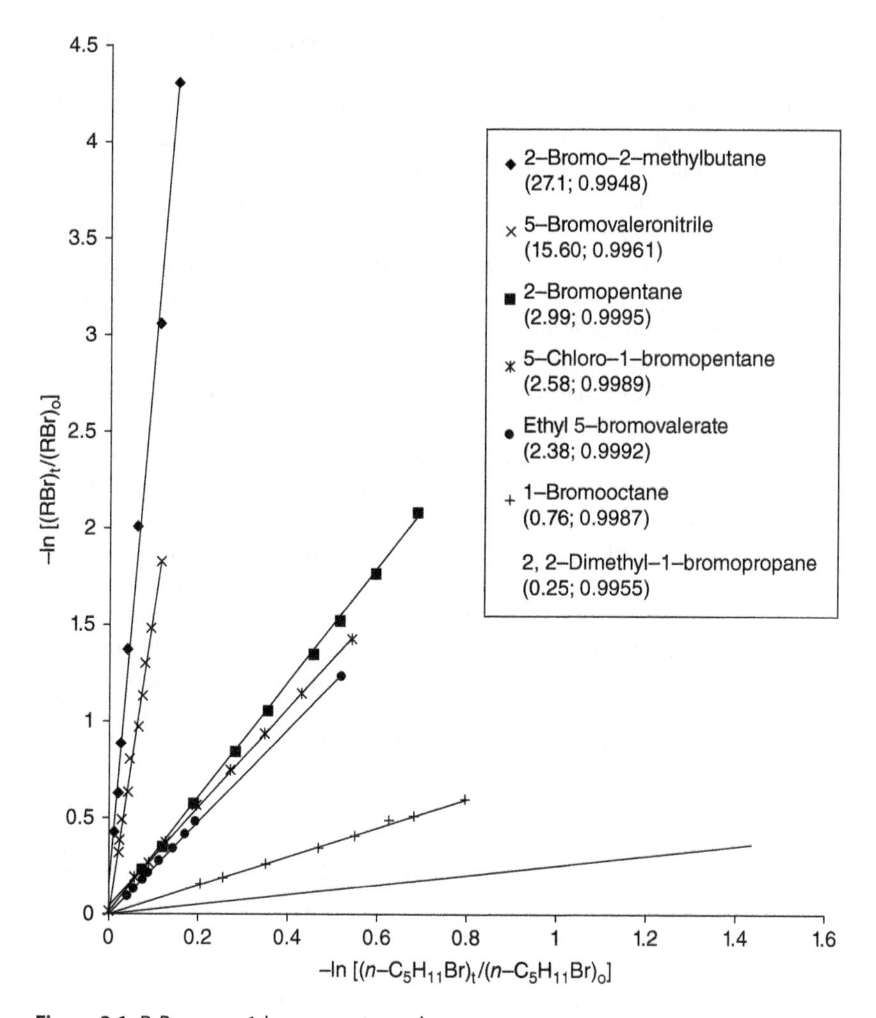

Figure 3.1 R_1Br versus 1-bromopentane plots.

Evaluation of the slope, employing linear regression analysis, yielded k_1/k_2 directly. Each line contains an average of ca. 9 points, with the linear regression coefficients being >0.99 in all cases for alkyl bromides and ca. 0.95 on average for aryl halides. This confirmed our hypothesis that the reaction was first order with respect to the organic halide, in agreement with kinetic data reported for magnesium and lithium [165b, 174]. Since the chemical yields of organozinc halides using this procedure are very high [168b, 170a], side reactions represent very small perturbations on the final results.

Alkyl Bromides

Table 3.48 summarizes the observed ratio of rate constants (k_1/k_2, THF, 0°C) for all the pairs of substrates assayed. The reactivity profile correlates with electron transfer (ET) being the rate-determining step of the reaction, in consonance to analogous processes with magnesium or lithium, as well as with other ET reactions. Other possible mechanisms of the oxidative addition will be discussed later. The reaction shows selectivity toward the nature of the organic moiety. The effect of the alkyl bromide on the rate of the ET is tertiary > secondary > primary (Table 3.48, entries 2, 3, 12, and 13), roughly in the order of 30 : 3 : 1. Cyclopentylbromide has a rate of 3.51 times faster than that for 2-bromopentane (Table 3.48, entry 5), and the 1-adamantyl bromide rate is 5.96 times slower than that for 2-bromooctane (and 77 times slower than that for 2-bromo-2-methylheptane) (Table 3.48, entries 10 and 13). These are consistent with the geometric requirements of radical intermediates [175]. The reaction also seems to be slightly sensitive to some degree of steric factors: neopentyl bromide was found to react 4.00 times slower than 1-bromopentane (Table 3.48, entry 4) [176]. The effect of a polar group in the chain is to increase the rate of the reaction (Table 3.48, entries 6 and 7). In the case of 5-bromovaleronitrile, this effect is notable (15.6 times faster than 1-bromopentane) and may be governed by binding interactions of the nitrile to the metal surface prior to ET or a reduced reduction potential (Table 3.48, entry 8). When we shorten the chain length by one methylene, much more increased rates were observed (131 times faster than 1-bromopentane, Table 3.48, entry 9). Comparative kinetic experiments using 1-bromopentane and 10 equiv of acetonitrile showed little effect on the reaction rate ($k_{THF\text{-}AN}/k_{THF} \approx 1.1–1.2$). Neither intramolecular nor intermolecular coordinative stabilization seems to be responsible for the increase in reaction rates observed. Purely inductive effects, which usually vanish rapidly in aliphatic chains, may not provide by themselves a satisfactory explanation.

Aryl, Vinyl, Benzyl, and Allyl Bromides

The study of these systems provided important information for both mechanistic and synthetic purposes. The reactivity of bromobenzene was evaluated against 1-bromooctane by following the same procedure used for alkyl bromides (THF, 0°C, GLC analysis). It reacts with active zinc ca. 20 times slower than this primary alkyl bromide and, subsequently, 27 times slower

Table 3.48 Relative rates (k_1/k_2).

	Relative rates (k_1/k_2) to 1- and 2-bromopentane		
	R_1Br	k_1/k_2	
1.	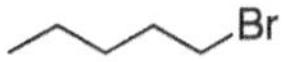 Br	*1*	*0.33*
2.	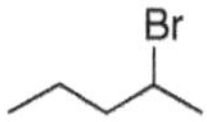 Br	2.99	*1*
3.	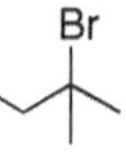 Br	27.1, *28.6*	9.58
4.	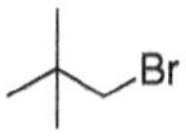 Br	0.25	*0.084*
5.	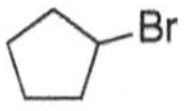 Br	*10.5*	3.51
6.	 Cl, Br	2.58	*0.86*
7.	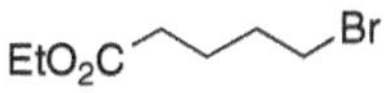 EtO_2C, Br	2.38	*0.80*
8.	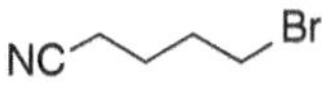 NC, Br	15.6	*5.22*
9.	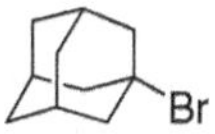 Br	*0.42*	0.17
10.	 Br	*1*	*0.39*
11.	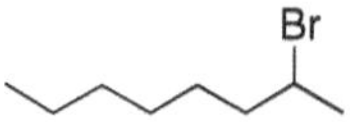 Br	2.52	*1*
12.	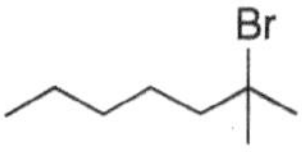 Br	31.8, *32.8*	13.02
13.	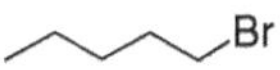 Br	1.32	*0.52*
		Primary	Secondary
		$C_8H_{17}Br$	

k_1/k_2 values for nondirectly measured pairs in italics.

than 1-bromopentane, arbitrarily chosen for comparative purposes (Table 3.49). As an example of a vinylic bromide, 2-bromo-3-methyl-2-butene was tested against bromobenzene in THF at 70°C. A rate ratio of 0.17 was observed, which corresponds to 160 times slower than 1-bromopentane (Table 3.49, entry 2).

Table 3.49 Relative rates of reaction of aryl, vinyl, benzyl, and allyl halides (k_1/k_2).[a]

	R_1Br	k_1/k_2	R_2Br	$k_1/k_{n\text{-bromopentane}}$[b]
1.	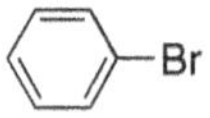 Br	0.049	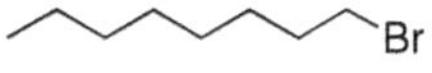 Br	*0.037*
2.	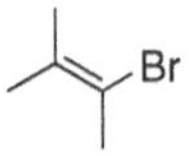 Br	0.17[c]	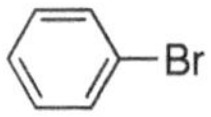 Br	*0.0062*[d]
3.	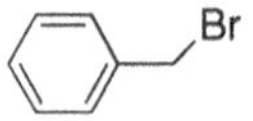 Br	≥14	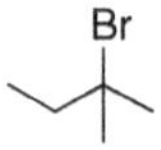 Br	*≥380*
4.	 Br	~1–10[e]	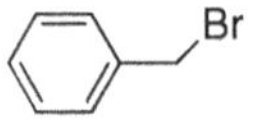 Br	*>380*
5.	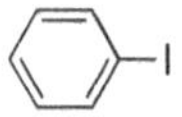 I	21	 Br	*16*

[a] Corresponding to the reaction of two alkyl bromides (R_1Br, R_2Br) with highly reactive zinc. All reactions were carried out in THF 0°C, unless otherwise stated. Relative rates were determined by the competitive kinetics technique.

[b] k_1/k_2 values for nondirectly measured pairs in italics.

[c] 70°C.

[d] Calculated from $k_{PhBr}/k_{n\text{-Bromopentane}}$ at 0°C.

[e] Approximate range (see text).

Benzyl and allyl bromides could not be studied using the normal experimental protocol. The reaction was too fast to adequately monitor by simply sampling at different time intervals. On the other hand, GLC response calibration of benzyl bromide gave inconsistent results [177]. We used a variation of the technique, in which several reactions—one for each point to be plotted—were set up. They were loaded with a previously prepared mixture of bromides/internal standard. A substoichiometric amount of active zinc, different for each reaction, was then injected. The reactions went to completion within minutes. Direct analysis of the unreacted bromides by ^{1}H NMR provided, if not an accurate measurement of the rates ratios, establishment of lower limits for them in these rapid reactions. Benzyl bromide was tested against the fastest organic bromide, whose kinetics were available at that point, that is, *tert*-pentyl bromide (2-bromo-2-methylbutane). $k_{BnBr}/k_{t\text{-PentBr}}$ was found to be ≥14 (THF, 0°C), the limit being dictated by the sensitivity of the analysis. Consistent with other ET reactions, benzyl bromide showed enhanced reactivity, the rate being at least 380 times faster than 1-bromopentane (Table 3.49). The measure of allyl bromide resulted in additional complications. Whereas side reactions were present, but unimportant in the formation of benzylzinc bromide, allyl bromide reacts with extensive homo- and cross-coupling product formation in competitive kinetic experiments with benzyl bromide. A study of the distribution of reaction products as well as unreacted starting materials only permits a rough assignment of reaction rates ratios. With reasonable reliance, we can state that allyl bromide reacts slightly faster than benzyl bromide but within the same order of magnitude (THF at 0°C). No attempts were made to determine whether the reaction rates observed for allyl and benzyl bromides are mass-transport limited or not. Though this may be possible, it does not alter the final conclusions of this work since most substrates assayed fall out of this range.

This set of kinetic data is compiled in Table 3.50, along with an extensive set of other ET reactions to organic halides, in which reactivity profiles were studied. These are as follows, starting from the left column: $Co(CN)_5^{3-}$ in MeOH—H_2O at 25°C (RI) [179]; $Cr(en)_2^{2+}$ (en = ethylenediamine) in dimethylformamide (DMF)—H_2O at 25°C (RBr) [180]; idem. (RCl) [180], active Zn in THF at 0°C (RBr) [181]; Bu_3Sn^* in Et_2O at 35°C (RCl) [165d]; lithium 4,4′-dimethylbenzophenone ketyl in THF at 24°C (RBr) [165d]; Mg in Et_2O at 0°C (RCl) [165d]; $(C_5Me_5)_2UCl{\cdot}THF$ in benzene at 25°C (RCl) [182]; Bu_3Sn^* at 45/80°C in chlorobenzene (RBr) [183]; electrochemically generated anthracene radical anion in DMF (0.1 M $(TBA)BF_4$) (TBA = tetrabutylammonium) at 25°C (RBr) [184]; sodium in gas phase (RCl) [185]; electrochemically generated benzophenones ketyl in DMF 0.1 M $(TBA)BF_4$) at 25°C (RBr) [184]; peak potentials (E_p) at glassy carbon electrode in DMF at 10°C (RBr) [186] and half-wave potentials ($E_{1/2}$) at a mercury electrode in different solvents [178]; and Mg in Et_2O at 0°C (RBr) [165a]. A comparative study of the processes will be discussed later.

Stereochemical Studies

The study of the stereochemical course of a reaction may provide definitive clues on its mechanistic nature. To further explore the oxidative addition step, the reaction of Zn* with an optically active alkyl bromide was investigated. (*R*)-(+)-2-bromooctane (**1**) [187] ($[\alpha]_d 25$ = +39.7°C, >95% op) (op = optical purity) was prepared and reacted with 4.0 equiv of Zn* at 0°C. The 2-octylzinc bromide generated was then reacted with 1.0 equiv of di-*tert*-butylazodicarboxylate (DBAD) at 0°C [60]. Deprotection and reduction [74] afforded (±)-2-octylamine (**3**), as shown by ^1H NMR spectroscopy using (*S*)-(+)-*O*-acetylmandelic acid as a chiral solvating agent [188]. The use of O_2 as an electrophile also resulted in formation of racemic 2-octanol (**4**, $^1H/^{19}F$ NMR spectra of the ester with (*R*)-(+)-Mosher acid) [189]. This is a less reliable reaction since it is possible that racemization might also have taken place during reaction with the oxygen [190]. According to the kinetics of the isotopic substitution of 2-bromooctane with radioactive bromide ion (Li*Br, acetone, 65.5°C) [191], racemization attributed to the bromide generated during the reaction may be excluded, without incurring an extensive error (Scheme 3.14, <5 min of reaction time to completion). On the other hand, configurational stability of a secondary carbon—zinc bond

Table 3.50 Relative rates of reaction of organic halides in various electron transfer processes.[a]

ET reag.[b]	Co(II)	Cr(II)	Cr(II)	Zn*	$Bu_3Sn^{\cdot}$	Ar_2COLi	Mg	U(III)	$Bu_3Sn^{\cdot}$	$Anthr^{\cdot-}$	$Na_{(g)}$	$Ph_2CO^{\cdot-}$	E_p ($E_{1/2}$)[c]	Mg
RX	RI	RBr	RCl	RBr	RCl	RBr	RCl	RCl	RBr	RBr	RCl	RBr	RBr	RBr
Me	0.22										0.43		(−1.96) [178b]	
n-	1 (Pr)	1 (Bu)	1 (Pr)	1 (Pent)	1 (Pent)	1 (Pent)	1 (Pr)	1 (Bu)	1 (Et)	1 (Bu)	1 (Pr)	1 (Bu)	−2.85 (−2.24) (Bu)	1 (Pr)
sec-	28 (Pr)	10 (Bu)	4.0 (Pr)	3.0 (Pent)	2.9 (Pent)	2.3 (Bu)	3.8 (Pr)	2.1 (Pr)	3.0 (Pr)	2.5 (Bu)	1.3 (Pr)	1.0 (Bu)	−2.63 (−2.34) (Bu)	1.1 (Pr)
tert-	210 (Bu)	41 (Bu)	32 (Bu)	27 (Pent)	19 (Bu)	21 (Hex)	7.7 (Bu)	7.5 (Bu)	7.0 (Bu)	4.4 (Bu)	2.9 (Bu)	2.6 (Bu)	−2.51 (−2.19) (Bu)	1.2 (Bu)
Bn	88 000	400[d]	9 300[d]	≥380			<92	12	34		4400		(−1.22)	≤1.4
1-Adam				0.32				5		0.023			(−2.38)	0.67
Neopent				0.25	0.52	0.68	0.22	0.5		0.014			(−2.46)	0.58
c-Pent				11	4.3	12	4.2		2.4				(−2.19)	1.2
Ph				0.037				0.05	0.063[e]				(−1.81) [178b][f]	0.58
Vinyl				0.0062[g]									(−2.46)[h]	0.231[i]
Allyl				>380									(−1.24) [178b][j]	≤1.8
Solvent	MeOH	DMF	DMF	THF	Et_2O	THF	Et_2O	PhH	PhCl	DMF	—	DMF	DMF	Et_2O

[a] All the rates are relative to the corresponding *n*-haloalkane.
[b] See text for details and references.
[c] Peak potentials, E_p (glassy C, volts) in DMF (0.1 M Bu_4NBF_4). Half-wave potentials, $E_{1/2}$ (Hg, V, in parentheses) in DMF (0.01 M Et_4NBr). Hg electrodes may be chemically involved in the electrochemical reduction of alkyl halides: Ref. [54].
[d] Estimate: see references in text.
[e] 80°C.
[f] DMF (0.02 M Et_4NBr).
[g] For 2-bromo-3-methyl-2-butene at 70°C.
[h] In 75% dioxane–water ($Bu_4$4NI).
[i] For 2-bromo-1-butene.
[j] In DMSO (0.1 M Et_4ClO_4).

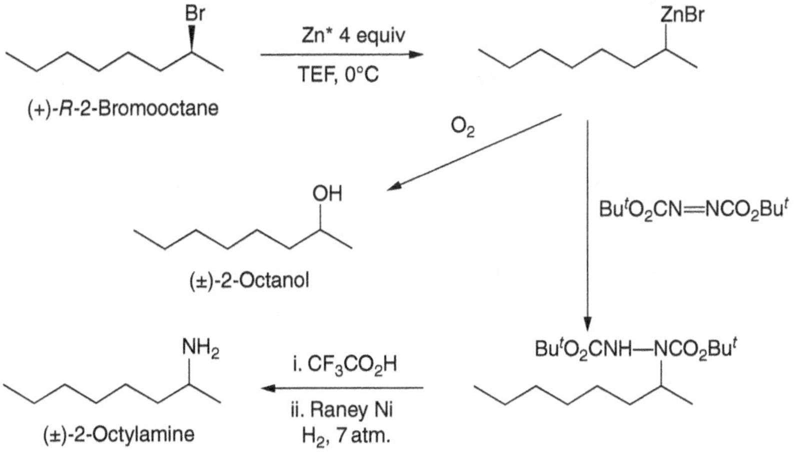

Br
(+)-R-2-Bromooctane
Zn* 4 equiv
TEF, 0°C
ZnBr
O_2
OH
(±)-2-Octanol
$Bu^tO_2CN{=}NCO_2Bu^t$
NH_2
(±)-2-Octylamine
i. CF_3CO_2H
ii. Raney Ni
H_2, 7 atm.
$Bu^tO_2CNH{-}NCO_2Bu^t$

Scheme 3.14 Reaction of Zn* with optically active bromides: intermediate radicals promote racemization.

has been previously observed [192]. Accordingly, racemization most likely occurred during the formation of the organozinc reagent [193].

Radical Detection

The rearrangement of radical probes (radical clocks) has been used extensively to elucidate the nature of the events following ET steps in which the formation of radicals is suspected [194]. Previous studies using 6-halo-1-hexene models suggested radicals as intermediates in the reaction of organic halides with zinc [195]. A difficulty inherent to these radical probes is that the corresponding organometal derivative also cyclizes, although at a much slower rate. It has been demonstrated recently that acyclic organozinc iodides can be prepared from several substituted 6-iodo-1-hexene probes by reaction with zinc, and they only cyclize upon warming [196]. In one example, using (*Z*)-8-iodo-3-octene and zinc in DMF at 80°C, cyclization of the corresponding organozinc iodide did not take place to any extent. This observation was clearly inconsistent with a scenario involving radical ring closure. The background depicted previously prompted us to examine with caution the study of the reaction of active zinc with radical clocks. Three different radical probes [197], covering a wide range of radical rearrangement rate constants (k_r, 25°C), were submitted to test. Bromomethylcyclopropane (**5**) ($k_r = 1.3 \times 10^8\,s^{-1}$), 6-bromo-1-hexene (**6**) ($k_r = 1.0 \times 10^5\,s^{-1}$), and 5-bromopentanenitrile (**7**) ($k_r = 3.9 \times 10^3\,s^{-1}$) were reacted with 2.0 equiv of Zn* at 0°C in THF. The analysis of products was performed directly on the crude reaction mixture after centrifugation and at the same temperature by 500 MHz ^1H NMR (Scheme 3.15). They reveal a complete rearrangement for the fastest radical clock (**5r**). In contrast, there was essentially

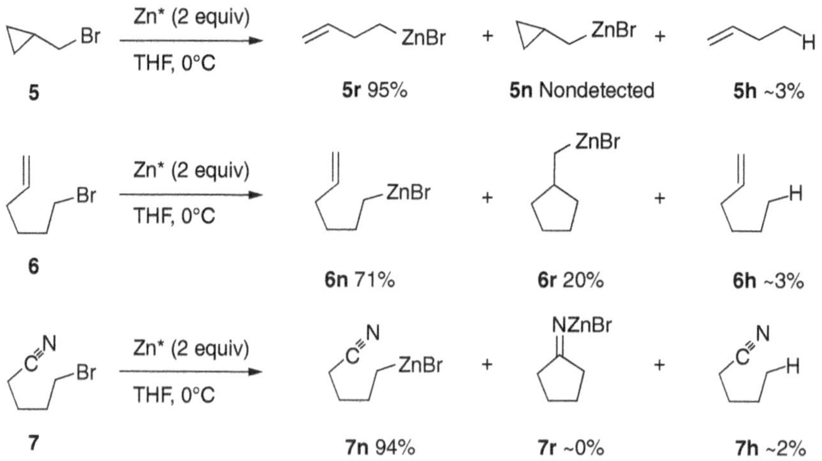

Zn* (2 equiv), THF, 0°C

5 → 5r 95% + 5n Nondetected + 5h ~3%

6 → 6n 71% + 6r 20% + 6h ~3%

7 → 7n 94% + 7r ~0% + 7h ~2%

Scheme 3.15 Indirect radical detection in the reaction of Zn* with radical probes.

total formation of the unrearranged organozinc bromide in the case of the 5-bromovaleronitrile (**7n**) [198]. A mixture of both rearranged and nonrearranged products was observed in the case of 6-bromo-1-hexene (6*n*:6*r* ratio = 3.6 : 1), along with small amounts of the hydrocarbons originated by hydrogen/proton abstraction from the media or during handling (1-hexene (**6h**) and methylcyclopentane (**6h**′) in ≈5 and 1–2%, respectively). The reaction was then allowed to warm to room temperature. After 24 h, no substantial change in the ratio of products was observed (3.5 : 1). In addition, the use of a higher amount of Zn* in the reaction (4 equiv) originated a mixture of compounds enriched in the olefinic component (8.7 : 1). From all of this, we can conclude that the cyclized products are originated during the formation of the organozinc reagent, via radical cyclization. No attempts to measure the reaction rate constant for this second ET were considered. Decay in the reactivity of any powdered metal as the reaction advances adds inherent difficulties to the experiment.

Mechanistic Considerations

Traditionally, we distinguish between one- and two-electron metal redox reagents. In practice, this distinction is only formal, merely reflecting the difficulty of stabilizing certain transient oxidation states. Zinc, with an intermediate oxidation state of low stability (Zn^+), is one of these reagents [199], its two-electron processes being better regarded as 2 one-electron steps. For the sake of simplicity and symmetry in the following schemes, after any kind of ET, the metallic surface will be referred to as $Zn^+(S)$. It does not imply the presence of an isolated monovalent zinc atom. The entire metallic cluster should be considered to be involved in the stabilization of this electron-deficient entity, until Zn^{2+} ions neutralize the charge by leaving the surface. Comparison of the reactivity profiles in Table 3.50 leaves little doubt about the nature of the reaction. However, a complete survey of other mechanisms involving oxidative addition will be considered [200].

Two-Electron Mechanisms: S_N2

This is a common oxidative addition mechanism. The order of reactivity for this S_N2-type reaction with alkyl halides is primary > secondary > tertiary, which is reversed to that actually observed [201]. Reaction with aryl and vinyl substrates would be also disallowed. The stereochemical outcome of the reaction does not sustain this reaction pathway either. Clearly, this route can be ruled out (Scheme 3.16).

Ate Complex

Similar to the halogen–lithium exchange, R′Br + RLi → R′Li + RBr, an ate complex could be either a transition state or an intermediate of the reaction. The reactivity pattern exhibited does not corroborate this mechanism. Aryl and vinyl bromides would be expected to be more reactive than alkyl bromides. The reverse order of reactivity for the alkyl bromides would also be expected, that is, primary > secondary > tertiary. Moreover, stereochemical studies do not show any degree of retention of the configuration at the carbon center either (Scheme 3.17).

S_N1

An S_N1-like transition state would consist of heterolytic cleavage of the carbon—carbon bond at the transition state of the reaction, followed by ET [202]. Neither the structure-reactivity profile exhibited nor the stereochemical outcome of the reaction is in complete disagreement with this mechanism [203]. Also, radicals might originate in subsequent steps after the rate-limiting step, rendering a positive test for radical presence. However, linear-free energy relationship studies with the aryl series, both for bromides and iodides, show that a negative charge is being developed at the transition state. Accordingly, a

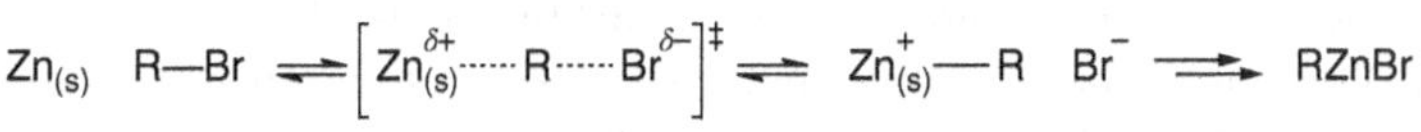

$$Zn_{(s)}\ \ R{-}Br \rightleftharpoons \left[Zn_{(s)}^{\delta+}\cdots R\cdots Br^{\delta-}\right]^{\ddagger} \rightleftharpoons Zn_{(s)}^{+}{-}R\ \ Br^{-} \longrightarrow RZnBr$$

Scheme 3.16 Oxidative addition via S_N2 transition state.

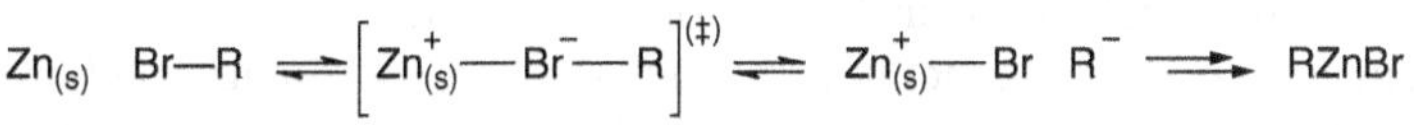

$$Zn_{(s)}\ \ Br{-}R \rightleftharpoons \left[Zn_{(s)}^{+}{-}\overset{-}{Br}{-}R\right]^{(\ddagger)} \rightleftharpoons Zn_{(s)}^{+}{-}Br\ \ R^{-} \longrightarrow RZnBr$$

Scheme 3.17 Oxidative addition via an ate complex transition state/intermediate.

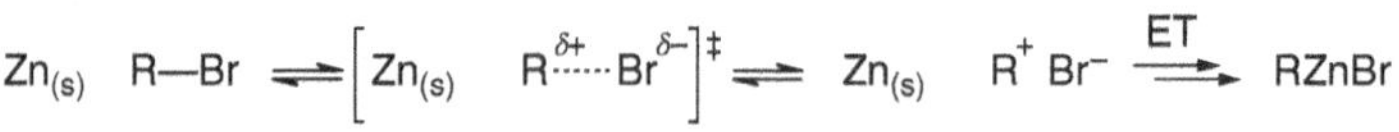

$$Zn_{(s)} \quad R{-}Br \rightleftharpoons \left[Zn_{(s)} \quad \overset{\delta+}{R}\cdots\overset{\delta-}{Br} \right]^{\ddagger} \rightleftharpoons Zn_{(s)} \quad R^{+}\,Br^{-} \xrightarrow{ET} RZnBr$$

Scheme 3.18 Oxidative addition via S_N1 transition state.

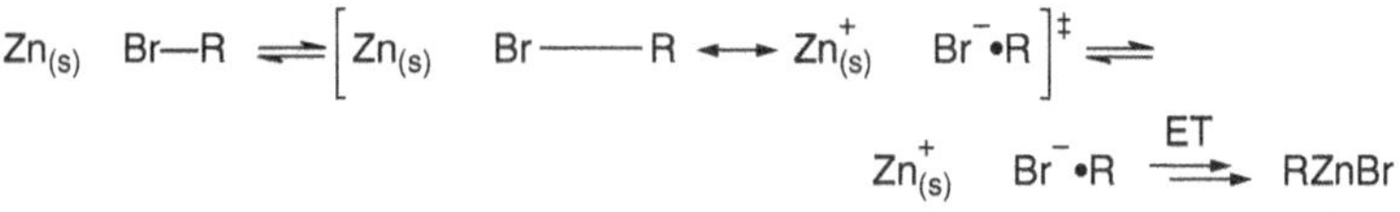

$$Zn_{(s)} \quad Br{-}R \rightleftharpoons \left[Zn_{(s)} \quad Br{-}R \longleftrightarrow \overset{+}{Zn}_{(s)} \quad Br^{-}\bullet R \right]^{\ddagger} \rightleftharpoons \overset{+}{Zn}_{(s)} \quad Br^{-}\bullet R \xrightarrow{ET} RZnBr$$

Scheme 3.19 Oxidative addition via an outer-sphere electron transfer.

$$Zn_{(s)} \quad Br{-}R \rightleftharpoons \left[Zn_{(s)}\cdots Br\cdots \overset{\delta\bullet}{R} \right]^{\ddagger} \rightleftharpoons Zn_{(s)}{-}Br\bullet R \xrightarrow{ET} RZnBr$$

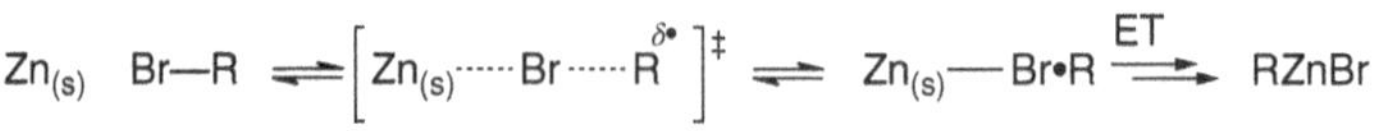

Scheme 3.20 Oxidative addition via an inner-sphere electron transfer.

positive slope (ρ) is observed in both cases (see succeeding text). These results clearly rule out an S_N1 in favor of ET processes (Scheme 3.18).

One-Electron Mechanisms

The radical clock experiments as well as the stereochemical outcome of the reaction along with the reactivity profiles observed pointed to an ET process as the operating mechanism. Linear-free energy relationships were also consistent with this mechanistic pathway (see succeeding text). ET may proceed in two ways, usually referred to as inner-sphere and outer-sphere ET, which can be contemplated as the two extremes of a continuous mechanism [204]. Both processes are dissociative in nature for alkyl halides and presumably do not involve a discrete radical anion, $RX^{-\cdot}$ [205]. The situation may, however, be different for aryl halides. Radical anions do exist, and aryl halides probably undergo a stepwise reaction with an electron donor to give rise to $RX^{-\cdot}$ [206].

Outer-Sphere Electron Transfer

Also called nonbonded ET or simply ET, the term emphasizes the low degree of interaction between reacting species in the transition state [207]. The inner coordination shells of the participating metal are intact in the transition state, no ligand-to-metal bond is broken or formed (Scheme 3.19).

Inner-Sphere Electron Transfer

Also called bonded or halogen transfer, denoting that the ET occurs within bonding distances between the reactants, through the first coordination sphere of the metal. The ET occurs through a bridged ligand, which is exchanged in the transition state (Scheme 3.20).

Discerning between both ET mechanisms is not an easy task [208]. Even though there are theoretical models for both processes [209], the limits of theoretical descriptions have not always been well substantiated by experimental evidence. Also in some systems, it is not clear what kind of theoretical model would be applicable [210]. Those ET reactions in which a single halide bridges the reactant centers at the moment of the ET have been very difficult to treat from a theoretical point of view, owing to ambiguity about the strength of the electronic interactions that accompany the activation process. Estimation of this (often based on spectroscopic data) permits one to estimate the inner-sphere contribution of the process, but this is especially difficult in a heterogeneous system. Table 3.50 contains data from a vast set of ET reactions in which some structure-reactivity profiles are available. For comparative purposes, we plotted the logarithm of the relative rates of tertiary and secondary versus primary alkyl halides for all of the reactions (Figure 3.2) [211]. The reactions are arranged in such a way that a decrease in selectivity is observed along the "*x*" axis. Among all the ET examples in Figure 3.2, only the first three reactions—involving Co(II) and Cr(II) complexes—are fully identified as inner-sphere ET processes [212]. The reactions involving Bu_3Sn^* have been ascribed as inner-sphere ET reactions (halogen transfer) [213], but this assumption has not been rigorously proved [214]. Whereas the rate-determining step of the reaction of Bu_3SnH with alkyl chlorides is the halogen transfer, this is not true for alkyl bromides. Hydrogen transfer becomes the determining step of the reaction instead [215]. In a similar way, the reduction of alkyl bromides with lithium 4,4′-dimethylbenzophenone ketyl, considered as an outer-sphere ET process [216], has by no means been strictly established [165d, 217]. It is noticeable that the electrolytically generated benzophenone ketyl shows a much lower degree of selectivity for the structure of the alkyl bromide, indicating that the lithium cation plays an important role in the reaction [217]. The reaction of

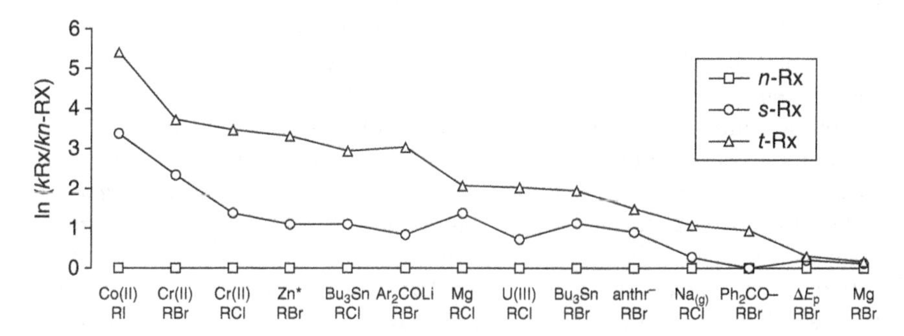

Figure 3.2 Structure-reactivity profiles for the reactions of a set of ET reagents with primary, secondary, and tertiary alkyl halides. Inner-sphere processes dominate the left side of the plot, whereas outer-sphere processes are dominant on the right side.

magnesium with alkyl chlorides, which is not a mass-transport-limited reaction, could not be assigned as either inner- or outer-sphere ET [163b, 165d]. Next we find the reduction of alkyl chlorides with a U(III) complex. Originally designated as a halogen-transfer reaction [182], it was later questioned. Since it was not possible to dismiss its outer-sphere counterpart, the process emerged as a possible borderline case of ET [218]. On the other side of the plot, we find the reduction of alkyl halides at an inert electrode and the reaction of alkyl bromides with radical anions, which are considered typical outer-sphere ET reactions [219]. $Na_{(g)}$ is also expected to be an outer-sphere ET reagent in its reaction with alkyl chlorides, based on ergonicity criteria [220]. As the rates of the reactions approach the diffusion-controlled limit, as they are known to for sodium atoms/RCl and Mg/RBr [165b, 185], the selectivity must become slight. Once the nature of these ET reactions has been assigned (tentatively sometimes), one can observe some general features in them. There is a general trend toward a decrease in the selectivity of the reactions, from the left, dominated by inner-sphere processes and, to the right, assumed to be outer-sphere processes. There is also an area of not well-defined processes in which one of them was referred to as borderline ET. The activation energy of these processes can thus be regarded as composed of two parts, the inner-sphere contribution and the outer-sphere contribution [204].

The transfer of an electron occurs under Franck–Condon restrictions [221]. Molecular vibrations are ca. 2 orders of magnitude slower than the transfer of the electron, and the position of the atoms remain frozen during the process. According to the definition of outer-sphere ET, the interaction between donor and acceptor is weak in the transition state [207]. Electronic interactions <1 kcal/mol define this ET, yet the inner-sphere ET must be assumed to possess a respectable donor/acceptor interaction, >5 kcal/mol. Stronger interactions with the donor in the transition state, that is, inner-sphere ET, should be accompanied with a higher degree of rehybridization (or C–X weakening) at the carbon center. It consequently may better accommodate the incipient radical which is being formed. It is at this stage that structure-reactivity relationships are displayed, following the normal order of thermodynamical stability of carbon-centered radicals. As we saw before, there are several cases of well-established inner-sphere ET. They all show pronounced dependence of the reactivity to the structure of the organic halide (or selectivity). However, in the case of outer-sphere ET, despite the fact that they follow the same general trend, none was found to show a high degree of selectivity. The presented results so far suggest, but do not prove, an important contribution of an inner-sphere ET in the oxidative addition of active zinc to alkyl bromides [222]. It is not clear that structure-reactivity studies may resolve this problem without the aid of supporting evidence [223]. The most straightforward proof of inner-sphere mechanism comes from the analysis of products [212]. In these processes, the halogen atom must remain bonded to the metal ion after the ET takes place. Presently, this can only be demonstrated when the metal ion is inert toward ligand substitution, which Zn^{2+} is not [224].

Linear-Free Energy Relationships (LFERs)

The use of LFERs for mechanistic assignment is a fundamental tool of physical organic chemistry and can be applied to the characterization of ET mechanisms as well. As previously mentioned, the ET to aryl halides may not be of a dissociative nature, in contrast to that of alkyl halides. A radical anion, $ArX^{-\cdot}$, is generated, which subsequently evolves to products (Scheme 3.21). The study of the electronic demand of the reaction Zn* + ArX can provide important clues about the structure of the transition state, linking it to an ET process or any other conceivable mechanism. A series of substituted aryl bromides, as well as aryl iodides—which are very convenient precursors of arylzinc halides when mild reaction conditions are needed—were submitted to competitive kinetic analysis in a similar way as described for alkyl bromides. Tables 3.51 and 3.52 summarize the observed ratio of rate constants (k_1/k_2) for all the substrates assayed. The most important fact derived from this set of data is that a notable grade of selectivity is again observed for the substrates examined, bearing in mind that a heterogeneous ET is, most likely, the key step of the reaction. Rates range from $k_1/k_2 = 700$ for 4-bromobenzonitrile/4-bromo-*N,N*-dimethylaniline in refluxing THF and go up to >1000 when 4-bromobenzonitrile is measured at 0°C (the reaction of 4-bromo-*N,N*-dimethylaniline is too slow to be conveniently monitored at 0°C) (Table 3.51). In the case of aryl iodides, this range is smaller, $k_1/k_2 = 300$ for 4-iodobenzonitrile/4-iodo-*N,N*-dimethylaniline in THF at 0°C, but still significant (Table 3.52). Comparison with other related processes leaves little doubt about the singularity of the reaction and its possible synthetic applications. Magnesium (and consequently lithium) shows little or no selectivity in its reaction with substituted aryl bromides and iodides [165c]. The rates summarized in Tables 3.51 and 3.52 were correlated with substituent effects using Hammett σ parameters [225]. Hammett plots for the reactions of Zn* with aryl bromides (THF, 70°C), aryl iodides (THF, 0°C), Mg with aryl bromides (Et_2O, THF, 0°C), and aryl iodides (Et_2O, 0°C) [226] are represented in Figure 3.3. The wide range of reactivity exhibited and σ values covered, as well as the good fitting of the lines for the Zn* reaction, permits an estimation

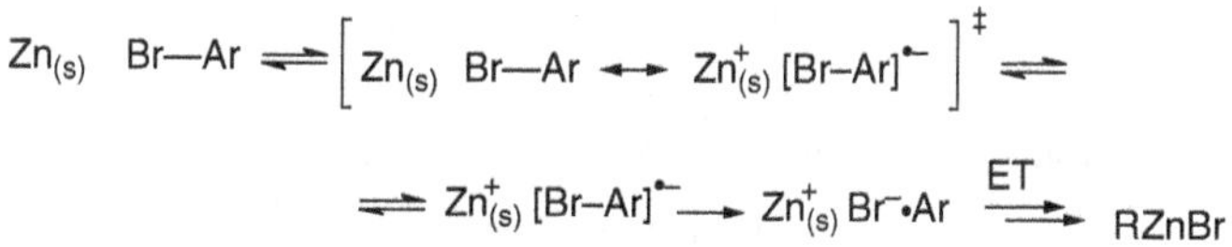

$$Zn_{(s)}\ \ Br{-}Ar \rightleftharpoons \left[Zn_{(s)}\ \ Br{-}Ar \leftrightarrow Zn^{+}_{(s)}\ [Br{-}Ar]^{\bullet -} \right]^{\ddagger} \rightleftharpoons$$

$$\rightleftharpoons Zn^{+}_{(s)}\ [Br{-}Ar]^{\bullet -} \longrightarrow Zn^{+}_{(s)}\ Br^{-}{\bullet}Ar \xrightarrow{ET} RZnBr$$

Scheme 3.21 Oxidative addition via a radical-anion intermediate.

Table 3.51 Relative rates of reaction of aryl bromides (k_1/k_2).[a]

	R-C$_6$H$_4$-Br	k_1/k_2	R_2Br	Temperature (°C)	σ[b]	k_1/k_{PhBr}[c]
1.	4-CN	7.8, 7.0	n-$C_8H_{17}Br$, 4-$EtO_2CC_6H_4Br$	0°, 70°	0.70	*160, 105*[d]
2.	4-COMe	37	PhBr	70°	0.48	37
3.	4-CO_2Et	15	PhBr	70°	0.45	15
4.	3-CO_2Et	8.3	PhBr	70°	0.35	8.3
5.	3-OMe	2.2	PhBr	70°	0.11	2.2
6.	H	*1*	PhBr	—	0	*1*
7.	4-Me	0.59	PhBr	70°	−0.17	0.59
8.	4-OMe	0.39	PhBr	70°	−0.28	0.39
9.	4-NMe_2	0.15	PhBr	70°	−0.63	−0.15
10.	1-bromonaphthalene (structure)	5.0	PhBr	70°		5.0
11.	2-bromonaphthalene (structure)	1.6	PhBr	70°		1.6

[a] Corresponding to the reaction of two organic bromides (R_1Br, R_2Br) with Zn^*. All reactions were carried out in THF at the given temperature. Relative rates were determined by the competitive kinetics technique.
[b] Substituent constants from Ref. [64].
[c] k_1/k_2 values for nondirectly measured pairs in italics.
[d] The difference reflects the difference in experiment temperature.

Table 3.52 Relative rates of reaction of aryl iodines (k_1/k_2).[a]

	R-C$_6$H$_4$-I R_1I	k_1/k_2	R_2I[b]	σ[c]	k_1/k_{PhI}[d]
1.	4-CN	4.4	4-$EtO_2CC_6H_4I$	0.70	*18.5*
2.	4-CO_2Et	4.2	PhI	0.45	4.2
3.	3-CO_2Et	3.1	PhI	0.35	3.1
4.	3-OMe	1.2	PhI	0.11	1.2
5.	H	*1*	PhI	0	*1*
6.	4-Me	0.41	PhI	−0.17	0.41
7.	4-OMe	0.25, 5.2	PhI, n-$C_8H_{17}Br$[b]	−0.28	0.25
8.	4-NMe_2	0.25	4-$MeOC_6H_4I$	−0.63	*0.062*

[a] Corresponding to the reaction of two aryl iodides (R_1I, R_2I) with Zn^* in THF at 0°C. Relative rates were determined by the competitive kinetics technique.
[b] 4-Bromoanisole was also measured versus n-bromooctane to link with the bromides series.
[c] Substituent constants from Ref. [64].
[d] k_1/k_2 values for nondirectly measured pairs in italics.

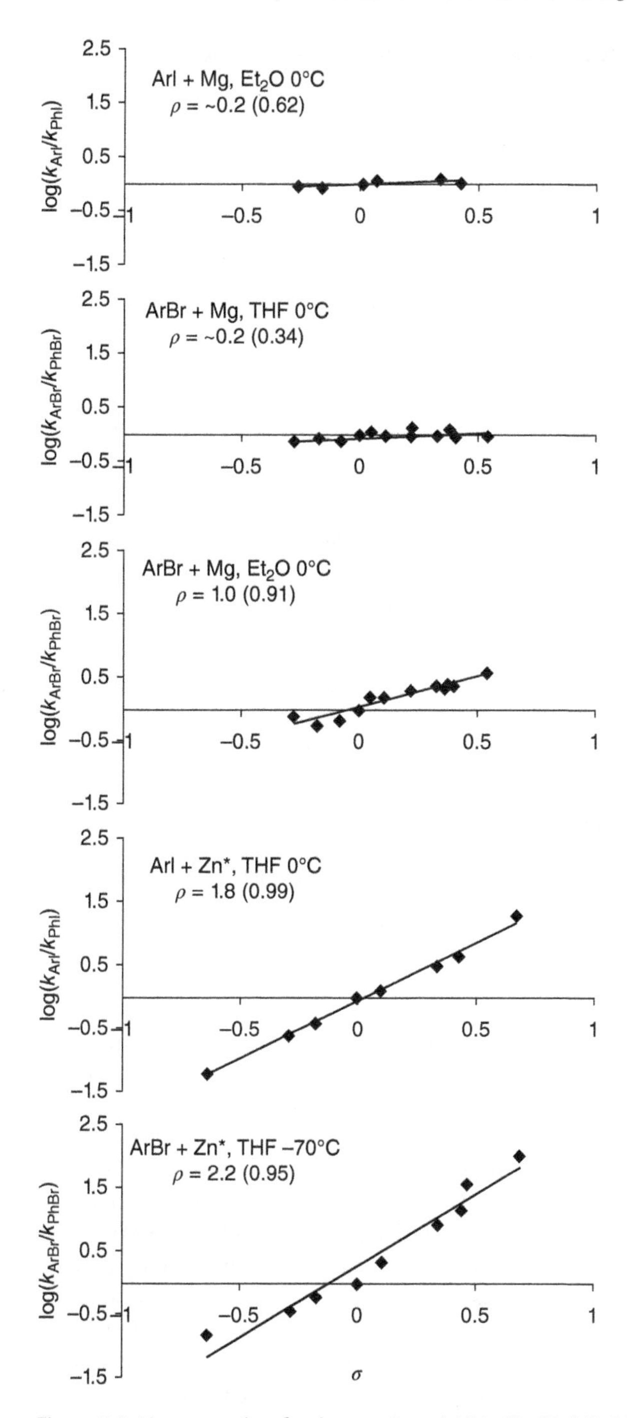

Figure 3.3 Hammett plots for the reactions ArI/Mg (Et_2O), ArBr/Mg (THF, Et_2O), ArI/Zn* (THF), and ArBr/Zn* (THF). Reaction constants, ρ; regression coefficients are in parentheses.

Table 3.53 Hammett constants (ρ) for some ET reactions (1–14) and one S_N (15) involving aryl halides.

	Reaction[a]	Solvent (T, °C)	ρ
1	ArBr + Zn*	THF (70)	2.2
2	ArI + Zn*	THF (0)	1.8
3	ArCl + Mg	Et_20 (23)	(~2)[b,c]
4	ArBr + Mg	THF (0)	(~0.2)[b,d]
5	ArBr + Mg	Et_2O (0)	1.0[e]
6	ArI + Mg	Et_2O (0)	(~0.2)[b,d]
7	ArBr + $Bu_3Sn^•$	Et_2O (0)	1.4
8	ArI + $Ph^•$	CCl_4 (60)	0.57
9	ArCl + e^-	DMF (25)	0.8[f]
10	ArBr + e^-	DMF (25)	0.57[f]
11	ArI + e^-	DMF (25)	0.35[f]
12	ArCl + $Ni(PEt_3)_4$	THF (25)	5.4[g]
13	ArBr + $Ni(PEt_3)_4$	THF (25)	4.4[g]
14	ArI + $Ni(PEt_3)_4$	THF (25)	2.0[g]
15	ArCl + NaOMe	MeOH (50)	8.47[h]

[a] See text for details and references.
[b] Nonlinear Hammett plots.
[c] Insufficient number of points and σ range covered.
[d] Mass transport limited.
[e] Obtained after data process of Ref. [6]. No statistical correction in dihalides was applied.
[f] Slopes of the plots $E_{1/2}$ versus σ.
[g] Only Me, Ph, H, and Cl substituents are included. *p*-MeO and *p*-MeOCO give deviations.
[h] σ^-.

a priori of the difference in reaction rates of two potentially reactive centers, by simply reading the plot. A hypothetical molecule bearing several aryl bromides in different electronic environments could thus be selectively metalated.

On the other hand, reaction constants for these plots, ρ, can be compared to a number of other reactions, heterogeneous or not, for which ρ values are available. For comparison, reaction constants, ρ, for these and other reactions involving halogen transfer (ArBr + $Bu_3Sn^•$ [165c], ArI + $Ph^•$) [227], heterogeneous ET to aryl halides (polarographic reduction of ArCl, Br, and I at a mercury electrode in DMF) [228], homogeneous ET (Ni(0) complexes + ArCl, –Br, and –I) [229], and nucleophilic substitutions (ArCl + MeO^-) [230] are included in Table 3.53. Unfortunately, linear-free energy relationships between log k and log K (the Hammett plots being one of these cases) cannot be taken as evidence for common outer-sphere or inner-sphere mechanisms for a series of related reactions. These plots may be linear for atom-transfer reactions as well as for ET reactions [231]. Indeed, the reaction of Mg with ArBr in ether (Table 3.53, entry 5) correlates equally well with rates of bromine abstraction by $Bu_3Sn^•$ [232] and polarographic reduction potentials [233]. The reaction of $Ni(PEt_3)_4$ with aryl halides (Table 3.53, entries 12–14) was interpreted as outer-sphere ET and is linearly correlated with $E_{½}$, as well as with σ [234]. σ correlates itself with the reaction $ArBr + Bu_3Sn^•$ [165c]. It is not surprising that the reaction of Zn* with aryl bromides also shows reasonable correlations versus $E_{½}$(ArBr) and $Bu_3Sn^• + ArBr$ (0.92 of correlation coefficients in both cases). These facts can be taken only as a reinforcement of the hypothesis of an ET process as the rate-determining step of the reaction but do not provide further evidence about its nature in a deeper level. However, comparison of the slopes for (supposedly) homogeneous halogen-transfer processes ($\rho = 1.4$, 0.57, Table 3.53, entries 7 and 8) and (supposedly) homogeneous outer-sphere ET processes ($\rho = 4.4$, 2.2; Table 3.53, entries 13 and 14) denotes a higher degree of charge separation in the transition state for the latter ones, in good agreement with the mechanism. The electronic demand of our reaction suggests the participation of an intermediate of the type $ArX^{-•}$, much like those proposed for the analogous reaction with other reducing agents (Scheme 3.21) [205a, 206]. Differences in selectivity can thus be explained in terms of an early ET transition state for Mg as a late ET transition state for Zn*, the latter having an increased ionic character and therefore displaying the observed reactivity. The proposed mechanism is equally valid for substituted ArBr and ArI, with $ArX^{-•}$ having the shortest lifetimes, since both series of halides are also correlated (0.96 of linear regression coefficient, eight different substituents), indicating only minor differences in their mechanistic pathways.

Other possibilities for the transition state, like carbanions, $[Ar^-]^{\ddagger}$ or concerted insertion of Zn in the C—Br bond, seem unlikely. The moderate ρ value does not indicate the development of a carbanion or a carbon—zinc bond in the transition state. In such a case, the carbon would bear a large negative charge, and a larger value for ρ would be expected [235].

To conclude with all of these mechanistic considerations, we have to point out that the experimental evidence collected, both for alkyl and aryl systems, goes against the possibility of multiple mechanisms [236]. Radicals have been detected on primary carbons. No sign of inversion (or retention) was detected in the stereochemical course of the reaction. For aryl systems, Hammett plots reveal also mechanistic homogeneity in all the range of σ covered by substituents.

Synthetic Applications

A superficial examination of the relative kinetic data gathered in Tables 3.48, 3.49, 3.51, and 3.52, along with the well-known functional group tolerance of organozinc reagents [169] and their potential in multiple processes mediated by transition metals [168], strongly stimulated the exploration of the synthetic scope of this promising scenario. To do so, a set of molecules carrying two different organic bromides (**8** [237], **9** [238], **10**, **12–13**) were synthesized (**11** was commercially available). The use of controlled reaction conditions and/or a stoichiometric amount of Zn* permitted the selective preparation of the monobrominated monoorganozinc bromides in good yields from the corresponding starting materials **8–13**. These intermediates showed remarkable stability (Table 3.54, entry 9, footnote *l*) and could be conveniently reacted either after transmetallation or directly (Table 3.54, entries 1, 3, 4, 6–8, 11 and 2, 5, 9, 10, respectively) with a variety of electrophiles ($CH_2{=}CClCH_2Cl$, D_2O, *tert*-$BuO_2CN{=}NCO_2Bu$-*tert*, PhCOCl, 2-cyclohexenone, AcOH).

A rough classification of the organic bromides, attending to their reactivities, resulted in three main groups: (i) benzyl/allyl bromides, (ii) alkyl bromides, and (iii) aryl and vinyl bromides. Benzylic and allylic bromides can be selectively reacted with Zn* in the presence of alkyl bromides. One example of this kind is presented in Table 3.54, entry 1. An estimation for this process is $k_{\text{benzyl}}/k_{n\text{-alkyl}} > 380$, provided that the reaction centers act independently. Presumably, this can also be extended to secondary alkyl bromides ($k_{\text{benzyl}}/k_{\text{ks-alkyl}} > 120$) and other bromides (Table 3.48). In a similar way, alkyl bromides can be selectively reacted in the presence of aryl bromides (provided that no electron-withdrawing groups are present on the aromatic ring) as well as vinyl bromides. As examples, compounds **9** and **10** were synthesized (Table 3.54, entries 2 and 4). $k_{n\text{-alkyl}}/k_{p\text{-tolyl}} = 45.8$ and $k_{n\text{-alkyl}}/k_{\text{vinyl}} = 160$ are the expected selectivities to be observed. Compounds **19** and **20** were prepared after selective reaction of the benzyl bromide in **11**. The selectivity expected is $k_{\text{benzyl}}/k_{p\text{-tolyl}} > 17\,000$.

Selective reactions are also possible inside the same group. Wide ranges of reactivity have already been mentioned for substituted aryl bromides and iodides. In the alkyl series, selective insertion of Zn* in a tertiary bromide, leaving intact a primary one, was achieved in compounds **12** and **13** with good yields (Table 3.54, entries 8–11). As we saw, the estimated selectivity for these processes is $k_{t\text{-alkyl}}/k_{n\text{-alkyl}} = 31.8$.

Finally, the overall spectrum of reactivity for the reaction of active zinc with organic bromides covers a range of $k_1/k_2 \approx 7 \times 10^5$, with allylic and benzylic bromides in one extreme and vinyl and aryl bromides (carrying electron-donation groups) in the other (Figure 3.4). Since preparation of the corresponding organozinc bromides can be accomplished for most substrates in good yields, including deactivated aryl bromides and vinyl bromides [60, 170a], sequential cycles of metalation reaction with an electrophile are theoretically feasible in multibrominated starting materials.

Conclusions

In conclusion, we have established that our highly reactive zinc exhibits unusual structure-dependent reactivity that can be used to elaborate complex molecules

Table 3.54 Synthetic applications.

Entry	Starting material[a]	No (%)[b]	Reaction conditions[c]		Product[f]	No (%)[g]
			Zn*[d]	Electrophile (catalyst)[e]		
1.	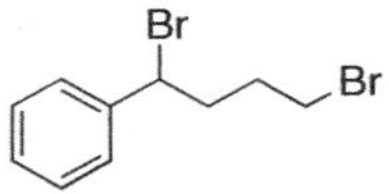	**8** (68)	1.1 0°C 10 min	(10% CuCN·2LiBr) $CH_2{=}CClH_2Cl$	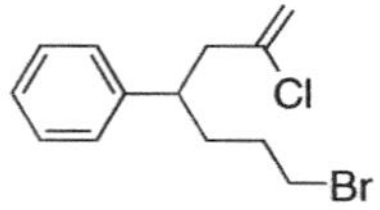	**14** (64)
2.	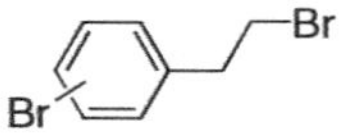	**9**[h] (80)	1 + 1 rt, 3 + 2 h	D_2O	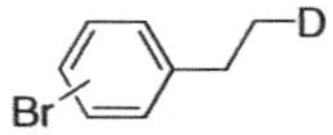	**15**[h,i] (96)
3.	**9**	**9**	"	(5% CuCN·2LiBr) $CH_2{=}CClCH_2Cl$	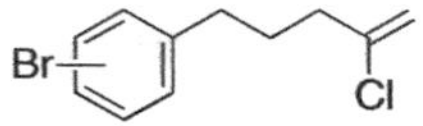	**16**[h] (65)
4.	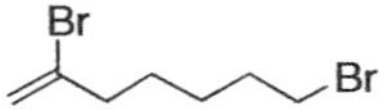	**10** (90)	3 rt, 3 h	(10% CuCN·2LiBr) $CH_2{=}CClCH_2Cl$	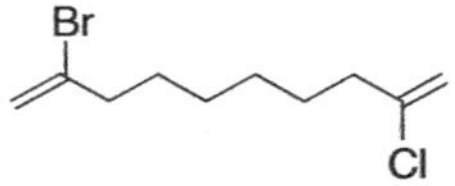	**17** (96)
5.	**10**	**10**	"	$^tBuO_2CN{=}NCO_2Bu^t$	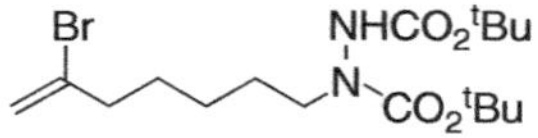	**18** (91)
6.	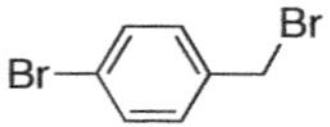	**11**[j] (–)	1.0 0°C, 1 h	(5% CuCN·2LiBr) $CH_2{=}CClCH_2Cl$	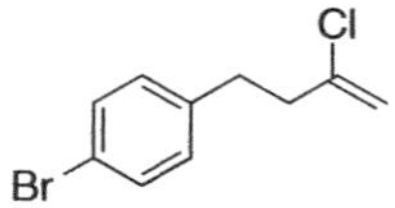	**19** (96)
7.	**11**	**11**	1.2 0°C, 1 h	(CuCN·2LiBr) PhCOCl	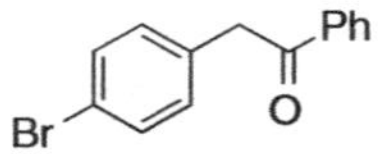	**20** (89)

(*Continued*)

Table 3.54 (Continued)

Entry	Starting material[a]	No (%)[b]	Reaction conditions[c] Zn*[d]	Electrophile (catalyst)[e]	Product[f]	No (%)[g]
8.	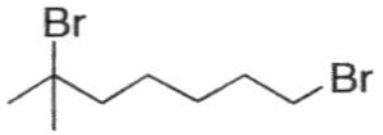 Br, Br	**12** (92)	1.1 rt, 10 h	(Li[(2-Th)CuCN]) 2-cyclohexenone 2 equiv TMSCl	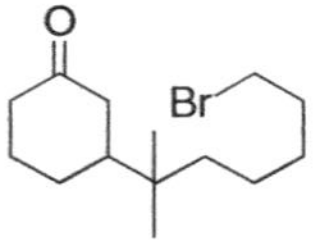 O, Br	**21** (65)
9.	**12**	**12**	1.1 rt, 12 h	AcOH[k]	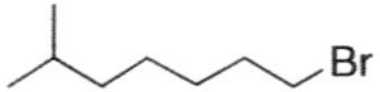 Br	**22** (81)[l]
10.	**12**	**12**	1.1 rt, 10 h	${}^{t}BuO_2CN{=}NCO_2Bu^{t}$	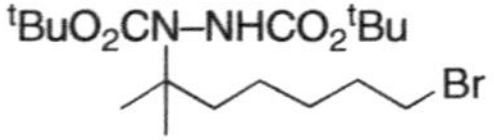 ${}^{t}BuO_2CN{-}NHCO_2{}^{t}Bu$, Br	**23** (65)
11.	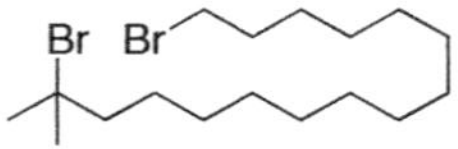 Br Br	**13** (95)	"	(Li[(2-Th)CuCN]) 2-cyclohexenone 2 equiv TMSCl	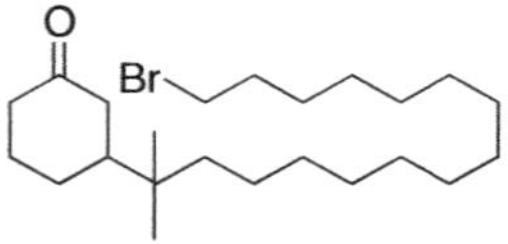 O, Br	**24** (63)

[a] See experimental for procedures.
[b] Overall yield for the synthesis of the starting material.
[c] All the reactions were carried out in THF.
[d] Equivalents of Zn*. Several additions and/or excess were used to minimize reaction times. Reaction time and temperature.
[e] 1.0 equiv unless otherwise noted.
[f] All products were pure (≥95%) and fully characterized by spectroscopical means (IR, ^{1}H and ^{13}C NMR, HRMS).
[g] Based on the dibrominated starting materials **8–13**.
[h] 61/39 *p*-/*o*-mixture of regioisomers (^{1}H NMR, GLC).
[i] >98% D incorporation (^{1}H NMR).
[j] Commercially available.
[k] Distilled under Ar.
[l] The organozinc intermediate did not decompose extensively after 12 h at room temperature.

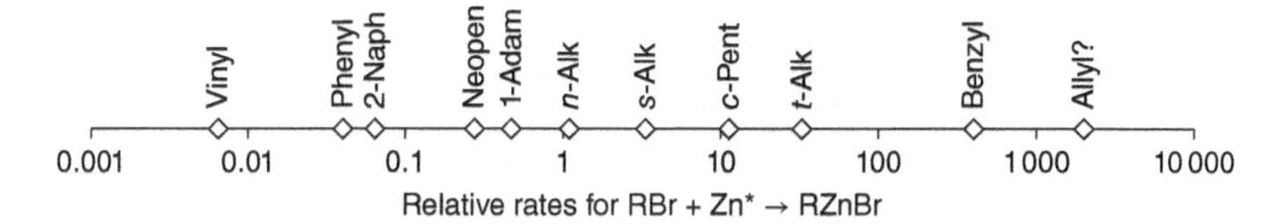

Figure 3.4 Relative rates of reaction.

containing more than one bromine atom. The kinetic and LFER studies suggest a mechanism in which ET is the rate-determining step of the reaction. Experiments carried out with radical clocks as well as the stereochemical outcomes of the reaction support the presence of radicals. The reactivity profiles suggest that the ET has an important component of inner-sphere processes in the reaction with alkyl bromides. In the case of aryl halides, Hammett plots are consistent with the participation of aryl halides radical anions as intermediates of the reaction. In addition, the use of the schemes described herein constitutes a new contribution to the current strategies in organic synthesis. Active zinc emerges out of this as a singular metal whose selectivity in these processes markedly distinguishes itself from other synthetically relevant metals.

3.19 Regiocontrolled Synthesis of Poly(3-Alkylthiophenes) Mediated by Rieke Zinc: A New Class of Plastic Semiconductors

Of the many conjugated polymers, poly(3-alkylthiophenes) (P3ATs) have been found to be an unusual class of polymers with good solubility, processability, environmental stability, electroactivity, and other interesting properties [239]. Practical and potential applications of P3ATs include rechargeable battery electrodes, electrochromic devices, chemical and optical sensors [240a], light-emitting diodes [240b, c], molecular-based devices, such as microelectrochemical amplifiers [240d] and field-effect transistors [240e], and nonlinear materials [240f–h]. With such a wide range of potential applications, they have attracted interest, both from synthetic considerations as well as from material science.

Since the discovery of insoluble polythiophene and the determination of its electroconductivity in 1980 [241], many studies have been completed to improve the synthetic feasibility and chemical and physical properties of this conjugated polymer. A milestone step in the development of polythiophene occurred in 1985 when the poly(3-alkylthiophene), which was a soluble, processable, and stable polymer, was achieved by introducing an alkyl group into the β-position of the thiophene ring [242]. Three general synthetic methods for polymer synthesis were developed. These included electrochemical polymerization [243], oxidative polymerization of 3-alkylthiophene by oxidants such as $FeCl_3$ [244], and polymerization by catalyzed dehalogenation of 2,5-dihalo-3-alkylthiophene, such as nickel-catalyzed coupling of thiophene Grignard reagents [242] or dihalothiophene monomers [245]. An important challenge remained, that being the development of a method which would yield regioregular P3ATs.

Little work has been done on the systematically structure-controlled synthesis of conjugated polymers. One elegant example is the conformation-controlled synthesis of polyacetylene. Shirakawa et al. [246] have succeeded in synthesizing the all-*cis* and all-*trans* isomers of polyacetylene and have developed techniques for controlling the ratio of *cis* to trans isomers in the product. The *trans*-polyacetylene polymer has quite different physical properties and conductivity than does the *cis*-isomer. Furthermore, the Naarmann's conformation-unique polyacetylene [247] has fewer sp^3 defects and less cross-linking in the polymer chain and leads to a large improvement in conductivity. It is clear that the higher-order and regular polymer chain resulted in higher conductivity, nonlinearity, and other improved physical properties of the material.

For poly(3-alkylthiophene), the 3-alkyl substituent in a thiophene ring can be incorporated into a polymer chain with two different regioregularities: head-to-tail (HT) and head-to-head (HH). Furthermore, this results in four triad regioisomers in the polymer chain: HT–HT, HT–HH, TT–HT, and TT–HH triad (Figure 3.5). Recent research on poly(3-alkylthiophene) and other conjugated polymers has concentrated on the regularity and structure of the polymer chain [248–252]. The HT regiospecific polymers have improved electroconductivities, optical nonlinearity, and magnetic properties over regiorandom polymers in which more sterically hindered HH linkages can cause defects in the conjugated polymer chain and reduce the desired physical properties of the material [249–251, 253, 254]. Our group and others have developed different methods for the synthesis of regioregular HT poly(3-alkylthiophenes). One approach involves the synthesis of HT regioregular P3ATs by a Ni-catalyzed coupling polymerization of a regiospecific thiophene Grignard reagent [253]. Our methodology involves a regiocontrolled synthesis for a series of regiospecific P3ATs, including HT regioregular P3ATs and completely regiorandom HT–HH-P3ATs, as well as other regioisopolymers with a variety of degrees of regioregularity on the percentage of the HT linkage in the polymer chain. The reactions involve Ni- or Pd-catalyzed coupling polymerizations of regiospecific thiophene organozinc reagents [254]. These organozinc reagents were formed by reaction of Rieke zinc (Zn*) with 2,5-dihalothiophenes (Scheme 3.22). Herein we report the detailed results of the regiocontrolled syntheses, mechanistic implications of these polymerizations, polymer characterizations, and solid-state properties, including spectroscopic studies, molecular weight determinations, polarizing microscopy, X-ray diffraction, and the crystallinity of the materials.

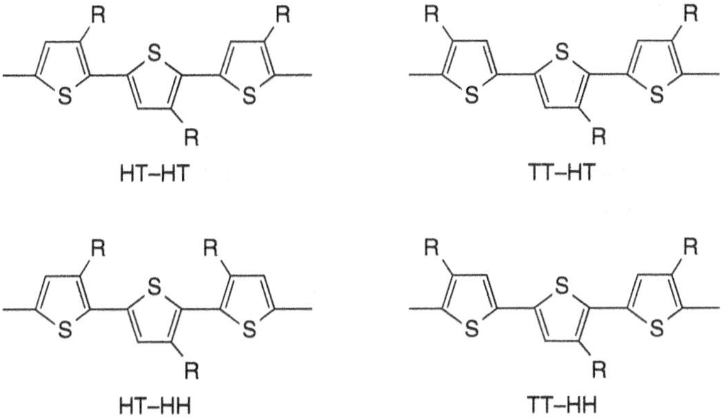

Figure 3.5 Orientation of 3-alkylthiophene units in polymers.

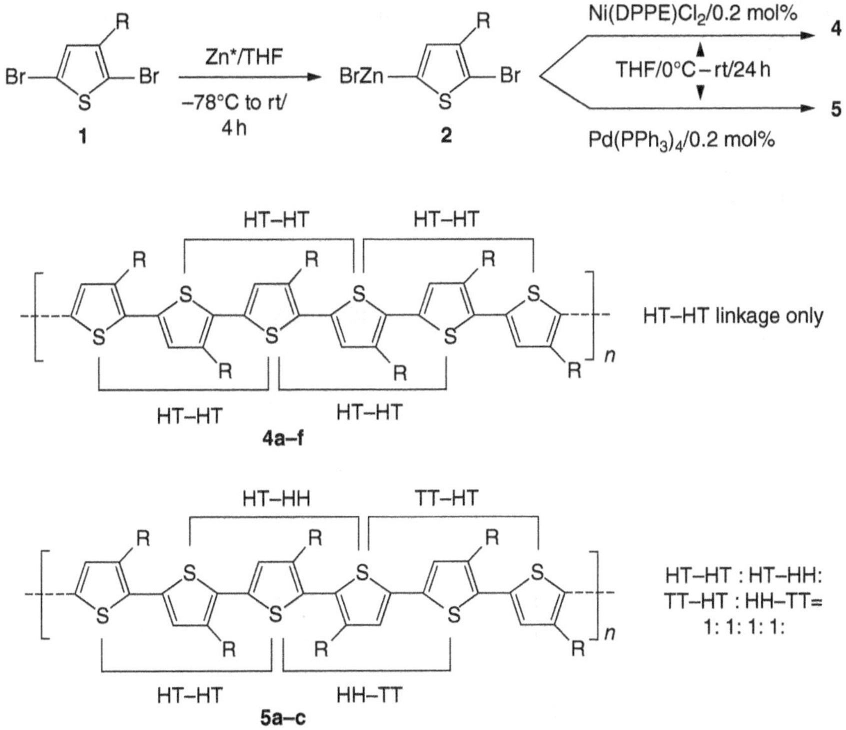

Yields (%) :	a: n-C_4H_9	b: n-C_6H_{13}	c: n-C_8H_{17}	d: n-$C_{10}H_{21}$	e: n-$C_{12}H_{25}$	f: n-$C_{14}H_{29}$
4	80	82	79	77	71	67
5	99	98	98			

Scheme 3.22 Regiocontrolled synthesis of poly(3-alkylthiophene) mediated by Zn*.

Results and Discussion

Regiocontrolled Synthesis of Poly(3-Alkylthiophenes) Mediated by Rieke Zinc

Rieke zinc (Zn*) [254] underwent direct oxidative addition to 2,5-dibromothiophene chemoselectively to afford 2-(bromozincio)-5-bromothiophene quantitatively (Equation 3.19) [254c]. Significantly, no bis(bromozincio)-thiophene was formed:

Br–(thiophene)–Br —(Zn*/THF, rt./1 h)→ BrZn–(thiophene)–Br (3.19)

We further examined the reactions of 2,5-dibromo-3-alkylthiophene with Zn* and found that the Zn* also underwent oxidative addition to 2,5-dibromo-3-alkylthiophene regioselectively, especially at low temperature, to afford the regioisomer 2-bromo-5-(bromozincio)-3-alkylthiophene (**2**) predominantly and 5-bromo-2-(bromozincio)-3-alkylthiophene (**3**) as a very minor regioisomer (Table 3.55).

Table 3.55 Regioselectivities of the reactions of Zn* with 2,5-dibromo-3-alkylthiophenes.

Br–(3-R-thiophene)–Br (**1**) —(Zn*/THF, *T*/*t*)→ BrZn–(3-R-thiophene)–Br (**2**) + Br–(3-R-thiophene)–ZnBr (**3**)

R	*T*/*t*	**2/3 (%)**[a]
n-Hexyl	rt/1 h	90:10
n-Hexyl	−45°C to rt/4 h	93:7
n-Hexyl	−78°C to rt/4 h	97:3
n-Butyl	−78°C to rt/4 h	94:6
n-Hexyl	−78°C to rt/4 h	97:3
n-Octyl	−78°C to rt/4 h	98:2
n-Decyl	−78°C to rt/4 h	98:2
n-Dodecyl	−78°C to rt/4 h	98:2
n-Tetradecyl	−78°C to rt/4 h	98:2

[a] The regioselectivity was identified by ^{1}H NMR analysis of the crude reaction mixture after quenching with a saturated NH_4Cl solution.

Zn* undergoes oxidative addition to the C—Br bond primarily at the 5-position of **1**. The selectivity of the oxidative addition is enhanced at lower temperature (Table 3.55); the lower the reaction temperature, the higher the regioselectivity. At −78°C the regioselectivity was as high as 97–98% for most cases. The total yield for two regioisomers **2** and **3** was larger than 99% by gas chromatography in all cases. This novel regioselective reaction of Zn* with dihalothiophenes and other dihaloarylenes [254c] represents a significant new approach for the synthesis of unsymmetrical heterocyclic compounds and arylene derivatives [254], particularly since organozinc reagents are stable and are formed under mild conditions with the tolerance of many functional groups [255]. The reaction with Zn* is an indispensable step to fulfill the regiospecific polymerizations. The polymerizations are performed in a simple one-pot reaction. **2** (and the very minor isomer **3**) was treated *in situ* either with Ni-(DPPE) Cl_2([1,2-bis(diphenylphosphino)ethane]nickel(II) chloride) to afford a completely regioregular HT P3AT **4** or with $Pd(PPh_3)_4$ to afford a totally regiorandom P3AT **5** [256]. The turnover number of the catalyst was quite high since only a very small amount (0.2 mol%) of catalyst was used and a high yield of product was achieved in these polymerizations. The use of the small amount of catalyst also facilitated the purification of the polymer products.

An alternate and efficient synthesis for **4** and **5** is shown in Scheme 3.23. 3-Alkyl-2-bromo-5-iodothiophenes **9** were synthesized by regiospecific bromination [257], then iodination [258], followed by the alkylation [259] of 3-bromothiophene. Rieke zinc oxidatively inserted into the C–I bond to afford the regiospecific 3-alkyl-2-bromo-5-(iodozincio)thiophenes **10** exclusively. **10** was treated either by $Ni(DPPE)Cl_2$ to afford (**4**) or by $Pd(PPh_3)_4$ to afford (**5**).

Mechanistic Implications of the Polymerizations

The polymerizations described in Schemes 3.22 and 3.23 not only resulted in two extremely regiospecific P3ATs **4** and **5**, but, moreover, the reactions can

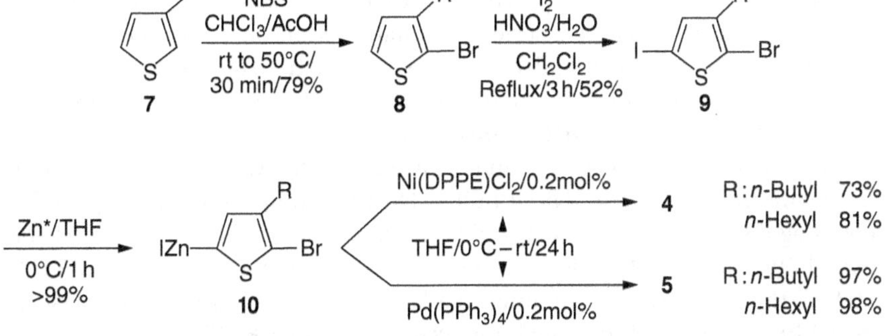

Scheme 3.23 Regiocontrolled synthesis of poly(3-alkylthiophenes) starting with 3-alkyl-2-bromo-5-iodothiophenes.

Table 3.56 Regioregularity of P3ATs controlled by different catalysts.[a]

Catalyst structure	Initialized species of the catalyst	Resulting P3ATs	Regioregularity (%) HT/HH
$Ni(DPPE)Cl_2$	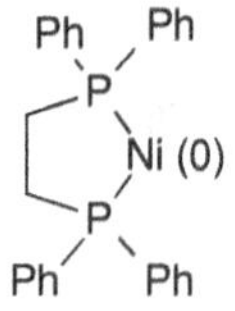 Ph Ph P Ni (0) P Ph Ph	**4**	HT > 98.5
$Pd(DPPE)Cl_2$	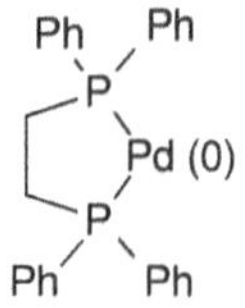 Ph Ph P Pd (0) P Ph Ph	**11**	70 : 30
$Ni(PPh_3)_4$	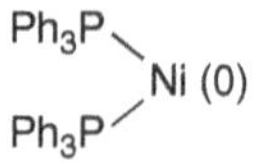 Ph_3P Ni (0) Ph_3P	**12**	65 : 35
$Pd(PPh_3)_4$	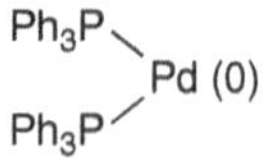 Ph_3P Pd (0) Ph_3P	**5**	50 : 50

[a] The reaction conditions are all the same as shown in Scheme 3.1 (catalyst amount: 0.2 mol%; 0°C to rt for 24 h).

also be expanded to a more diversified regiocontrolled syntheses for a series of different regiospecific poly(3-alkylthiophenes) by choosing a different catalyst (Table 3.56).

The mechanism of Ni- and Pd-catalyzed cross-coupling organozinc compounds with organic halides has been extensively investigated [256c, 260]. Yamamoto and coworkers [261] have examined the mechanism of the Ni- and Pd-catalyzed cross-coupling polymerization of dihaloarylenes. The polymerizations were performed by a series of oxidative addition, transmetallation (or disproportionation [261c]), and the reductive elimination steps. Transmetallation (or the disproportionation) for these reactions was the rate-determining step. In our research, it was found that the degree of regioregularity of P3AT was a function of both the metal (Ni vs. Pd) and the ligands (DPPE vs. PPh_3) of the catalysts in polymerizations. Using $Ni(DPPE)Cl_2$ led to a completely regioregular HT coupling P3AT. Switching the ligand of the catalyst from DPPE to a more liable ligand PPh_3 [as $Ni(PPh_3)_4$] led to a reduction in the regioregularity (65 : 35 HT/HH) of the P3AT, while switching to a larger size atom Pd in the catalyst (as $Pd(DPPE)Cl_2$) also led to a much reduced regioregular P3AT (70 : 30 HT/HH). A totally regiorandom P3AT (50 : 50 HT/HH) was obtained when both the metal and the ligand were changed in the catalyst [$Pd(PPh_3)_4$]. The smaller ionic radius of Ni^{2+} versus Pd^{2+} along with the higher steric demands for DPPE indicated that the degree of stereoregularity was controlled by steric congestion of the transmetallation (or disproportionation) step. The polymerizations were also performed by a series of transmetallation (or disproportionation) reductive elimination, followed by oxidative addition.

Spectroscopic Studies and Other Characterization

NMR Spectroscopy

All polymers were studied by solution 500-MHz ^{1}H NMR and 125-MHz ^{13}C NMR spectroscopy. The ^{1}H and ^{13}C NMR spectra of P3ATs provide sensitive probes for the substitution pattern in the polymer backbone [249, 262, 263]. The proton in the 4-position of the thiophene ring bears four different chemical environments in a mixture of the four possible triad regioisomers (Figure 3.6a). These four chemical shift distinct protons are uniquely distinguishable by NMR spectroscopy. This is clearly demonstrated in the ^{1}H NMR spectra of **5** (e.g., Figure 3.6a for polymer **5b**, regiorandom P3HT). The observed spectra are consistent with a totally random (1 : 1 : 1 : 1 HT–HT (δ, 6.98)/TT–HT (7.00)/HT–HH (7.02)/TT–HH (7.05) linkages based on NMR integration) mixture of the four triad structures depicted. Previous reports [262, 264] showed that the α-methylene protons of the alkyl group could be resolved by two different diads: HT and HH. We found that even β-methylene protons could be resolved by these diads. The expanded ^{1}H NMR spectrum of polymer **5b** (Figure 3.8a) also showed the polymer had a regiorandom chain structure with an equal distribution of HT and HH linkages (for α-methylene protons, 1 : 1 HT (δ, 2.80/HH (2.58)); for β-methylene protons, 1 : 1 HT (δ, 1.72/HH (1.63)). All of the thiophene carbons (16 peaks are theoretically possible) can be resolved in the mixture of the four triad regioisomers. This is clearly demonstrated, again, in the ^{13}C NMR spectra of regiorandom P3AT **5** (see Figure 3.9a for **5a**). The observed spectra are consistent with an equal distribution of the four triad structures (HT–HT, TT–HT, HT–HH, TT–HH) [254a] (Table 3.57).

In contrast, only one sharp band for the thiophene proton, which denotes the HT–HT structure, is observed in the ^{1}H NMR spectra of **4** (see Figure 3.6b for regioregular P3HT **4b**). The spectra did not indicate the presence of other irregular linkages as observable within the NMR resolution in this region. The expanded ^{1}H NMR spectrum in the α-methylene proton region was indicative of $>$98.5% for the HT linkage, and in the β-methylene proton region only the HT linkage was present (Figure 3.8b). In the ^{13}C NMR spectra, only four sharp bands for the thiophene carbon atoms (one band for each carbon), which also denoted the HT–HT structure, are observed (Figure 3.9b for **4b**, and Table 3.57). The NMR data analyses and the signal-to-noise ratio characterized that the polymer **4b** was a structurally homogeneous, regioregular poly(3-hexylthiophene) containing at least 98.5% HT linkage in the polymer chain.

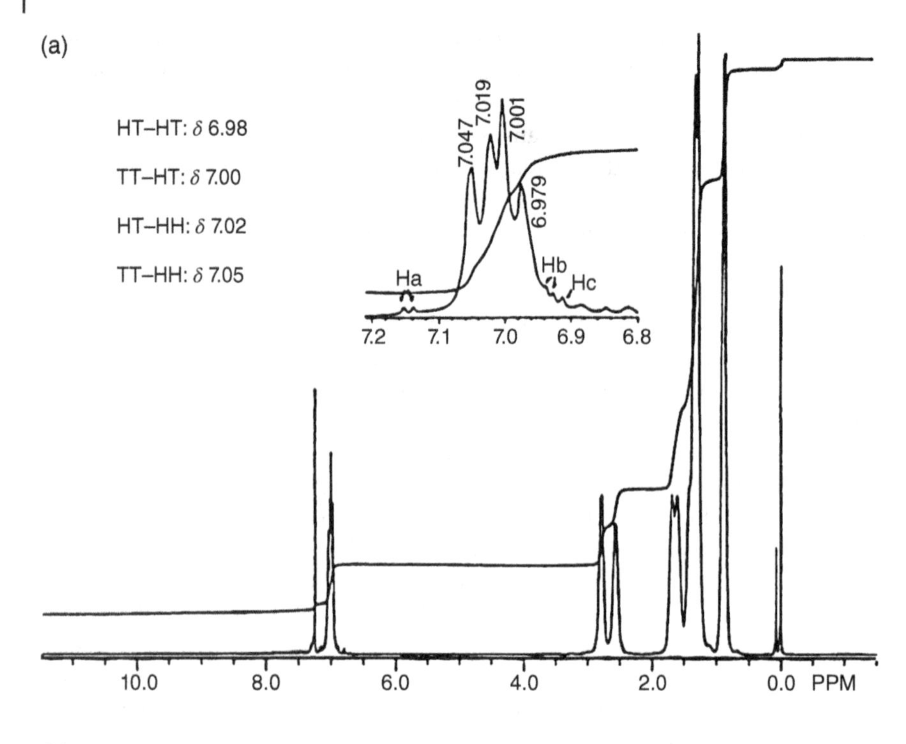

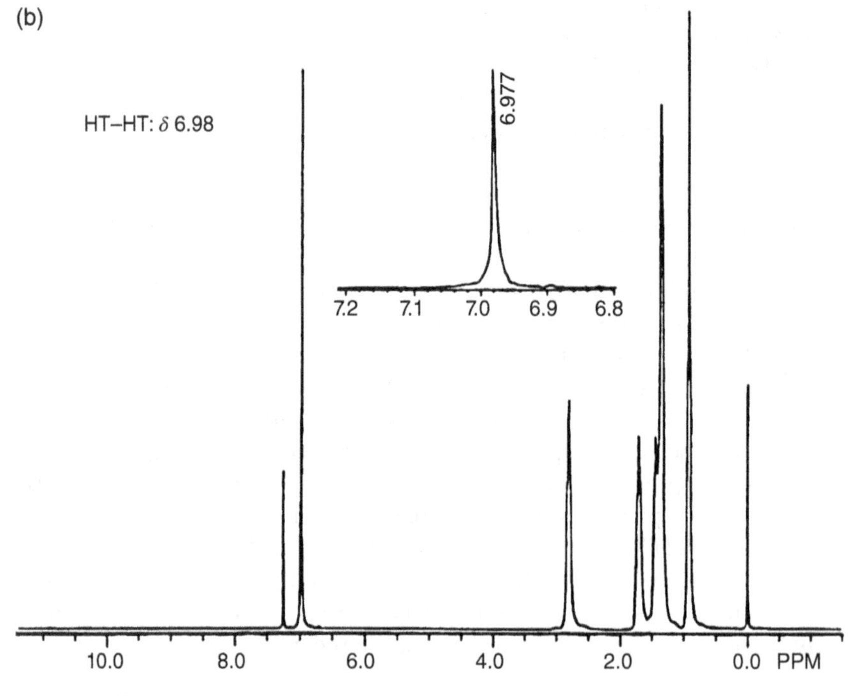

Figure 3.6 ^{1}H NMR spectra of (a) regiorandom P3HT **5b** (1 : 1 : 1 : 1 HT–HT/HT–HH/TT–HT/TT–HH) and (b) regioregular P3HT **4b** (HT linkage >98.5%).

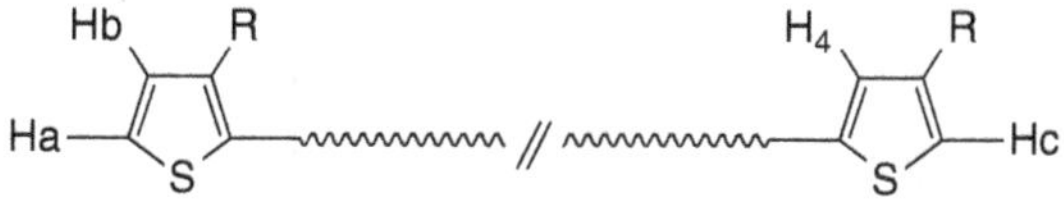

Figure 3.7 Terminal polymer groups.

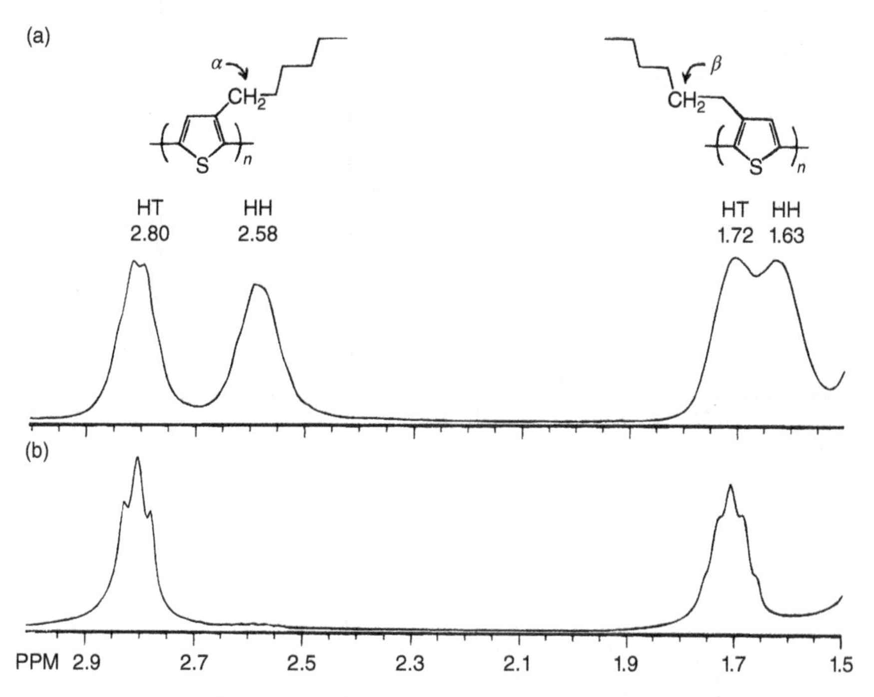

Figure 3.8 Expanded ^{1}H NMR spectra of (a) regiorandom P3HT **5b** (1 : 1 HT/HH) and (b) regioregular **4b** (HT linkage >98.5%).

By analyses of NMR data, polymers **4**, which are catalyzed by $Ni(DPPE)Cl_2$, are regioregular HT P3ATs with at least 98.5% of HT linkage when the alkyl group is hexyl **4b**, octyl **4c**, decyl **4d**, dodecyl **4e**, and tetradecyl **4f** and with 97% of the HT linkage when the alkyl group is butyl **4a**, in the polymer chain. Polymers **5**, formed via $Pd(PPh_3)_4$ catalysts, are regiorandom P3ATs with an equal distribution of different regioisomers in the polymer chain (1:1:1:1 HT–HT/TT–HT/HT–HH/TT–HH or 1:1 HT/HH). Polymer **11**, catalyzed by $Pd(DPPE)Cl_2$, is a P3HT with 70% HT and 30% HH linkage in the polymer chain. Polymer **12**, catalyzed by $Ni(PPh_3)_4$, contains 65% HT and 35% HH. The NMR analyses confirm that our methodology is a regiocontrolled synthesis for a series of regiospecific P3ATs.

The two doublet peaks (7.15, 6.93 ppm) and one singlet peak at 6.91 ppm (Figure 3.6a) are assigned as the terminal ring protons Ha, Hb, and Hc in the polymer chain (Figure 3.7) [251a]. The percentage of these terminal protons in the polymer chain is <1% based on the NMR integration.

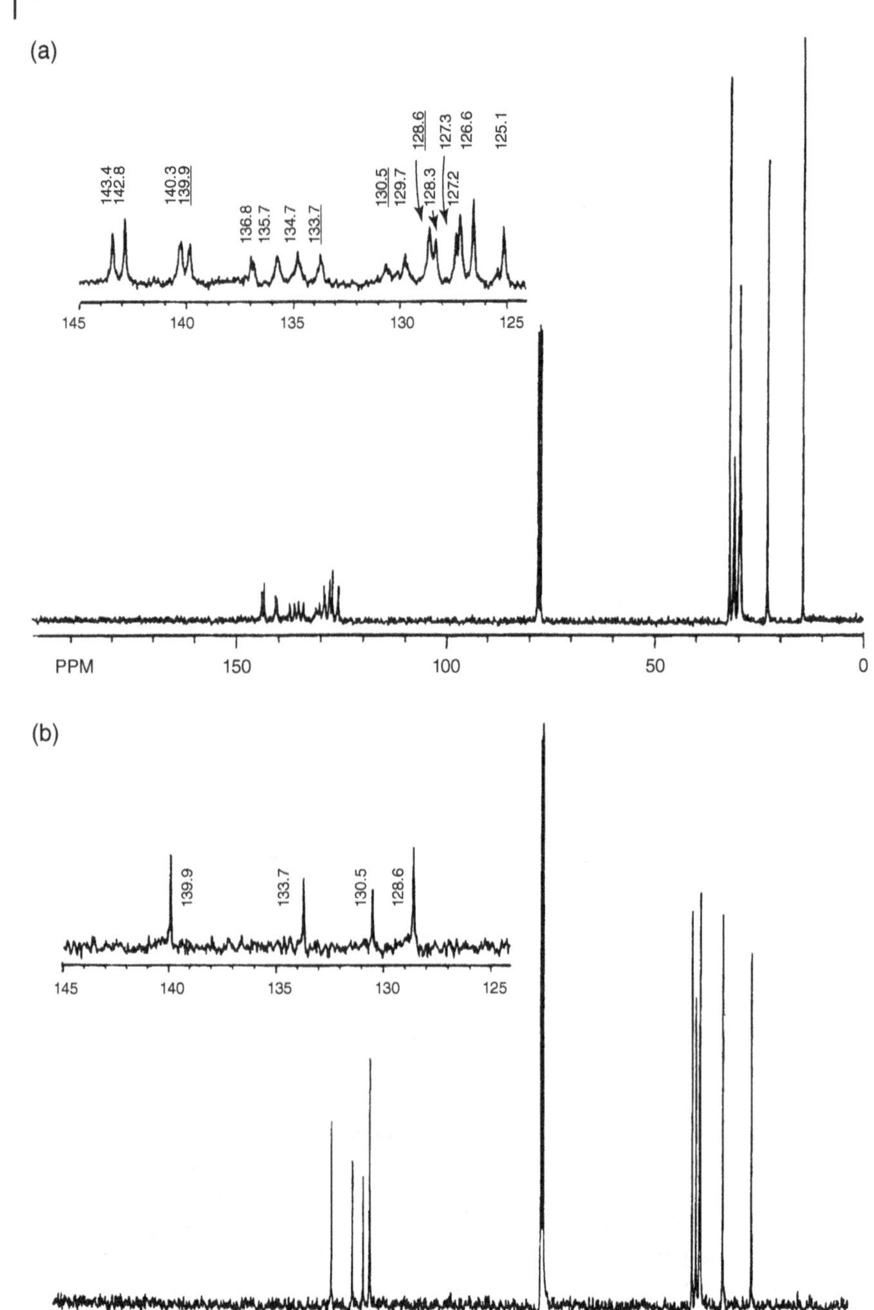

Figure 3.9 ^{13}C NMR spectra of (a) regiorandom P3HT **5b** (1 : 1 : 1 : 1 HT–HT/HT–HH/TT–HT/TT–HH) and (b) regioregular P3HT **4b** (HT linkage only).

Table 3.57 NMR chemistry shift (ppm) of ^1H and ^{13}C in different regiostructures.

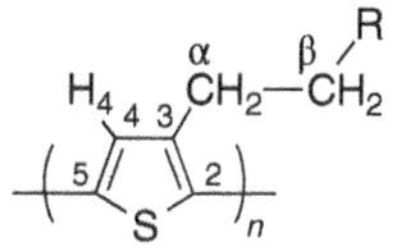

	HT–HT	TT–HT	HT–HH	TT–HH
C2	130.5	128.3	129.7	127.2
C3	139.9	142.8	140.3	143.4
C4	128.6	126.6	127.3	125.1
C5	133.7	135.7	134.7	136.8
H4	6.98	7.00	7.02	7.05
		Head-to-tail		Head-to-head
α-Methylene-H		2.80		2.58
β-Methylene-H		1.72		1.63

Conclusion

We have demonstrated a systematically regiocontrolled synthesis of poly(3-alkylthiophenes) mediated by Rieke zinc. Starting with the same compounds, a series of P3ATs with different regioregularities were obtained, among which, a completely regioregular HT and a totally regiorandom poly(3-alkylthiophene) were most significant. The regioregularity of the polymers was controlled by the structure of different catalysts in the polymerization. The regioregular HT P3ATs were consistently characterized as a class of polymers with regiospecific HT conformation, significantly extended conjugation length, self-organized structure of polymer chain, low bandgap, and polycrystalline texture. The regioregular HT P3ATs had considerably higher neutral and doping conductivities than those of regiorandom P3ATs. These regioregular polymers are being used worldwide in the development of photovoltaic devices, plastic electronics, and organic light-emitting diode devices.

General Preparation of Regioregular HT Poly(3-Alkylthiophenes) from 2,5-Dibromo-3-Alkylthiophenes: Preparation of Regioregular HT Poly(3-Hexylthiophene) (4b)

2,5-Dibromo-3-hexylthiophene (3.26 g, 10.0 mmol, in 20 ml of THF) was added via a cannula to the newly prepared Rieke Zn* (11.0 mmol, in 40 ml of THF) at −78°C. The mixture was stirred for 1 h at this temperature and allowed to warm to 0°C in ca. 3 h; 0.2 mol% of Ni(DPPE)Cl_2 (11.0 mg, 0.02 mmol, in 20 ml of THF) was added via cannula at 0°C. The mixture was stirred for 24 h at room temperature. A dark-purple precipitate was formed gradually in this period. The mixture was poured into a solution of MeOH (100 ml) and 2 N HCl (50 ml);

the resulting dark precipitate was filtered and washed with MeOH and 2 N HCl solution and then dried. Reprecipitation of polymer from chloroform solution upon addition of MeOH and drying under vacuum gave 1.62 g of dark polymer of regioregular HT poly(3-hexylthiophene) (98% yield). The product had apparently the same NMR, IR, UV-Vis, and fluorescence characterization as the product after extractions (see succeeding text). The polymer had average molecular weight, $M_w = 15\,000$ and $M_n = 4\,130$ with a polydispersity index (PD) of 3.63 determined by GPC. Anal. Calcd for $(C_{10}H_{14}S)_n$: C, 72.22; H, 8.49; S,19.29. Found: C, 71.91; H, 8.36; S, 18.67.

Repurification of the polymer by Soxhlet extractions with MeOH for 24 h and then with hexane for 24 h afforded 1.37 g (82% final yield) of regioregular HT poly(3-hexylthiophene). The polymer was characterized as a regioregular poly(3-hexylthiophene) with at least 98.5% HT linkage in the polymer chains by NMR analysis. The polymer after extractions had much higher average molecular weight than did the polymer before extractions. The $M_w = 37\,680$ and $M_n = 25\,500$ corresponding to a PD of 1.48. Anal. Calcd for $(C_{10}H_{14}S)_n$: C, 72.22; H, 8.49; S, 19.29. Found: C, 71.90; H, 8.48; S, 18.73.

General Preparation of Regiorandom Poly(3-Alkylthiophenes) from 2,5-Dibromo-3-Alkylthiophenes: Preparation of Regiorandom Poly(3-Hexylthiophene) (5b)

2,5-Dibromo-3-hexylthiophene (3.26 g, 10.0 mmol, in 20 ml THF) was added to the flask with newly prepared Zn* (11.0 mmol in 40 ml of THF) via cannula at 0°C, and the mixture was stirred for 1 h at room temperature. A 0.2 mol% amount of $Pd(PPh_3)_4$ (23.1 mg, 0.02 mmol, in 20 ml of THF) was added via cannula. The mixture was then stirred for 24 h at room temperature (or reflux for 6 h). The polymer was precipitated with a solution of MeOH (100 ml) and 2 N HCl (50 ml) and purified by reprecipitation from the polymer solution of chloroform upon addition of MeOH. After drying under vacuum, 1.63 g (98% yield) of red–brown rubber-like polymer of regiorandom P3HT was obtained. The polymer was characterized as a regiorandom P3HT (1 : 1 : 1 : 1 HT–HT/HT–HH/TT–HT/TT–HH or 50 : 50 HT–HH) by the NMR analysis: $M_w = 24\,400$ and $M_n = 5\,650$ with a PD of 4.32. Anal. Calcd for $(C_{10}H_{14}S)_n$: C, 72.22; H, 8.49; S, 19.29. Found: C, 71.50; H, 8.28; S, 19.77.

References

1 Frankland, E. *Ann. Chem. Pharm.* 1849, **71**, 171.
2 Frankland, E. *Justus Liebigs Ann. Chem.* 1856, **99**, 333.
3 Frankland, E. *Ann. Chem. Pharm.* 1849, **71**, 213.
4 (a) Frankland, E. *Justus Liebigs Ann. Chem.* 1853, **85**, 329. (b) Theyer, J. *J. Chem. Educ.* 1969, **46**, 474.

5 Gladstone, J.H. *J. Chem. Soc.* 1891, **59**, 290.
6 Job, A.; Reich, R. *Bull. Soc. Chim. Fr.* 1923, **33**, 1414.
7 Kurg, R.C.; Tang, R.J.C. *J. Am. Chem. Soc.* 1954, **76**, 2262.
8 Renshaw, R.R.; Greenlaw, C.E. *J. Am. Chem. Soc.* 1920, **42**, 1472.
9 Noller, C.R. *Org. Synth.* 1932, **12**, 86.
10 Shriner, R.L. *Org. React.* 1942, **1**, 1.
11 Cornforth, J.W.; Cornforth, R.H.; Popjak, G.; Gore, I.Y. *Biochem. J.* 1958, **69**, 146.
12 (a) Knochel, P.; Yeh, M.C.P.; Berk, S.C.; Talbert, J. *J. Org. Chem.* 1988, **53**, 2390. (b) Yeh, M.C.P.; Knochel, P.; Santa, L.E. *Tetrahedron Lett.* 1988, **29**, 3887. (c) Yeh, M.C.; Knochel, P.; Butler, W.M.; Berk, S.C. *Tetrahedron Lett.* 1988, **29**, 6693. (d) Majid, T.N., Yeh, M.C.; Knochel, P. *Tetrahedron Lett.* 1989, **30**, 5069. (e) Majid, T.N.; Knochel, P. *Tetrahedron Lett.* 1990, **32**, 4413.
13 (a) Knochel, P.; Yeh, M.C.; Berk, S.C.; Talbert, J. *J. Org. Chem.* 1988, **53**, 2390. (b) Yeh, M.C.; Knochel, P.; Santa, L. *Tetrahedron Lett.* 1988, **29**, 3887. (c) Yeh, M.C.; Knochel, P.; Butler, M.W.; Berk, S.C. *Tetrahedron Lett.* 1988, **29**, 6693. (d) Knochel, P.; Yeh, M.C.P.; Berk, S.C.; Talbert, J. *Tetrahedron Lett.* 1988, **29**, 2395. (e) Majid, T.N.; Knochel, P. *Tetrahedron Lett.* 1990, **31**, 4413.
14 Zhu, L.; Wehmeyer, R.M.; Rieke, R.D. *J. Org. Chem.* 1991, **56**, 1445.
15 (a) Knochel, P.; Yeh, M.C.; Berk, S.C.; Talbert, J. *J. Org. Chem.* 1988, **53**, 2390. (b) Yeh, M.C.; Knochel, P.; Santa, L. *Tetrahedron Lett.* 1988, **29**, 3887. (c) Yeh, M.C.; Knochel, P.; Butler, M.W.; Berk S.C. *Tetrahedron Lett.* 1988, **29**, 6693. (d) Knochel, P.; Yeh, M.C.P.; Berk, S.C.; Talbert, J. *Tetrahedron Lett.* 1988, **29**, 2395. (e) Majid, T.N.; Knochel, P. *Tetrahedron Lett.* 1990, **31**, 4413.
16 Negishi, E.; Bagheri, V.; Chatterjee, S.; Luo, F.; Miller, J.A.; Stoll, A.T. *Tetrahedron Lett.* 1983, **24**, 5181.
17 Hanson, M.; Rieke, R.D. *Synth. Commun.* 1995, **25**, 101.
18 Petrier, C.; Dupuy, C.; Luche, J.L. *Tetrahedron Lett.* 1986, **27**, 3149.
19 Isobe, M.; Kondo, S.; Nagasawa, N.; Goto, T. *Chem. Lett.* 1997, 679.
20 Kjonaas, R.A.; Vawter, E.J. *J. Org. Chem.* 1986, **51**, 3993.
21 Watson, R.A.; Kjonaas, R.A. *Tetrahedron Lett.* 1986, **27**, 1437.
22 (a) Knochel, P.; Yeh, M.C.P.; Berk, S.C.; Talbert, J. *J. Org. Chem.* 1988, **53**, 2392. (b) Knochel, P.; Yeh, M.C.P.; Berk, S.C.; Talbert, J. *Tetrahedron Lett.* 1988, **29**, 2395. (c) Yeh, M.C.P.; Knochel, P.; Butler, W.M.; Berk, S.C. *Tetrahedron Lett.* 1988, **29**, 6693.
23 Lipschutz, B.H.; Parker, D.A.; Nguyen, S.L.; McCarthy, K.E.; Barton, J.C.; Whitney, S.E.; Kotsuki, H. *Tetrahedron* 1986, **42**, 2873.
24 For example, allylic alcohols, allylic esters, allylic halides, and allylic tosylates.
25 Magid, R.M. *Tetrahedron* 1980, **36**, 1901. And reference cited therein.
26 See examples in Refs. 31 and 32.
27 Miyaura, N.; Itoh, M.; Suzuki, A. *Bull. Chem. Soc. Jpn.* 1977, **50**, 2199.
28 Mourik, G.L.V.; Pabon, H.J.J. *Tetrahedron Lett.* 1978, **30**, 2705.
29 Ochiai, H.; Tamaru, Y.; Tsubaki, K.; Yoshida, Z. *J. Org. Chem.* 1987, **52**, 4418.

30 (a) Nakamura, E.; Aoki, S.; Sekiya, K.; Oshino, H.; Kuwajima, J. *J. Am. Chem. Soc.* 1987, **109**, 8056. (b) Sekiya, K.; Nakamura, E. *Tetrahedron Lett.* 1988, **29**, 5155.

31 Yamamoto, Y.; Yamamoto, S.; Yatagai, H.; Maruyama, K. *J. Am. Chem. Soc.* 1980, **102**, 2318. And references cited therein.

32 Zhu, L.; Rieke, R.D. *Tetrahedron Lett.* 1991, **32**, 2865.

33 Zhu, L.; Shaughnessy, K.H.; Rieke, R.D. *Synth. Commun.* 1993, **23**, 525.

34 (a) Dieter, R.K. *Tetrahedron* 1999, **55**, 4177. (b) *Comprehensive Organic Transformations*; Larock, R.C., Ed., 2nd ed.; Wiley-VCH: Weinheim, 1999. (c) Diederich, F.; Stang, P.J. *Metal-Catalyzed Cross-Coupling Reactions*; Wiley-VCH: Weinheim, 1998.

35 (a) Maeda, H.; Okamato, J.; Ohmori, H. *Tetrahedron Lett.* 1996, **37**, 5381. (b) Malanga, C.; Aronica, L.A.; Lardicci, L. *Tetrahedron Lett.* 1995, **36**, 9185.

36 (a) *Modern Organocopper Chemistry*; Krause, N., Ed.; Wiley-VCH: Weinheim, 2001. (b) *Organocopper Reagents: A Practical Approach*; Taylor, R.J.K., Ed.; Oxford University Press: New York, 1994.

37 (a) Rieke, R.D., Suh, Y.S.; Kim, S.H. *Tetrahedron Lett.* 2005, **46**, 5961. (b) Kim, S.H.; Rieke, R.D. *J. Org. Chem.* 2000, **65**, 2322. (c) Kim, S.H. Rieke, R.D. *Tetrahedron Lett.* 1999, **40**, 4391. (d) Cahiez, G.; Laboue, B. *Tetrahedron Lett.* 1989, **30**, 7169. (e) Cahiez, G.; Laboue, B. *Tetrahedron Lett.* 1989, **30**, 3545.

38 (a) Wakefield, B.J. *Organolithium Methods in Organic Synthesis*; Academic: London, 1990. (b) Tucker, C.E.; Majid, T.N.; Knochel, P. *J. Am. Chem. Soc.* 1992, **114**, 3893.

39 (a) Haddach, M.; McCarthy, J.R. *Tetrahedron Lett.* 1999, **40**, 3109. (b) Bumagin, N.A.; Korolev, D.N. *Tetrahedron Lett.* 1999, **40**, 3057. (c) Goosen, L.J.; Ghosh, K. *Angew.Chem. Int. Ed.* 2001, **40**, 3458.

40 (a) Stille, J.K. *Angew. Chem. Int. Ed. Engl.* 1986, **25**, 508. (b) Lerebours, R.; Camacho-Soto, A.; Wolf, C. *J. Org. Chem.* 2005, **70**, 8601.

41 (a) Karzmierski, I.; Bastienne, M.; Gosmini, C.; Paris, J.M.; Perichon, J. *J. Org. Chem.* 2004, **69**, 936 and references cited therein. (b) Knochel, P.; Yeh, M.C.P.; Berk, S.C.; Talbert, J. *J. Org. Chem.* 1988, **53**, 2390.

42 Fillon, H.; Gosmini, C.; Perichon, J. *Tetrahedron* 2003, **59**, 8199.

43 Erdik, E.; Pekel, Ö.Ö. *Tetrahedron Lett.* 2009, **50**, 1501.

44 Xu, H.; Ekoue-Kovi, K.; Wolf, C. *J. Org. Chem.* 2008, **73**, 7368.

45 (a) Bercot, E.A.; Rovis, T. *J. Am. Chem. Soc.* 2005, **127**, 247. (b) Zhang, Y.; Rovis, T. *J. Am. Chem. Soc.* 2004, **126**, 15964. (c) Bercot, E.A.; Rovis, T. *J. Am. Chem. Soc.* 2001, **124**, 174.

46 For the general procedure of the preparation of active zinc and organozinc reagents, see: Rieke, R.D.; Hanson, M.V. *Tetrahedron* 1997, **53**, 1925.

47 Kim, S.H.; Rieke, R.D. *Tetrahedron Lett.* 2011, **52**, 1523.

48 A ligand-less Pd system: Iwai, T.; Nakai, T.; Mihara, M.; Ito, T.; Mizuno, T.; Ohno, T. *Synlett* 2009, **7**, 1091.

49 Rieke, R.D. *Aldrich Chim. Acta* 2000, **33**, 52.

50 Bronk, B.S.; Lippard, S.J.; Danheiser, R.L. *Organometallics* 1993, **12**, 3340.
51 Meyer, C.; Marek, I.; Courtemanche, G.; Normant, J.-F. *Synlett* 1993, 266.
52 Meyer, C.; Marek, I.; Courtemanche, G.; Normant, J.-F. *Tetrahedron Lett.* 1993, **34**, 6053.
53 Hanson, M.V.; Rieke, R.D. *J. Am. Chem. Soc.* 1995, **117**, 10775.
54 Kim, S.; Lee, J.M. *Tetrahedron Lett.* 1990, **31**, 7627.
55 (a) Isobe, M.; Kondo, S.; Nagasawa, N.; Goto, T. *Chem. Lett.* 1977, 679. (b) Tuckmantel, W.; Oshima, K.; Nozaki, H. *Chem. Ber.* 1986, **119**, 1581. (c) Kjonaas, R.A.; Vawter, E.J. *J. Org. Chem.* 1986, **51**, 3993.
56 (a) Knochel, P.; Singer, R.D. *Chem. Rev.* 1993, **93**, 2117. (b) Klement, I.; Knochel, P.; Chau, K.; Cahiez, G. *Tetrahedron Lett.* 1994, **35**, 1177. (c) Rozema, M.J.; Eisenberg, C.; Lutjens, H.; Ostwald, R.; Belyk, K.; Knochel, P. *Tetrahedron Lett.* 1993, **34**, 3115. (d) Rozema, M.J.; AchyuthaRao, S.; Knochel, P. *J. Org. Chem.* 1992, **57**, 1956. (e) Knochel, P.; Yeh, M.C.P.; Berk, S.C.; Talbert, J. *J. Org. Chem.* 1988, **53**, 2390.
57 Lipshutz, B.H.; Keil, R. *J. Am. Chem. Soc.* 1992, **114**, 7919.
58 Lipshutz, B.H.; Wood, M.R.; Tirado, R. *J. Am. Chem. Soc.* 1995, **117**, 6126.
59 (a) Hanson, M.V.; Brown, J.D.; Niu, Q.J.; Rieke, R.D. *Tetrahedron Lett.* 1994, **35**, 7205. (b) Zhu, L.; Wehmeyer, R.M.; Rieke, R.D. *J. Org. Chem.* 1991, **56**, 1445.
60 Velarde-Ortiz, R.; Guijarro, A.; Rieke, R.D. *Tetrahedron Lett.* 1998, **39**, 9157.
61 Erdik, E., Ay, M. *Chem. Rev.* 1989, **89**, 1947.
62 Coleman, G.H.; Andersen, H.P.; Hermanson, J.L. *J. Am. Chem. Soc.* 1934, **56**, 1381.
63 Coleman, G.H.; Hermanson, J.L.; Johnson, H.L. *J. Am. Chem. Soc.* 1937, **59**, 1896.
64 Genari, C.; Colombo, L.; Bertolini, G. *J. Am Chem. Soc.* 1986, **108**, 6394.
65 Evans, D.A.; Britton, T.C.; Dorow, R.L.; Dellaria, J.F. *J. Am. Chem. Soc.* 1986, **108**, 6395.
66 Trimble, L.A.; Verderas, J.C. *J. Am. Chem. Soc.* 1986, **108**, 6397.
67 Gmeiner, P.; Bollinger, B. *Liebigs Ann. Chem.* 1992, 273.
68 Greck, C.; Bischoff, L.; Ferreira, F.; Pinel, C.; Piveteau, E.; Genet, J.P. *Synlett* 1993, 475.
69 Page, P.C.B.; Allin, S.M.; Collington, E.W.; Carr, R.A.E. *Tetrahedron Lett.* 1994, **35**, 2427.
70 Oppolzer, W.; Moretti, R. *Helv. Chim. Acta* 1986, **69**, 1923.
71 Mellor, J.M. Smith, N.M. *J. Chem. Soc. Perkin Trans.* **1**. 1984, 2927.
72 Shvo, Y. Conformational Analysis of Hydrazines. In *The Chemistry of Hydrazo, Azo and Azoxy Groups*; Patai, S., Ed.; Interscience: New York, 1972; *part 2*, p 1017.
73 Oki, M. *Applications of Dynamic NMR Spectroscopy to Organic Chemistry*; VCH Publishers: Deerfield Beach, FL, 1985; **2**, p 3.
74 Evans, D.A.; Britton, T.C.; Dorow, R.L.; Dellaria, J.F. *Tetrahedron* 1988, **44**, 5525.

75 Fitzsimmons, B.J.; Leblanc, Y.; Rokach, J. *J. Am. Chem. Soc.* 1987, **109**, 285.
76 Frankenfeld, J.W.; Werner, J.J. *J. Org. Chem.* 1969, **34**, 3689 and references cite therein.
77 Rathke, M.W.; Lindert, A. *J. Org. Chem.* 1970, **35**, 3966.
78 Ruppert, J.; White, J.D. *J. Org. Chem.* 1974, **39**, 269.
79 Rieke, R.D.; Uhm, S.J.; Hudnall, P.M. *J. Chem. Soc. Chem. Comm.* 1973, 269.
80 Rieke, R.D.; Bales, S.E. *J. Amer. Chem. Soc.* 1974, **96**, 1775.
81 Rieke, R.D.; Li, P.T.; Burns, T.P.; Uhm, S.T. *J. Org. Chem.* 1981, **46**, 4323.
82 Duddu, R.; Eckhardt, M.; Furlong, M.; Knoess, H.P.; Berger, S.; Knochel, P. *Tetrahedron* 1994, **50**, 2415.
83 Lemoucheux, L.; Rouden, J.; Lasne, M.-C. *Tetrahedron Lett.* 2000, **41**, 9997.
84 Lemoucheux, L.; Seitz, T.; Rouden, J.; Lasne, M.-C. *Org. Lett.* 2004, **6**, 3703.
85 (a) Balas, L.; Jousseaume, B.; Shin, H.; Verlhac, J.-B; Wallian, F. *Organometallics* 1991, **10**, 366. (b) Jousseaume, B.; Kwon, H.; Verlhac, J.-B.; Denat, F.; Dubac, J. *Synlett* 1993, 117. (c) Murakami, M.; Hoshino, Y.; Ito, H.; Ito, Y. *Chem. Lett.* 1998, **27**, 163.
86 (a) Lysēn, M.; Kelleher, S.; Begtrup, M.; Kristensen, J.L. *J. Org. Chem.* 2005, **70**, 5342. (b) Duan, Y.-Z.; Deng, M.-Z. *Synlett* 2005, 355. (c) Yasui, Y.; Tsuchida, S.; Miyabe, H.; Takemoto, Y. *J. Org. Chem.* 2007, **72**, 5898. (d) Krishnamoorthy, R.; Lam, S.Q.; Manley, C.M.; Herr, R.J. *J. Org. Chem.* 2010, **75**, 1251.
87 Cunico, R.F.; Pandey, R.K. *J. Org.Chem.* 2005, **70**, 9048.
88 (a) Martinelli, J.R.; Watson, D.A.; Freckmann, D.M.M.; Barder, T.E.; Buchwald, S.L. *J. Org. Chem.* 2008, **73**, 7102. (b) Martinelli, J.R.; Freckman, D.M.M.; Buchwald, S.L. *Org. Lett.* 2006, **8**, 4843. (c) Zhuang, L.; Wai, J.S.; Embrey, M.W.; Fisher, T.E.; Egbertson, M.S.; Payne, L.S.; Guare, J.P., Jr.; Vacca, J.P.; Hazuda, D.J.; Felock, P.J.; Wolfe, A.L.; Stillmock, K.A.; Witmer, M.V.; Moyer, G.; Schleif, W.A.; Gabryelski, L.J.; Leonard, Y.M.; Lynch, J.J., Jr.; Michelson, S.R.; Young, S.D. *J. Med. Chem.* 2003, **46**, 453. (d) Deagostino, A.; Larini, P.; Occhiato, E.G.; Pizzuto, L.; Prandi, C.; Venturello, P. *J. Org. Chem.* 2008, **1941**, 73.
89 Meijere, A.; Diederich, F. *Metal-Catalyzed Cross-Coupling Reactions*; 2nd ed.; Wiley-VCH: Weinheim, 2004; Vol. **2**.
90 Rieke, R.D.; Kim, S.-H. *Tetrahedron Lett.* 2012, **53**, 3478.
91 Organozinc reagents used in this study were prepared by the direct insertion of highly active zinc to the corresponding aryl halide. For details, see: Rieke, R.D.; Hanson, M.V. *Tetrahedron* 1997, **1925**, 53.
92 (a) A limited number of examples of pyridinyl amide via aminocarbonylation appeared in Buchwald's report (see: Ref. 88a). (b) For the low reactivity of 3-pyridylboronic acid in the coupling reaction with carbamoyl chloride, see: Ref. 86d.
93 A transition metal-catalyzed ring opening of THF, see: Frior, G.; Alexakis, A.; Cahiez, G.; Normant, J. *Tetrahedron* 1984, **40**, 683.
94 Kim, S.-H.; Rieke, R.D. *J. Org. Chem.* 2013, **78**, 1984.

95 (a) Keay, B.A.; Dibble, P.W. In *Comprehensive Heterocyclic Chemistry II*; Bird, C.W., Ed.; Elsevier: New York, 1996; Vol. **2**, Chapter 2.08, pp 395–436. (b) Kea B.A. *Chem. Soc. Rev.* 1999, **28**, 209. (c) Lipshutz, B.H. *Chem. Rev.* 1986, **86**, 795. (d) Hou, X.L.; Cheung, H.Y.; Hon, T.Y.; Kwan, P.L.; Lo, T.H.; Tong, S.Y.; Wong, H.N.C. *Tetrahedron* 1998, **54**, 1955. (e) Konig, B. In *Science of Synthesis: Houben-Weyl Methods of Molecular Transformations, Category 2*; Maas, G.; Regitz, M., Eds.; Georg Thiem Verlag: New York, 2001; Vol. **9**, pp 183–286.

96 (a) Dunlop, A.P.; Peters, F.N. *The Furans*; Reinhold Publishing Corporation: New York, 1953. (b) Donnelly, D.M.S.; Meegan, M.J. In *Comprehensive Heterocyclic Chemistry*; Katrizky, A.R., Rees, C.W., Eds; Pergamon Press: New York, 1984; Vol. **4**, Selection 3.12.

97 Li, J.J.; Gribble, G.W. *Palladium in Heterocyclic Chemistry, A Guide for the Synthetic Chemist*; 2nd ed.; Elsevier: New York, 2007; Chapter 6.

98 (a) Balachari, D.; Quinn, L.; O'Doherty, G.A. *Tetrahedron Lett.* 1999, **40**, 4769. (b) Balachari, D.; O'Doherty, G.A. *Org. Lett.* 2000, **2**, 863. (c) Balachari, D.; O'Doherty, G.A. *Org. Lett.* 2000, **2**, 4033.

99 Gauthier, D.R., Jr.; Szumigala, R.H., Jr.; Dormer, P.G.; Armstrong, J.D., III; Volante, R.P.; Reider, P.J. *Org. Lett.* 2002, **4**, 375.

100 (a) McClure, M.S.; Glover, B.; McSorley, E.; Millar, A.; Osterhout, M.H.; Roschangar, F. *Org. Lett.* 2001, **3**, 1677. (b) McClure, M.S.; Roschangar, F.; Hodson, S.J.; Millar, A.; Osterhout, M.H. *Synthesis* 2001, **11**, 1681. (c) Other example of direct coupling: Liegault, B.; Lapointe, D.; Caron, L.; Vlassava, A.; Fagnou, K. *J. Org. Chem.* 2009, **74**, 1826.

101 Hanar, J.; Jack, K.; Nagireddy, J.; Raheem, M.A.; Durham, R.; Tam, W. *Synthesis* 2011, **5**, 731.

102 Piller, F.M.; Knochel, P. *Synthesis* 2011, **11**, 1751.

103 For examples of using 5-bromo-2-furaldehyde for coupling reaction, see: (a) Karpov, A.S.; Rominger, F.; Muller, T.J. *J. Org. Chem.* 2003, **68**, 1503. (b) Lewis, T.A.; Bayless, L.; Eckman, J.B.; Ellis, J.L.; Grewal, G.; Libertine, L.; Nicolas, J.M.; Scannell, R.T.; Wels, B.F.; Wenberg, K.; Wypij, D.M. *Bioorg. Med. Chem. Lett.* 2004, **14**, 2265.

104 For general procedure for organozincs, see:(a) Rieke, R.D.; Hanson, M.V. *Tetrahedron* 1997, **53**, 1925. (b) Rieke, R.D.; Hanson, M.V.; Brown, J.D. *J. Org. Chem.* 1996, **61**, 2726.

105 (a) Manolikakes, G.; Hernandez, C.M.; Schade, M.A.; Metzger, A.; Knochel, P. *J. Org.Chem.* 2008, **73**, 8422. (b) Manolikakes, G.; Schade, M.A.; Hernandez, C.M.; Mayr, H.; Knochel, P. *Org. Lett.* 2008, **10**, 2765. (c) Kim, S.H.; Rieke, R.D. *Tetrahedron Lett.* 2009, **50**, 6985.

106 An attempt for the preparation of 3.4.5-trimethoxyphenylzinc bromide using the direct insertion of active zinc (Zn*) to the corresponding organic halide was unsuccessful. Instead, an unidentified mixture was obtained.

107 (a) Murakami, A.; Gao, G.; Omura, M.; Yano, M.; Ito, C.; Furukawa, H.; Takahashi, D.; Koshimizu, K.; Ohigashi, H. *Bioorg. Med. Chem. Lett.* 2000,

10, 59. (b) Maier, W.; Schmidt, J.; Nimtz, M.; Wray, Y.; Strack, G. *Phytochemistry* 2000, **54**, 473. (c) Garcia-Argaez, A.N.; Ramirez Apan, T.O.; Delgado, H.P.; Velazquez, G.; Martinez-Vazquez, M. *Planta Med.* 2000, **66**, 279. (d) Zhou, P.; Takaishi, Y.; Duan, H.; Chen, B.; Honda, G.; Itoh, M.; Takeda, Y.; Kodzhimatov, O.K.; Lee, K.-H. *Phytochemistry* 2000, **53**, 689.

108 (a) Jones, G., II; Jackson, W.R.; Choi, C. *J. Phys. Chem.* 1985, **89**, 294. (b) Jones, G., II; Ann, J.; Jimenez, A.C. *Tetrahedron Lett.* 1999, **40**, 8551. (c) Raboin, J.-C.; Beley, M.; Kirsch, G. *Tetrahedron Lett.* 2000, **41**, 1175.

109 (a) De la Hoz, A.; Moreno, A.; Vazquez, E. *Synlett* 1999, 608. (b) Aracadi, A.; Cacchi, S.; Fabrizi, G.; Marinelli, F.; Pace, P. *Synlett* 1996, 568. (c) Donnely, D.M.X.; Finet, J.-P.; Guiry, P.J.; Hutchinson, R.M. *J. Chem. Soc. Perkin Trans.* **1**. 1990, 2851. (d) Britto, N.; Gore, V.G.; Mali, R.S.; Ranade, A.C. *Synth. Commun.* 1989, **19**, 1899. (e) Sato, K.; Inour, S.; Ozawa, K.; Kobayashi, T.; Ota, T.; Tazaki, M. *J. Chem. Soc. Perkin Trans.* **1**. 1987, 1753. (f) Awasthi, A.K.; Tewari, R.S. *Synthesis* 1986, 1061.

110 (a) Wu, J.; Liao, Y.; Yang, Z. *J. Org Chem.* 2001, **66**, 3462. (b) Wu, J.; Yang, Z. *J. Org. Chem.* 2001, **66**, 7875.

111 (a) Wattanasin, S. *Synth. Commun.* 1988, **18**, 1919. (b) Schio, L.; Chatreaux, F.; Klich, M. *Tetrahedron Lett.* 2000, **41**, 1543.

112 (a) Boland, G.M.; Donnelly, D.M.X.; Finet, J.-P.; Rea, M.D. *J. Chem. Soc. Perkin Trans.* **1** 1996, 2591. (b) Yao, M.-L.; Deng, M.-Z. *Heteroat. Chem.* 2000, **11**, 380.

113 Rieke, R.D.; Kim, S.-H. *Tetrahedron Lett.* 2011, **52**, 3094.

114 Representative procedures: (a) Preparation of 4-coumarinylzinc bromide (**I**); In an oven-dried 50 ml round-bottomed flask equipped with a stir bar was added 1.40 g of active zinc (Zn*, 22.0 mmol). 4-Bromocoumarin (4.50 g, 20.0 mmol) dissolved in 20 ml of THF was then cannulated neat into the flask at room temperature. The resulting mixture was stirred for 1 h at room temperature. The whole mixture was settled down, and then the supernatant was used for the subsequent coupling reactions. (b) Representative Pd-catalyzed cross-coupling reaction procedure: Into a 25 ml round-bottomed flask were added $Pd(PPh_3)_4$ (0.025 g, 1 mol%) and 4.0 ml of 4-coumarinylzinc bromide (I) (0.5 M in THF, 2.0 mmol) under an argon atmosphere. Next, 4-methoxybenzoyl chloride (0.27 g, 160 mmol) was slowly added via a syringe while being stirred at room temperature. The resulting mixture was stirred at room temperature for 30 min. Quenched with 3 M HCl solution, it was then extracted with ethyl ether (10 ml × 3). Washed with saturated $NaHCO_3$, $Na_2S_2O_3$ solution, and brine, it was then dried over anhydrous $MgSO_4$. Purification by column chromatography on silica gel (20% ethyl acetate/80% heptane) afforded 0.39 g of **1e** in 92% isolated yield as white solid (mp 107–108°C). MS (EI) *m/z* (relative intensity): 280 (m^+, 80), 252 (25), 135 (100).

115 All chemical shifts of the reduced product(coumarin) were consistent with the literature values; The Aldrich Library of ^{13}C and ^{1}H FT NMR Spectra.

116 Recent examples of using $Pd(OAc)_2$ /SPhos-system for the Negishi-type coupling reaction with haloaromatic amine: see, (a) Kim, S.H.; Rieke, R.D. *Tetrahedron* 2010, **66**, 3135. (b) Manolikakes, G.; Schade, M.A.; Hernandez, C.M.; Mayr, H.; Knochel, P. *Org. Lett.* 2008, **10**, 2765.

117 For recent examples, (a) Denton, T.T.; Zhang, X.; Cashman, J.R. *J. Med. Chem.* 2005, **48**, 224. (b) Davies, J.R.; Kane, P.D.; Moody, C.J.; Slawin, A.M. *J.Org. Chem.* 2005, **70**, 5840. (c) Yang, C.-G.; Huang, H.; Jiang, B. *Curr. Org. Chem.* 2004, **8**, 1691. (d) Che, D.; Wegge, T.; Stubbs, M.T.; Seitz, G.; Meier, H.; Methfessel, C. *J. Med. Chem.* 2001, **44**, 47.

118 (a) Trecourt, F.; Gervais, B.; Mallet, M.; Queguiner, C. *J. Org. Chem.* 1996, **61**, 1673. (b) Trecourt, F.; Gervais, B.; Mongin, O.; Le Gal, C.; Mongin, F.; Queguiner, G. *J. Org. Chem.* 1998, **63**, 2892.

119 (a) Hargreaves, S.L.; Pilkington, B.L.; Russell, S.E.; Worthington, P.A. *Tetrahedron Lett.* 2000, **41**, 1653. (b) Zimmermann, J.; Buchdunger, E.; Mett, H.; Meyer, T.; Lydon, N.B. *Bioorg. Med. Chem. Lett.* 1997, **7**, 187. (c) Zimmermann, J.; Buchdunger, E.; Mett, H.; Meyer, T.; Lydon, N.B. *Bioorg. Med. Chem. Lett.* 1996, **6**, 1221.

120 Getmanenko, Y.A.; Twieg, R.J. *J.Org. Chem.* 2008, **73**, 830.

121 For recent examples, see: (a) Deng, J.Z.; Paone, D.V.; Ginnetti, A.T.; Kurihara, H.; Dreher, S.D.; Weissman, S.A.; Stauffer, S.R.; Burgey, C.S. *Org. Lett.* 2009, **11**, 345. (b) Yang, D.X.; Colletti, S.L.; Wu, K.; Song, M.; Li, G.Y.; Shen, H.C. *Org. Lett.* 2009, **11**, 381. (c) Voisin-Chiret, A.S.; Bouillon, A.; Burzicki, G.; Celant, M.; Legay, R.; El-Kashef, H.; Rault, S. *Tetrahedron* 2009, **65**, 607. (d) Hodgson, P.B.; Salingue, F.H. *Tetrahedron Lett.* 2004, **45**, 685.

122 (a) Schwab, P.F.H.; Fleischer, F.; Michl, J. *J. Org. Chem.* 2002, **67**, 443. (b) Zhang, N.; Thomas, L.; Wu, B. *J. Org. Chem.* 2001, **66**, 1550. (c) Schubert, U.C.; Eschbaumer, C.; Heller, M. *Org. Lett.* 2000, **2**, 3373. (d) Gronowitz, S.; Bjork, P.; Malm, J.; Hornfeldt, A.-B. *J. Organomet. Chem.* 1993, **460**, 127.

123 (a) Sugimoto, O.; Yamada, S.; Tanji, K. *J. Org. Chem.* 2003, **68**, 2054. (b) Song, J.J.; Yee, N.K.; Tan, Z.; Xu, J.; Kapadia, S.R.; Senanayake, C.H. *Org. Lett.* 2004, **6**, 4905. (c) Duan, X.-F.; Ma, X.-Q.; Zhang, F.; Zhang, Z.-B. *J. Org. Chem.* 2009, **74**, 939.

124 (a) Savage, S.A.; Smith, A.P.; Fraser, C.L. *J. Org. Chem.* 1998, **63**, 10048. (b) Fang, Y.-Q.; Hanan, G.S. *Synlett* 2003, 852. (c) Lutzen, A.; Hapke, M.; Staats, H.; Bunzen, J. *Eur. J. Org. Chem.* 2003, 3948.

125 Billingsley, K.L.; Buchwald, S.L. *Angew. Chem. Int. Ed.* 2008, **47**, 4695.

126 Berman, A.M.; Lewis, J.C.; Bergman, R.G.; Ellman, J.A. *J. Am. Chem. Soc.* 2008, **130**, 14926.

127 Li, M.; Hua, R. *Tetrahedron Lett.* 2009, **50**, 1478.

128 Cho, S.H.; Hwang, S.J.; Chang, S. *J. Am. Chem. Soc.* 2008, **130**, 9254.

129 Campeau, L.-C.; Rousseaux, S.; Fagnou, K. *J. Am. Chem. Soc.* 2005, **127**, 18020.

130 Andersson, H.; Almqvist, F.; Olsson, R. *Org. Lett.* 2007, **9**, 1335.

131 (a) Furukawa, N.; Shibutani, T.; Fujihara, H. *Tetrahedron Lett.* 1987, **28**, 5845. (b) Berillon, L.; Lepretre, A.; Turck, A.; Ple, N.; Queguiner, G.; Cahiez, G.; Knochel, P. *Synlett* 1998, 1359. (c) Trecourt, F.; Breton, G.; Mongin, F.; Marsais, F.; Queguiner, G. *Tetrahedron Lett.* 1999, **40**, 4339. (d) Abardi, M.; Dehmel, F.; Knochel, P. *Tetrahedron Lett.* 1999, **40**, 7449. (e) Abarbri, M.; Thibonnet, J.; Berillon, L.; Dehmel, F.; Rottlander, M.; Knochel, P. *J. Org. Chem.* 2000, **65**, 4618. (f) Trecourt, F.; Breton, G.; Bonnet, V.; Mongin, F.; Marsais, F.; Queguiner, G. *Tetrahedron* 2000, **56**, 1349. (g) Dumouchel, S.; Mongin, F.; Trecourt, F.; Queguiner, G.; *Tetrahedron Lett.* 2003, **44**, 3887.

132 Simkovsky, N.M.; Ermann, M.; Roberts, S.M.; Parry, D.M.; Baxter, A.D. *J. Chem. Soc. Perkin Trans.* **1** 2002, 1847.

133 (a) Font-Sanchis, E.; Cespedes-Guirao, F.J.; Sastre-Santos, A.; Fernandez-Lazaro, F. *J. Org. Chem.* 2007, **72**, 3589. (b) Lee, P.H.; Seomoon, D.; Lee, K. *Org. Lett.* 2005, **7**, 343. (c) Chen, Y.-H.; Knochel, P. *Angew. Chem. Int. Ed.* 2008, **47**, 7648.

134 (a) Molander, G.A.; Canturk, B.; Kennedy, L.E. *J. Org. Chem.* 2009, **74**, 973. (b) Billingsley, K.; Buchwald, S.L. *J. Am. Chem. Soc.* 2007, **129**, 3358. (c) Thompson, A.E.; Hughes, G.; Batsanov, A.S.; Bryce, M.R.; Parry, P.R.; Tarbit, B. *J. Org. Chem.* 2005, **70**, 388. (d) Li, W.; Nelson, D.P.; Jensen, M.S.; Hoerrner, R.S.; Cai, D.; Larsen, R.D.; Reider, P.J. *J. Org. Chem.* 2002, **67**, 5394. (e) Cai, D.; Larsen, R.D.; Reider, P.J. *Tetrahedron Lett.* 2002, **43**, 4285.

135 For the side reaction of lithiation of 3-halopyridines, see:(a) Mallet, M.; Branger, G.; Marsais, F.; Queguiner, G. *J. Organomet. Chem.* 1990, **382**, 319 and also see; Ref. 131c, f.

136 (a) Sakamoto, T.; Kondo, Y.; Murata, N.; Yamanaka, H. *Tetrahedron* 1993, **49**, 9713. (b) Sakamoto, T.; Kondo, Y.; Murata, N.; Yamanaka, H. *Tetrahedron Lett.* 1992, **33**, 5373. (c) Pryor, L.; Kiessling, A. *Am. J. Undergrad. Res.* 2002, **1**, 25. (d) Bell, A.S.; Roberts, D.A.; Ruddock, K.S. *Synthesis* 1987, 843. (e) Sugimoto, O.; Yamada, S.; Tanji, K. *Tetrahedron Lett.* 2002, **43**, 3355.

137 Kim, S.-H.; Rieke, R.D. *Tetrahedron* 2010, **66**, 3135.

138 Rieke, R.D. *Science* 1989, **246**, 1260.

139 For Examples, (a) Rieke, R.D.; Hanson, M.V. *Tetrahedron* 1997, **53**, 1925. (b) Krasovskiy, A.; Malakhov, V.; Gavryushin, A.; Knochel, P. *Angew. Chem. Int. Ed.* 2006, **45**, 6040.

140 For review; Mitsche, U.; Bauerle, P. *J. Mater. Chem.* 2000, **10**, 1471. See also Ref. 120.

141 For a review, see: Kaes, C.; Katz, M.; Hosseini, M.W. *Chem. Rev.* 2000, **100**, 3553 and see also Ref. 122c cited therein.

142 (a) Manolikakes, G.; Schade, M.A.; Hernandez, C.M.; Mayr, H.; Knochel, P. *Org. Lett.* 2008, **10**, 2765. For scattered examples of Suzuki coupling, see: (b) Parry, P.A.; Wang, C.; Batsanov, A.S.; Bryce, M.R.; Trabit, B. *J. Org. Chem.* 2002, **67**, 7541. (c) Jiang, B.; Wang, Q.-F.; Yang, C.-G.; Xu, M. *Tetrahedron Lett.* 2001, **42**, 4083. (d) For an example of the coupling reaction

of 2-pyridylzinc bromide with bromoaromatic amine, see Charifson, P.S.; Grillot, A.-L.; Grossman, T.H.; Parsons, J.D.; Badia, M.; Bellon, S.; Deininger, D.D.; Drumm, J.E.; Gross, C.H.; LeTiran, A.; Liao, Y.; Mani, N.; Nicolau, D.P.; Perola, E.; Ronkin, S.; Shannon, D.; Swenson, L.L.; Tang, Q.; Tessier, P.R.; Tian, S.-K.; Trudeau, M.; Wang, T.; Wei, Y.; Zhang, H.; Stamos, D. *J. Med. Chem.* 2005, **51**, 5243.

143 Manolikakes, G.; Hernandez, C.M.; Schade, M.A.; Metzger, A.; Knochel, P. *J. Org. Chem.* 2008, **73**, 8422 Slow cannulation (over 90 min) of organozinc (2.4 mmol) via syringe pump was required.

144 pK_a values; range between 15 and 20 for phenols and 20–30 for anilines, source from http://www.chem.wisec.edu/areas/reich/pkatable/index.htm

145 (a) Eweiss, N.F.; Katritzky, A.R.; Nie, P.L.; Ramsden, C.R. *Synthesis* 1997, 634. (b) Boyer, J.H.; McCane, D.I.; McCarville, W.J.; Tweedie, A.T. *J. Am. Chem. Soc.* 1953, **75**, 5298.

146 (a) Kato, Y.; Okada, S.; Tomimoto, K.; Mase, T. *Tetrahedron Lett.* 2001, **42**, 4849. (b) Sugimoto, O.; Mori, M.; Tanji, K.-I. *Tetrahedron Lett.* 1999, **40**, 7477.

147 Kim, S.-H.; Rieke, R.D. *Tetrahedron Lett.* 2009, **50**, 5329.

148 (a) Mahatsekake, C.; Catel, J.M.; Andrieu, C.G.; Ebel, M.; Mollier, Y.; Tourillon, G. *Phosphorus Sulfur Silicon Relat. Elem.* 1990, **47**, 35. (b) Tarhouni, R.; Kirschleger, B.; Rambaud, M.; Villieras, J. *Tetrahedron Lett.* 1984, **25**, 835. (c) Reutrakul, V.; Kanghae, W. *Tetrahedron Lett.* 1977, **14**, 1225.

149 Lee, J.G.; Ha, D.S. *Tetrahedron Lett.* 1989, **30**, 193.

150 Kimpe, N.D.; Cock, W.D.; Schamp, N. *Synthesis* 1987, 188.

151 Rieke, R.D.; Brown, J.D.; Wu, X. *Synth. Commun.* 1995, **25**, 3923.

152 Knochel, P.; Yeh, M.L.P.; Berk, S.C.; Talbert, J. *J. Org. Chem.* 1988, **53**, 2390.

153 Kroemer, J.; Kirkpatrick, C.; Moride, B.; Gawrych, R.; Mosher, M.D.; Kaufman, D. *Tetrahedron Lett.* 2006, **47**, 6339.

154 Furstner, A. *Angew. Chem. Int. Ed. Engl.* 1993, **32**, 164.

155 Chou, W.-N.; Clark, D.L.; White, J.B. *Tetrahedron Lett.* 1991, **32**, 299.

156 Olah, G.A.; Molnar, A. *Hydrocarbon Chemistry*; John Wiley & Sons, Inc.: New York, 1995, 472.

157 House, H.O. *Modern Synthetic Reactions*; 2nd ed.; W.A. Benjamin: Menlo Park, 1972, 205.

158 Kaufman, D.; Johnson, E.; Mosher, M.D. *Tetrahedron Lett.* 2005, **46**, 5613.

159 See, for instance; Durant, A.; Delplancke, J.L.; Libert, V.; Reisee, J. *Eur. J. Org. Chem.* 1999, **11**, 2845.

160 Grignard, V. *Compt. Rend.* 1900, **130**, 1322.

161 Ziegler, K.; Colonius, H. *Liebig's Ann. Chem.* 1930, **479**, 135.

162 (a) Wakefield, B.J. In *Comprehensive Organic Chemistry*; Barton, D.; Ollis, W.D., Eds.; Pergamon Press: New York, 1979; Vol. **3**, pp 944, 974. (b) Wardell, J.L.; Lindsell, W.E. In *Comprehensive Organometallic Chemistry*; Wilkinson, G.; Stone, F.G.A.; Abel, E.W., Eds.; Pergamon Press: New York, 1982; Vol. **1**, pp 52 and 156, respectively.

163 (a) Walborsky, H.M.; Hamdouchi, C. *J. Am. Chem. Soc.* 1993, **115**, 6406. (b) Walborsky, H.M. *Acc. Chem. Res.* 1990, **23**, 286. (c) Walborsky, H.M.; Zimmermann, C. *J. Am. Chem. Soc.* 1992, **114**, 4996. (d) Garst, J.F.; Ungvary, F.; Baxter, J.T. *J. Am. Chem. Soc.* 1997, **119**, 253. (e) Garst, J.F.; Deutch, J.E.; Whitesides, G.M. *J. Am. Chem. Soc.* 1986, **108**, 2490.

164 (a) Blomberg, C. In *The Barbier Reaction and Related One-Step Processes*; Springer-Verlag: Berlin/Heidelberg, 1993; pp 146, 150. (b) References 170a, b. (c) Walborsky, H.M.; Ollman, J.; Hamdouchi, C.; Topolski, M. *Tetrahedron Lett.* 1992, **33**, 761 and references therein.

165 (a) Rogers, H.R.; Hill, C.L.; Fugiwara, Y.; Rogers, R.J.; Mitchell, H.L.; Whitesides, G.M. *J. Am. Chem. Soc.* 1980, **102**, 217. (b) Rogers, H.R.; Deutch, J.; Whitesides, G.M. *J. Am. Chem. Soc.* 1980, **102**, 226. (c) Rogers, H.R.; Rogers, R.J.; Mitchell, H.L.; Whitesides, G.M. *J. Am. Chem. Soc.* 1980, **102**, 231. (d) Barber, J.J.; Whitesides, G.M. *J. Am. Chem. Soc.* 1980, **102**, 239.

166 (a) Curran, D.P.; Lin, C.-H.; DeMello, N.; Junggebauer, J. *J. Am. Chem. Soc.* 1998, **120**, 342. (b) Sibi, M.P.; Ji, J. *Angew. Chem. Int. Ed. Engl.* 1996, **35**, 190. (c) Curran, D.P.; Qi, H.; DeMello, N.C.; Lin, C.-H. *J. Am. Chem. Soc.* 1994, **116**, 8430. (d) Beckwith, A.L. *J. Chem. Soc. Rev.* 1993, **22**, 143. (e) Hoffmann, H.M.R. Angew, Chem. *Int. Ed. Engl.* 1992, **31**, 1332.

167 Frankland, E. *Liebigs Ann. Chem.* 1849, **71**, 171.

168 For a recent review, see: (a) Knochel, P.; Almena-Perea, J.J.; Jones, P. *Tetrahedron* 1998, **54**, 8275. For some articles, see: (b) Rieke, R.D.; Hanson, M.V.; Brown, J.D. *J. Org. Chem.* 1996, **61**, 2726. (c) Chen, T.-A.; Wu, X.; Rieke, R.D. *J. Am. Chem. Soc.* 1995, **117**, 233. (d) Hanson, M.D.; Rieke, R.D. *J. Am. Chem. Soc.* 1995, **117**, 10775.

169 (a) Thorn, S.N.; Gallagher, T. *Synlett* 1996, 185. (b) Lipshutz, B.H.; Wood, M.R.; Tirado, R. *J. Am Chem. Soc.* 1995, **117**, 6126. (c) Brunner, M.; Maas, G. *Synthesis* 1995, 957. (d) Knochel, P.; Chou, T.-S.; Jubert, C.; Rajagopal, J. *J. Org. Chem.* 1993, **58**, 588. (e) Reference 178a.

170 (a) Zhu, L.; Wehmeyer, R.M.; Rieke, R.D. *J. Org. Chem.* 1991, **56**, 1445. For earlier forms of Zn* with lower activity, see: (b) Rieke, R.D.; Li, P.T.; Burns, T.P.; Uhm, S.T. *J. Org. Chem.* 1981, **46**, 4323. (c) Rieke, R.D.; Uhm, S.J.; Hudnall, P.M. *J. Chem. Soc. Chem. Commun.* 1973, 269.

171 Boersma, J. In *Comprehensive Organometallic Chemistry*; Wilkinson, G., Stone, F.G.A., Abel, E.W., Eds.; Pergamon Press: New York, 1982; Vol. **2**, p 832.

172 (a) Guijarro, A.; Rosenberg, D.M.; Rieke, R.D. *J. Am. Chem. Soc.* 1999, **121**, 4155. (b) Guijarro, A.; Rieke, R.D. *Angew. Chem. Int. Ed. Engl.* 1998, **37**, 1679.

173 Stile, J.K. In *The Chemistry of Metal-Carbon Bond*; Hartley, F.R., Patai, S., Eds.; John Wiley & Sons, Ltd: Chichester, 1985; Vol. **2**, pp 629–630.

174 Yao, C.-Y. *Diss. Abstr.* 1964, **24**, 4414; *Chem. Abstr.* **1964**, *61*, 6443g.

175 Rehybridization of the carbon in the transition state releases strain in cyclopentyl derivatives: Brown, H.C. *Boranes in Organic Chemistry*; Cornell

University Press: London, 1972; pp 107, 126. In 1-adamantyl derivatives the situation is opposite, due to the rigidity of the backbone.
176 This factor is, for example, $k_{Et}/k_{neo} = 10^5$ for a S_N2 reaction: $RBr + {}^*Br^- \rightleftharpoons RBr^* + Br^-$. Isaacs, N.S. *Physical Organic Chemistry*; Longman Scientific & Technical: Harlow Essex, 1987; pp 380.
177 Intuitively attributed to some degree of nonreproducible thermal decomposition through the stationary phase.
178 (a) Casanova, J.; Eberson, L. In *The Chemistry of the Carbon-Halogen Bond*; Patai, S., Ed.; John Wiley & Sons, Inc.: London/New York, 1973; Part 2, pp 979–1047. (b) Meites, L.; Zuman, P. *CRC Handbook Series in Organic Electrochemistry*; CRC Press: Cleveland, 1977; Vol. **I**.
179 Chock, P.B.; Halpern, J. *J. Am. Chem. Soc.* 1969, **91**, 582.
180 (a) Kochi, J.K.; Powers, J.W. *J. Am. Chem. Soc.* 1970, **92**, 137. (b) Bakac, A.; Butkovic, V.; Espenson, J.H.; Orhanovic, M. *Inorg. Chem.* 1993, **32**, 5886.
181 Present book chapter 3.18
182 Finke, R.G.; Schiraldi, D.A.; Hirose, Y. *J. Am. Chem. Soc.* 1981, **103**, 1876.
183 Kuivila, H.G. *Adv. Organomet. Chem.* 1964, **1**, 78.
184 Lund, T.; Lund, H. *Acta Chem. Scand.* 1986, **B40**, 470.
185 From Ref. 179. Original paper: Fox, R.J.; Evans, F.W.; Szwarc, M. *Trans. Faraday Soc.* 1961, **57**, 1915.
186 Andrieux, C.P.; Gallardo, I.; Saveant, J.-M.; Su, K.-B. *J. Am. Chem. Soc.* 1986, **108**, 638.
187 Mukaiyama, T.; Hogo, K. *Chem. Lett.* 1976, 619.
188 Parker, D.; Taylor, R.J. *Tetrahedron* 1987, **43**, 5451.
189 (a) Dale, J.A.; Dull, D.L.; Mosher, H.S. *J. Org. Chem.* 1969, **34**, 2543. (b) Sullivan, G.R.; Dale, J.A.; Mosher, H.S. *J. Org. Chem.* 1973, **38**, 2143.
190 Radicals on the carbon have been proposed in the reaction of Grignard reagents with oxygen: Lamb, R.C.; Ayers, P.W.; Toney, M.K.; Garst, J.F. *J. Am. Chem. Soc.* 1966, **88**, 4261.
191 Evans, C.C.; Sudgen, S. *J. Chem. Soc.* 1949, 270.
192 Micounin, L.; Oestreich, M.; Knochel, P. *Angew. Chem. Int. Ed. Engl.* 1997, **36**, 245.
193 (a) The participation of radicals in the reaction of zinc with alkyl iodides has been suggested: Luche, J.L.; Allavena, C.; Petrier, C.; Dupuy, C. *Tetrahedron Lett.* 1988, **29**, 5373. (b) Partial loss of the stereochemical integrity was reported for the reaction of zinc with diasteromerical secondary iodides followed by deuteration: Turecek, F.; Veres, K.; Kocovsky, P.; Pouzar, V.; Fajkos, J. *J. Org. Chem.* 1983, **48**, 2233. NaI S_N2 fast exchange and "retention of the configuration in the reduction step" were then proposed. However, substrates nonsusceptible to S_N2 equilibration such as *exo*- and *endo*-7-iodonorcarane were found to react with complete loss of their stereochemical integrity: Duddu, R.; Eckhardt, M.; Furlong, M.; Knoess, P.; Berger, S.; and Knochel, P. *Tetrahedron* **1994**, *50*, 2415. This and the evidence

accumulated in this work point to a radical-induced loss of stereochemistry, rather than retention in the reduction step.
194 Eberson, L. In *Electron Transfer Reactions in Organic Chemistry*; Springer-Verlag: Berlin/Heidelberg, 1987; Chapter V.
195 (a) Brace, N.O.; Van Elswyk, J.E. *J. Org. Chem.* 1976, **41**, 766. (b) Samat, A.; Vacher, B.; Chanon, M. *J. Org. Chem.* 1991, **56**, 3524.
196 (a) Meyer, C.; Marek, I., Courtemanche, G.; Normant, J.-F. *Synlett* 1993, 266. (b) Meyer, C.; Marek, I.; Courtemanche, G.; Normant, J.-F. *Tetrahedron Lett.* 1993, **34**, 6053.
197 (a) Griller, D.; Ingold, K.U. *Acc. Chem. Res.* 1980, **13**, 317. (b) Bowry, V.W.; Lusztyk, J.; Ingold, K.U. *J. Am. Chem. Soc.* 1991, **113**, 5687.
198 No cyclopentanone could be detected by GLC analysis of the hydrolyzed (2 M HCI) crude reaction mixture.
199 (a) Very unstable Zn(I) complexes have been produced by the pulse radiolysis of solutions of Zn(II) complexes: Weddel, J.K.; Allred, A.L.; Meyerstein, D. *J. Inorg. Nucl. Chem.* 1980, **42**, 219. (b) Buxton, G.V.; Sellers, R.M. *Coord. Chem. Rev.* 1977, **22**, 195.
200 Collmann, J.P.; Hegedus, L.S. *Principles and Applications of Organotransition Metal Chemistry*; University Science Books: Mill Valley, 1980; Chapter 4.
201 See, for example: Schrauzer, G.N.; Deutsch, E. *J. Am. Chem. Soc.* 1969, **91**, 3341.
202 Alignment of the C—X dipole to the potential field gradient near the surface might induce ionization of the bond. This mechanism, suggested in electrochemical reductions, was later discarded: Ref. 178a, p 1016.
203 Bridgehead bicyclic halides, among others, would be expected to be far less reactive: Ref. 178a, p 1017.
204 (a) Sastry, G.N.; Danovich, D.; Shaik, S. *Angew. Chem. Int. Ed. Engl.* 1996, **35**, 1098. (b) Fawcett, W.R.; Opallo, M. *Angew. Chem., Int. Ed. Engl.* 1994, **33**, 2131.
205 (a) Bertran, J.; Gallardo, I.; Moreno, M.; Savēant, J.-M. *J. Am. Chem. Soc.* 1992, **114**, 9576. (b) Savēant, J.-M. *Acc. Chem. Res.* 1993, **26**, 455. (c) Reference 194, Chapter viii. (d) Grimshaw, J.; Langan, J.R.; Salmon, G.A. *J. Chem. Soc. Faraday Trans. 1* 1994, **90**, 75.
206 (a) Andrieux, C.P.; Blocman, C.; Dumas-Bouchiat, J.-M.; M'Halla, F.; Saveant, J.-M. *J. Am. Chem. Soc.* 1980, **102**, 3806. (b) Andrieux, C.P.; Blocman, C.; Dumas-Bouchiat, J.-M.; Savēant, J.-M. *J. Am. Chem. Soc.* 1979, **101**, 3432. (c) Reference 194, p 122. (d) Pierini, A.B.; Duca, J.S., Jr. *J. Chem. Soc. Perkin Trans.* **2** 1995, 1821.
207 Eberson, L. *New J. Chem.* 1992, **16**, 151.
208 (a) Cannon, R.D. *Electron Transfer Reactions*; Butterworth: London, 1980; Chapter 5.
209 (a) Marcus, R.A. *Pure App. Chem.* 1997, **69**, 13. (b) Sutin, N. *Prog. Inorg. Chem.* 1983, **30**, 441.

210 For example, the Marcus theory for outer-sphere ET was applied to alkyl halide reductions by inner- and outer-sphere reagents with reasonable success: Eberson, L. *Acta Chem. Scand.* 1982, **B36**, 533.

211 For electrochemical reduction of alkyl bromides, $\Delta E_p = E_p(\text{RBr}) - E_p(n\text{-RBr})$.

212 Kochi, J.K. *Organometallic Mechanisms and Catalysis*; Academic Press: New York, 1978; pp 139–146, 153.

213 (a) Neumann, W.P. *The Organic Chemistry of Tin*; Interscience: London/New York, 1970; pp 114–115. (b) Kuivila, H.G. *Acc. Chem. Res.* 1968, **1**, 299.

214 (a) Polar resonant structures [$RBr^- + SnBu_3^{\cdot}$] were used to describe the transition state: Ref. 213. (b) An outer-sphere ET cannot be totally ruled out: Ref. 165d.

215 The measure of the disappearance of a pair of halides in a competitive process, however, yields the ratio of rate constants of halogen abstraction by $Bu_3Sn\cdot$: (a) Carlsson, D.J.; Ingold, K.U. *J. Am. Chem. Soc.* 1968, **90**, 7047. (b) Reference 165d. This is true provided that the halogen transfer, Bu3Sn. $Bu_3Sn\cdot + RBr \rightarrow Bu_3SnBr + R^{\cdot}$, is an irreversible step. An equilibrium seems reasonable for this step. Indeed, ΔH is only slightly exothermic in these halogen transfers: Nash, G.A.; Skinner, H.A.; Stack, W.F. *Trans. Faraday Soc.* 1965, **61**, 640, 2122. This would explain why this reaction is abnormally positioned toward the right in Figure 3.2.

216 (a) Garst, J.F.; Smith, C.D. *J. Am. Chem. Soc.* 1976, **98**, 1520. (b) Holy, N.L. *Chem. Rev.* 1974, **74**, 243.

217 Lund, H.; Daasbjerg, K.; Lund, T.; Occhiliani, D.; Pederson, S.U. *Acta Chem. Scand.* 1997, **51**, 135. Reaction of RBr with $Ar_2CO^-Li^+$ might involve inner-sphere transfer through bridging halogen and metal ion: (Ar_2CO...Li...BrR). This was suggested in Ref. 165d and is now reinforced with the previous observation.

218 Eberson, L. *Acta Chem. Scand.* 1982, **B36**, 533.

219 (a) Andrieux, C.P.; Gallardo, I.; Savēant, J.-M. *J. Am. Chem. Soc.* 1989, **111**, 1620. (b) Reference 186.

220 Reference 194, Chapter vii.

221 Reynolds, W.L.; Lumry, R.W. *Mechanisms of Electron Transfer*; Ronald Press: New York, 1966.

222 "The formations of Grignard reagents and organolithium compounds are examples of the reduction of alkyl halides directly by metals...It is not clear whether *inner-* or *outer-sphere* processes are involved since the reactions occurring on a metal surface offer only a limited number of probes": Kochi, J.K. *Free Radicals*; John Wiley & Sons, Inc.: New York, 1973; 663.

223 The rate-structure profiles for the reaction of Mg with alkyl chlorides, Bu_3SnH with alkyl chlorides, $Ar_2Co\text{-}Li+$ with alkyl bromides, and $E_{1/2}$ (C, alkyl chlorides) are all linearly correlated: Ref 165d. It was also pointed that in better defined inorganic ET reactions, inner- and outer-sphere processes may exhibit linearly correlated rate-structure profiles. Linearly correlated

reactions can have different selectivities, however, the slope of the log k_{rel} plot for both reactions being a measure of the relative selectivity.
224 (a) Prince, R.H. In *Comprehensive Coordination Chemistry*; Wilkinson, G., Ed.; Pergamon Press: Oxford/New York, 1987; Vol. **5**, pp 983–985. (b) For a conductrimetric study of the equilibrium $ZnCl_2 + Cl^- \rightleftharpoons ZnCl_3^-$ in THF: Folest, J.-C.; Troupel, M.; Perichon, J. *Bull. Soc. Chim.* 1980, I–181.
225 σ values; (a) Hine, J. *Physical Organic Chemistry*; McGraw-Hill: New York, 1962. (b) Isaacs, N.S. *Physical Organic Chemistry*; Longman Scientific & Technical: Harlow, Essex, 1987.
226 Calculated from kinetic data in Ref. 165c.
227 Dannen, W.C.; Saunders, D.G. *J. Am. Chem. Soc.* 1969, **91**, 5924.
228 Sease, J.W.; Burton, F.G.; Nickol, S.L. *J. Am. Chem. Soc.* 1968, **90**, 2595.
229 Tsou, T.T.; Kochi, J.K. *J. Am. Chem. Soc.* 1979, **101**, 6319.
230 Exner, O. *Correlation Analysis of Chemical Data*; Plenum Press: New York, 1988; pp 72–73.
231 Chipperfield, J.R. In *Advances in Linear Free Energy Relationships*; Chapman, N.B., Shorter, J., Eds.; Plenum Press: London/New York, 1972; p 340.
232 Studies of tin hydride reductions of bridgehead aliphatic halides have been interpreted as indicating little or no charge separation in the transition state: Fort, R.C., Jr.; Hiti, J. *J. Org. Chem.* 1977, **42**, 3968. However, positive slope of the Hammett plot $\rho = 1.4$ points toward some participation of ET with partial negative charge generation at the transition state.
233 Electrochemical reductions have been referred to as "compulsory ET processes," and for inert smooth Pt and some C, electrodes are assumed to be outer-sphere ET. For a dropping Hg electrode, which is the case considered here, the situation has been discussed: Ref. 194, p 79.
234 With some restrictions, see Table 3.53, footnote g.
235 Compare with $ArH + NH_2^-/NH_{3(1)} \rightleftharpoons [Ar^-] \rightarrow ....$ ($\rho \approx 6$): Shatenshtein, A.I. *Adv. Phys. Org. Chem.* 1963, **1**, 155, and reaction in Table 3.53, entry 15.
236 Depending on the nature of the substrate, for example, S_N2 for primary bromides and ET for hindered systems: Zhou, D.-L.; Walder, P.; Scheffold, R.; Walder, L. *Helv. Chim. Acta.* 1992, **75**, 995.
237 Truce, W.E.; Lindy, L.B. *J. Org. Chem.* 1961, **26**, 1463. It was also prepared by reduction of γ-phenylbutyrolactone with $LiA1H_4$ and bromination using the procedure described for **12** (68%).
238 Grinberg, S.; Shaubi, E. *Tetrahedron* 1991, **47**, 2895. The crude reaction mixture was distilled under vacuum, and an intermediate fraction (61% *p*-isomer/39% *o*-isomer) was used in the reaction with active zinc.
239 For reviews, see: (a) Patil, A.O.; Heeger, A.J.; Wudl, F. *Chem. Rev.* 1988, **88**, 183. (b) Roncali, J. *Chem. Rev.* 1992, **92**, 711. (c) Kaner, R.B.; MacDiarmid, A.G. *Sci. Am.* 1988 (February), 106. (d) Kanatzidis, M.G. *Chem. Eng. News.* 1990 (December 3), 36–54. (e) Kovacic, P.; Jones, M.B. *Chem. Rev.* 1987, **87**,

357. (f) Baughman, R.H.; Brédas, J.L.; Chance, R.R.; Elsenbaumer, R.L.; Shacklette, L.W. *Chem. Rev.* 1982, **82**, 209. (g) Bredas, J.L.; Street, G.B. *Acc. Chem Res.* 1985, **18**, 309. (h) Skotheim, T.A. *Handbook of Conducting Polymers*; Marcel Dekker: New York, 1986. (i) Sandman, D.J. *Trends Polym. Sci.* 1994, **2**, 44. (j) Clery, D. *Science* 1994, **263**, 1700.

240 (a) Scrosati, B. *Applications of Electroactive Polymers*; Chapman & Hall: London, 1994. (b) Greenham, N.C.; Morratt, S.C.; Bradley, D.D.C.; Friend, R.H.; Holmes, A.B. *Nature* 1993, **365**, 628. (c) Gustafsson, G.; Cao, Y.; Treacy, G.M.; Klavetter, F.; Colaneri, N.; Heeger, A.J. *Nature* 1992, **357**, 477. (d) McCoy, C.H.; Wringhton, M.S. *Chem. Mater.* 1993, **5**, 914. (e) Garnier, F.; Yassar, A.; Hajlaoui, R.; Horowitz, G.; Deloffre, F.; Servet, B.; Ries, S.; Alnot, P. *J. Am. Chem. Soc.* 1993, **115**, 8716. (f) Prasad, P.N.; Williams, D.J. *Introduction to Nonlinear Optical Effects in Molecules & Polymers*; John Wiley & Sons, Inc.: New York, 1991. (g) Robitaille, L.; Leclerc, M.; Callender, C.L. *Chem. Mater.* 1993, **5**, 1755. (h) Chittibabu, K.G.; Li, L.; Kamath, M.; Kumar, J.; Tripathy, S.K. *Chem. Mater.* 1994, **6**, 475.

241 (a) Yamamoto, T.; Sanechika, K.; Yamamoto, A.J. *Polym. Sci. Polym. Lett. Ed.* 1980, **18**, 9. (b) Lin, J.W.-P.; Dudek, L.P. *J. Polym. Sci. Polym. Chem. Ed.* 1980, **18**, 2869.

242 (a) Jen, K.-Y.; Miller, G.G.; Elsenbaumer, R.L. *J. Chem. Soc. Chem. Commun.* 1986, 1346. (b) Elsenbaumer, R.L.; Jen, K.-Y.; Oboodi, R. *Synth. Met.* 1986, **15**, 169.

243 (a) Sato, M.; Tanka, S.; Kaeriyama, K. *J. Chem. Soc. Chem. Commun.* 1986, 873. (b) Lemaire, M.; Roncali, J.; Garnier, F.; Garreau, R.; Hannecart, E. French Patent 86.04744, April 4, 1986.

244 (a) Yoshino, K.; Nakajima, S.; Onoda, M.; Sugimoto, R. *Synth. Met.* 1989, **28**, C349. (b) Osterholm, J.E.; Laakso, J.; Nyholm, P.; Isotalo, H.; Stubb, H.; Inganäs, O.; Salaneck, W.R. *Synth. Met.* 1989, **28**, C435.

245 Yamamoto, T.; Morita, A.; Miyazaki, Y.; Maruyama, T.; Wakayama, H.; Zhou, Z.-H.; Nakamura, Y.; Kanbara, T.; Sasaki, S.; Kubota, K. *Macromolecules* 1992, **25**, 1214.

246 (a) Ito, T.; Shirakawa, H.; Ikeda, S. *J. Polym. Sci. Polym. Chem. Ed.* 1974, **12**, 11. (b) Ito, T.; Shirakawa, H.; Ikeda, S. *J. Polym. Sci. Polym. Chem. Ed.* 1975, **13**, 1943. (c) Shirakawa, H.; Louis, B.J.; MacDiarmid, A.G.; Chiang, C.K.; Heeger, A.J. *J. Chem. Soc. Chem. Commun.* 1977, 578. (d) Shirakawa, H.; Zhang, Y.-X.; Akagi, K. *Polym. Prepr.* 1994, **35**, 189.

247 Basescu, N.; Liu, Z.-X.; Moses, D.; Heeger, A.J.; Naarmann, H.; Theophilou, N. *Nature* 1987, **327**, 403.

248 (a) Roncali, J.; Carreau, R.; Yassar, A.; Marque, P.; Garnier, F.; Lemaire, M. *J. Phys. Chem.* 1987, **91**, 6706. (b) Roncali, J.; Marque, P.; Garreau, R.; Garnier, F.; Lemaire, M. *Macromolecules* 1990, **23**, 1347.

249 Souto Maior, R.M.; Hinkelmann, K.; Eckert, H.; Wudl, F. *Macromolecules* 1990, **23**, 1268.

250 Wei, Y.; Chan, C.-C.; Tian, J.; Jang, G.-W.; Hsueh, K.F. *Chem. Mater.* 1991, **3**, 888.

251 (a) Mao, H.; Holdcroft, S. *Macromolecules* 1992, **25**, 554. (b) Mao, H.; Xu, B.; Holdcroft, S. *Macromolecules* 1993, **26**, 1163, 4457.

252 (a) Tour, J.M.; Wu, R.; Schumm, J.S. *J. Am. Chem. Soc.* 1991, **113**, 7064. (b) Pearson, D.L.; Schumm, J.S.; Tour, J.M. *Macromolecules* 1994, **27**, 2348. (c) Tour, J.M.; Wu, R. *Macromolecules* 1992, **25**, 1901.

253 (a) McCullough, R.D.; Lowe, R.D. *J. Chem. Soc. Chem. Commun.* 1992, 70. (b) McCullough, R.D.; Lowe, R.D. Jayaraman, M.; Anderson, D.L. *J. Org. Chem.* 1993, **58**, 904. (c) McCullough, R.D.; Tristram-Nagle, S.; Williams, S.P.; Lowe, R.D.; Jayarman, M. *J. Am. Chem. Soc.* 1993, **115**, 4910. (d) McCullough, R.D.; Williams, S.P. *J. Am. Chem. Soc.* 1993, **115**, 11608.

254 (a) Chen, T.-A.; Rieke, R.D. *J. Am. Chem. Soc.* 1992, **114**, 10087. (b) Chen, T.-A.; Rieke, R.D. *Synth. Met.* 1993, **60**, 175. (c) Chen, T.-A.; O'Brien, R.A.; Rieke, R.D. *Macromolecules* 1993, **26**, 3462. (d) Chen, T.-A.; Wu, X.; Rieke, R.D. *J. Am. Chem. Soc.* 1995, **117**, 233.

255 (a) Zhu, L.; Wehmeyer, R.M.; Rieke, R.D. *J. Org. Chem.* 1991, **56**, 1445. (b) Zhu, L.; Rieke, R.D. *Tetrahedron Lett.* 1991, **32**, 2865.

256 For a review of catalyzed cross-coupling reactions of organozinc compounds, see: (a) Negishi, E.; King, A.O.; Okukado, N. *J. Org. Chem.* 1977, **42**, 1821. (b) Negishi, E.; *Organometallics in Organic Synthesis*; John Wiley & Sons, Inc.: New York, 1980; Vol. **1**. (c) Negishi, E. *Acc. Chem. Res.* 1982, **15**, 340.

257 (a) Kellogg, R.M.; Schaap, A.P.; Harper, E.T.; Wynberg, H. *J. Org. Chem.* 1968, **33**, 2902. (b) Fournari, P.; Guilard, R.; Person, M. *Bull. Soc. Chim. Fr.* 1967, **11**, 4115.

258 Barker, J.M.; Huddleston, P.R.; Wood, M.L. *Synth. Commun.* 1975, **5**, 59.

259 (a) Tamao, K.; Kodama, S.; Nakajima, I.; Kumada, M.; Minato, A.; Suzuki, K. *Tetrahedron* 1982, **38**, 3347. (b) Pham, C.V.; Mark, H.B., Jr.; Zimmer, H. *Synth. Commun.* 1986, **16**, 689.

260 For a general review of the subject, see: Bumagin, N.A.; Beletskaya, I.P. *Russ. Chem. Rev.* 1990, **59**, 1174 (translated from *Uspekhi Khimii* **1990**, *59*, 2003) and references cited therein.

261 (a) Yamamoto, A.; Yamamoto, T.; Ozawa, F. *Pure Appl. Chem.* 1985, **57**, 1799. (b) Ozawa, F.; Hidaka, T.; Yamamoto, T.; Yamamoto, A. *J. Organomet. Chem.* 1987, **330**, 253. (c) Yamamoto, T.; Wakabayashi, S.; Osakada, K. *J. Organomet. Chem.* 1992, **428**, 223.

262 (a) Sato, M.-A.; Morii, H. *Macromolecules* 1991, **24**, 1196–120. (b) Sato, M.-A.; Morii, H. *Polym. Commun.* 1991, **32**, 42.

263 Barbarella, G.; Bongini, A.; Zambianchi, M. *Macromolecules* 1994, **27**, 3039.

264 Ekeblad, P.O.; Inganäs, O. *Polym. Commun.* 1991, **32**, 436.

4

Magnesium

4.1 General Background and Mechanistic Details of Grignard Reaction

Among the most powerful tools in the synthetic organic chemists' arsenal are organometallic compounds and metal-mediated transformations. The literature involving organometallic intermediates has literally exploded in the past 30 years, and the rate of publication in this area continues to accelerate. While many transformations involve heterogeneous conditions (metal–solution interface), the majority of these reactions involve discrete organometallic species.

In most cases, each organometallic reagent has its own characteristic chemo-, regio-, or stereoselectivity. Many times the desired organometallic reagent is prepared by a chemical transformation of one organometallic reagent into the desired reagent. The final organometallic reagent will, however, only be able to contain functional groups that would tolerate all of these manipulations. The final organometallic reagent often would be able to tolerate a number of functional groups. Hence, any approach that could directly prepare this final reagent would be quite useful. One such approach would be the direct reaction of an organic molecule with a metal. An example is the reaction of an organic halide with magnesium to generate a Grignard reagent. The reaction

$$RX + Mg \rightarrow RMgX$$

is a classic example of an oxidative addition reaction; however, this reagent will tolerate only a very limited number of functional groups (e.g., ketones, esters, aldehydes, nitriles, and epoxides will react with Grignard reagents). Grignard reagents are extremely useful because of their ability to add to a variety of functional groups forming carbon–carbon bonds. One other primary use of Grignard as well as organolithium reagents is the generation of other organometallic reagents by a metathesis reaction. For the majority of the metals shown here,

Chemical Synthesis Using Highly Reactive Metals, First Edition. Reuben D. Rieke.

$$RMgX(RLi) + MX_2 \rightarrow RMX + MgX_2(LiX)$$

$$M = Cd, Zn, Ni, Pt, Fe, In, \; Co, Cr, Mo, W, Cu, Pd, Tl$$

it would not be possible to prepare the corresponding organometallic species by the direct reaction of the metal with an organic halide due to the low reactivity of the metal. Accordingly, since Frankland [1] first reported the oxidative addition of zinc to alkyl iodides back in 1849, chemists have been searching for ways to enhance the direct reaction of zerovalent metals with organic substrates. Early approaches generally included grinding or milling the metal to a finely divided state in an inert atmosphere; addition of a catalytic amount of I_2, which is highly successful with magnesium; entrainment methods, which make use of the slow addition of highly reactive organic substrates such as CH_2BrCH_2Br throughout the reaction; use of polar solvents; high-temperature sealed tube reactions; or addition of inorganic salts [2]. These approaches were developed in many cases simply by trial and error. A logical approach to increase the reactivity of organic molecules at a metal surface would appear to be found in an understanding of the detailed mechanism of the reaction such as the oxidative addition reaction. Unfortunately, for the vast majority of metals mentioned in the preceding text, few if any data are available. The only reaction that has been studied in detail is that of organic halides at a magnesium surface (the Grignard reaction and the corresponding reaction with zinc (see Chapter 3)). Whitesides and coworkers [3] have carried out a series of elegant studies from which several significant factors were delineated. The rate of reaction with organic iodides and bromides is proportional to the organic halide concentration and to the magnesium surface area, which suggests that the reaction occurs at the metal–solution interface. Also, the data suggested that the most plausible rate-determining step was electron transfer from the metal to the organic halide. As the lifetime of the resulting radical anion would be expected to be very short, this is also tantamount to carbon–halogen bond cleavage. Lack of steric effects on rates also suggested that preadsorption of the alkyl halide onto the magnesium surface seems improbable. Other studies with single crystals have attempted to answer the question of whether the electron transfer is inner or outer sphere [4]; however, no definitive conclusion was reached. Whitesides and coworkers also argued that the low susceptibility of Grignard formation to poisoning argues against competitive adsorption by other solution components. However, Burns and Rieke [5] demonstrated that when highly reactive magnesium is used, the Grignard formation can be totally inhibited at low temperatures (−78°C) by the presence of certain Lewis bases such as alkyl nitriles. This result would strongly support a mechanism in which preadsorption of the organic halide occurs first, followed by the rate-limiting electron transfer. This was supported by the observation that, for several reactive halides that contained functional groups with lone-pair electrons or π-clouds, no reaction was observed at −78°C, suggesting that adsorption of the Lewis bases on the metal surface prevented oxidative addition [5].

Another controversial aspect of the mechanism of Grignard formation is the fate of the organic radicals formed after the rate-limiting single-electron transfer from the magnesium surface. There is in general agreement that the formation of Grignard reagents involves the intermediacy of free radicals [6–8]. Whether these radicals are adsorbed on the magnesium surface, exist free in the solution close to the magnesium surface, or are involved in an equilibrium between these two species is not entirely settled. An analytical model has been presented by Garst et al. [9], which assumes that the radicals "diffuse freely in solution at all times." Ashby and Oswald [10] have presented data indicating that ~25% of the Grignard reagent is formed from radicals that diffuse into the solvent phase and then return to the magnesium surface to receive a second electron and form the Grignard reagent. Recently, however, Walborsky and Rachon [11] have presented data demonstrating that several secondary and vinyl chiral halides lead to Grignard reagents with retention of configuration as high as 71%. As some of these model systems involve radicals with known inversion rates as high as 10^{11} s^{-1}, the results strongly support surface-adsorbed radicals at least for these species. Recently, Walborsky and Rachon [11] have reported the reaction of Rieke magnesium with (*s*)-(+)-1-bromo-methyl-2,2-diphenylcyclopropane at −65°C to yield a chiral Grignard reagent that is 33–43% optically pure.

It is interesting to note, however, that most of the systems studied supporting the "diffusion-free in solution" theory involve primary halides with no functionality next to the halogen (most of these systems have remote double bonds). Most of Walborsky's model compounds have π-clouds next to (or close to) the halogen, and hence surface adsorption may be more favored with these systems prior to or simultaneous with electron transfer.

Two primary conclusions can be drawn from these studies. First, a reaction at a metal surface can be expected to be a function of surface area. Thus, any process that increases the effective surface area should have a rate-enhancing effect. In conjunction with this is the obvious desirability of removing all passivating surface oxide coatings. The second conclusion is that for oxidative addition reactions, the rate-limiting step is single-electron transfer for magnesium, and it is likely to be the rate-limiting step for most metals.

Accordingly, any factor that can facilitate electron transfer from the metal surface to the carbon–halogen bond should accelerate oxidative addition. We have, in fact, pointed out in the 1970s that for certain metals (i.e., zinc), the addition of certain Lewis bases (i.e., amines and phosphines and also the addition of metal salts such as LiCl, LiBr, or LiI) will accelerate oxidative addition. A likely explanation is that the adsorption of the Lewis bases changes the work function of the metal atoms (lowers the work function) in the vicinity of the

Lewis base and hence reduces the barrier for electron transfer. The alkali salts can also serve as electrolytes and facilitate electron transfer at an electrode.

General Methods of Metal Activation

In recent years, three major approaches have been developed to increase the reactivity of metal powders: metal vaporization, use of ultrasound, and reduction of metal salts. An in-depth review of each approach is beyond the scope of this book, and only the reduction of metal salts is treated here in detail.

The work initiated by Skell and coworkers [12] in the early 1960s on carbine chemistry proved to have a profound influence on the chemistry of metals and metal powders. These were the first studies in which high temperatures were used to generate unstable gaseous species under vacuum inside a liquid-nitrogen-cooled reaction vessel. It was shown that highly reactive vaporized carbon species could be generated from a carbon arc. A number of novel reactions were carried out by condensing the highly reactive carbon species with organic compounds on the liquid-nitrogen-cooled walls of the vessel. It was only a matter of time before this technique was applied to main group as well as transition metals [13–19]. These studies clearly demonstrated that all metal atoms can be considered to be highly reactive. The cocondensation of metal atoms with a wide variety of organic and inorganic substrates proved to be a powerful new approach to many heretofore unknown organometallic compounds. A variation on this method developed by Klabunde [19] was the cocondensation of metal atoms with solvents such as benzene, tetrahydrofuran (THF), and hexane. Upon melting, a slurry of finely divided metal particles was obtained that exhibited high reactivity, particularly toward oxidative addition reactions. Although the metal slurries prepared in this manner were highly reactive, they were not as reactive toward oxidative addition as the metal powders prepared by reduction of metal salts (discussed in the succeeding text). The basic metal-vaporization approach was remarkably successful in the synthesis of novel arene and π-complexes but was of limited value in organic synthesis. One of the principal reasons for this is that the organic substrate must be volatile; this would prove fatal for many high-boiling organic materials. However, the technique of reacting metal atoms with ligands in cold solutions has been reported. This suggests the possibility of performing such reactions with nonvolatile organic molecules [20, 21].

The use of ultrasound to accelerate heterogeneous reactions is growing rapidly [22–25]. This technique has generated a field of study commonly labeled sonochemistry. Much of the rapid rise in popularity of sonochemistry stems from the fact that ultrasound can be delivered safely, conveniently, and with inexpensive equipment. Ultrasound can affect chemical reactions at a metal–solution interface when the sound waves induce “cavitation” or the rapid growth and sudden collapse of bubbles at the metal–solution interface.

The rapid collapse leads to intense localized pressure and temperature differentials as well as electric discharges. Principal uses have involved the generation of organolithium, organozinc, and organocopper reagents. The majority of the examples reported involved organic halides, which can be considered to be "quite reactive" toward standard activated metals. The primary advantages include reduced reaction times and temperatures, increased yields, and in some cases superior regioselectivity. Luche et al. [25] have demonstrated that less reactive alkyl and aryl bromides can be utilized by employing a metathesis step. Sonication of the organic halide, lithium, and metal salt of the desired organometallic reagent ultimately yields the desired organometallic reagent. The process involves the initial formation of an organolithium reagent. Thus, this procedure precludes the presence of most functional groups in the organic halide.

4.2 General Methods for Preparation of Rieke Magnesium

In 1972, Rieke and associates [2, 5, 26–81] reported a general approach for preparing highly reactive metal powders by reducing metal salts in ethereal or hydrocarbon solvents using alkali metals as reducing agents. Several basic approaches are possible and each has its own particular advantages. For some metals, all approaches lead to metal powders of identical reactivity. However, for some metals, one method can lead to far superior reactivity. High reactivity, for the most part, refers to oxidative addition reactions. Since our initial report, several other reduction methods have been reported including metal-graphite compounds, magnesium-anthracene complex, and dissolved alkalides [82].

Although our initial entry into this area of study involved the reduction of $MgCl_2$ with potassium biphenylide, our early work concentrated on reductions without the use of electron carriers. In this approach, reductions are conveniently carried out with an alkali metal and a solvent whose boiling point exceeds the melting point of the alkali metal. The metal salt to be reduced must also be partially soluble in the solvent, and the reductions are carried out under an argon

$$MX_n + nK \rightarrow M^* + nKX$$

atmosphere. The reductions are exothermic and generally are complete within a few hours. In addition to the metal powder, one or more moles of alkali salt are generated. Convenient systems of reducing agents and solvents include potassium and THF, sodium and 1,2-dimethoxyethane (DME), and sodium or potassium with benzene or toluene. A second general approach is to use an alkali metal in conjunction with an electron carrier such as naphthalene.

The electron carrier is normally used in less than stoichiometric proportions, generally 5–10% by mole based on the metal salt being reduced. This procedure allows reductions to be carried out at ambient temperatures or at least at lower temperatures compared to the previous approach, which requires refluxing. A convenient reducing metal is lithium. Not only is the procedure much safer when lithium is used rather than sodium or potassium, but also in many cases the reactivity of the metal powders is greater.

A third approach is to use a stoichiometric amount of preformed lithium naphthalide. This approach allows for very rapid generation of the metal powders in that the reductions are diffusion controlled. Very low to ambient temperatures can be used for the reduction. In some cases, the reductions are slower at low temperatures because of the low solubility of the metal salts. This approach frequently generates the most active metals as the relatively short reduction times at low temperatures leads to reduced sintering (or growth) of the metal particles. This approach has been particularly important for preparing active copper. Fujita et al. [83] have shown that lithium naphthalide in toluene can be prepared by sonicating lithium, naphthalene, and *N,N,N′,N′*-tetramethylethylenediamine (TMEDA) in toluene, which allows reductions of metal salts in hydrocarbon solvents. This proved to be especially beneficial with cadmium [58]. An extension of this approach is to use the solid dilithium salt of the dianion of naphthalene. Use of this reducing agent in a hydrocarbon solvent is essential in the preparation of highly reactive uranium [57, 65].

For many of the metals generated by one of the prementioned three general methods, the finely divided black metals will settle after standing for a few hours leaving a clear, and in most cases colorless, solution. This allows the solvent to be removed via cannula. Thus, the metal powder can be washed to remove the electron carrier as well as the alkali salt, especially if it is a lithium salt. Moreover, a different solvent may be added at this point, providing versatility in solvent choice for subsequent reactions.

An important aspect of the highly reactive metal powders is their convenient preparation. The apparatus required is very inexpensive and simple. The reductions are usually carried out in a two-necked flask equipped with a condenser (if necessary), septum, heating mantle (if necessary), magnetic stirrer, and argon atmosphere. A critical aspect of the procedure is that anhydrous metal salts are to be used. Alternatively, anhydrous salts can sometimes be easily prepared as, for example, $MgBr_2$ from Mg turnings and 1,2-dibromoethane. In some cases, anhydrous salts can be prepared by drying the hydrated salts at high temperatures in a vacuum. This approach must be used with caution as many hydrated salts are very difficult to dry completely by this method or lead to mixtures of metal oxides and hydroxides. This is the most common problem when metal powders of low reactivity are obtained. The introduction of the metal salt and reducing agent to the reaction vessel is best done in a dry box or glove bag; however, nonhygroscopic salts can be weighed out in the air and then introduced into the reaction vessel. Solvents, freshly distilled from suitable drying agents under argon, are then added to the flask with a syringe. While it varies from metal to metal, the reactivity will diminish with time, and the metals are best used within a few days of preparation.

Rieke and coworkers have never had a fire or explosion caused by the activated metals; however, extreme caution should be exercised when working with these materials. Until a person becomes familiar with the characteristics of the metal powder involved, careful consideration should be taken at every step. To date, no metal powder except for magnesium the author and associates have generated will spontaneously ignite if removed from the reaction vessel while wet with solvent. They do, however, react rapidly with oxygen and with moisture in the air. Accordingly, they should be handled under an argon atmosphere. If the metal powders are dried before being exposed to the air, many will begin to smoke and/or ignite, especially active magnesium. Perhaps the most dangerous step in the preparation of the active metals is the handling of sodium or potassium. This can be avoided for most metals by using lithium as the reducing agent. In rare cases, heat generated during the reduction process can cause the solvent to reflux excessively. For example, reductions of $ZnCl_2$ or $FeCl_3$ in THF with potassium are quite exothermic. This is generally only observed when the metal salts are very soluble and the molten alkali metal (Method 1) approach is used. Sodium–potassium alloy is very reactive and difficult to use as a reducing agent; it is used only as a last resort in special cases.

4.3 Grignard Reagent Formation and Range of Reactivity of Magnesium

Perhaps the most widely known and used organometallic reagent is the Grignard reagent. Although commonly thought to be completely general, there are many organic halides that do not form Grignard reagents. Also, in some cases the Grignard reagents are not stable under the reaction conditions that commonly involve refluxing in THF. Prior to our studies, there were several modifications that greatly expanded the range of organic halides that were amenable to Grignard preparation. These modifications included (i) use of more strongly coordinated ethereal solvents [84–88], (ii) reduction of particle size by grinding or milling magnesium turnings [89], (iii) chemical activation such as the Gilman catalyst (I_2) [90] or the addition of ethylene bromide or ethyl bromide in molar quantities as an entrainer [91], (iv) sonication [23, 25, 92], (v) high-purity magnesium, and (vi) magnesium slurries produced by cocondensation metal-vaporization methods [19]. While these approaches were successful in advancing Grignard chemistry, a wide range of Grignard reagents were still not possible. Use of highly reactive magnesium prepared by the reduction method dramatically extends the Grignard reaction to many organic halides considered unreactive, including organic chlorides and even fluorides. The authors' early reports primarily involved reduction of $MgCl_2$ or $MgBr_2$ with K [2, 26, 27, 40, 50]. However, recent studies have shown that reductions with lithium and naphthalene as an electron carrier are not only safer and more convenient, but also the metal is more reactive [5, 44, 73, 81]. The high reactivity of the magnesium is demonstrated by its ability to form a Grignard reagent with bromobenzene in a few minutes at −78°C. Most dramatic is the ability to generate a Grignard reagent from fluorobenzene. Most organic textbooks today still state that one cannot form Grignard reagents from aryl fluorides. High yields of Grignard reagents have been prepared from a wide variety of alkyl, vinyl, and aryl halides. In many of these cases, standard conditions with ordinary magnesium turnings either do not work or give low yields.

A simple example of the advantage of generating Grignard reagents at low temperatures is the reaction of 3-halophenoxypropane [49]. Although the Grignard reagent is easy to generate at room temperature or above, it eliminates the phenoxy group by an S_N2 reaction to generate cyclopropane. This reaction is, in fact, a standard way to prepare cyclopropanes, and the cyclization cannot be stopped; however, by using the highly reactive magnesium, the Grignard reagent can be prepared at −78°C, and it does not cyclize at these low temperatures. It then can be added to a variety of other substrates in standard Grignard reactions. We have attempted the low-temperature formation of Grignard reagents containing nitrile, ketone, and ester functional groups [5]. These reactions were largely unsuccessful, except in the case of 8-bromooctanenitrile. Alkyl nitriles were largely unsuccessful, except in the case of 8-bromooctanenitrile. Aromatic nitriles were reduced to radical anions. In the case of the alkyl system, ethyl 3-bromopropionate, 4-bromobutyronitrile, and 5-bromovaleronitrile, the heteroatoms coordinated strongly to the magnesium surface at the temperatures required to prevent attack of Grignard reagents on the functionality present. Thus, Grignard reagent formation was inhibited by the occupation of the active sites on the magnesium surface, although 8-bromooctanenitrile reacted rapidly with Mg* at −78°C to give 74% conversion to the Grignard reagent. Carbonation gave 8-cyclooctanoic acid in 35% yield based on Grignard reagent. This appears to be the first example of a Grignard reagent which contains the nitrile functionality. It is interesting to note that the nitrile functionality coordinates so strongly to the Mg surface that reaction with organic halides is completely inhibited until 20°C is reached. 4-Bromobutyronitrile did not react at all with Mg* at −78°C. When the temperature was slowly raised, no reaction was observed until 20°C, at which time polymerization occurred. In confirmatory experiments, 1-bromopentane reacted quantitatively with Mg* in 1 min at −78°C. The resulting Grignard reagent did not react with subsequently added butyronitrile even after 1 h at −78°C.

If butyronitrile was added to Mg* at −78°C, followed by addition of bromopentane, no Grignard reagent was formed after 2 h.

The active magnesium can be used to readily prepare Grignard reagents from a wide variety of primary, secondary, and tertiary halides (see Table 4.1) [30]. Significantly, allyl halides also give good yields. For example, 3-chloro-2-methylpropene was converted to 3-methyl-3-butenoic acid in 82% yield after 1 h of reaction at room temperature. Vinyl halides are also readily converted to the corresponding Grignard reagent; many times the reaction proceeds in a few minutes at room temperature [30].

Previous efforts to prepare dimagnesium derivatives of benzene have been successful only with dibromo- or bromoiodobenzene and required forcing conditions. Moreover, they usually resulted in the monomagnesium derivative as the main product. This work has been surveyed by Ioffe and Nesmayanov [93]. Using the $MgCl_2$-KI-K-THF system, we have prepared the di-Grignard of *p*-dibromobenzene in 100% yield in 15 min at room temperature. In earlier work, only one halogen atom of dichloro derivatives of benzene and naphthalene reacted with magnesium [94–97], and the chlorine of *p*-chlorobromobenzene (**2**) was found to be completely unreactive [98]. Using our method, we obtained a 100% yield of the mono-Grignard and a 10% yield of the di-Grignard of **2** in 15 min at room temperature. After 2 h, the yield of di-Grignard was 100%. With *p*-dichlorobenzene (**3**) the yield of di-Grignard was 30% in 2 h. The use of a lower Mg/halide ratio permitted the rapid and selective formation of the mono-Grignard of **3**, which was converted to 4-chlorobenzoic acid in 89% yield by CO_2 quench (Table 4.1).

Ashby et al. [99] have shown that alkyl fluorides can in fact be converted to the corresponding Grignard reagents in high yields. They found that refluxing THF and magnesium activated with the Gilman catalyst (I_2) gave the best results. We have tried our "activated magnesium" and "KI-activated magnesium" with *p*-fluorotoluene (**10**) and 1-fluorohexane (**11**) under a variety of conditions. The results are summarized in Table 4.2. Perhaps this table best demonstrates the high reactivity of the "KI-activated magnesium." Prior to our work, all efforts to prepare the Grignard reagent from fluorobenzene and magnesium failed. Some of these attempts included reflux times of weeks or months. Using the $MgCl_2$-KI-K system, refluxing THF, and a Mg/**10** ratio of 4 gave 69% yield of the fluoro Grignard in 1 h. Quenching the Grignard with CO_2 gave a 63% yield of the expected *p*-toluic acid. Using a lower Mg/**10** ratio resulted in lower yields. Using diglyme but no KI gave only 8% yield after 1 h at 162°C (solvent refluxing). Continued refluxing in THF or diglyme for periods of 24–48 h gave no appreciable increase in Grignard formation compared to reflux times of 2–3 h.

Reaction with (**11**) using the normal procedure gave 54% yield in 3 h at room temperature and 71% yield in 3 h at 66°C. When the KI procedure was employed, the yield was 89% after 3 h at room temperature. Quenching the Grignard of **11**

Table 4.1 Reactions of activated magnesium with various halides.[a]

Halide	Mg/halide	Mg/KI	Reaction temp. (°C)	Reaction time (min)	% yield Grignard[b] Mono	Di	% yield[c] —CO_2H
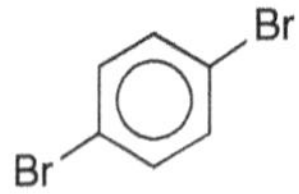 Br, Br	4	2	25	15		100	
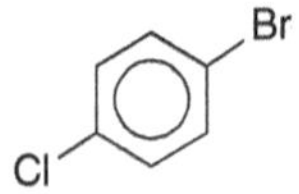 Br, Cl	4	2	25	15	100	10	
				60	100	57	
				120	100	100	
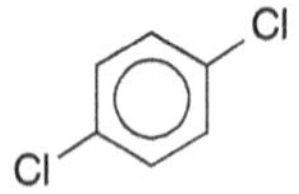 Cl, Cl	4	2	25	15	100	15	
				120	100	30	
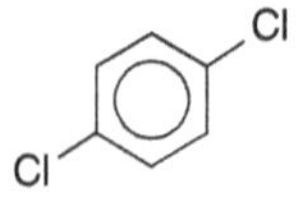 Cl, Cl	2	2	25	5	70	0	
				60	86	0	
				180	90	0	89[d]
 Br	2		25	5	100		81[e]
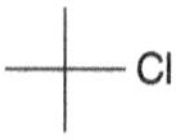 Cl	2		25	10	100		
				60			52[e]
 Cl	1.7	2	66	15	11		
				90	42		
				360	74		63[d]
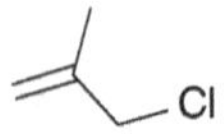 Cl	2		25	60			82[e]
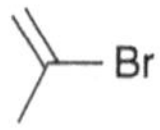 Br	2		25	5	100		71[e]
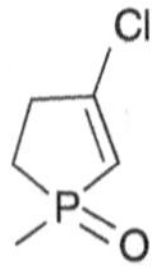 Cl, P, O	5		66	390			40

[a] Mg source was either $MgCl_2 + K$ or $MgCl_2 + KI + K$, THF solvent, refluxed 2–3 h prior to halide addition.
[b] Yield by vpc after hydrolysis.
[c] Yield of isolated carboxylic acid.
[d] Yield based on Grignard.
[e] Yield based on starting halide.

Table 4.2 Reaction of *p*-fluorotoluene and 1-fluorohexane with activated magnesium.[a,b]

Fluoride	Mg/fluoride	Mg/KI	Reaction temp. (°C)	Solvent	% yield Grignard at reaction time (min)			
					10	30	60	180
10[c]	1		162	Diglyme			8	5
10	2	2	66	THF		26	27	55
10	3	2	66	THF	40	53	59	58
10	4	2	66	THF	58	57	69	66
11	2		25	THF	10	26	27	54
11	2		66	THF	53	48	72	71
11	2	2	25	THF	38	53	73	89

[a] Magnesium source for all reactions was $MgCl_2 + K$.
[b] All THF reactions refluxed 3 h prior to fluoride.
[c] Refluxed 18 h prior to fluoride addition.

with CO_2 gave the expected heptanoic acid in 88% yield, based on the Grignard. These results compare favorably to those obtained by Ashby et al. [99] for the preparation of the Grignard of **11**.

Another example of the remarkable reactivity of Mg* is its reaction with nitriles. In this respect, the Mg* resembles an alkali metal more than an alkaline earth. Benzonitrile reacts with Mg* overnight in refluxing DME to give 2,4,6-triphenyl-1,3,5-triazine and 2,4,5-triphenylimidazole in 26% and 27% yield, respectively, based on magnesium. The imidazole was shown to partly arise from the action of Mg* on the triazine. The trimerization of aromatic nitriles to give symmetrical triazines is not unknown, but generally the reactions are catalyzed by strong acids [100–102], less often by strong bases, or least frequently, a weak base and extremely high pressure [103]. The action of phenylmagnesium bromide on benzonitrile gives 2,3,4,6-tetraphenyl-1,2-dihydro-1,3,5-triazine along with a small amount of symmetrical triazine [104]. Organoalkalis react with benzonitrile to give 2,2,4,6-tetraphenyl-1,2-dihydro-1,3,5-triazine [105, 106]. To our knowledge, no one has previously demonstrated a direct reaction of magnesium to give the symmetrical triazine we observed.

Neat butyronitrile reacted under reflux to give 4-amino-5-ethyl-2,6-dipropylpyrimidine isolated in 98% yield based on Mg*. Similarly, acetonitrile gave 4-amino-2,6-dimethylpyrimidine in 52% yield, based on Mg, after reacting in a sealed tube at 130°C for 13 h. The reaction appears catalytic with respect to Mg*. Treatment of alkyl nitriles with metallic sodium [107] or Grignard reagents [108] is known to produce aminopyrimidines, but there have been no reports of such results by the direct action of metallic magnesium on alkyl nitriles.

We have also found that our Mg* reacts with tosylates to give good yields of the coupled products. For example, benzyl tosylate reacts in refluxing THF to give the expected bibenzyl in 84% yield after 12 h. Neat hexane-1,6-ditosylate reacted with Mg* after 36 h at 90–100°C to give a polymer with an average carbon chain length of 26.5. Mass spectrometry (EI) revealed masses up to 1135 amu, but much of the sample was unable to be volatilized. Butyl tosylate did not react easily with the Mg*.

Dibenzyl ether reacted with Mg* after 5 days of reflux in THF to give phenylacetic acid in 42% yield after carbonation. Maercker [109] was able to obtain a 15% yield of 3-butenoic acid from allyl methyl ether after 56 h of reflux.

4.4 1,3-Diene-Magnesium Complexes and Their Chemistry

Metallic magnesium is known to react with certain 1,3-butadienes yielding halide-free organomagnesium compounds [110]; however, there is a primary problem associated with the preparation of these reagents. The reaction of ordinary magnesium with 1,3-dienes such as 1,3-butadiene or isoprene is usually slow and is accompanied by dimerization, trimerization, and oligomerization. This reaction may be catalyzed by alkyl halides or transition metal salts but is generally accompanied by a variety of by-products. Consequently, the utilization of these reagents in organic synthesis has been quite limited [111], except for 1,3-butadiene-magnesium, which has found considerable application in organometallic synthesis [112].

The author and associates published a communication [73] demonstrating that substituted (2-butene-1,4-diyl)magnesium complexes can be conveniently prepared from the reaction of highly reactive magnesium with the corresponding 1,3-dienes. Significantly, these bis-Grignard reagents react with α,ω-alkylene dihalides to give complex carbocycles. Polyfunctionalized organic compounds can be obtained in this manner. The following are typical examples. The magnesium prepared by lithium reduction of $MgCl_2$ reacts with (*E*,*E*)-1,4-diphenyl-1,3-butadiene (**1a**), 2,3-dimethyl-1,3-butadiene (**1b**), isoprene (**1c**), myrcene (**1d**), and 2-phenyl-1,3-butadiene (**1e**) in THF at ambient temperature to give the corresponding substituted (2-butene-1,4-diyl)magnesium complexes. The structures of these complexes have not been determined to date except for (1,4-diphenyl-2-butene-1,4-diyl)magnesium, which has been shown to be a five-membered ring metallocycle [113]. The most probable structure for the magnesium complex of 1,3-butadiene is an oligomer. Accordingly, the most likely structures for these complexes are five-membered metallocycles or oligomers. It is also possible that an equilibrium exists between these various forms.

Cyclizations of (1,4-Diphenyl-2-butene-1,4-diyl)magnesium with α,ω-Alkylene Dihalides

Reaction of (*E,E*)-1,4-diphenyl-1,3-butadiene with newly generated magnesium in THF gave (1,4-diphenyl-2-butene-1,4-diyl)magnesium (**2a**). This reagent acted as a dinucleophile. Reactions of **2a** with α,ω-alkylene dibromides resulted in either cyclization or reduction of the electrophile, depending on the initial dibromides. The results are summarized in Table 4.3. Significantly, these cyclizations are always stereospecific and completely regioselective. For example, cyclization proceeds rapidly at −78°C in the reaction of **2a** with 1,3-dibromopropane and 1,3-dichloropropane, yielding a single product, *trans*-1-phenyl-2-((*E*)-2-phenylethenyl)cyclopentanes (**5**), in 65 and 81% yield, respectively (Table 4.3, entries 3 and 4).

In contrast to the 1,2-cyclizations of **2a** with 1,3-dihalopropane and 1,4-dihalobutane, treatment of **2a** with 1,2-dichloroethane resulted in 1,4-addition, producing a cyclohexene derivative, *cis*-3,6-diphenylcyclohexene [77, 114].

4.5 Regioselectivity of Reaction of Complexes with Electrophiles

The reactions discussed in the following illustrate the unusual reactivity of these magnesium-diene complexes. While one might expect the complexes to react as bis-Grignard reagents, it is clear that they are much more powerful nucleophiles. Remarkably, they react with alkyl bromides and chlorides at temperatures as low as −80°C. They appear to react via a standard S_N2 mechanism. However, they can serve as electron transfer agents with certain reagents such as metal salts. With most electrophiles possessing good leaving groups, clean S_N2 chemistry is observed. Soft electrophiles such as organic halides demonstrate complete regioselectivity for the 2-position as shown in Equation 4.1. The resulting primary organomagnesium intermediate can then be reacted with a wide range of second electrophiles. Harder electrophiles such as

Equation 4.1 Reaction of soft electrophile with (2,3-dimethyl-2-butene-1,4-diyl)magnesium.

Mg; $\xrightarrow{E_1\text{ (soft)}}$ E_1, MgX; $\xrightarrow{E_2}$ E_1, E_2 (4.1)

monohalosilanes, monohalostannanes, and so on, react completely regioselectively in the 1-position as shown in Equation 4.2. Addition of a second electrophile occurs primarily in the 2-position.

Table 4.3 Cyclization of (1,4-diphenyl-2-butene-1,4-diyl)magnesium with α,ω-alkylene dihalides.[a]

Entry	Dihalide	Product[b]	% iso. yield
1	$Br(CH_2)_4Br$	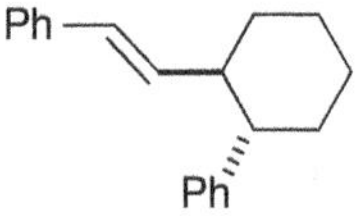 Ph, Ph **4**	40
2	$Cl(CH_2)_4Cl$	4	51
3	$Br(CH_2)_3Br$	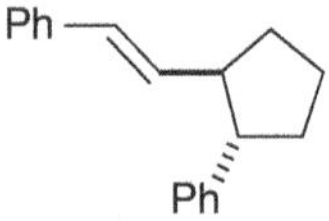 Ph, Ph **5**	65
4	$Cl(CH_2)_3Cl$	5	81
5	$Br(CH_2)_2Br$	 Ph, Ph	–
6	$Cl(CH_2)_2Cl$	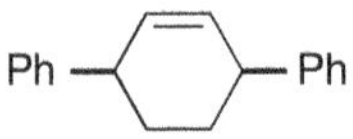 Ph, Ph **6**	59
7	$BrCH_2Br$	 Ph, Ph	–
8	$ClCH_2Cl$	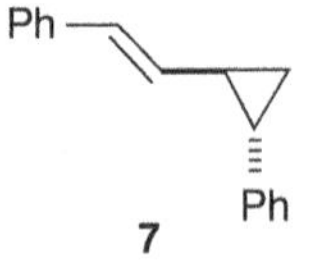 Ph, Ph **7**	76
9	$CH_3(CH_2)_3Br$	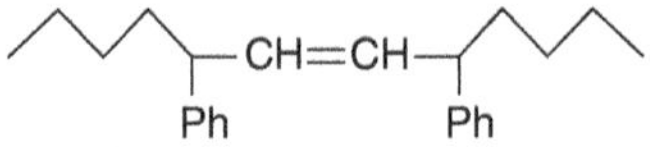 CH=CH, Ph, Ph **8** (*cis:trans* = 56:44)c	93
10	$CH_3(CH_2)_3Cl$	**8** (*cis* : *trans* = 28 : 72)[c]	87

[a] Reactions were typically done at –78°C, and then the reaction mixtures warmed to room temperature prior to workup.
[b] All compounds were fully characterized by ^1H NMR, ^{13}C NMR, IR, and mass spectra.
[c] The ratio of *cis* to *trans* was based on the ^1H NMR spectra of the crude products. The individual isomers were separated by chromatography.

Equation 4.2 Reaction of hard electrophile with (2,3-dimethyl-2-butene-1,4-diyl) magnesium.

$$\text{Mg-diene} \xrightarrow{E_1 \text{ (hard)}} \text{MgX}\ldots E_1 \xrightarrow{E_2} \ldots E_2, E_1 \quad (4.2)$$

The overall result is that one can effect a net 2,1-addition or a 1,2-addition by choosing appropriate electrophiles. Complex, highly functionalized molecules have been prepared using this approach and are shown in Table 4.4 [81].

4.6 Carbocyclization of (1,4-Diphenyl-2-butene-1,4-diyl) magnesium with Organic Dihalides

(*E,E*)-1,4-Diphenyl-1,3-butadiene has been found to be one of the most reactive conjugated dienes toward metallic magnesium. Reaction of this diene with Rieke magnesium affords (1,4-diphenyl-2-butene-1,4-diyl)magnesium (**2**) which can be treated with various electrophiles (Table 4.3). Reactions of **2** with 1,*n*-dibromoalkanes resulted in either cyclization or reduction of the electrophile, depending on the initial dibromide. For example, cyclization proceeded rapidly at −78°C in the reaction of **2** with 1,3-dibromopropane or 1,3-dichloropropane, yielding a single product, *trans*-1-phenyl-2-((*E*)-2-phenylethenyl)cyclopentanes (**4**) in 65 or 81% isolated yield, respectively (Scheme 4.1). Similar cyclizations were obtained with (2,3-dimethyl-2-buten-1,4-diyl)magnesium (Table 4.5).

(1,4-Diphenyl-2-butene-1,4-diyl)magnesium can be easily prepared using Rieke magnesium. Treatment of the diene-magnesium reagent with 1,*n*-dihaloalkanes provides a convenient method for the generation of substituted three-, five-, and six-membered carbocycles. Significantly, the cyclizations are always stereoselective and completely regioselective.

It should be noted that the (1,4-diphenyl-2-butene-1,4-diyl)barium complex has been prepared in our laboratories and exhibits higher reactivity than its magnesium-diene counterpart.

4.7 1,2-Dimethylenecycloalkane-Magnesium Reagents

Based upon the bis-nucleophilicity of 1,3-diene-magnesium intermediates, reactions of these intermediates with bis-electrophiles can lead to spiro or fused bicyclic molecules, depending upon the regioselectivity of the cyclization. It has been shown that treatment of magnesium complexes of 1,2-bis(methylene) cyclohexane with 1,*n*-dibromoalkanes results in overall 1,2-cyclizations of the original dienes, affording spirocarbocycles in good to excellent yields [115].

Table 4.4 Stepwise reactions of (2,3-dimetyl-2-butene-1,4-diyl)magnesium with electrophiles.

Entry	First electrophile[a]	Second electrophile[b]	Product[c]	% iso. yield
1	$Br(CH_2)_4Br$	MeCOCl	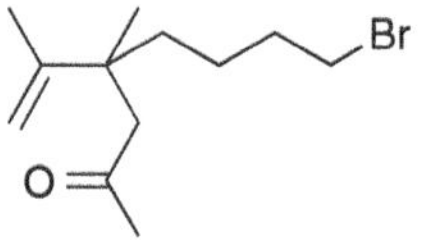 Br, O **17**	62
2	$Br(CH_2)_4Br$	PhCOCl	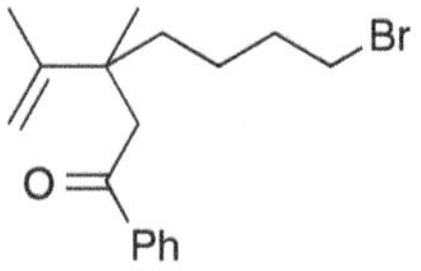 Br, O, Ph **18**	60
3	$Me(CH_2)_3Br$	MeCOCl	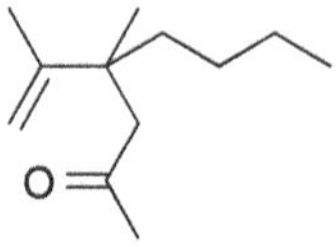 O **19**	61
4	$Me(CH_2)_3Br$	PhCOCl	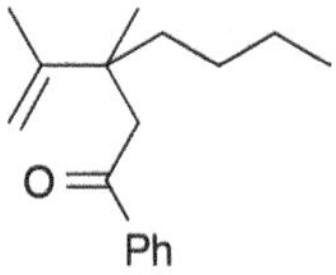 O, Ph **20**	82
5	Me_3SiCl	MeCOCl	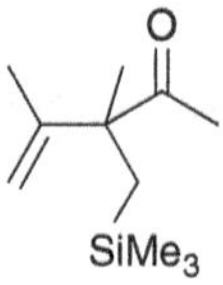 O, $SiMe_3$ **21**	73
6	Me_3SiCl	PhCOCl	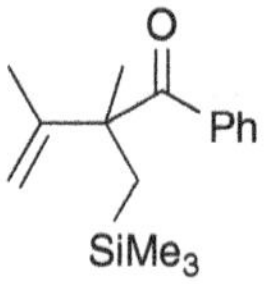 O, Ph, $SiMe_3$ **22**	79
7	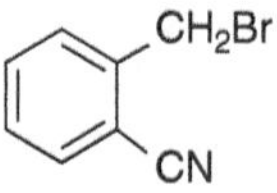 CH_2Br, CN	H_3^+O	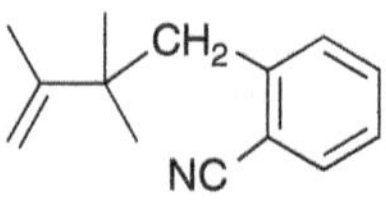 CH_2, NC **23**	35[d]

Table 4.4 (Continued)

Entry	First electrophile[a]	Second electrophile[b]	Product[c]	% iso. yield
8	2-(CH_2Br)benzonitrile (CH_2Br, CN)		24	30[e]
9	$Br(CH_2)_3CN$	H_3^+O	25 (CN)	58[f]
10	$Br(CH_2)_3CN$		26 (O)	31[g]
11	$Br(CH_2)_2CN$		27 (O)	42[h]

[a] The first electrophile was added to the THF solution of (2,3-dimethyl-2-butene-1,4-diyl) magnesium at −78°C. The reaction mixture was then warmed to room temperature prior to the addition of the second electrophile.
[b] The second electrophile was added at 0°C.
[c] All new compounds were completely characterized spectroscopically.
[d] Protonation at −78°C resulted in the survival of the cyano group.
[e] Cyclization was achieved at reflux.
[f] Acidic hydrolysis at −40°C gave the nitrile.
[g] The cyclic ketone was obtained at reflux followed by acidic hydrolysis.
[h] Cyclization completed at room temperature.

Scheme 4.2 depicts a general route for the spiro-olefin synthesis. Typically, a 1,*n*-dibromoalkane was added to the THF solution of the diene-magnesium complex at −78°C, producing an organomagnesium intermediate accommodating a bromo group. The intermediate cyclized upon warming, affording the corresponding spirocarbocycles containing an exocyclic double bond.

The spiroannulation approach described in Scheme 4.2 has been easily extended to other 1,2-dimethylenecycloalkanes. A wide variety of spirocycles were synthesized from the reactions of magnesium complexes of 1,2-dimethylenecyclopentane and 1,2-dimethylenecycloheptane with 1,*n*-dibromoalkanes

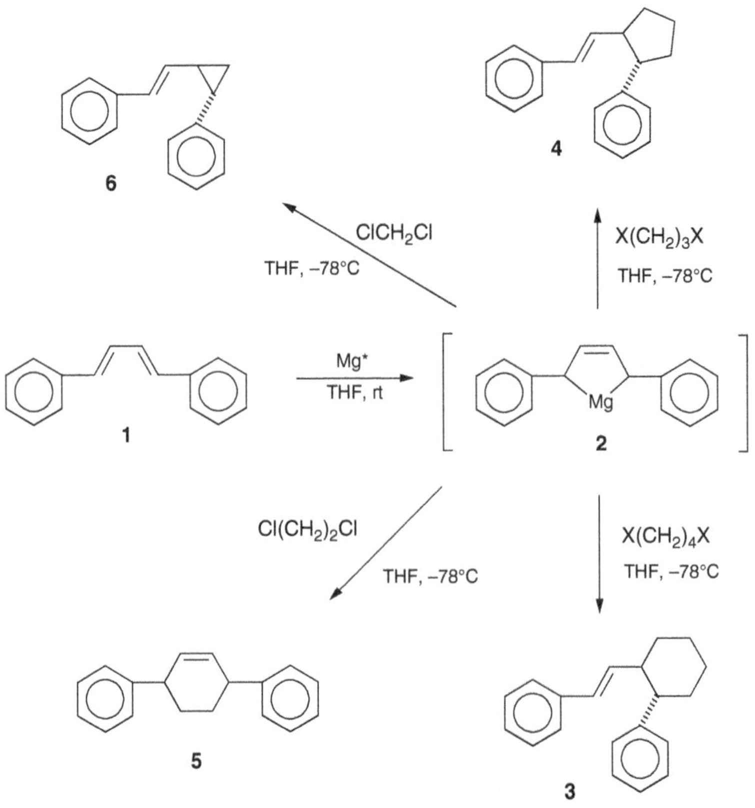

Scheme 4.1 Carbocyclization of (1,4-diphenyl-2-butene-1,4-diyl)magnesium with organic dihalides.

as shown in Table 4.6. Therefore, this spiroannulation method provides a very general approach to a large number of spirocarbocycles with different combinations of ring sizes.

4.8 Synthesis of Fused Carbocycles, β-γ-Unsaturated Ketones, and 3-Cyclopentenols from Conjugated Diene-Magnesium Reagents

With the aid of Rieke magnesium, fused carbocyclic enols can be effortlessly synthesized from 1,3-diene-magnesium complexes by reacting with carboxylic esters [116]. By controlling the reaction temperature, β,γ-unsaturated ketones can also be produced.

Table 4.5 Reactions of (2,3-dimethyl-2-butene-1,4-diyl)magnesium with organodihalides.

Entry	Dihalide[a]	Product[b]	% iso. yield[c]
1	$Br(CH_2)_4Br$	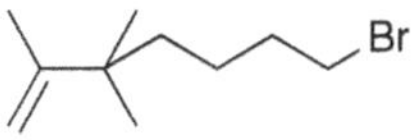 **9**	79
2	$Br(CH_2)_4Br$	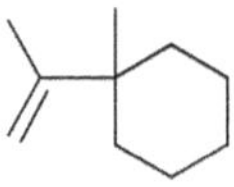 **10**	53[d] (69)
3	$Br(CH_2)_3Br$	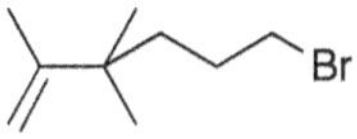 **11**	72[e]
4	$Br(CH_2)_3Br$	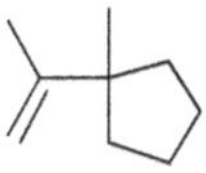 **12**	– (75)
5	$Cl(CH_2)_3Cl$	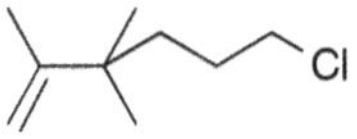 **13**	81
6	$Br(CH_2)_2Br$	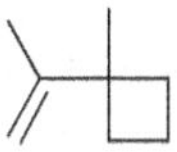 **14**	– (49)[f]
7	$Cl(CH_2)_2Cl$	**14**	– (61)[f]
8	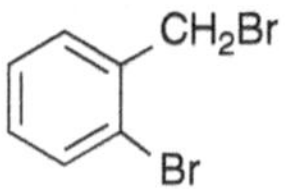 2-Bromobenzyl bromide (CH_2Br, Br)	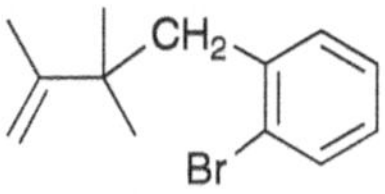 **15** (CH_2, Br)	62[g]
9	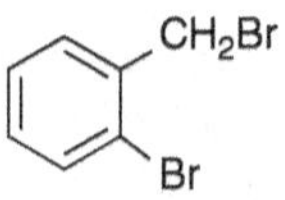 2-Bromobenzyl bromide (CH_2Br, Br)	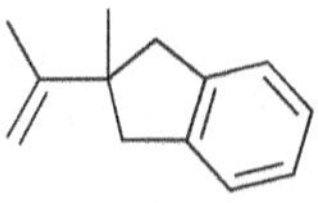 **16**	~30[h]

[a] Organodihalides were added to the THF solution of (2,3-dimethyl-2-butene-1,4-diyl) magnesium at −78°C. The reaction mixture was stirred at −78°C for 2 h and then typically warmed to room temperature (unless specified) prior to workup.
[b] All compounds have satisfactory spectral data including ^{1}H NMR, ^{13}C NMR, IR, and mass spectral data.
[c] GC yields are given in parentheses.
[d] Cyclization completed after the mixture was refluxed for 5 h.
[e] Uncyclized product was obtained by controlling the reaction temperature below −35°C.
[f] Product was isolated by preparative gas chromatography.
[g] Monoalkylated product was obtained by protonation at −78°C.
[h] Cyclization was achieved at reflux.

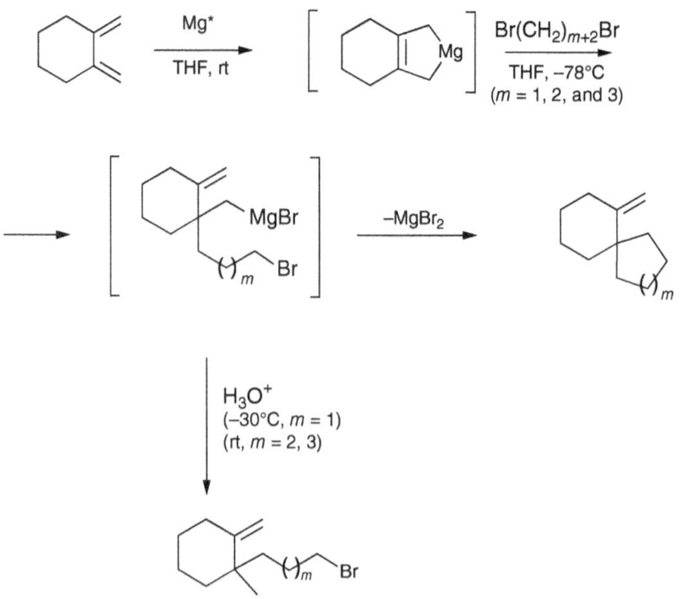

Scheme 4.2 Synthesis of spiro-olefins.

As shown in Scheme 4.3, reaction of the 1,2-bis(methylene)cycloalkane-magnesium, **1**, with ethyl acetate at −78°C, followed by acidic quenching at −10°C, afforded the β,γ-unsaturated ketone, (2-methyl-1-cyclohexenyl)propan-2-one (**6**). Alternatively, upon refluxing the initially formed intermediate **2**, followed by acidic workup, the fused bicyclic enol 2,3,4,5,6,7-hexahydro-2-methyl-1*H*-indene-2-ol (**4**) was obtained. It was found that the initial product generated upon treating the 1,2-bis(methylene)cycloalkane-magnesium, **1**, with ethyl acetate was **2**, a magnesium salt of a spiroenol containing a cyclopropane ring. This intermediate underwent ring expansion upon warming to produce the fused carbocyclic product **4**. Alternatively, protonation of the intermediate **2** at −10°C generated the corresponding spiroenol **5**, which underwent an *in situ* rearrangement to afford the β,γ-unsaturated ketone **6**. The identity of the initially formed versatile intermediate, **2**, was validated by trapping with acetyl chloride to give 1-methyl-4-methylenespiro[2.5]oct-1-yl acetate, **7**. Upon basic workup, the spiroacetate **7** also afforded the β,γ-unsaturated ketone product **6**.

It is important to note that the ring enlargement for **2** to **3** involves a vinylcyclopropane–cyclopentene ring expansion which has also been observed for the lithium salts of 2-vinylcyclopropanol systems [117]. On the other hand, the rearrangement of **5** to **6** is formally a 2-vinylcyclopropanol ring opening with a proton transfer. To our knowledge this is the first report of

Table 4.6 Reactions of magnesium complexes of 1,2-dimethylenecycloalkanes with bis-electrophiles.

Entry	Diene[a]	Electrophile	Product	% yield[b]	Note[c]
1	**1a**	$Br(CH_2)_5Br$	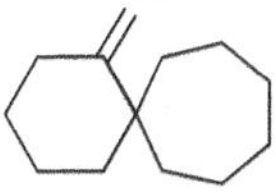	45	A
2	**1a**	$Br(CH_2)_5Br$	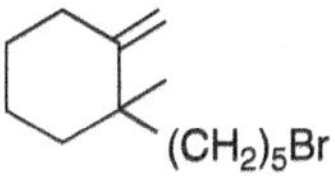 $(CH_2)_5Br$	79	B
3	**1a**	$Br(CH_2)_4Br$	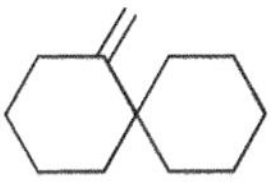	75 (81)	C
4	**1a**	$Br(CH_2)_4Br$	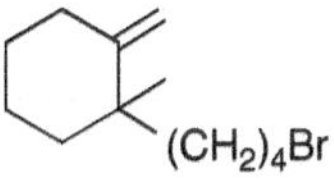 $(CH_2)_4Br$	81	B
5	**1a**	$Br(CH_2)_3Br$	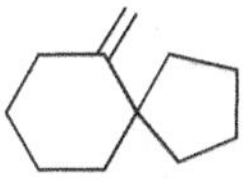	75 (87)	D
6	**1a**	$Br(CH_2)_3Br$	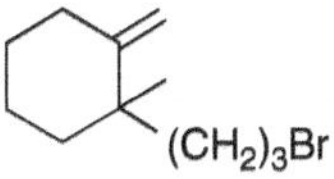 $(CH_2)_3Br$	78	E
7	**1a**	$Cl(CH_2)_3Cl$	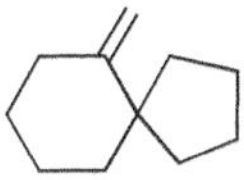	– (78)	F
8	**1a**	$Br(CH_2)_2Br$	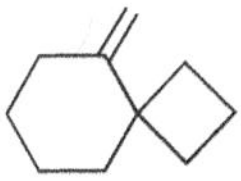	– (15)	B
9	**1a**	$Cl(CH_2)_2Cl$	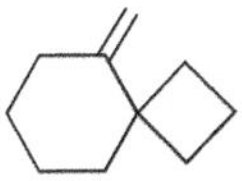	– (40)	B
10	**1a**	$(T_8OCH_2)_2$	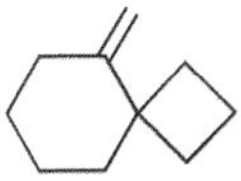	52 (67)	B
11	**1b**	$Br(CH_2)_3Br$	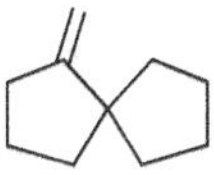	60 (70)	D

(*Continued*)

Table 4.6 (Continued)

Entry	Diene[a]	Electrophile	Product	% yield[b]	Note[c]
12	**1c**	$Br(CH_2)_4Br$	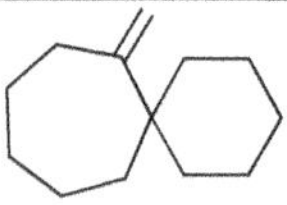	73	C
13	**1c**	$Br(CH_2)_3Br$	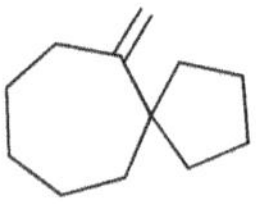	77 (86)	D
14	**1c**	$(T_8OCH_2)_2$	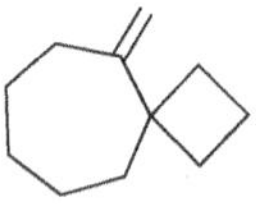	46 (59)	B
15	**1a**	Ph_2SiCl_2	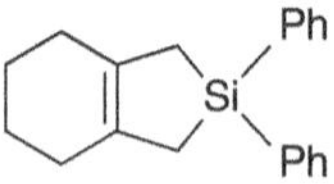 Si, Ph, Ph	89	B

[a] **1a**: 1,2-bis(methylene)cyclohexane; **1b**: 1,2-bis(methylene)-cyclopentane; **1c**: 1,2-bis(methylene) cycloheptane.

[b] Isolated overall yields were based on 1,2-bis(methylene)cycloalkanes. GC yields are shown in parentheses.

[c] Bis-electrophiles were added to the THF solution of the diene–magnesium reagent at −78°C. The reaction mixture was then stirred at −78°C for 1 h prior to warming to the specified temperature. A, reflux (15 h); B, room temperature (30 min); C, reflux (10 h); D, room temperature (10 h); E, −30°C; F, room temperature (30 h).

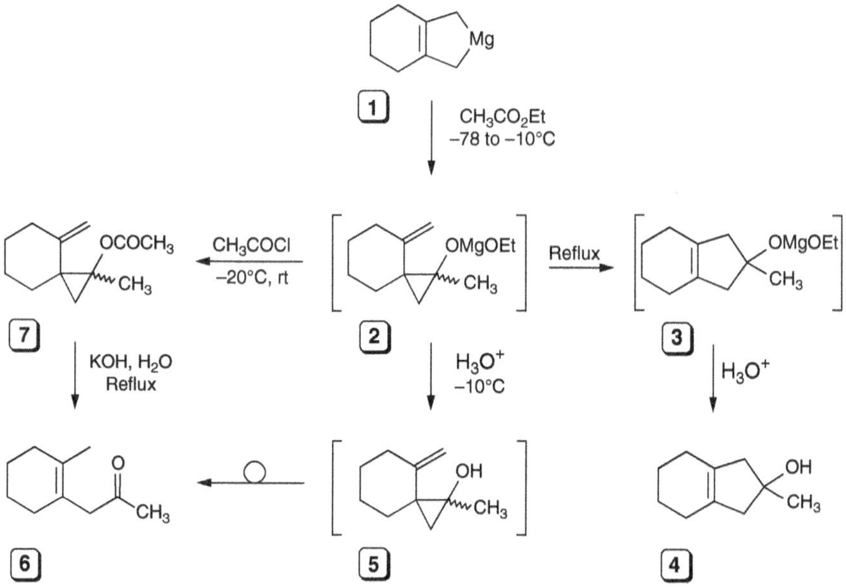

Scheme 4.3 Reaction of the 1,2-bismethylenecyclohexane-magnesium complex with an ester.

such a rearrangement, although 1-vinylcyclopropanol-cyclobutanone rearrangements have been well documented [118].

The absence of the formation of a five-membered ring at low temperature suggests that the initial attack of ethyl acetate occurred at the 2-position of the 1,2-bis(methylene)cycloalkane-magnesium (**1**), followed by an intramolecular cyclization, generating the spiro intermediate (**2**).

This new methodology is quite general and can be applied to the magnesium complexes of 1,2-dimethylenecyclopentane, 1,2-dimethylenecycloheptane, and 2-methyl-3-phenyl-1,3-butadiene. Likewise, other carboxylic esters, such as butyl and ethyl benzoates, can also be used to make various fused carbocyclic enols or β,γ-unsaturated ketone products. Significantly, the overall synthetic process to form fused carbocyclic enols from the corresponding 1,2-bis(methylene)cycloalkanes represents a formal [4+1] annulation.

Table 4.7 shows representative examples of the high-temperature pathway affording cyclopentenols, and Table 4.8 depicts the low-temperature pathway which generates β,γ-unsaturated ketones from intermediate **2**.

We have extended this chemistry by incorporating a lactone as the electrophile and generated β,γ-unsaturated ketone alcohols, as well as cyclopentenols containing both 3° and 1° alcohols within the same molecule depending on the reaction temperature [119]. We have also broadened this chemistry to

Table 4.7 Reactions of diene-magnesium reagents with carboxylic esters: formation of cyclopentenols.

Entry	Diene	Ester	Product	% yield[a]
1	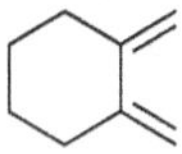	CH_3COOEt	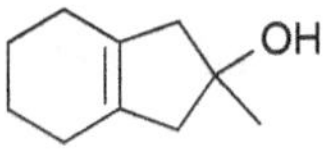 OH	91
2	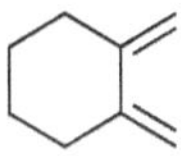	$CH_3(CH_2)_2COOEt$	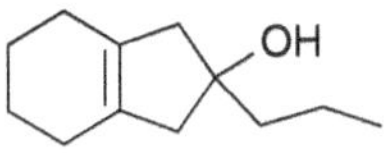 OH	96
3	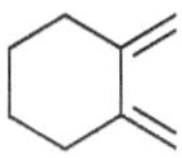	PhCOOEt	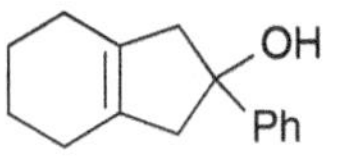 OH Ph	55
4	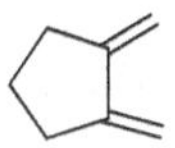	$CH_3(CH_2)_2COOEt$	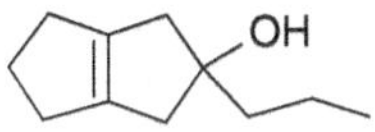 OH	59
5	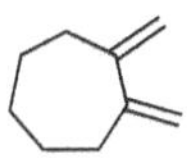	$CH_3(CH_2)_2COOEt$	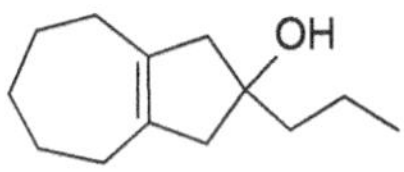 OH	74

[a] Fused bicyclic product was obtained at reflux.

Table 4.8 Reactions of diene-magnesium reagents with carboxylic esters: formation of β,γ-unsaturated ketones.

Entry	Diene	Ester	Product	% yield[a]
1	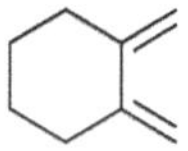	CH_3COOEt	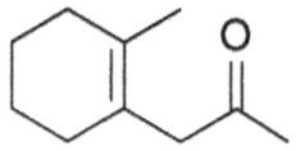 O	72
2	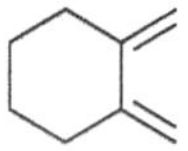	$CH_3(CH_2)_2COOEt$	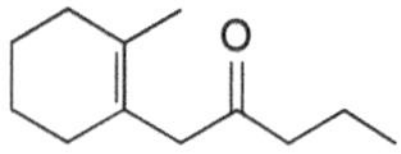 O	81
3	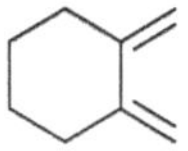	PhCOOEt	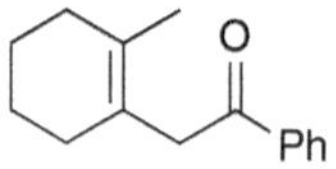 O Ph	62
4	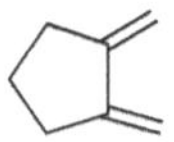	$CH_3(CH_2)_2COOEt$	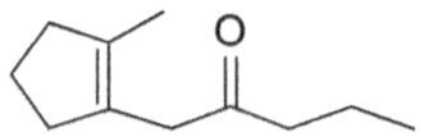 O	76
5	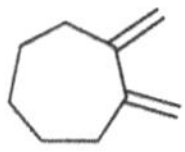	$CH_3(CH_2)_2COOEt$	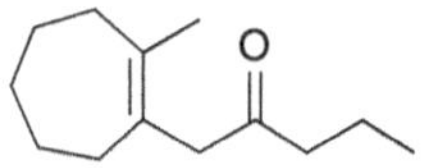 O	84

[a]Quenching the reaction at −10°C gave the β,γ-unsaturated ketone.

include magnesium-diene intermediates derived from isoprene, myrcene, and 2,3-dimethyl-1,3-butadiene in addition to the 1,2-bis(methylene) cyclohexane complex.

4.9 Synthesis of Spiro-γ-Lactones and Spiro-δ-Lactones from 1,3-Diene-Magnesium Reagents

The generation of a quaternary carbon center and the introduction of functional groups, which are present in the process of lactonization, are some of the difficulties associated with the formation of spirolactones. There have been many reports in the literature regarding the development of synthetic strategies to overcome these difficulties [120]. However, many require the use of complex reagents and multiple synthetic steps to achieve the overall lactonization. One of the more effective methods is the reaction of bis(bromomagnesio) alkanes with dicarboxylic anhydrides [121].

We have developed a direct synthetic method for the one-pot synthesis of lactones, spirolactones, di-spirolactones, tertiary alcohols, and even 1,2-diols from the corresponding conjugated diene-magnesium reagents mediated by Rieke magnesium [122]. The overall lactonization procedure can be considered as a molecular assembling process in which three distinct individual species—a conjugated diene, a ketone, and carbon dioxide—are used to build a complex organic molecule in a well-controlled manner. Likewise, the generation of a quaternary center and the incorporation of both a hydroxyl and a carboxyl group required for lactonization are achieved in one synthetic procedure.

An example of the formation of a spiro-γ-lactone from the corresponding 1,2-bis(methylene)cyclohexane-magnesium complex is shown in Scheme 4.4. The reaction between 1,2-bis(methylene)cyclohexane-magnesium (**1**) and acetone at −78°C yielded a 1,2-adduct, **2**, resulting from the incorporation of one molecule of acetone with the diene complex. Acidic workup of the 1,2-adduct at −78°C gave a tertiary alcohol, **3**, containing both a quaternary center and a vinyl group at the β-position. Upon warming, the 1,2-adduct can further undergo a nucleophilic addition to a second electrophile. Carbon dioxide introduced into the reaction mixture at 0°C as the second electrophile reacted immediately with **2**, presumably giving a magnesium salt of γ-hydroxy acid (**4**). After acidic hydrolysis and mild warming, the spiro-γ-lactone product 4,4-dimethyl-6-methylene-3-oxaspiro[4.5]decan-2-one (**6**) was obtained. Table 4.9 shows representative spiro-γ-lactones that have been prepared this way.

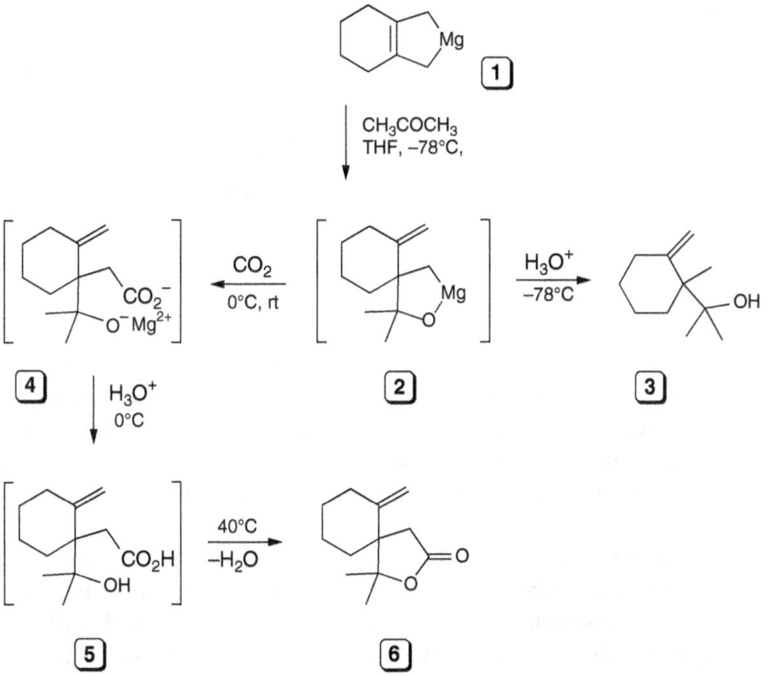

Mg
1
CH_3COCH_3
THF, −78°C,
CO_2^-
O^- Mg^{2+}
CO_2
0°C, rt
Mg
O
H_3O^+
−78°C
OH
4
H_3O^+
0°C
2
3
CO_2H
OH
40°C
$-H_2O$
O
O
5
6

Scheme 4.4 Synthesis of spiro-γ-lactones.

Table 4.9 Synthesis of spiro-γ-lactones from conjugated diene, ketones, and CO_2.

Entry	Diene	Ketone	Product	% yield[a]
1	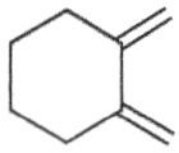	Acetone	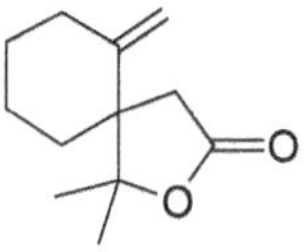	68
2	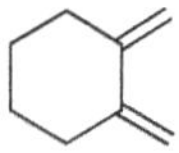	Cyclopentanone	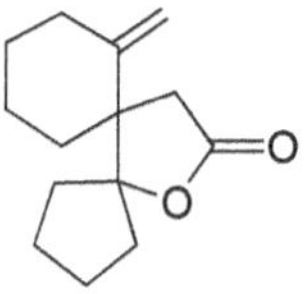	66
3	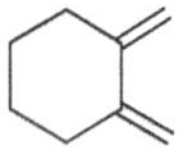	Cyclohexanone	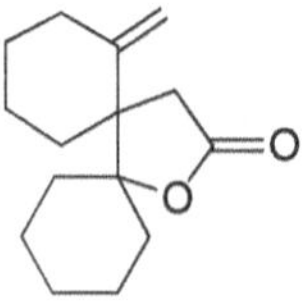	60
4		Cyclopentanone	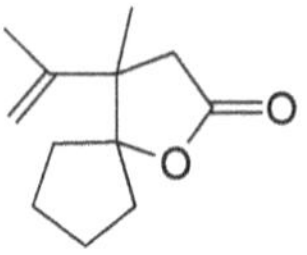	68
5		Cyclohexanone	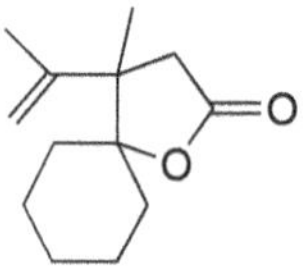	61

[a] Isolated yields.

We have reported an extension of this methodology in the production of spiro-δ-lactones [123]. Much work has been done to elucidate novel synthetic routes for these types of molecules [124]. In particular, δ-substituted δ-lactones have recently attracted considerable attention, mainly because molecules of this class include many natural products that exhibit significant biological activity [125].

Scheme 4.5 illustrates a route for the synthesis of spiro-δ-lactones from the magnesium complex of 1,2-bis(methylene)cyclohexane **1** [126]. Initially, treatment of the 1,2-bis(methylene)cyclohexane-magnesium reagent **2** with an excess of ethylene oxide at −78°C resulted in the formation of the 1,2-adduct of **3** by the incorporation of one equivalent of epoxides with the diene complex. Significantly, the bis-organomagnesium reagent **2** reacted with only one mole

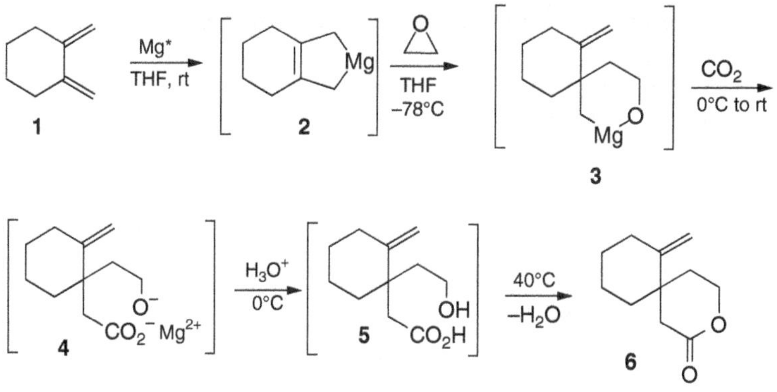

Scheme 4.5 Synthesis of spiro-δ-lactones.

of epoxide with 100% regioselectivity in the 2-position to give the intermediate **3**. Upon warming to 0°C, **3** reacted with CO_2 to yield the magnesium salt of a δ-hydroxy acid (**4**). Acidic hydrolysis followed by warming to 40°C generated the spiro-δ-lactone 7-methylene-3-oxaspiro[5.5]undecan-2-one (**6**) in 69% yield (Table 4.10, entry 1).

Importantly, the approach described in Scheme 4.5 can be used to prepare both bicyclic and tricyclic spiro-δ-lactones, and representative examples are listed in Table 4.10 (entries 3 and 4, respectively). For example, 1,2-bis(methylene) cyclohexane-magnesium reagent was treated with cyclohexane oxide at −78°C followed by the introduction of CO_2 at 0°C and acidic hydrolysis with subsequent warming to 40°C. Workup afforded the tricyclic spiro-δ-lactone hexahydro-2′-methylene-spiro[4*H*-1-benzopyran-4,1′-cyclohexan]-2(3*H*)-one in 63% isolated yield as a 1:1 mixture of diastereomers (Table 4.10, entry 4).

This method also exhibited good regioselectivity when an unsymmetrical epoxide was used as the primary electrophile. The formation of 6-butyltetrahydro-4-methyl-4-(1-methylethylene)-2*H*-pyran-2-one (Table 4.10, entry 5) demonstrates that the attack of the asymmetric epoxide occurs at the less sterically hindered carbon.

The generation of a quaternary carbon center is not a trivial undertaking in organic synthesis, and multiple synthetic steps are often required. In work related to our studies of the synthesis of spiro-δ-lactones [123], the preparation of alcohol and *vic*-diols has also been realized, utilizing this one-pot methodology [127]. In the process the formation of a quaternary carbon center is achieved. Scheme 4.6 shows that the initial adduct **3**, when treated with dilute acid at low temperature, affords the alcohol **4** containing a quaternary carbon center. It is important to note that in all cases where asymmetric epoxides were used, the bis-organomagnesium reagent **2** reacted with good regioselectivity at the least hindered carbon atom (Table 4.11, entries 3–6). Also, the organomagnesium

Table 4.10 Reactions of conjugated diene-magnesium reagents with epoxides followed by carbon dioxide.

Entry	Diene	Epoxide[a]	Product[b]	% yield[c]
1	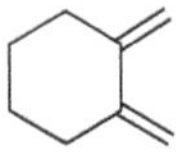		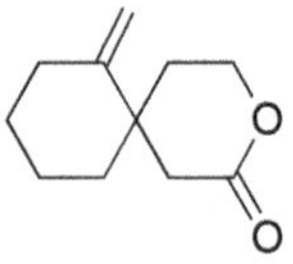	69
2			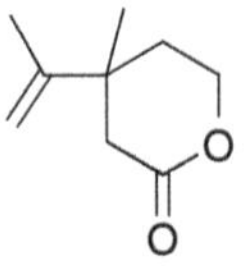	39[d]
3		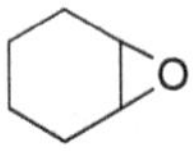	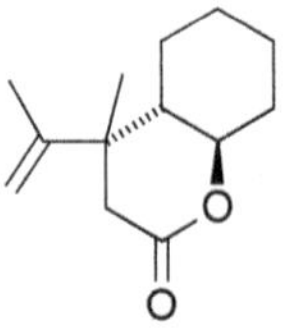	69[e]
4	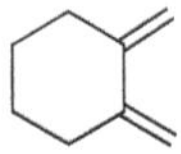	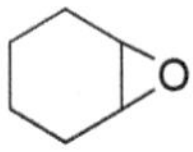	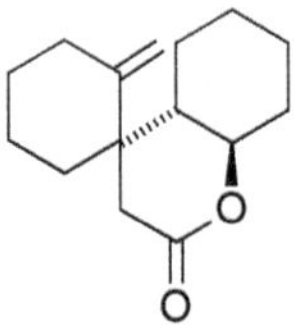	63[e]
5		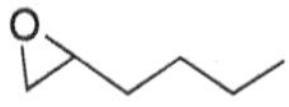	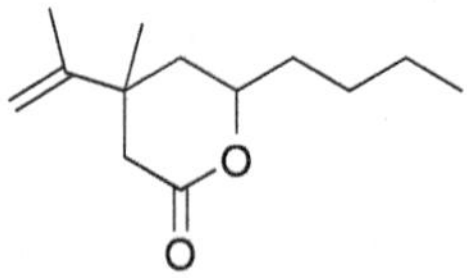	72[e]

[a] The epoxide was added to the diene-magnesium complex at −78°C, and the reaction mixture was stirred at −78°C for 30 min and then gradually warmed to 0°C followed by the bubbling of CO_2.
[b] Elemental analysis, mass spectra, ^{1}H NMR, ^{13}C NMR, and FTIR were all consistent with the indicated formulation.
[c] Isolated yields.
[d] Yield was based on the amount of active magnesium.
[e] A 1 : 1 mixture of diastereomers as determined by ^{1}H NMR.

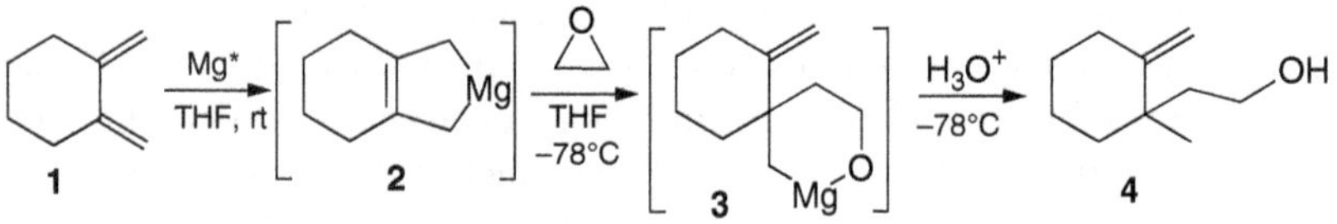

1 —Mg*, THF, rt→ [2] —THF, −78°C→ [3] —H_3O^+, −78°C→ 4

Scheme 4.6 Preparation of alcohols containing a quaternary carbon center.

Table 4.11 Reactions of conjugated diene-magnesium reagents with epoxides followed by acidic hydrolysis.

Entry	Diene	Epoxide	Product[a]	% yield[b]
1	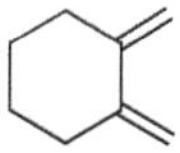		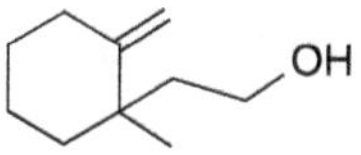 OH	86
2			 OH	60[c]
3		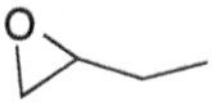	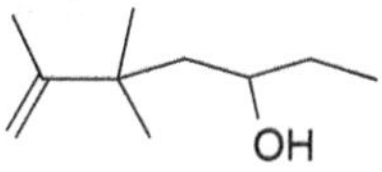 OH	64
4		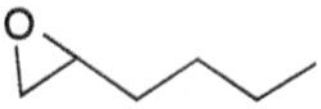	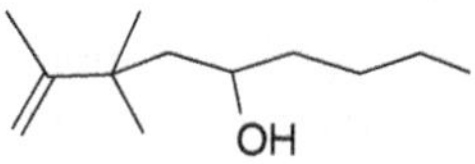 OH	90
5		 OH	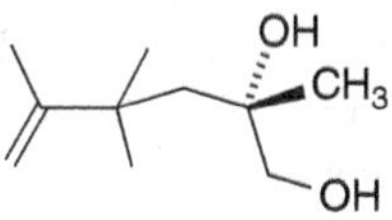 OH, CH_3, OH	47
6	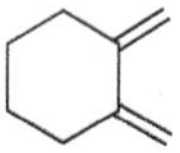	 OH	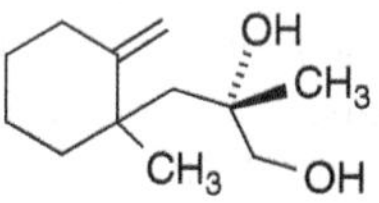 OH, CH_3, CH_3, OH	62[d]

[a] Elemental analysis, mass spectra, ^{1}H NMR, ^{13}C NMR, and FTIR were all consistent with the indicated formulation.
[b] Isolated yields.
[c] Yield was based on the amount of active magnesium.
[d] A 1:1 mixture of diastereomers as determined by ^{1}H NMR.

reagent **2** reacted with only one mole of epoxide in cases in which an excess of epoxide was used (Table 4.11, entries 1 and 2). Low-temperature protonation of the initial adduct **3** afforded a primary alcohol containing a quaternary center, 2-(β-hydroxyethyl)-2-methyl-1-methylene-cyclohexane (**4**), in excellent yield (Table 4.11, entry 1). An asymmetric chiral epoxide was utilized as the primary electrophile, followed by protonation which afforded a 1,2-diol. It was hoped that the attack of the organomagnesium intermediate **2** could be selectively induced to produce only one diastereomer. Unfortunately, the chiral epoxide did not exhibit any influence on the diastereoselectivity of the attack (Table 4.11, entry 6). However, the reactions proved that the organomagnesium reagent **2** will attack the epoxide even in the presence of the unprotected proximal hydroxyl functional group with no epimerization of the chiral center present in the epoxide.

The process presented here is a facile means for the preparation of alcohols and *vic*-diols in one pot in high yields.

4.10 Synthesis of γ-Lactams from Conjugated Diene-Magnesium Reagents

γ-Lactams are important intermediates in synthetic routes to five-membered heterocyclic compounds. Moreover, tetramic acids and 3-pyrrolin-2-ones represent a diverse and profoundly important family of biologically active secondary metabolites, many of which have potential uses in medicine and agriculture [128]. Synthetic interest in this class of molecules has been intense, particularly in the past decade [129].

Most approaches to γ-lactams have relied on cyclization via acyl-nitrogen bond formation [130]. Cyclization involving carbon–carbon bond formation is an alternative route; however, until recently this approach has received little attention. Mori and coworkers reported a palladium-catalyzed cyclization of *N*-allyl iodoacetamides, in which the intramolecular addition reaction of the carbon–iodine bond to an olefinic linkage is a key step [131].

Itoh and coworkers developed a new route to γ-lactams by the ruthenium-catalyzed cyclization of *N*-allyltrichloroacetamides [132]. Also, Stork and Mah have reported on the radical cyclization of *N*-protected haloacetamides to yield *N*-protected lactams. The protecting groups can then be easily removed under a variety of conditions [133]. This efficient radical cyclization route to *cis*-fused pyrrolidones and piperidones is interesting because of the widespread occurrence of related systems in natural products [134]. We have developed a direct method for the one-pot synthesis of γ-lactams and secondary amines utilizing conjugated diene-magnesium complexes [135].

Scheme 4.7 illustrates a route for the synthesis of a γ-lactam from the (2,3-dimethyl-2-butene-1,4-diyl)magnesium intermediate (**2**). Initially, treatment of **2** with *N*-benzylidenebenzylamine at −78°C resulted in the formation of the 1,2-adduct (**3**). Significantly, the bis-organomagnesium reagent (**2**) reacted with 100% regioselectivity in the 2-position to give the intermediate **3**. Upon warming to 0°C, **3** reacted with CO_2 to yield the magnesium salt of a γ-amino acid (**5**). Acidic hydrolysis followed by warming to 40°C generated β-(1-methylethenyl)-β-methyl-γ-phenyl-*N*-benzyllactam (**7**) in 67% isolated yield as a 75:25 mixture of diastereomers (Table 4.12, entry 2). Importantly, the generation of a highly substituted γ-lactam was accomplished in a one-pot process and in good overall chemical yield. This approach was equally applicable to 1,2-bis(methylene)cyclohexane and provided a facile route to a spiro-γ-lactam, β-((2-methylene)cyclohexyl)-γ-phenyl-*N*-benzyllactam, in 36% isolated yield.

Scheme 4.7 also illustrates a facile route to secondary amines containing a quaternary carbon center and a vinyl group in the β-position. When the

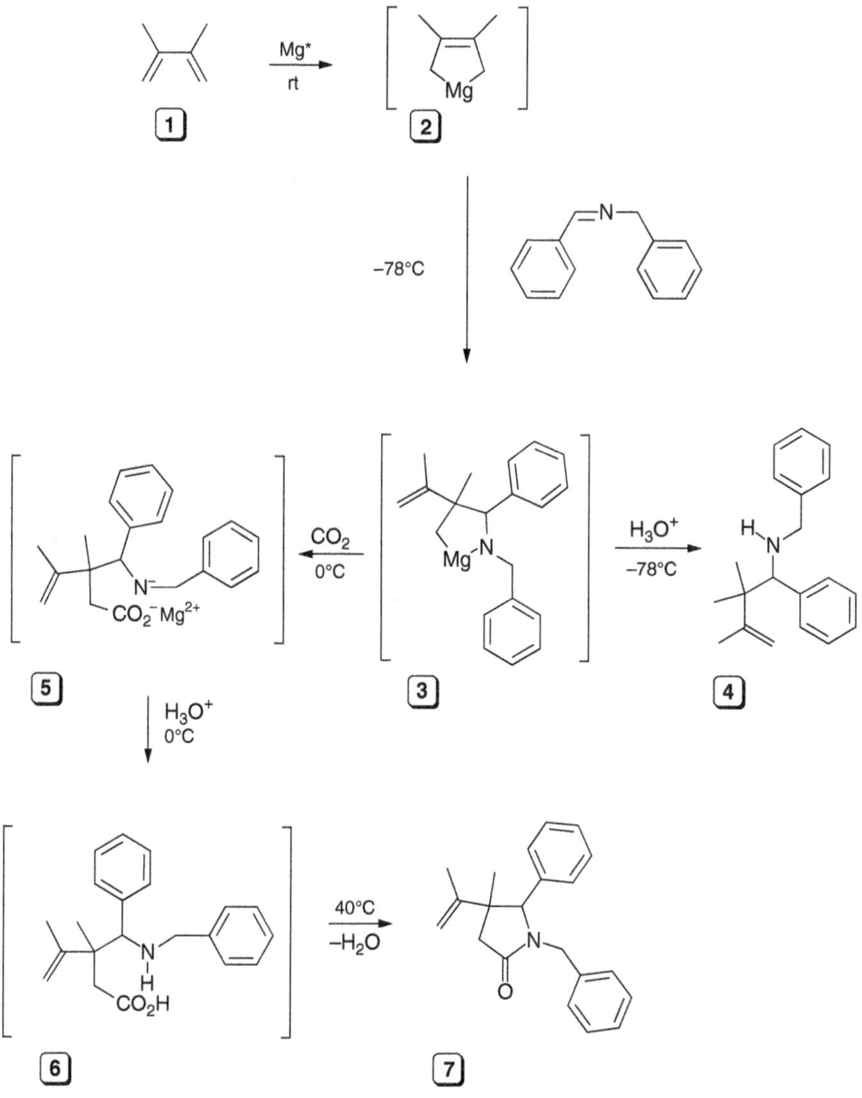

Scheme 4.7 Formation of spiro-γ-lactams.

magnesium-diene intermediate (**2**) was treated with *N*-benzylidenebenzylamine at −78°C, the initially formed adduct (**3**) could be easily hydrolyzed at 0°C to afford *N*-(1-phenyl-2,2,3-trimethylbut-3-en-1-yl)benzylamine (**4**) in 92% isolated yield (Table 4.13, entry 2). Other secondary amines are shown in Table 4.13 and all were prepared in high isolated yields.

Table 4.12 Lactamization of conjugated diene-magnesium reagents with imines and CO_2.

Entry	Diene	Imine	Product	% yield[a]
1	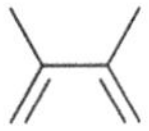	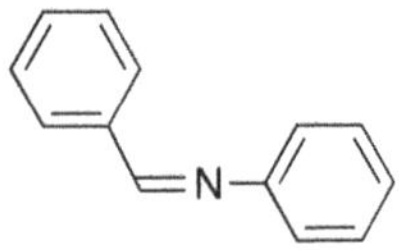	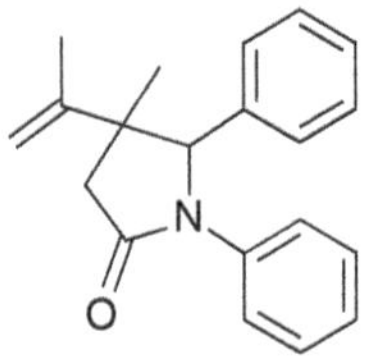	47 (80 : 20)
2	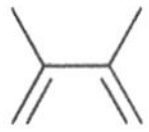	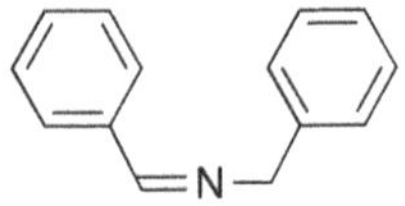	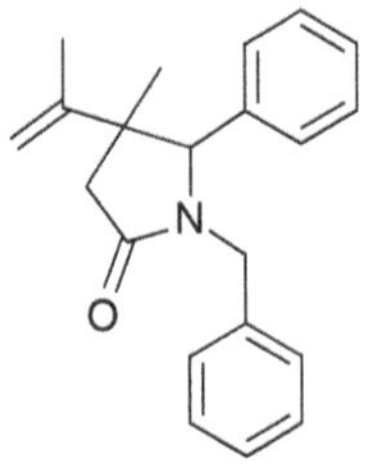	67 (75 : 25)
3	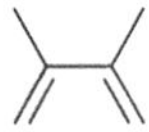	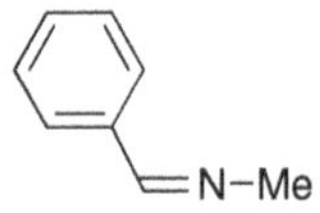	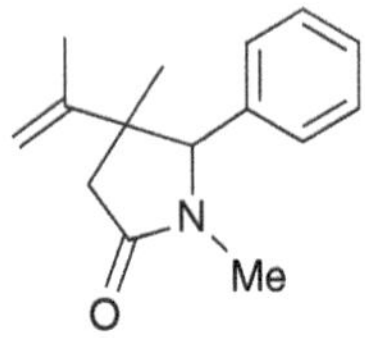	58 (70 : 30)

[a] Yield and diasternomeric ratio.

This facile one-pot transformation of a 1,3-diene-magnesium intermediate provides a direct route to the formation of both secondary amines and β,γ,*N*-trisubstituted γ-lactams in good to high isolated yields. The overall procedure of the γ-lactam synthesis can be thought of as a molecular assembling process in which three independent species, namely, a conjugated diene, an imine, and carbon dioxide, mediated by Rieke magnesium, are transformed into a complex organic molecule in a well-controlled fashion. In the process, the construction of a quaternary center and the introduction of both the amino and carboxyl groups required for lactamization are achieved in one synthetic step.

4.11 Low-Temperature Grignard Chemistry

Despite being over 100 years old, the Grignard reagent still plays a central role in synthetic chemistry today [136]. However, few functionalized Grignard reagents have been prepared due to the low functional group tolerance [137]. In 1998, several functionalized aromatic and vinylic Grignard reagents were prepared by

Table 4.13 Reactions of conjugated diene-magnesium reagents with imines followed by acidic hydrolysis.

Entry	Diene	Imine	Product	% yield[a]
1		=N–	H–N	87
2		=N–	H–N	92
3		=N–Me	H–N–Me	96
4		=N–	H–N	85

[a] Isolated yields based on the imine.

iodine–magnesium exchange using an excess of diisopropylmagnesium as metallating agent at a low temperature (−40°C). The yields were excellent to moderate after reaction with allyl bromide or benzaldehyde as electrophiles [138]. However, the thermodynamic nature of the equilibrium of exchange limits the range of substrates to acrylic and vinylic halides. Moreover, the transmetallation failed to afford functionalized Grignard reagents from brominated substrates. Chromium salts can form functionalized organochromium compounds, but the chromium salts needed a catalyst for the oxidative addition step, and the functionalized organochromium compounds reacted only with aldehydes so far [139]. We [140] reported the use of activated magnesium (Rieke magnesium) to prepare functionalized Grignard reagents from aryl bromides at low temperatures (−78°C). The Grignard reagents were reacted with several electrophiles such as benzaldehyde, benzoyl chloride, or allyl iodide. The reaction of the Grignard reagents with acid chlorides yielded ketones. Because the reactions were carried out at low temperatures, we observed little or no addition to the ketones [141]. The use of the functionalized Grignard reagents represents a straightforward method to synthesize polyfunctionalized organic molecules.

Results and Discussion

In general, the low functional group tolerance of Grignard reagents precludes the use of most groups in the same molecule. However, if the oxidative addition reaction is carried out at low temperatures (−78°C), the functionalized Grignard reagents are stable for a limited time. Functionalized Grignard reagents were successfully prepared by direct oxidative addition to aryl bromide substrates containing a nitrile, ester, or chloride group using highly active magnesium (Rieke magnesium) at low temperature (−78°C). The oxidative addition was rapid even at this temperature and was completed in 15 min (Scheme 4.8). The reaction with the electrophiles (PhCHO, allyl iodide, and PhCOCl using 10% CuI) was carried out at low temperatures (−78 and −40°C) within about 1 h, and the yields were good to moderate (Table 4.14).

When ethyl or methyl arylcarboxylates were used, that is, ethyl 3-bromobenzoate, the Grignard reagent did not form at −78°C. A possible explanation might be the coordination of the ester group to the magnesium surface, blocking the active sites of the metal and thus preventing the oxidative addition step. Raising the temperature to −50°C permits the formation of the organomagnesium reagents. However, at the higher temperatures, more by-products were formed, which limited the scope of the reaction. Other bidentate substrates containing groups of increased polarity, such as amides, failed to react with active magnesium at low temperature. A highly temperature-dependent equilibrium of adsorption–desorption of the functionalized group on the metal surface and active sites seems to be responsible for the capricious reactivity displayed by the halogenated substrates.

tert-Butyl arylcarboxylates proved to be good candidates for the preparation of functionalized Grignard reagents. The bulky ester group most likely prevented the coordination of the ester to the magnesium surface. The reactivity of the three isomers of the Grignard reagents also displayed important differences. Only the *para*-isomer gave moderate yields in the reaction with benzaldehyde. Entry 4 demonstrates that the highly reactive magnesium displays unusual selectivity in moderate yield.

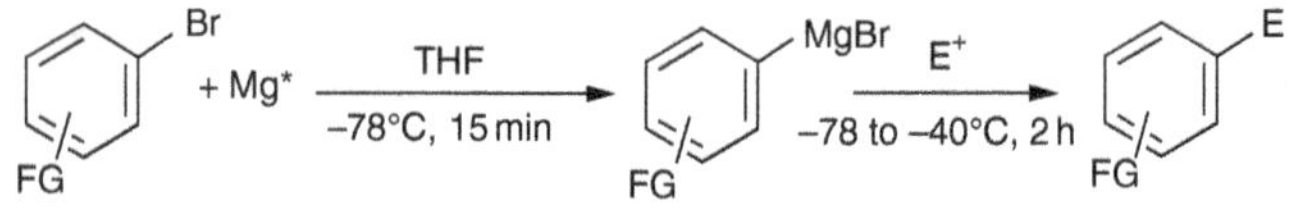

Br + Mg* —THF, −78°C, 15 min→ MgBr —E⁺, −78 to −40°C, 2 h→ E (FG-substituted aryl)

FG = −CN, $-CO_2R$, −OCOR, −Cl
R = *t*-butyl

Scheme 4.8 Low-temperature magnesium chemistry.

Table 4.14 Formation of the functionalized Grignard reagents and their reactions with electrophiles.

Entry	Substrate	Electrophile	Product	Yield (%)[a]
1	4-Br-C_6H_4-$CO_2{}^tBu$ **1a**	PhCHO	4-Ph(OH)HC-C_6H_4-$CO_2{}^tBu$ **2a**	86
2	4-Br-C_6H_4-OCO^tBu **1b**	PhCHO	4-Ph(OH)HC-C_6H_4-OCO^tBu **2b**	85
3	3-Br-C_6H_4-OCO^tBu **1c**	PhCHO	3-CH(OH)Ph-C_6H_4-OCO^tBu **2c**	76
4	2,4-Br_2-C_6H_3-OCO^tBu **1d**	PhCHO	4-Ph(OH)HC-2-Br-C_6H_3-OCO^tBu **2d**	53
5	4-Br-C_6H_4-CN **1e**	PhCHO	4-Ph(OH)HC-C_6H_4-CN **2e**	65
6	**1e**	PhCOCl	4-PhOC-C_6H_4-CN **2f**	62[b]
7	**1e**	CH_2–$CHCH_2I$	4-H_2C=HCH_2C-C_6H_4-CN **2g**	65[b]
8	4-Br-C_6H_4-Cl **1f**	PhCHO	4-Ph(OH)HC-C_6H_4-Cl **2h**	72

[a] Isolated yields.
[b] CuI 10 mol% (based on the substrate) was used.

The coupling reactions with the benzoyl chloride and allyl iodide required the addition of 10% CuI as catalyst to yield the corresponding ketone and allyl derivatives. In the absence of copper iodide catalyst, the prementioned two electrophiles gave only low yields of the expected products. 1,2-Dibromoethane was used to remove the excess of magnesium when benzoyl chloride or ally iodide was used as electrophiles (entries 6 and 7), and it did not react with the Grignard reagents under the given reaction conditions.

Finally, bromochlorobenzene is also an adequate substrate for this low-temperature chemistry (entry 8). We tested several additives, such as Me_2AlCl, LiBr, or TMSCl, on an attempt to enhance the reactivity of the resulting Grignard reagent. However, the presence of these additional Lewis acids failed to increase the yields.

When ethyl alkylcarboxylates were used, that is, ethyl 6-bromohexylbenzoate, some Grignard reagent was formed at −78°C along with two major nonidentified by-products, and some of the starting material still remained. From these data, we can conclude that simple alkyl esters are not suitable candidates for functionalized Grignard reagent preparation. Even when the functionalized Grignard reagents were made, they reacted with the ester group either in an intra- or intermolecular way. *tert*-Butyl 11-bromoundecanoate reacted with active magnesium to afford excellent yields of the corresponding Grignard reagent at −78°C within 15 min, but the yield of the reaction with the benzaldehyde was poor (≈30%).

In summary, several functionalized Grignard reagents containing an ester, nitrile, or chloride have been prepared at low temperature, and their reactivities were studied. The reported procedures involving the use of highly reactive magnesium expands the well-known methodologies of Grignard reagent preparation.

Typical Procedure for the Preparation of the Corresponding Grignard Reagents

To an oven-dried 50 ml centrifuge tube, equipped with a Teflon-coated magnetic stirring bar and septum, was added, under an atmosphere of prepurified argon, Mg* (3.0 mmol) in dry THF (5 ml) via syringe. This mixture was cooled to −78°C, and the organic substrate (1 mmol) was added dropwise with stirring via syringe. After 15 min the electrophile (1.1 mmol) was added in the same fashion at −78°C. The temperature was increased to −40°C for the reaction of the nitrile substrate with the benzoyl chloride and allyl iodide.

Reaction progress was monitored by gas chromatography (GC) of hydrolyzed reaction aliquots. After 2 h, the reaction was quenched with saturated aqueous NH_4Cl. The aqueous layers were extracted with diethyl ether three times. The organic layers were then combined, dried with $MgSO_4$, and filtered. The solvent was removed under vacuum. The crude product obtained was purified by flash chromatography.

4.12 Typical Procedures for Preparation of Active Magnesium and Typical Grignard Reactions as Well as 1,3-Diene Chemistry

Anhydrous Magnesium Salts

The preparation of Rieke magnesium is relatively straightforward, but a few factors must be strictly adhered to. It is absolutely critical that anhydrous magnesium salts be used as one of the starting materials. The use of wet or hydrated magnesium salts will lead to incomplete reductions and low reactivity. Moreover, incomplete reduction will mean leftover reducing metals which can be dangerous upon workup of reactions. We have had excellent success using anhydrous magnesium chloride and bromide purchased from Cerac, Inc., P.O. Box 1178, Milwaukee, Wisconsin 53201. Anhydrous magnesium chloride cannot, however, be prepared by heating the hexahydrate under vacuum, since hydrogen chloride is released before dehydration is complete. We have prepared active magnesium from anhydrous magnesium bromide and iodide; however, highly insoluble magnesium salts such as the fluoride or sulfate are not reduced. A small excess of magnesium chloride is used in this procedure to ensure that the potassium is completely consumed.

One can prepare anhydrous magnesium chloride or bromide very easily and at little cost by the following straightforward procedure. The following procedure is suitable for preparing both anhydrous magnesium chloride and bromide. The magnesium turnings and 1,2-dibromoethane used were purchased from Aldrich Chemical Company, Inc., Milwaukee, WI, USA. A 200 ml three-necked round-bottomed flask equipped with a magnetic stirring bar, two stoppers, and a condenser connected to an argon inlet is charged with 0.35 g of magnesium turnings, 50 ml of THF, and 3.0 g (0.016 mol) of 1,2-dibromoethane. The suspension is warmed gently, initiating the reaction. After the initially exothermic reaction subsides, the mixture is heated at reflux for 50 min. The solvent is evaporated under a reduced pressure of argon or nitrogen, leaving a white solid. The flask is then evacuated and heated in an oil bath at 150°C for 1 h. The dry magnesium bromide is ready for preparing active magnesium in the same flask.

Preparation of Rieke Magnesium Using Potassium or Sodium as Reducing Agent

A 200 ml three-necked round-bottomed flask equipped with a Teflon-coated magnetic stirring bar, rubber septum, and condenser connected to an argon inlet is charged with 1.5 g (0.038 g.-atom) of freshly cut potassium, 2.01 g (0.0211 mol) of anhydrous magnesium chloride, 3.55 g (0.0214 mol) of anhydrous potassium iodide, and 50 ml of THF. The mixture is stirred vigorously and heated to reflux with an electric heating mantle. A black precipitate starts to form within a few minutes. After 3 h at reflux temperature, the reduction should be complete, producing active magnesium as a black powder that settles very slowly when the stirring is stopped.

Purified grade potassium from J. T. Baker Chemical Company, Center Valley, PA, USA has been found to give the most consistent results. Very impure potassium or sodium generally gives magnesium powder with much reduced reactivity. Sodium may be used in place of potassium provided that the boiling point of the solvent chosen is higher than the melting point of the metal.

The potassium is usually cut into two or three pieces under hexane or heptane and placed wet in a tared flask that has been purged with argon. The flask is evacuated, removing the hydrocarbon, filled again with argon, and weighed to determine the exact amount of potassium. The amount of potassium used varied from 1.4 to 1.6 g, the weights of the other reagents being adjusted proportionately. With this procedure the pieces of potassium are shiny and relatively free from oxide coating. Alternatively, the potassium cuttings may be wiped free of solvent, quickly weighed in air, and placed in the flask.

Potassium iodide (>99% purity) from Allied Chemical Corporation or Mallinckrodt Chemical Works, St. Louis, MO, USA is finely ground with a mortar and pestle, dried overnight in an oven at 120°C, and stored in a desiccator. The molar ratio of potassium iodide to magnesium chloride is not highly critical and may vary from 0.05 to 2.0. However, the optimum ratio is 1:1. If the potassium iodide is omitted, the black magnesium powder produced reacts with bromobenzene at −78°C. However, since the magnesium prepared in this way does not react with fluorobenzene in refluxing THF, it is evidently less reactive than that produced in the presence of potassium iodide. We have found that diglyme and DME are also effective solvents. The reactivity of the magnesium obtained with DME as solvent is slightly reduced. Hydrocarbons, amines, and dioxane proved to be ineffective solvents, owing to the insolubility of the magnesium salts and consequent incomplete reduction.

Efficient stirring is essential for the generation of highly reactive magnesium. If the stirring is not effective, the reduction may not be complete after the 3 h reaction time. The remaining unreacted potassium is a fire hazard during the isolation of the product. If the scale of the reaction is increased, measures should be taken to ensure that effective stirring can be maintained throughout the reaction period. The mildly exothermic reduction may result in excessive foaming which carries potassium particles up into the condenser. This problem is avoided by using a relatively large flask (in this case, 200 ml instead of 100 ml) and by carefully controlling the temperature at the beginning of the reduction. The reduction appears to be essentially complete in 30–45 min. However, a reaction time of 3 h is recommended to ensure complete consumption of the potassium.

Preparation of a Grignard Reagent Using Rieke Magnesium Prepared Using Potassium–Potassium Iodide: 1-Norbornanecarboxyl Acid

The mixture of active magnesium metal and potassium salts as prepared before is allowed to cool to room temperature, after which 1.25 g (0.00958 mol) of 1-chloronorborane is injected with a syringe through the septum into the flask. The solid 1-chloronorborane was melted by warming on a steam bath and drawn into a syringe that had been warmed briefly in an oven. The reaction mixture is heated under reflux for 6 h and cooled to room temperature. A large excess of freshly sublimed dry ice chunks is added quickly to the Grignard reagent through the extra neck of the flask. The mixture is stirred vigorously, warmed to room temperature, acidified with 50 ml of 20% hydrochloric acid, and extracted with three 100 ml portions of diethyl ether. The combined ether layers are extracted with 100 ml of 10% aqueous sodium hydroxide. The alkaline solution is acidified with concentrated hydrochloric acid, and the acidic solution is extracted with two 100 ml portions of ether. The ether extracts are combined, washed with two 50 ml portions of water, dried over anhydrous sodium sulfate, and evaporated, giving 0.80–0.94 g (60–70%) of 1-norbornanecarboxylic acid as a slightly yellow crystalline solid, mp 106–109°C.

Preparation of Rieke Magnesium Using Lithium and Naphthalene as an Electron Carrier

We have developed the following procedure, which is suitable for preparing highly reactive magnesium powder, using lithium as a reducing agent. This procedure avoids potassium and produces a magnesium powder equal in reactivity to that obtained using potassium as the reducing agent. A 50 ml two-necked round-bottomed flask equipped with a Teflon-coated magnetic stirring bar, rubber septum, and condenser connected to an argon inlet is charged with 0.224 g (0.0325 g.-atom) of freshly cut lithium, 1.57 g (0.0165 mol) of anhydrous magnesium chloride, 0.436 g (0.00341 mol) of naphthalene, and 10 ml of THF. The mixture is stirred vigorously at room temperature for 24 h. After complete reduction, the highly reactive magnesium appears as a dark gray to black powder which slowly settles after stirring is stopped. In some cases, the THF has a slight olive green color due to a small amount of lithium naphthalide. This can be ignored when the highly reactive magnesium is reacted. If desired, this can be removed by withdrawing the THF with a syringe and adding fresh dry THF or other solvent.

Lithium (99.9%, rod, 1.27 cm dia.) from Alfa or a battery quality lithium ribbon lithium from FMC has been used extensively in our studies. The lithium rod is cut under oil, rinsed in hexane, and transferred to a tared 24/40 adapter with a stopcock and rubber septum which has been filled with argon. The adapter is evacuated, removing the hexane, filled with argon and weighed.

The lithium is then transferred to the reaction vessel under an argon stream. If lithium ribbon is used, it can be cut and weighed in a dry box under argon and then transferred to the reaction vessel.

It is important that the reaction be stirred vigorously and that the lithium make frequent contact with the stirring bar, as the lithium has a tendency to become coated with magnesium and stop the reduction from continuing. If the reduction does stop, it can be initiated again by gently rubbing the piece of lithium against the wall of the flask with a metal spatula; the rubber septum can be temporarily removed under a stream of argon to carry out this procedure.

Although we have never had a fire or explosion caused by active magnesium or other activated metals, we suggest extreme caution in working with these reactive materials, especially while the worker familiarizes him- or herself with the characteristics of each step in the procedure. We advise that the magnesium powder be kept under an argon atmosphere at all times as it can burst into flames when exposed to air.

Chemistry of (2-Butene-1,4-diyl)magnesium: Preparation of Activated Magnesium (Mg*)

Activated magnesium was prepared by the reduction of anhydrous magnesium chloride with lithium using naphthalene as an electron carrier. Highly reactive magnesium can also be prepared from the reduction of magnesium chloride by preformed lithium naphthalenide. In a typical preparation, lithium (10.0 mmol) and naphthalene (10.8 mmol) in freshly distilled THF (15 ml) were stirred under argon until the lithium was completely consumed (ca. 2 h). The resulting dark green lithium naphthalenide was then transferred dropwise via a cannula into a THF solution (10 ml) of anhydrous magnesium chloride (4.8 mmol). The mixture was stirred at room temperature for 30 min. The newly formed magnesium slurry was allowed to settle for at least 3 h, and then the supernatant was drawn off via a cannula. Freshly distilled THF was added, followed by the appropriate 1,3-diene. (Note: The number of millimoles of Mg* cited refers to the theoretical amount possible, based on the original amount of magnesium chloride.)

Typical Cyclization of (1,4-Diphenyl-2-butene-1,4-diyl)magnesium

(*E,E*)-1,4 Diphenyl-1,3-butadiene (0.825 g, 4.00 mmol) dissolved in 10 ml of THF was added via a cannula to the activated magnesium (4.80 mmol) in THF (20 ml) at room temperature. The reaction mixture turned purple immediately and then became red after the addition was completed. The mixture was stirred at room temperature for 2 h. The resulting dark red solution was cooled to −78°C. 1,3-Dichloropropane (0.507 g, 4.48 mmol) was added via a disposable syringe at −78°C. Stirring was continued at −78°C for 2 h. Then the reaction mixture was gradually warmed to room temperature and stirred for 30 min. An aqueous solution of 3 N HCl (10 ml) was added at 0°C, giving a clear solution. The reaction mixture was washed with diethyl ether (30 ml). The aqueous layer was extracted with diethyl ether (2 × 20 ml). The combined organic portions were washed with saturated aqueous $NaHCO_3$ (2 × 20 ml) and water (20 ml) and then dried over anhydrous $MgSO_4$. After evaporation of the solvent, the residue was flash chromatographed on silica gel, eluting sequentially with hexanes and 100:1 hexanes/Et_2O. *trans*-1-Phenyl-2-((*E*)-2-phenylethenyl) cyclopentane (0.807 g) was obtained in 81% isolated yield. This compound was also prepared in 65% isolated yield from **2a** and 1,3-dibromopropane.

Typical Reaction of (2,3-Dimethyl-2-butene-1,4-diyl)magnesium

An excess of freshly distilled 2,3-dimethyl-1,3-butadiene (1.5 ml) was added to activated magnesium (6.10 mmol) in THF (10 ml). The mixture was stirred at room temperature for 8 h, giving a pale orange solution. The THF solution of newly formed complex **2b** was cooled to −78°C, and 1,3-dibromopropane (1.079 g, 5.34 mmol) mixed with 10 ml of THF was added via a cannula. *n*-Undecane was added as an internal standard via a disposable syringe. The reaction was monitored by GC with an OV-17 column, and GC yield was based on the analyses of reaction quenches. The mixture was stirred at −78°C for 2 h and then gradually warmed to room temperature and stirred overnight. An aqueous solution of 3 N HCl (10 ml) was added at 0°C. The reaction mixture was washed with diethyl ether (30 ml). The aqueous layer was extracted with diethyl ether (2 × 20 ml), and the combined organic phases were washed with saturated aqueous $NaHCO_3$ (2 × 20 ml) and water (20 ml) and dried over anhydrous $MgSO_4$. Evaporation of solvents using a rotary evaporator at 0–5°C and flash chromatography (eluted by pentane only) gave 1-methyl-1-(1-methylethenyl)cyclopentane (**12**): 75% GC yield.

Typical Stepwise Reaction of (2,3-Dimethyl-2-butene-1,4-diyl) magnesium

Freshly formed complex **2b**, prepared from freshly distilled 2,3-dimethyl-1,3-butadiene (2 ml) and activated magnesium (7.32 mmol), in 10 ml of THF was cooled to −78°C. 1,4-Dibromobutane (1.299 g, 6.01 mmol) in 10 ml of THF was added dropwise via a cannula at −78°C. After being stirred at −78°C for 30 min, the reaction mixture was allowed to warm gradually to room temperature. Stirring was continued for 1 h at room temperature. The reaction flask was cooled to 0°C with an ice bath, and an excess of benzoyl chloride (1.111 g, 7.90 mmol) was added via a disposable syringe. The mixture was stirred for 1 h at 0°C and 1 h at room temperature. An aqueous solution of HCl (1.5 N, 15 ml) was added at 0°C. The mixture was washed with diethyl ether (20 ml).

The aqueous layer was extracted with diethyl ether (2 × 20 ml), and the combined organic portions were washed with brine (2 × 20 ml) and dried over $MgSO_4$. Evaporation of solvents and flash column chromatography (eluted by hexanes/Et_2O, 100:3) gave 7-bromo-2,3-dimethyl-3-((phenylcarbonyl)methyl)-1-heptene (**18**): 1.165 g, 60% yield.

Typical Regioselective Reaction of Unsymmetrical (2-Butene-1,4-diyl)magnesium

A THF solution of complex **2d** (20 ml), prepared from myrcene (0.281 g, 2.06 mmol, technical grade) and activated magnesium (3.44 mmol), was cooled to −78°C, and Me_3SiCl (0.171 g, 1.57 mmol) was added via a disposable syringe. Stirring was continued at −78°C for 1 h, and the reaction mixture was then gradually warmed to 0°C. Excess cyclohexanone (0.278 g, 2.83 mmol) was added at 0°C. The reaction mixture was warmed to room temperature and stirred for 1 h. An aqueous solution of HCl (1.5 N, 10 ml) was added at 0°C. The mixture was washed with diethyl ether (20 ml), and the aqueous layer was extracted with diethyl ether (2 × 20 ml). The combined organic phases were washed with a saturated aqueous solution of $NaHCO_3$ (2 × 15 ml) and brine (20 ml) and dried over $MgSO_4$. Removal of solvents and flash column chromatography (eluted by hexanes/Et_2O, 98:2) gave **32** (0.372 g, 77%) and **33** (0.025 g, 5%) (compound **33** was eluted out before compound **32**) in 82% total yield.

Typical Reaction of Unsymmetrical (2-Butene-1,4-diyl)magnesium with $SiCl_4$

Newly formed complex **2c**, prepared from isoprene (0.250 g, 3.67 mmol) and excess activated magnesium, in 20 ml of THF was cooled to −78°C. $SiCl_4$ (0.256 g, 1.50 mmol) was added via a disposable syringe. After being stirred at −78°C for 1 h, the mixture was gradually warmed to 0°C, and an aqueous solution of 1.5 N HCl (15 ml) was added. The reaction mixture was washed with diethyl ether (20 ml). The aqueous layer was extracted with diethyl ether (2 × 20 ml), and the combined organic parts were washed with saturated aqueous $NaHCO_3$ (2 × 20 ml) and brine (15 ml) and dried over anhydrous Na_2SO_4. Evaporation of solvents and flash column chromatography afforded 2,7-dimethyl-5-silaspiro[4.4]nona-2,7-diene (**42**): 0.185 g, 75%.

Typical Reaction with 1,2-Dimethylenecyclohexane

In a typical preparation, 1,2-dimethylenecyclohexane (2.0 mmol) was added via a disposable syringe to the newly prepared activated Mg* (3.0 mmol) in THF (15 ml). The mixture was stirred for 3–4 h at room temperature under argon. The yellowish gold THF solution of the complex was separated from the excess magnesium either by filtration or by cannulating the solution to another flask after the mixture had settled and the solution became transparent (ca. 2 h). The freshly prepared complex was then treated with appropriate electrophiles.

References

1 Frankland, E. *Justus Liebigs Ann. Chem.* 1849, **71**, 171.
2 Rieke, R.D. *Top. Curr. Chem.* 1975, **59**, 1; *Acc. Chem. Res.* **1977**, *10*, 301, and references therein.
3 (a) Hill, C.L.; Vander Sande, J.B.; Whitesides, G.M. *J. Org. Chem.* 1980, **45**, 1020. (b) Rogers, H.R.; Hill, C.L.; Fujiwara, Y.; Rogers, R.J.; Mitchell, H.L.; Whitesides, G.M. *J. Am. Chem. Soc.* 1980, **102**, 217. (c) Rogers, H.R.; Deutch, J.; Whitesides, G.M. *J. Am. Chem. Soc.* 1980, **102**, 226. (d) Rogers, H.R.; Rogers, R.J.; Mitchell, H.L.; Whitesides, G.M. *J. Am. Chem. Soc.* 1980, **102**, 231. (e) Barber, J.J.; Whitesides, G.M. *J. Am. Chem. Soc.* 1980, **102**, 239.
4 Nuzzo, R.G.; Dubois, L.H. *J. Am. Chem. Soc.* 1986, **108**, 2881.
5 Burns, T.P.; Rieke, R.D. *J. Org. Chem.* 1987, **52**, 3674.
6 (a) Gomberg, M.; Bachmann, W.E. *J. Am. Chem. Soc.* 1927, **49**, 236. (b) Kharasch, M.S.; Reinmuth, C. *Grignard Reactions of Nonmetallic Substances*; Prentice-Hall: New Brunswick, NJ, 1954.
7 (a) Walborsky, H.M.; Young, A.E. *J. Am. Chem. Soc.* 1961, **83**, 2595. (b) Walborsky, H.M. *Rec. Chem. Prog.* 1962, **23**, 75.
8 Grotveld, H.H.; Blomberg, C.; Bickelhaupt, F. *Tetrahedron Lett.*, 1991, **31**, 1971.
9 Garst, J.F.; Deutsch, J.E.; Whitesides, G.M. *J. Am. Chem. Soc.* 1986, **108**, 2490.
10 Ashby, E.C.; Oswald, J. *J. Org. Chem.* 1988, **53**, 6068.
11 (a) Walborsky, H.M.; Rachon, J. *J. Am. Chem. Soc.* 1989, **111**, 1897. (b) Rachon, J.; Walborsky, H.M. *Tetrahedron Lett.* 1989, **30**, 7345.
12 (a) Skell, P.S.; Westcott, L.D. *J. Am. Chem. Soc.* 1963, **85**, 1023. (b) Skell, P.S.; Westcott, D.; Goldstein, J.P.; Engel, R.R. *J. Am. Chem. Soc.* 1965, **87**, 2829.
13 Andrews, L.; Pimentel, G.C. *J. Chem. Phys.* 1966, **44**, 2527.
14 Mile, B. *Angew. Chem. Int. Ed. Engl.* 1968, **7**, 507.
15 Timms, P.L. *Chem. Commun.* 1969, 1033.
16 Timms, P.L. *Adv. Inorg. Chem. Radiochem.* 1972, **14**, 121.
17 Moskovits, M.; Oziss, G.A. *Cryochemistry*; Wiley-Interscience: New York, 1976.
18 Blackborow, J.R.; Young, D. *Metal Vapor Synthesis in Organometallic Chemistry*; Springer-Verlag: Berlin, 1979.
19 Klabunde, K.J. *Chemistry of Free Atoms and Particles*; Academic Press: New York, 1980.
20 Francis, C.G.; Timms, P.L. *J. Chem. Soc. Dalton Trans.* 1980, **14**, 1401.

21 Francis, C.G.; Ozin, G.A. *J. Macromol. Sci. A Chem.* 1981, **16**, 167.
22 (a) Suslick, K.S. *Adv. Organomet. Chem.* 1986, **25**, 73. (b) Suskick, K.S. *ACS Symp. Ser.* 1987, **333**, 191.
23 Boudjouk, P. *ACS Symp. Ser.* 1987, **333**, 209.
24 Abdulla, R.F. *Aldrichim. Acta* 1988, **21**, 31.
25 (a) Luche, J.L.; Petrier, C.; Lansard, J.P.; Greene, A.E. *J. Org. Chem.* 1983, **48**, 3837. (b) Petrier, C.; Luche, J.L.; Dupuy, C. *Tetrahedron Lett.* 1986, **25**, 3463.
26 Rieke, R.D.; Burns, T.P.; Wehmeyer, R.M.; Kahn, B.E. *ACS Symp. Ser.* 1987, **333**, 223, and references therein.
27 Rieke, R.D.; Hudnall, P.M. *J. Am. Chem. Soc.* 1972, **94**, 7178.
28 Rieke, R.D.; Hudnall, P.M.; Uhm, S. *J. Chem. Soc. Chem. Commun.* 1973, 269.
29 Rieke, R.D.; Bales, S.E. *J. Chem. Soc. Chem. Commun.* 1973, 269.
30 Rieke, R.D.; Bales, S.E. *J. Am. Chem. Soc.* 1974, **96**, 1775.
31 Rieke, R.D.; Chao, L. *Synth. React. Inorg. Met.-Org. Chem.* 1974, **4**, 101.
32 Rieke, R.D.; Ofele, K.; Fischer, E.O. *J. Organomet. Chem.* 1974, **76**, C19.
33 Rieke, R.D.; Wolf, W.J.; Kujundzic, N.; Kavaliunas, A.V. *J. Am. Chem. Soc.* 1977, **99**, 4159.
34 Chao, L.; Rieke, R.D. *J. Organomet. Chem.* 1974, **67**, C64.
35 Chao, L.; Rieke, R.D. *Synth. React. Inorg. Met.-Org. Chem.* 1974, **4**, 373.
36 Chao, L.; Rieke, R.D. *Synth. React. Inorg. Met.-Org. Chem.* 1975, **5**, 165.
37 Chao, L.; Rieke, R.D. *J. Org. Chem.* 1975, **40**, 2253.
38 Rieke, R.D.; Uhm, S.J. *Synthesis* 1975, 452.
39 Rieke, R.D.; Kavaliunas, A.V.; Rhyne, L.D.; Fraser, D.J. *J. Am. Chem. Soc.* 1979, **101**, 246.
40 Rieke, R.D.; Bales, S.E.; Hudnall, P.M.; Poindexter, G.S. *Org. Synth.* 1979, **59**, 85.
41 Rieke, R.D.; Kavaliunas, A.V. *J. Org. Chem.* 1979, **44**, 3069.
42 Rieke, R.D.; Rhyne, L.D. *J. Org. Chem.* 1979, **44**, 3445.
43 Kavaliunas, A.V.; Rieke, R.D. *J. Am. Chem. Soc.* 1980, **102**, 5944.
44 Rieke, R.D.; Li, P.T.; Burns, T.P.; Uhm, S.T. *J. Org. Chem.* 1981, **46**, 4323.
45 Inaba, S.; Matsumoto, H.; Rieke R.D. *Tetrahedron Lett.* 1982, **23**, 4215.
46 Kavaliunas, A.V.; Taylor, A.; Rieke, R.D. *Organometallics* 1983, **2**, 377.
47 Matsumoto, H.; Inaba, S.; Rieke, R.D. *J. Org. Chem.* 1983, **48**, 840.
48 Inaba, S.; Rieke, R.D. *Tetrahedron Lett.* 1983, **24**, 2451.
49 Burns, T.P.; Rieke, R.D. *J. Org. Chem.* 1983, **48**, 4141.
50 Rieke, R.D.; Bales, S.E.; Hudnall, P.M.; Burns, T.P.; Poindexter, G.S. *Org. Synth. Coll.* 1988, **6**, 845.
51 Rochfort, G.L.; Rieke, R.D. *Inorg. Chem.* 1984, **23**, 787.
52 Inaba, S.; Matsumoto, H.; Rieke, R.D. *J. Org. Chem.* 1984, **49**, 2093.
53 Inaba, S.; Rieke, R.D. *Chem. Lett.* 1984, 25.
54 Inaba, S.; Rieke, R.D. *Synthesis*, 1984, 842.
55 Inaba, S.; Rieke, R.D. *Synthesis*, 1984, 844.
56 Inaba, S.; Rieke, R.D. *J. Org. Chem.* 1985, **50**, 1373.

57 Inaba, S.; Rieke, R.D. *Tetrahedron Lett.* 1985, **26**, 155.
58 Burkhardt, E.; Rieke, R.D. *J. Org. Chem.* 1985, **50**, 416.
59 Ebert, G.W.; Rieke, R.D. *J. Org. Chem.* 1984, **49**, 5280.
60 Rochfort, G.L.; Rieke, R.D. *Inorg. Chem.* 1986, **25**, 348.
61 Wehmeyer, R.M.; Rieke, R.D. *J. Org. Chem.* 1987, **52**, 5056.
62 Wu, T.-C.; Wehmeyer, R.M.; Rieke, R.D. *J. Org. Chem.* 1987, **52**, 5057.
63 Inaba, S.; Wehmeyer, R.M.; Forkner, M.W.; Rieke, R.D. *J. Org. Chem.* 1988, **53**, 339.
64 Rieke, R.D.; Kavaliunas, A.V. *Organomet. Synth.* 1988, **4**, 319.
65 Kahn, B.E.; Rieke, R.D. *Organometallics* 1988, **7**, 463.
66 Kahn, B.E.; Rieke, R.D. *Chem. Rev.* 1988, **88**, 733.
67 Kahn, B.E.; Rieke, R.D. *J. Organomet. Chem.* 1988, **346**, C45.
68 Wu, T.-C.; Rieke, R.D. *J. Org. Chem.* 1988, **53**, 2381.
69 Wehmeyer, R.M.; Rieke, R.D. *Tetrahedron Lett.* 1988, **29**, 4513.
70 Ebert, G.W.; Rieke, R.D. *J. Org. Chem.* 1988, **53**, 4482.
71 Wu, T.-C.; Rieke, R.D. *Tetrahedron Lett.* 1988, **29**, 6753.
72 Rieke, R.D.; Wehmeyer, R.M.; Wu, T.-C.; Ebert, G.W. *Tetrahedron* 1989, **45**, 443 (Symposium-in-Print on Organcopper Chemistry).
73 Xiong, H.; Rieke, R.D. *J. Org. Chem.* 1989, **54**, 3247.
74 Rieke, R.D.; Wu, T.-C.; Stinn, D.E.; Wehmeyer, R.M. *Synth. Commun.* 1989, **19**, 1833.
75 Rieke, R.D. *Science* 1989, **246**, 1260.
76 O'Brien, R.A.; Rieke, R.D. *J. Org. Chem.* 1990, **55**, 788.
77 Wu, T.-C.; Xiong, H.; Rieke, R.D. *J. Org. Chem.* 1990, **55**, 5045.
78 Rieke, R.D.; Dawson, B.T.; Stack, D.E.; Stinn, D.E. *Synth. Commun.* 1990, **20**, 2711.
79 Zhu, L.; Wehmeyer, R.M.; Rieke, R.D. *J. Org. Chem.* 1991, **56**, 1445.
80 Stack, D.E.; Dawson, B.T.; Rieke, R.D. *J. Am. Chem. Soc.* 1991, **113**, 4672.
81 Rieke, R.D.; Xiong, H. *J. Org. Chem.* 1991, **56**, 3109.
82 (a) Csuk, R.; Glanzer, B.L. Furstner, A. *Adv. Organomet. Chem.* 1988, **28**, 85. (b) Savoia, D.; Tombini, C.; Umani-Ronchi, A. *Pure Appl. Chem.* 1985, **57**, 1877. (c) Bogdanovic, B. *Acc. Chem. Res.* 1988, **21**, 261. (d) Marceau, P.; Gautreau, L.; Beguin, F. *J. Organomet. Chem.* 1991, **21**, 403. (e) Tsai, K.-L; Dye, J. *J. Am. Chem. Soc.* 1991, **113**, 1650.
83 Fujita, T.; Watanaba, S.; Suga, K.; Sugahara, K.; Tsuchimoto, K. *Chem. Ind. (Lond.)* 1983, **4**, 167.
84 Normant, H. *Compt. Rend.* 1955, **240**, 1111.
85 Normant, H. *Bull. Soc. Chim. Fr.* 1957, **144**.
86 Ramsden, H.E.; Balint, A.E.; Whitford, W.R.; Walburn, J.J.; Serr, R.C. *J. Org. Chem.* 1957, **22**, 1201.
87 Ransden, H.E.; Leebrick, J.R.; Rosenberg, S.D.; Miller, E.H.; Walburn, J.J.; Balint, A.E.; Serr, R.C. *J. Org. Chem.* 1957, **22**, 1602.
88 Marvel, C.S.; Wollford, R.G. *J. Org. Chem.* 1958, **23**, 1658.

89 (a) Fuson, R.C.; Hammann, W.C.; Jones, P.R. *J. Am. Chem. Soc.* 1957, **79**, 928. (b) Baker, K.V.; Brown, J.M.; Hughes, N.; Skarnulis, A.J.; Sexton, A. *J. Org. Chem.* 1991, **56**, 698.

90 (a) Gilman, H.; St. John, N.B. *Recl. Trav. Chim. Pays-Bas* 1930, **49**, 717. (b) Gilman, H.; Kirby, R.H. *Recl. Trav. Chim. Pays-Bas* 1935, **54**, 577.

91 Pearson, E.; Cowan, D.; Becker, J.D. *J. Org. Chem.* 1959, **24**, 504.

92 Luche, J.-L.; Damiano, J.-C. *J. Am. Chem. Soc.* 1980, **102**, 7926.

93 For a review, see Ioffe, S.T.; Nesmayanov, A.N. *Methods of Elemento-organic Chemistry*; North Holland Publishing. Co.: Amsterdam, 1967; Vol. **2**, p. 18.

94 Gomberg, M.; Cove, L.H. *Berichte* 1906, **39**, 3274.

95 Pink, H.S. *J. Chem. Soc.* 1923, **123**, 3418.

96 John, E.; John, N. *Recl. Trav. Chim. Pays-Bas* 1936, **55**, 585.

97 Normant, H. *C. R. Acad. Sci.* 1954, **239**, 1510.

98 Krause, E.; Weinberg, K. *Berichte* 1929, **62**, 2235.

99 (a) Ashby, E.C.; Yu, S.H.; Beach, R.G. *J. Am. Chem. Soc.* 1970, **92**, 433. (b) Yu, S.H.; Ashby, E.C. *J. Org. Chem.* 1971, **36**, 2123.

100 Cook, A.H.; Jones, D.G. *J. Am. Chem. Soc.* 1941, **278**.

101 Scholl, R.; Norr, W. *Chem. Ber.* 1900, **33**, 1054.

102 Yanagida, S.; Yokoe, M.; Katagiri, I.; Ohoka, M.; Komori, S. *Bull. Chem. Soc. Jpn.* 1973, **46**, 306.

103 Cairns, T.L.; Larchar, A.W.; McKusick, B.C. *J. Am. Chem. Soc.* 1952, **74**, 5633.

104 Anker, R.M.; Cook, A.H. *J. Chem. Soc.* 1941, 329.

105 Hofmann, C. *Chem. Ber.* 1868, **1**, 198.

106 Anker, R.M.; Cook, A.H. *J. Chem. Soc.* 1941, 324.

107 Meyer, E. *J. Prakt. Chem.* 1888, **37**(2), 397.

108 Baerts, F. *Chem. Zentralbl.* 1923, **111**, 124.

109 Maercker, A.J. *Organomet. Chem.* 1969, **18**, 249.

110 (a) Fujita, K.; Ohnuma, Y.; Yasuda, H.; Tani, H. *J. Organomet. Chem.* 1976, **113**, 201. (b) Yang, M.; Yamamoto, K.; Otake, N.; Ando, M.; Takase, K. *Tetrahedron Lett.* 1970, **30**, 3843. (c) Nakamo, Y., Natsukawa, K.; Yasuda, K.; Tani, H. *Tetrahedron Lett.* 1972, **32**, 2833. (d) Yasuda, H.; Nakano, Y.; Natsukawa, K.; Tani, H. *Macromolecules* 1978, **11**, 586.

111 (a) Herberich, G.E.; Boveleth, W.; Hessner, B.; Hostalek, M.; Koeffer, D.P.J.; Ohst, H.; Soehnen, D. *Chem. Ber.* 1986, **119**, 420. (b) Richter, W.J. *Chem. Ber.* 1983, **116**, 3293.

112 (a) For review articles see Erker, G.; Kruger, C.; Muller, G. *Adv. Organomet. Chem.* 1985, **24**, 1. (b) Yasuda, H.; Tasumi, K.; Nakamura, A. *Acc. Chem. Res.* 1985, **18**, 120. (c) Walther, D.; Pfuetrenreuter, C. *Naturwiss. Reihe*, 1985, **34**, 789.

113 Kai, Y.; Kanehisa, N.; Miki, K.; Kasai, N.; Mashima, K.; Yasuda, H.; Nakamura, A. *Chem. Lett.* 1982, 1277.

114 Mandrou, A.-M.; Potin, P.; Wylde-Lachazette, R. *Bull. Soc. Chim. Fr.* 1962, 1546.

115 Rieke, R.D.; Xiong, H. *J. Org. Chem.* 1992, **57**, 6560.
116 Xiong, H.; Rieke, R.D. *J. Am. Chem. Soc.* 1992, **114**, 4415.
117 (a) Danheiser, R.L.; Martinex-Davila, C.; Morin, J.M. *J. Org. Chem.* 1980, **45**, 1340. (b) Danheiser, R.L., Martinez-Davila, C.; Auchus, R.J.; Kadonaga, J.T. *J. Am. Chem. Soc.* 1981, **103**, 2443.
118 (a) For a review, see Goldschmidt, Z.; Crammer, B. *Chem. Soc. Rev.* 1988, **17**, 229. For representative references, see (b) Wasserman, H.H.; Cochoy, R.E.; Baird, M.S. *J. Am. Chem. Soc.* 1969, **91**, 2375. (c) Trost, B.M.; Lee, D.C. *J. Am. Chem. Soc.* 1988, **110**, 6556. (d) Ollivier, J.; Salaun, J. *Tetrahedron Lett.* 1984, **25**, 1269. (e) Trost, B.M.; Mao, M.K. *J. Am. Chem. Soc.* 1983, **105**, 6753.
119 Rieke, R.D.; Sell, M.S.; Xiong, H. *J. Am. Chem. Soc.* 1995, **117**, 5429.
120 (a) Alonso, D.; Font, J.; Ortuno, R.M. *J. Org. Chem.* 1991, **56**, 5567. (b) Mudryk, B.; Shook, C.A.; Cohen, T. *J. Am. Chem. Soc.* 1990, **112**, 6389. (c) Fristad, W.E.; Hershberger, S. *J. Org. Chem.* 1985, **50**, 1026. (d) Trost, B.M.; Mao, M.K. *J. Am. Chem. Soc.* 1983, **105**, 6753.
121 Canonne, P.; Belanger, D.; Lemay, G. *J. Org. Chem.* 1982, **47**, 3953.
122 (a) Rieke, R.D.; Sell, M.S.; Xiong, H. *J. Org. Chem.* 1995, **60**, 5143. (b) Sell, M.S.; Xiong, H.; Rieke, R.D. *Tetrahedron Lett.* 1993, **34**, 6007, *Tetrahedron Lett.* **1993**, *34*, 6011. (c) Xiong, H.; Rieke, R.D. *J. Org. Chem.* 1992, **57**, 7007.
123 Sell, M.S.; Xiong, H.; Rieke, R.D. *Tetrahedron Lett.* 1993, **34**, 6007.
124 Set, L.; Cheshire, D.; Clive, D.L.; *J. Chem. Soc. Chem. Commun.* 1985, 1205.
125 (a) Barua, N.C.; Schmidt, R.R. *Synthesis* 1986, 1067. (b) Thompson, C.M. *Tetrahedron Lett.* 1987, **28**, 4243. (c) Canonne, P.; Belanger, G.; Lemay, G.; Foscolos, G.B. *J. Org. Chem.* 1981, **46**, 3091.
126 Le, N.; Jones, M.; Bickelhaupt, F.; deWolf, W.H. *J. Am. Chem. Soc.* 1989, **111**, 8691.
127 Sell, M.S.; Xiong, H.; Rieke, R.D. *Tetrahedron Lett.* 1993, **34**, 6011.
128 Laskin, A.I.; Lechevalier, H.A. *CRC Handbook of Microbiology*, 2nd ed.; CRC: New York, 1984; *Vol.* **5**, pp. 575–581, and references therein.
129 Paquette, L.A.; MacDonald, D.; Anderson, L.G.; Wright, J. *J. Am. Chem. Soc.* 1989, **111**, 8037, and references therein.
130 Frank, R.L.; Schmitz, W.R.; Zeidman, B. *Org. Synth.* 1955, **3**, 328.
131 Mori, M.; Oda, I.; Ban, Y. *Tetrahedron Lett.* 1982, **23**, 5315.
132 Nagashima, H.; Wakamatsu, H.; Ozaki, N.; Ishii, T.; Watanabe, M.; Tajima, T.; Itoh, K. *J. Org. Chem.* 1992, **57**, 1682.
133 Stork, G.; Mah, T. *Heterocycles* 1989, **28**, 723.
134 (a) Bachi, M.D.; Frolow, F.; Hornaert, C. *J. Org. Chem.* 1983, **48**, 1841. (b) Choi, J.K.; Hart, D.J. *Tetrahedron* 1985, **41**, 3959. (c) Danishefsky, S.J.; Panek, J.S. *J. Am. Chem. Soc.* 1987, **109**, 917.
135 Sell, M.S.; Klein, W.R.; Rieke, R.D. *J. Org. Chem.* 1995, **60**, 1077.
136 Grignard, V.C.R. *Acad. Sci.* 1900, **130**, 1322.
137 (a) Silverman, G.S.; Rakita, P.E. *Handbook of Grignard Reagents*; Marcel Dekker: New York, 1996. (b) Burns, T.P.; Rieke, R.D. *J. Org. Chem.* 1983, **48**, 4141.

138 Knochel, P.; Cahiez, G.; Boymond, L.; Rottlander, M. *Angew. Chem. Int. Ed.* 1998, **37**, 1701.

139 (a) Takai, K.; Nitta, K.; Fijimura, O.; Utimoto, K. *J. Org. Chem.* 1988, **54**, 4732. (b) Chen, C.; Tagami, K.; Kishi, Y. *J. Org. Chem.* 1995, **60**, 5386. (c) Furstner, A.; Shi, N. *J. Am. Chem. Soc.* 1996, **118**, 2533.

140 Lee, J.; Velarde-Ortiz, R.; Guijarro, A.; Wurst, J.; Rieke, R.D. *J. Org. Chem.* 2000, **65**, 5428.

141 (a) Sato, F.; Inoue, M.; Oguro, K.; Sato, M. *Tetrahedron Lett.* 1979, **44**, 4304. (b) Eberle, M.K.; Kahle, G.G. *Tetrahedron Lett.* 1980, **21**, 2303.

5

Copper

5.1 Background of Copper and Organocopper Chemistry

Copper metal is one of the oldest metals known to man. Copper beads found in Iraq date back to 9000 BC. Methods for refining copper from ores were discovered around 5000 BC. The pure metal has little strength and was of little use for tools or weapons. Around 5000 BC, it was discovered that copper could form an alloy with zinc to yield brass, a much stronger metal. Around this time, a second alloy was discovered made with copper and tin, yielding bronze. The use of copper in organic chemistry had to wait well into the twentieth century before it made any significant impact. The majority of the early studies on generating organocopper reagents relied primarily on transmetallation chemistry. The reactions generally involved the addition of a Grignard or organolithium reagent to a copper(I) salt. The use of organocopper reagents in organic synthesis has grown dramatically over the past decades. Both their ease of preparation and their ability to react with other substrates have led to their widespread use and acceptance. However, ordinary copper metal is not sufficiently reactive to add oxidatively to organic halides. This is evidenced by the Ullmann biaryl synthesis, which involves the reaction of copper bronze with aryl iodides to form biaryl compounds [1]. Typically, these reactions are carried out in sealed tubes at temperatures of 100–300°C for long periods of time ranging from hours to days. Of much more synthetic utility are organocopper, lithium diorganocuprate, and various heterocuprate reagents developed over the past years. The vast majority of organocopper reagents are prepared by a transmetallation reaction involving an organometallic reagent and a copper(I) salt. A variety of organometallic reagents, derived from metals more electropositive than copper, have been utilized in the preparation of these organocopper reagents. The most common organometallic precursors have been organomagnesium and organolithium reagents. However, this approach severely limits the functionalities that may be incorporated into the organocopper reagent.

Chemical Synthesis Using Highly Reactive Metals, First Edition. Reuben D. Rieke.

The use of the traditional organolithium or Grignard precursors can be circumvented by using Rieke's highly reactive zerovalent copper, which undergoes direct oxidative addition to a wide variety of organic halides [2]. Significantly, the organocopper reagents prepared in our laboratories with this form of active copper may incorporate a wide variety of functionalities, such as allyl, nitrile, chloride, fluoride, epoxide, ketone, and ester moieties. The functionalized organocopper reagents undergo many of the same reactions as other organocopper species, including cross-coupling reactions with acid chlorides, 1,4-conjugate additions with α,β-unsaturated carbonyl compounds, and intermolecular and intramolecular epoxide-opening reactions.

5.2 Development of Rieke Copper

Initial attempts to prepare highly reactive copper in our laboratories were only partially successful due to rapid sintering of the finely divided copper [3]. This method involved the reduction of cuprous iodide with a stoichiometric amount of potassium metal, along with a catalytic amount of naphthalene (10 mol%) as an electron carrier, under refluxing condition in 1,2-dimethoxyethane (DME) under argon. The reductions are usually complete within 8–12 h at room temperature. While the heterogeneous copper metal formed by this process was found to be more reactive than ordinary copper metal, the active copper could not undergo oxidative addition at temperatures low enough to form stable organocopper reagents. Consequently, the active copper formed by this method was prone to sinter, giving larger particles upon extended stirring or prolonged periods of standing. As a result, the reactivity of the copper metal was reduced. It was also found that substitution of the potassium metal by lithium results in much longer reduction times, giving a copper metal that was relatively unreactive. The reactivity of the copper metal was determined by measuring its ability to homocouple organic halides, analogous to the Ullmann reaction. It was evident from this initial work that copper could be activated by the Rieke method. Indeed, pentafluoroiodobenzene (C_6F_5I) easily formed the homocoupled product, $C_{12}F_{10}$, under very mild conditions in high yield when compared to the classical Ullmann reaction. Similarly, allyl iodide also homocoupled in high yield. However, the copper was not reactive enough to give good yields of homocoupled products for C_6F_5Br, *p*-$NO_2C_6H_4Br$, or *n*-butyl iodide. Although this copper was found to be more reactive than the ordinary copper bronze used in the traditional Ullmann reaction, it lacked the reactivity necessary to react with the majority of organic halides.

When compared with other metals produced by the Rieke method of metal activation, copper suffers from the disadvantage of sintering into larger particles, thereby reducing its reactivity. It was apparent that the main difficulty associated with the copper metal sintering was the exceptionally long reduction time. Previous reports from our laboratory have shown a correlation between particle size and reactivity [4]. It was anticipated that the reduction of a soluble copper complex would result in a short reduction time, thereby giving a finely divided copper metal. It was later discovered that a highly reactive zerovalent copper solution could be prepared by the reduction of a soluble copper(I) salt complex ($CuI{\cdot}PEt_3$) by a stoichiometric amount of preformed lithium naphthalenide in an ethereal solvent under inert, ambient conditions [5]. By using this method the reduction time was substantially shortened. The formation of lithium naphthalenide required 2 h, and the subsequent reduction of the copper(I) salt complex was usually completed within 5 min, giving an active copper that was much more reactive than previous forms of copper. The resulting copper solutions appeared homogeneous and could be stirred for prolonged periods without sintering. Moreover, these new active copper species could undergo oxidative addition with functionalized alkyl and aryl organic halides to form the corresponding functionalized organocopper reagents at temperatures conducive to the formation of organocopper compounds. Since then many other copper(I) salt complexes have been reduced by preformed lithium naphthalenide. The most common include copper(I) halide phosphine complexes, copper cyanide–lithium halide complexes, and lithium 2-thienylcyanocuprate. The choice of ligand used in preparing the active copper is a critical factor influencing the reactivity of the resulting active copper species. Although all of the complexes used to form active copper are capable of undergoing the transformations previously mentioned, each has its inherent advantages and disadvantages.

5.3 Phosphine-Based Copper

The choice of ligand used in preparing the active copper is a critical factor affecting the reactivity of the active copper species. The use of phosphine ligands significantly enhanced the reactivity of active copper. In general, the more electron donating the phosphine, the more reactive the copper was toward oxidative addition. Also, the resulting organocopper reagent was generally more nucleophilic. The general trend of reactivities for phosphine-based copper is $P(NMe_2)_3 > PEt_3 > P(CH_2NMe_2)_3 > P(cyclohexyl)_3 > PBu_3 > PPh_3 >$ Dip hos $> P(OEt)_3$, as shown in Table 5.1. Phosphites were of little use, but nitrogen-containing phosphine ligands had the advantage that phosphine-containing impurities could be removed by a simple dilute acid workup. The active copper prepared utilizing trialkylphosphines, such as PEt_3, PBu_3, or $P(Cy)_3$, gave 10–30% of homocoupled products when using alkyl bromides (2 RBr → R—R). Higher proportions of the homocoupled product were found when using alkyl iodides. The tendency for homocoupling was highly dependent upon the ratio

Table 5.1 Reactions of activated copper with alkyl halides.

Equivalent octyl halide	Copper complex	Temp. (°C)	Time (min)	Product yields (%)		
				R—H	R—X	R—R
0.45 I	CuI-PEt_3	0	2	46	0	49
0.45 I	CuI-PBu_3	0	2	68	0	49
0.45 I	CuI-PBu_3	0	2	68	0	27
0.50 I	CuI-PBu_3	−70	15	35	0	64
0.50 I	CuI-PBu_3	−50	20	44	0	52
0.45 I	CuI-$P(OEt)_3$	0	140	0	78	0
0.40 I	CuI-PPh_3	−78	30	67	9	23
0.40 I	CuI-$P(CH_2NMe_2)_3$	−78	75	64	1	34
0.43 I	CuI-DIPHOS	−78	10	49	13	23
0.50 I	CuBr-PBu_3	0	5	58	0	42
0.50 I	$CuBrSMe_2$	0	60	0	99	0
0.50 I	CuCN	0	60	7	86	8
0.50 I	CuCN-PEt_3	−78	30	18	56	14
0.50 I	CuCN-PEt_3	0	60	30	31	16
0.46 I	CuCn-PBu_3	0	10	40	8	27
0.45 I	CuCN-$P(OEt)_3$	0	60	30	28	2
0.45 I	CuCN-Bipy	0	60	40	2	2
0.40 Br	CuI-PBu_3	−50	110	71	3	30
0.40 Br	CuI-PBu_3	−78	20	65	5	25
0.50 Br	CuI-PBu_3	−50	110	58	24	30
0.50 Br	CuI-PBu_3	−50	105	44	22	35
0.50 Br	CuI-PBu_3	−50	75	60	27	13
0.40 Br	CuI-PBu_3	−78	120	65	3	26
0.40 Br	CuI-$PÜh_3$	−78	60	53	42	1
0.40 Br	CuI-PPh_3	−30	150	67	31	1
0.41 Br	Cu-(NMe_2)	−78	15	68	6	21
0.41 Br	CuI-PCy_3	−78	20	70	0	25
0.44 Br	CuI-$PCy_2(CH_2)_2NMe_2$	−78	65	37	31	26
0.40 Br	CuI-$P(CH_2NMe_2)_3$	−78	130	64	28	8
0.41 Br	CuILiSCN	24	170	0	28	0
0.37 Br	CuCN-PBu_3	0	5	19	50	3

Table 5.1 (Continued)

Equivalent octyl halide	Copper complex	Temp. (°C)	Time (min)	Product yields (%)		
				R—H	R—X	R—R
0.30 Br	CuCNLi(2-thienyl)	-60	30	91	0	5
0.35 Br	$CuCn(LiBr)_2$	−78	5	70	25	6
0.35 Br	$CuCN(LiBr)_2$	−78	5	84	3	2
0.37 Br	$CuSCN(LiBr)_2$	−35	10	29	60	0
0.50 Cl	$CuI\text{-}PBu_3$	−78	100	19	76	0
0.50 Cl	$CuI\text{-}PBu_3$	−50	80	23	71	0
0.50 Cl	$CuI\text{-}PBu_3$	−30	50	26	70	0
0.50 Cl	$CuI\text{-}PBu_3$	24	35	32	61	0

of copper to alkyl halide used, the temperature of addition, and the manner in which the alkyl halide was added. In general, rapid low-temperature addition of the alkyl bromide to the active copper gives the best yield of the organocopper species while minimizing the formation of homocoupled products. Conversely, dropwise addition of the alkyl bromide to active copper at low temperature is not desirable and usually forms considerable amounts of homocoupled products. The active copper prepared utilizing triphenylphosphine gave less than 1% of homocoupled product with alkyl bromides. However, the resulting organocopper reagent was less nucleophilic and not efficient in reactions that require a higher degree of nucleophilicity such as intermolecular epoxide ring-opening reactions.

Although the choice of phosphine ligand was found to be crucial for the formation and subsequent reactivity of the organocopper species, the presence of malodorous phosphine ligands was often found to interfere with product isolation. In our search for a more reactive copper that did not require the use of phosphorus ligands, we discovered that the lithium napthalenide reduction of a solution of a commercially available lithium 2-thienylcyanocuprate or CuCN·LiCl complex produced a highly reactive zerovalent copper complex.

Alkylcopper reagents derived from copper(I) iodide trialkylphosphine complexes are very reactive in conjugate addition reactions. Ligands such as PBu_3, PEt_3, PCy_3, and HMPT have been used quite effectively in preparing organocopper species that undergo 1,4-conjugate additions. The 1,4-conjugate addition of these alkylcopper species with 2-cyclohexenone proceeds readily at −78°C, with the enone being consumed upon warming to −50°C. In contrast, organocopper species derived using triphenylphosphine gave little or no conjugate addition products. The amount of phosphine ligands present in the reaction mixture has a pronounced effect upon the yield of the conjugate adduct in these reactions. These results are in agreement with those reported by Noyori, in which a significant enhancement in the yields of the conjugate adducts was observed when using excess PBu_3 with the organocopper reagents [6]. Chlorotrimethylsilane is also compatible with the reaction conditions [7]. The low-temperature addition of chlorotrimethylsilane to the organocopper reagents followed by addition of the enone and subsequent acidic workup results in the formation of the 3-alkylated cyclohexanone products in good yields.

The triphenylphosphine-based copper reacts rapidly with functionalized alkyl bromides to give stable alkylcopper reagents, followed by subsequent trapping with acid chlorides to give good yields of the functionalized ketone products. An excess of the acid chloride must be used in order to trap the organocopper reagent [8], since otherwise the unreacted copper species will further react with the acid chloride; for example, the use of benzoyl chloride gives *cis*-α,α′-stilbenediol dibenzoate. Homocoupling of primary alkyl bromides does not occur when using triphenylphosphine-based active copper, but some homocoupled products are formed when using primary alkyl iodides. Although triphenylphosphine alkylcopper reagents are less nucleophilic, they give higher yields in cross-coupling reactions with acid chlorides since none of the primary alkyl bromide is lost as the homocoupled product. In the aryl cases, the choice of the phosphine ligand is not crucial in cross-coupling reactions with acid chlorides because aryl halides do not homocouple to the same extent as alkyl halides. Trialkylphosphine-based copper has also been used to form ketone-functionalized, remote ester-functionalized, stable *ortho*-halophenylcopper reagents and in the cyclization of α,ω-dihaloalkanes [9].

The functionalized organocopper reagents prepared from organic halides and the trialkylphosphine derived copper species also undergo epoxide-opening reactions. The reactions of primary alkylcopper reagents with 1,2-epoxybutane gave a single regioisomer, with the alkylation taking place at the least hindered position of the epoxide, proceeding to completion at −15°C or lower. The epoxides are found to be reasonably stable in the presence of activated copper under these conditions; otherwise the alkylcopper reagents decompose above −10°C. Aryl halides form arylcopper compounds at 25°C or lower. Arylcopper compounds also undergo epoxide-opening reactions with 1,2-epoxybutane to form a single regioisomer in good yield at room temperature or with moderate heating.

Molecules containing both an epoxide and a nucleophile which can undergo an intramolecular cyclization are synthetically useful and have increasingly been reported [10]. Epoxyalkylcopper species, using trialkylphosphine-based copper, are able to undergo intramolecular cyclization via an epoxide-opening process upon warming, thereby generating new carbocycles. It should be noted that triphenylphosphine-based organocopper reagents may also be used for these intramolecular epoxide-opening reactions. In general, the regioselectivity of intramolecular epoxide-opening reactions usually follows Baldwin's rules [11]. Exo-mode ring formations are favored in medium-sized ring cyclizations when both termini of the epoxide are equally substituted. For a shorter connecting chain, such as four-carbon bromoepoxides, reactions always prefer the exo-mode ring closures.

Parham cyclialkylations [12] and cycliacylations [13] involving an anionic cyclization of *ortho*-functionalized aryllithium compounds have been found to be synthetically useful [14]. The intramolecular cyclizations of epoxyaryl-copper compounds, mediated by Rieke copper, undergo *exo*-mode ring closures to form 2,3-dihydrobenzofuran and *endo*-ring closures to give 3-chromanol, as shown in Table 5.2. Both trialkylphosphine- and triphenylphosphine-based copper may be used for the intramolecular epoxide-opening reactions of these arylcopper reagents. The regioselectivity of these cyclizations is also affected by the substitution pattern, the reaction solvent, and the $CuI{\cdot}PR_3$ complex used to generate the active copper. Table 5.3 contains several more examples of cyclization of haloepoxides using activated copper.

Cyclization of 6-bromo-1,2-epoxyhexane with activated copper in THF gave cyclohexanol and cyclopentylmethanol in a ratio of 6:1 in a 56% combined yield. Solvent effects were found to be drastic for this reaction. The regiochemistry of this reaction was reversed by carrying out the reaction in toluene.

It should be pointed out that the homocoupling of alkyl halides could be completely eliminated when the CuI/PPh_3 complex was used. The cycloalkylation of 5-bromo-1,2-epoxypentane with highly reactive copper, which was generated from lithium naphthalenide and CuI/PPh_3, gave a 37:1 mixture of cyclopentanol and cyclobutylmethanol in excellent yield (87%) with no evidence of any homocoupling reaction.

Methyl substitution at the internal position of the epoxides gave only the endo-mode cyclized products for the medium-sized connecting chains. *cis*-6-Bromo-2,3-epoxyhexane, with a methyl substituent at the external position of the epoxide, surprisingly reversed the regioselectivity to undergo preferential exo-mode ring closure to furnish 1-cyclobutylethanol and *cis*-2-methylcyclopentanol in a 29:1 ratio in a 61% combined yield. For these medium-sized ring closures, ring strain imposed by the connecting chain and nonbonding interactions arising from the substitution are both equally important. Therefore, the exo-mode ring closure will be favored when both termini of the epoxide are equally substituted. Surprisingly, 7-bromo-1,2-epoxyheptane gave only cycloheptanol in 12% yield.

Reaction of 4-bromo-1,2-epoxybutane and 4-bromo-2-methyl-1,2-epoxybutane with the activated copper gave only the 3-membered ring alcohols cyclopropylmethanol and 1-methyl-1-(hydroxymethyl)cyclopropane in 89% and 96% yields, respectively. For the small-sized ring closure, ring strain imposed by the connecting chain becomes crucial. The geometry of

Table 5.2 Intramolecular epoxide-opening reactions of epoxy aryl halides using phosphine-based active copper.

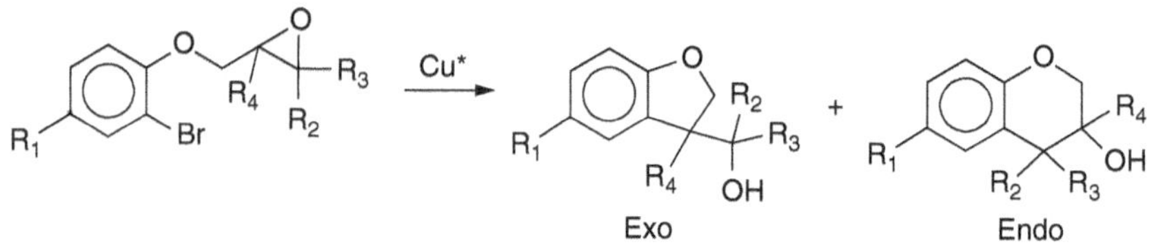

Entry	No.	R1	R2	R3	R4	PR_3	Solvent	Temp. (°C)	*exo*:*endo*	Yield (%)
1	3	H	H	H	H	PBu_3	THF	0→rt	82:18	61
2	3	H	H	H	H	PPh_3	THF	0→rt	88:12	63
3	3	H	H	H	H	PPh_3	Toluene	0→rt	58:42	29
4	3	H	H	H	H	PBu_3	THF/DMF	0→rt	92:8	25
5	3	H	H	H	H	PPh_3	THF/DMF	0→rt	100:0	5
6	4	CO_2Et	H	H	H	PBu_3	THF	0	87:13	80
7	5	Cl	H	H	H	PBu_3	THF	0→rt	83:17	53
8	5	Cl	H	H	H	PPh_3	THF	0→rt	89:11	63
9	6	Me	H	H	H	PBu_3	THF	0→rt	82:18	59
10	6	Me	H	H	H	PPh_3	THF	0→rt	87:13	56
11	7	H	Me	H	H	PBu_3	THF	0	100:0	86
12	7	H	Me	H	H	PPh_3	THF	0	100:0	85
13	8	CO_2Et	Me	H	H	PBu_3	THF	0	100:0	74
14	8	CO_2Et	Me	H	H	PPh_3	THF	0	100:0	68
15	9	Cl	Me	H	H	PBu_3	THF	0	100:0	96
16	10	Cl	Ph	H	H	PBu_3	THF	0	84:16	86
17	10	Cl	H	H	Me	PBu_3	THF	0	0:100	53
18	12	CO_2Et	Me	Me	H	PBu_3	THF	0	100:0	79

endo-mode ring closure is more difficult to reach for these cycloalkylation reactions with a shorter connecting chain, and the exo-mode cyclization is consequently preferred.

As most of the products were volatile with relatively low boiling points, the isolated yields were usually low. Trapping the cyclized alkoxide with an acid chloride to form a higher boiling point derivative helps prevent the evaporation of volatile cycloproducts and facilitates the isolation. For example, cyclopropylmethyl benzoate was isolated in 77% yield by this method.

A bromoepoxide containing a remote cyano group gave only the 5-membered ring alcohol in 83% isolated yield (Table 5.3, entry 20). With an appropriate

Table 5.3 Intramolecular cyclizations of bromoepoxides with activated copper.

Entry	Haloepoxides	PR_3[a]	Solvent	Products[b]	*exo*:*endo*	Yield (%)[c]
1	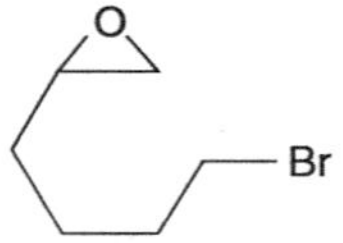 1	PBu_3	THF	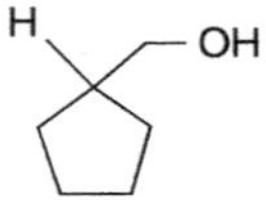 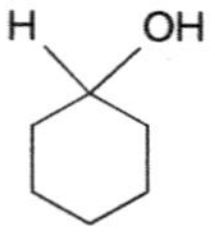 3, 4	1:6	56
2		PBu_3	Toluene[d]		1:0	45
3		PPh_3	THF		1:4	37
4		PPh_3	Toluene[d]		1:1.5	62
5	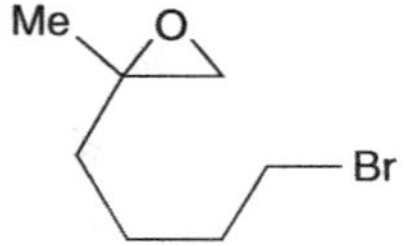 5	PBu_3	THF	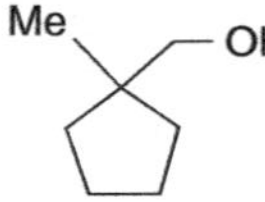 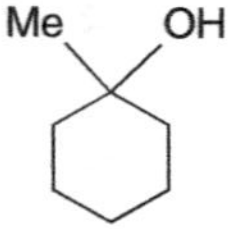 6, 7	0:1	35
6		PPh_3	THF		0:1	31
7	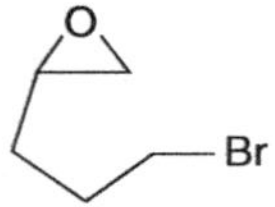 8	PBu_3	THF	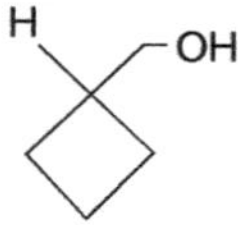 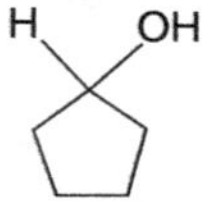 9, 10	1:35	57
8		PPh_3	THF		1:37	87 (78)[e]
9		PPh_3	Toluene[d]		1:7	—[f]
10	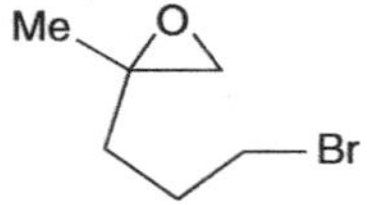 11	PPh_3	THF	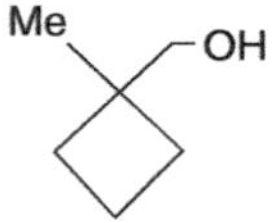 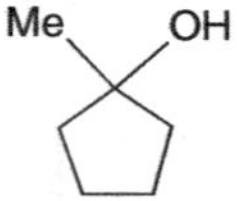 12, 13	0:1	95

(*Continued*)

Table 5.3 (Continued)

Entry	Haloepoxides	PR_3[a]	Solvent	Products[b]	*exo*:*endo*	Yield (%)[c]
11	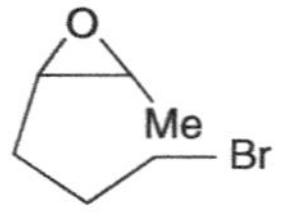 14	PPh_3	THF	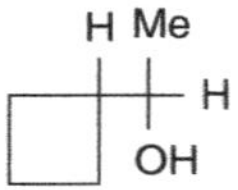 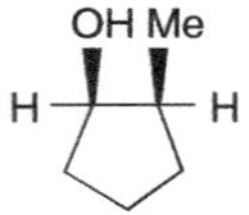 15, 16	25:1	41
12		PPh_3	Toluene[d]		29:1	61
13	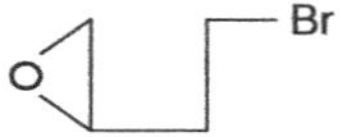 17	PPh_3	THF	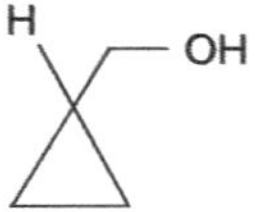 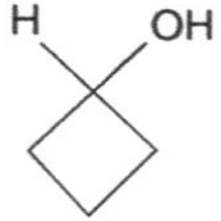 18, 19	1:0	89 (77)[e]
14		PBu_3	THF		1:0	51
15		PPh_3	Toluene[d]		6:1	—[f]
16	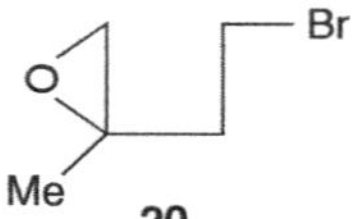 20	PPh_3	THF	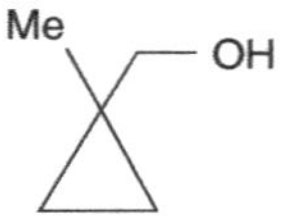 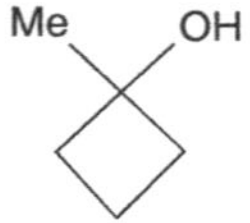 21, 22	1:0	96 (81)[e]
17		PPh_3	Toluene[d]		1:0	—[f]
18	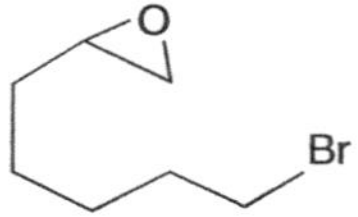 23	PBu_3	THF	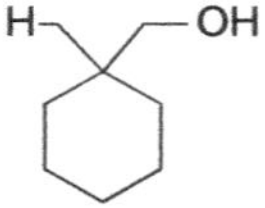 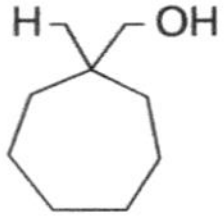 24, 25	0:1	10
19		PBu_3	Toluene[d]		0:1	(12)

20	$NC(CH_2)_6$ epoxide–Br **26**	PPh_3	THF	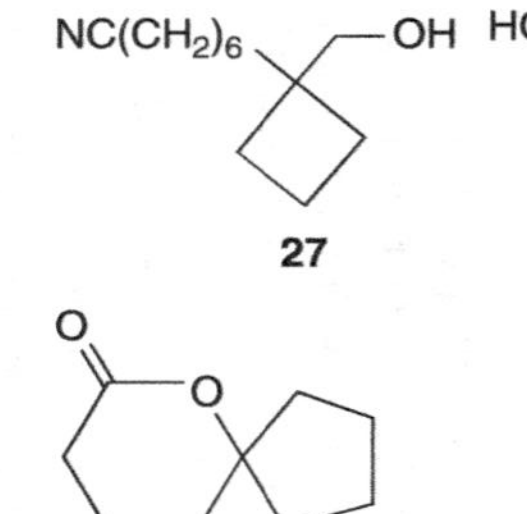 **27** **28**	0:1	(83)
21	$EtO_2C(CH_2)_3$ epoxide–Br **29**	PPh_3	THF	**30**		(50)

[a] $P(n\text{-}Bu)_3$ (2.3 equiv) or PPh_3 (2.0 equiv) was often used.
[b] The spectra data of cyclized products are identical to those obtained from authentic samples. All new substances have satisfactory spectroscopic data.
[c] Yields reported were mostly determined by gas chromatography analysis. Isolated yields are shown in parenthesis.
[d] Activated copper was first generated in THF as usual. Upon the removal of THF in vacuum, toluene was then added.
[e] Isolated as a benzoate derivative.
[f] The GC yield could not be determined due to the overlapping of the product peak with the solvent peak.

linkage between epoxide and ester groups, 4-bromo-2-(3-carboethoxypropyl)-1,2-epoxybutane underwent tandem epoxide-opening and lactonization reactions to form the spirolactone in 50% yield (Table 5.3, entry 21).

In conclusion, we have demonstrated that difunctional molecules, containing both an epoxide and a halide, can undergo intramolecular cyclization mediated by the activated copper to generate new carbocycles. Significantly, the haloepoxides can contain other functional groups leading to highly functionalized carbocycles.

5.4 Lithium 2-Thienylcyanocuprate-Based Copper

The reduction of lithium 2-thienylcyanocuprate with lithium naphthalenide at −78°C generates a highly reactive form of active copper [15]. This clearly demonstrates that highly reactive copper can also be produced from copper salts that do not contain phosphines (Equation 5.1). As a result, product isolation and purification is greatly facilitated by avoiding the use of malodorous phosphine ligands. The resulting zerovalent copper species will react oxidatively with alkyl halides under very mild conditions at −78°C to form stable alkylcopper compounds in high yields. The formation of the alkylcopper reagent is generally accompanied by trace amounts (<2%) of eliminated and homocoupled by-products. A minor amount of the thienyl ligand transfer product has been found. However, these thienyl transfer products only affected product isolation in a few cases.

$$1.05\ Li^{+}\ [C_{10}H_8]^{\bullet -} + 1.0\ Li\ (2\text{-thienyl})CuCN \xrightarrow[-78^{\circ}C]{THF} 1.0\ Cu^{*} \quad (5.1)$$

The ability to directly form various organocopper reagents utilizing this thienyl-based active copper solution allows easy preparation of functionalized organocopper reagents containing ester, nitrile, chloride, fluoride, epoxide, amine, and ketone moieties. In turn, these stable alkylcopper reagents are able to cross-couple with benzoyl chloride at −78°C to generate ketone products in excellent isolated yields, as shown in Table 5.4. Significantly, this thienyl-based active copper adds oxidatively to allyl chlorides and acetates at −78°C to allow the direct formation of allylic organocopper reagents, accompanied by less than 10% of the Wurtz-type homocoupling products. In turn, these allylic organocopper reagents are capable of undergoing cross-coupling reactions with electrophiles such as benzoyl chloride and benzaldehyde to give the corresponding ketone and alcohol products, respectively, in good isolated yields, as shown in Table 5.5. Another common, although tedious, route to allylic organocopper reagents is the transmetallation of allylic stannanes with an appropriate organocopper reagent, itself derived from a transmetallation of an organolithium or Grignard precursor.

Table 5.4 Cross-coupling reactions of thienyl-based organocopper reagents with acid chlorides.

Entry	Halide	Acid chloride	Product	Yield (%)
1	$Br(CH_2)_7CH_3$	PhCOCl	$PhCO(CH_2)_7CH_3$	73
2	$Br(CH_3)_3CO_2Et$	PhCOCl	$PhCO(CH_2)_3CO_2Et$	47
3	$Br(CH_2)_3CN$	PhCOCl	$PhCO(CH_2)_3CN$	61
4	$Br(CH_2)_6Cl$	PhCOCl	$PhCO(CH_2)_6Cl$	42
5	$Br(CH_2)_6CH\text{–}CH_2$ (epoxide, O bridging CH–CH$_2$)	PhCOCl	$PhCO(CH_2)_6CH\text{–}CH_2$ (epoxide, O bridging CH–CH$_2$)	65
6	$BrC_6H_4(p\text{-}CH_3)$	PhCOCl	$PhCOC_6H_4(p\text{-}CH_3)$	86
7	$BrC_6H_4(p\text{-}OCH_3)$	PhCOCl	$PhCOC_6H_4(p\text{-}OCH_3)$	87
8	$IC_6H_4(p\text{-}OCH_3)$	*n*-BuCOCl	$n\text{-}BuCOC_6H_4(p\text{-}OCH_3)$	63
9	$BrC_6H_4(p\text{-}CN)$	PhCOCl	$PhCOC_6H_4(p\text{-}CN)$	75
10	$BrC_6H_4(m\text{-}CN)$	PhCOCl	$PhCOC_6H_4(m\text{-}CN)$	62
11	$BrC_6H_4(p\text{-}NMe_2)$	PhCOCl	$PhCOC_6H_4(p\text{-}NMe2)$	81
12	$IC_6H_4(p\text{-}Cl)$	PhCOCl	$PhCOC_6H_4(p\text{-}Cl)$	90
13	$BrC_6H_4(p\text{-}F)$	PhCOCl	$PhCOC_6H_4(p\text{-}F)$	93
14	$BrC_6H_4(p\text{-}COPh)$	PhCOCl	$PhCOC_6H_4(p\text{-}COPh)$	44

Conjugate additions of thienyl-based organocopper reagents to α,β-unsaturated ketones proceed in excellent yields, and these are the organocopper reagents of choice. The addition of TMSCl to the organocopper reagent prior to the addition of the enone allows the facile formation of the 1,4-conjugate product at −78°C in excellent isolated yields, as shown in Table 5.6. Competitive 1,2-addition products were not observed for these conjugate addition reactions. The addition of TMSCl significantly increased product formation as much as fivefold and is considered essential. The addition of Lewis acids, such as $BF_3{\cdot}Et_2O$, did not have as dramatic an effect upon product formation as did TMSCl. These thienyl-based organocopper reagents are very efficient for conjugate addition reactions, the ideal ratio of organocopper reagent to enone being as low as 2 : 1. The use of larger excesses of the organometallic reagent (10 equiv) has been reported in the literature [16]. Both cyclic and acyclic enones worked well with this formulation of active copper. However, more sterically hindered enones, such as isophorone, carvone, and 2,4,4-trimethyl-2-cyclohexen-1-one, were not amenable for these 1,4-conjugate additions.

The thienyl-based organocopper reagents were also found to be nucleophilic enough to undergo intermolecular epoxide-opening reactions with 1,2-epoxybutane to form single regioisomers in good isolated yields, as shown in

Table 5.5 Reactions of thienyl-based allylic organocopper reagents with electrophiles.

Entry	Allyl chloride	Electrophile	Product	Yield (%)
1	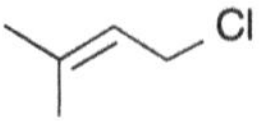 Cl	PhCHO	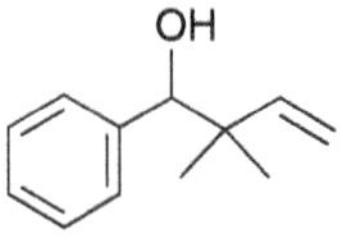 OH	80
2	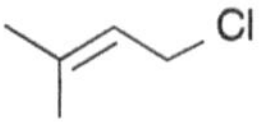 Cl	PhCOCl	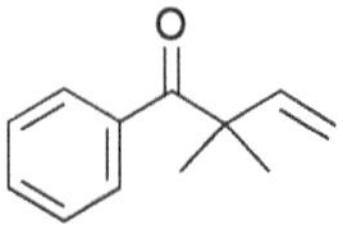 O	56
3	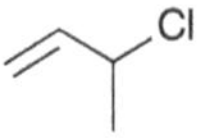 Cl	PhCHO	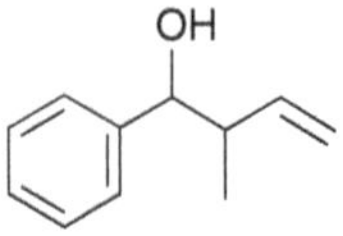 OH	70
4	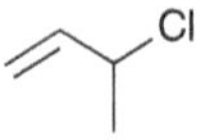 Cl	PhCOCl	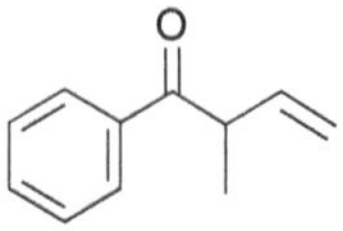 O	40
5	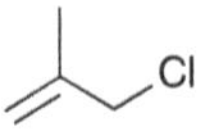 Cl	PhCHO	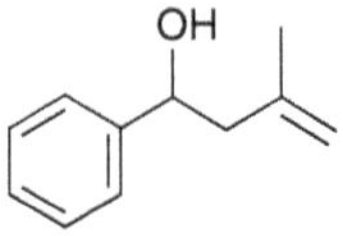 OH	78
6	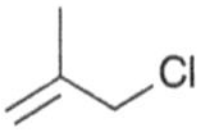 Cl	PhCOCl	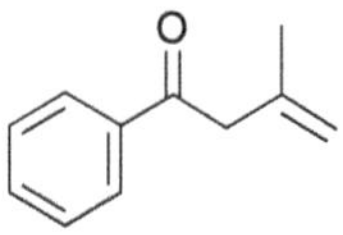 O	55
7	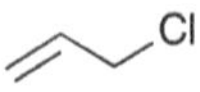 Cl	PhCHO	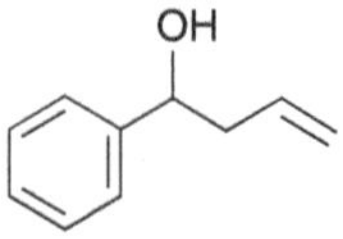 OH	61
8	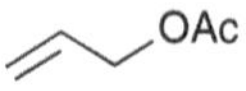 OAc	PhCHO	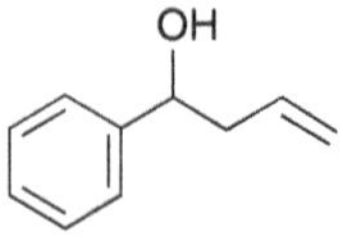 OH	26

Table 5.7. We have attempted to activate the epoxide using Lewis acids to accelerate the organometallic epoxide-opening reaction, but to no avail. Analogously, intramolecular cyclizations via an epoxide cleavage process can also be realized. For example, treatment of 6-bromo-1,2-epoxyhexane with thienyl-based copper resulted in intramolecular cyclization; subsequent

Table 5.6 1,4-Conjugate addition reactions with thienyl-based organocopper reagents.

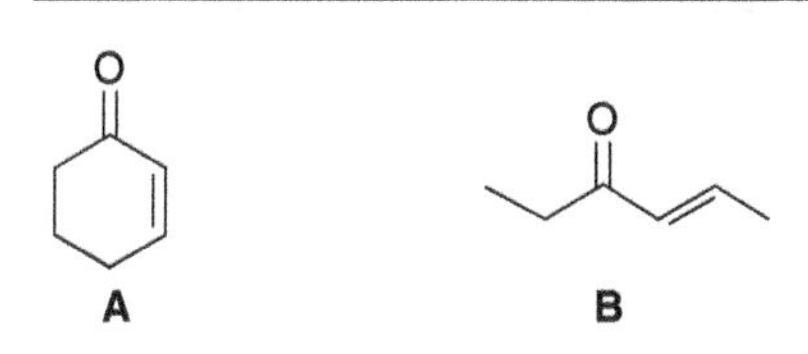

Entry	Eq. halide	Eq. additive	Eq. enone	Product	Yield (%)
1	0.5 $Br(CH_2)_7CH_3$	2.0 TMSCl	0.25 A	$(CH_2)_7CH_3$	91
2	0.5 $Br(CH_2)_7CH_3$	1.0 TMSCl	0.125 A	"	97
3	0.5 $Br(CH_2)_7CH_3$	None	0.25 A	"	20
4	0.5 $Cl(CH_2)_7CH_3$	2.0 TMSCl	0.25 A	$(CH_2)_7CH_3$	71
5	0.5 $Cl(CH_2)_7CH_3$	None	0.25 A	"	16
6	0.5 $Cl(CH_2)_7CH_3$	1.0 TMSCl	0.167 A	"	87
7	0.5 $Br(CH_2)_7CH_3$	1.0 TMSCl	0.125 B	$(CH_2)_7CH_3$	88
8	0.5 $Cl(CH_2)_7CH_3$	1.0 TMSCl	0.125 A	$(CH_2)_7CH_3$	87
9	0.5 $Br(CH_2)_6Cl$	1.0 TMSCl	0.25 A	$(CH_2)_6Cl$	80
10	0.5 $Br(CH_2)_6Cl$	1.0 TMSCl	0.125 A	$(CH_2)_6Cl$	83
11	0.5 cyclohexyl–Br	1.0 TMSCl	0.125 A		72

(*Continued*)

Table 5.6 (Continued)

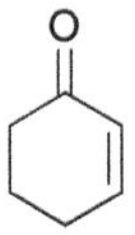

A (cyclohexenone) B (hex-4-en-3-one)

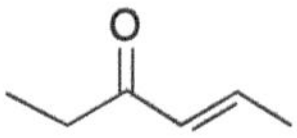

Entry	Eq. halide	Eq. additive	Eq. enone	Product	Yield (%)
12	0.5 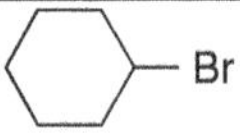 cyclohexyl-Br	1.0 TMSCl	0.125 B	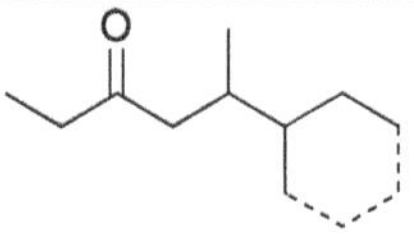	80
13	0.5 $Br(CH_2)_3CO_2Et$	1.0 TMSCl	0.125 A	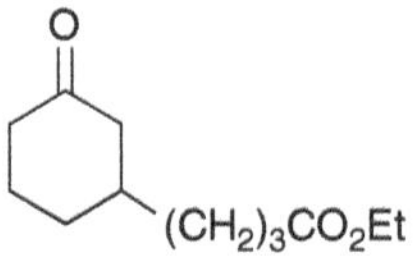 $(CH_2)_3CO_2Et$	79
14	0.5 $Br(CH_2)_3CO_2Et$	1.0 TMSCl	0.125 B	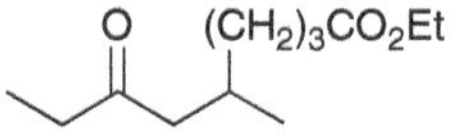 $(CH_2)_3CO_2Et$	81

Table 5.7 Reactions of thienyl-based organocopper reagents with 1,2-epoxybutane.

Entry	Halide	Product	Yield (%)
1	$Br(CH_2)_7CH_3$	$CH_3CH_2CH(OH)(CH_2)_6CH_3$	71
2	$Br(CH_2)_5OPh$	$CH_3CH_2CH(OH)(CH_2)_6OPh$	60
3	$Br(CH_2)_6Cl$	$CH_3CH_2CH(OH)(CH_2)_7Cl$	68
4	$IC_6H_4(p\text{-}CH_3)$	$CH_3CH_2CH(OH)CH_2C_6H_4(p\text{-}CH_3)$	64
5	$IC_6H_4(p\text{-}OCH_3)$	$CH_3CH_2CH(OH)CH_2C_6H_4(p\text{-}OCH_3)$	78
6	$IC_6H_4(p\text{-}Cl)$	$CH_3CH_2CH(OH)CH_2C_6H_4(p\text{-}Cl)$	62

trapping of the intermediate with benzoyl chloride gave a 64 : 36 mixture of cyclohexyl benzoate and cyclopentylmethyl benzoate in 52% isolated yield.

5.5 Copper Cyanide-Based Active Copper

Initial attempts to make active copper from CuCN·*n*LiX complexes (X = Br, Cl) met with only limited success. However, exploiting the low-temperature reduction effects seen with other Cu(I) complexes led to a successful approach for the formation of active copper from CuCN·*n*LiX [17]. Lowering the reduction temperature to −110°C allowed the active copper solution to react with organic iodides, bromides, and to some extent chlorides to produce organocopper species in high yields. Like the triphenylphosphine and thienyl-based copper species, active copper derived from CuCN·*n*LiX produced very little of the homocoupled products when using alkyl bromides. Better product yields were obtained when two equivalents of the lithium salt were used to solubilize CuCN. Both LiBr and LiCl gave comparable results. Also, the manner in which the CuCN·*n*LiX and lithium naphthalenide are added together affects the reactivity of the resulting active copper solution. The best results were obtained when the CuCN·*n*LiX was added into the preformed lithium naphthalenide. It should be noted that the phosphine-based active copper is more reactive than the active copper derived from CuCN·*n*LiX. However, active copper derived from CuCN·*n*LiX offers several advantages. First, CuCN is an inexpensive and stable source of Cu(I), which can be used as received without further purification. Both active copper chemistry and traditional organocopper chemistry involving Cu(I) have been shown to be effected by the purity of the Cu(I) used [18]. Second, product isolation is facile, and product purity is greatly improved using this phosphine-free source of Cu(I). The lithium salts are easily removed during aqueous workup. However, this form of active copper is not as nucleophilic, as evidenced by its inability to undergo epoxide opening.

Table 5.8 shows the results of reacting various functionalized organocopper reagents, produced from both alkyl and aryl bromides, with benzoyl chloride

Table 5.8 Cross-coupling of benzoyl chloride with organocopper reagents derived from CuCN·2 LiBr-based active copper.

Entry	Halide (equiv.)	Product	Yield (%)
1	$Br(CH_2)_7CH_3$ (0.25)	$PhCO(CH_2)_7CH_3$	82
2	$Br(CH_2)_6Cl$ (0.25)	$PhCO(CH_2)_6Cl$	80
3	$Br(CH_2)_3CO_2Et$ (0.25)	$PhCO(CH_2)_3CO_2Et$	81
4	$Br(CH_2)_2CO_2Et$ (0.25)	$PhCO(CH_2)_2CO_2Et$	43
5	$Br(CH_2)_3CN$ (0.25)	$PhCO(CH_2)_3CN$	86
6	Bromobenzene (0.20)	PhCOPh	87
7	*p*-BrC_6H_4CN (0.20)	*p*-NCC_6H_4COPh	60
8	*o*-BrC_6H_4CN (0.20)	*o*-NCC_6H_4COPh	74
9	*o*-$BrC_6H_4CO_2Et$ (0.20)	$EtO_2CC_6H_4COPh$	51
10	*p*-BrC_6H_4Cl (0.20)	*p*-ClC_6H_4COPh	83

Table 5.9 Conjugate additions with organocopper reagents derived from CuCN·2 LiBr-based active copper.

A B C

Entry	Halide	Enone (equiv.)	Yield (%)
1	$Br(CH_2)_7CH_3$	A (0.17)	92
2	$Cl(CH_2)_7CH_3$	A (0.16)	42
3	$Br(CH_2)_3CO_2Et$	A (0.17)	70
4	$Br(CH_2)_3CO_2Et$	A (0.12)	90
5	$Br(CH_2)_3CO_2Et$	B (0.11)	94
6	$Br(CH_2)_3CO_2Et$	C (0.11)	87
7	$Br(CH_2)_3CN$	A (0.12)	87
8	$Br(CH_2)_3CN$	B (0.11)	92
9	$Br(CH_2)_6Cl$	A (0.12)	82
10	BrC_6H_{11}	A (0.12)	80
11	BrC_6H_5	A (0.11)	45
12	$ClCH_2CH{=}C(CH_3)_2$	A (0.10)	81

to form the corresponding functionalized ketone products, which were obtained in good isolated yields. Chloride, nitrile, and ester functionalities can be incorporated into the organocopper reagent, but if the functional group is proximal to the carbon–bromine bond, the yields are lower.

The essential addition of TMSCl to organocopper reagents made from CuCN·2LiBr-based active copper allowed 1,4-conjugate additions to occur at −78°C in good isolated yields, as shown in Table 5.9. Both cyclic and acyclic enones can be used, the ideal organocopper to enone ratio being 2.5 : 1. As with other species of active copper, competitive 1,2-addition products were not observed.

The reactions of several nonfunctionalized allyl chlorides and acetates with active copper followed by cross-coupling with various electrophiles at −100°C gave ketone and alcohol products in good isolated yields. Although this species of active copper is not nucleophilic enough to undergo intramolecular or intermolecular epoxide-opening reactions, the addition of MeLi enhances the nucleophilicity of the allylic organocopper reagents and allows substitution

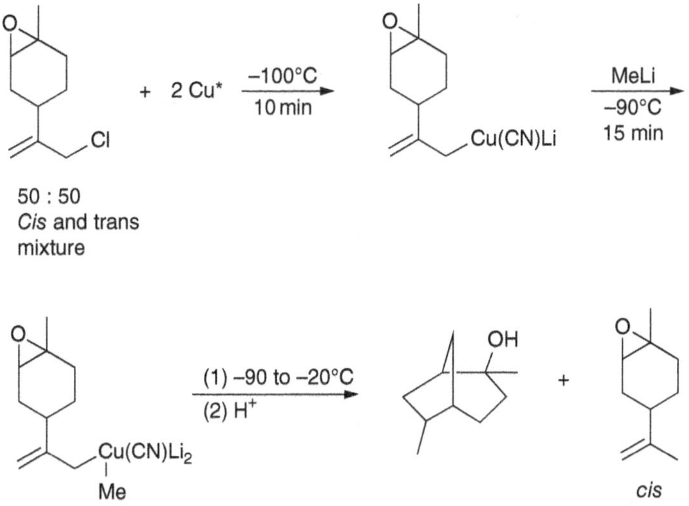

Scheme 5.1 Intramolecular epoxide opening to afford a bicyclic product.

reactions with epoxides. The MeLi is believed to act as a "dummy" nontransferable ligand, presumably forming a higher-order cuprate.

Primary and secondary allyl chlorides containing diverse functionalities are tolerated by the CuCN-based copper. Allyl organocopper reagents containing ketone, α,β-unsaturated ketone, epoxide, nitrile, alkyl acetate, ester, alkyl chloride, and carbamate functionalities have been prepared. The ability of this CuCN-based active copper to tolerate a wide variety of functionalities allows the facile formation of highly functionalized homoallylic alcohols, β,γ-unsaturated ketones, and amines in excellent isolated yields. The organocopper reagents derived from primary allyl chlorides showed a remarkable thermostability with little decomposition at 0°C, unlike secondary allyl organocopper reagents, which decompose at a significant rate. When MeLi was added to these functionalized organocopper species derived from allyl chlorides, an intramolecular opening of the epoxide produced a bicyclic product, as shown in Scheme 5.1.

This highly reactive copper derived from CuCN·2LiCl reacts directly with 2,3-dichloropropene to yield a new bis-organocopper species which contains both a nucleophilic allylic and a vinylic moiety [19]. This novel bis-organocopper reagent undergoes a selective one-pot addition to two different electrophiles in good to excellent yields. The more reactive allyl carbon–copper bond adds to the first electrophile, followed by the incorporation of the second electrophile into the vinyl carbon–copper bond. Table 5.10 shows the one-pot reactions of the bis-organocopper reagent with various combinations of electrophiles.

Table 5.10 Reaction of CuCN·2 LiBr-derived copper with 2,3-dichloropropene.

Entry	E1	E2	Product	Yield (%)
1	PhCHO	H^+	OH Ph	88
2	PhCHO	I_2	I OH Ph	69
3	PhCHO	Br	OH Ph	71
4	PhCHO	Cl	OH Ph	83
5	PhCHO	MeI	OH Ph	63
6	$PhCH_2COCH_3$	Cl	OH Ph	54
7	O CN	Br	OH CN	79
8	NCH_2PH ‖ PhCH	Cl	$HNCH_2Ph$ Ph	22

Aldehydes, ketones, and imines are all suitable electrophiles for the reaction with the allylic terminus, whereas allyl chlorides, allyl bromides, alkyl iodides, and iodine add to the vinyl–copper bond.

5.6 Formal Copper Anion Preparation and Resulting Chemistry

A new, highly reactive copper species made from the reduction of copper(I) complexes with two equivalents of preformed lithium naphthalenide leads to a reagent that behaves chemically as a formal copper anion solution [20]. Lithium naphthalenide provides a suitable reducing agent for a variety of metals and offers unique advantages over other techniques used in metal activation, such

Table 5.11 Reaction of copper anion with organohalides and subsequent cross-coupling reactions with benzoyl chloride.

$$\text{CuI complex} + 2\text{LiNp} \xrightarrow[\text{THF}]{^\circ\text{C}} \xrightarrow[-35^\circ\text{C}]{\text{RX}} \xrightarrow[-35^\circ\text{C}]{\text{PhCOCl}} \text{RCOPh}$$

				Yields (%)		
Entry	**Halide**	**Complex**	**Temp. (°C)**	**RCOPh**	**R–X**	**R–R**
1	$CH_3(CH_2)_7Cl$	$CuI \cdot PPh_3$	0	40	39	0
2	$CH_3(CH_2)_7Cl$	$CuI \cdot PPh_3$	−78	77	12	0
3	$CH_3(CH_2)_7Cl$	$CuI \cdot PPh_3$	−107	96	0	0
4	$CH_3(CH_2)_7Br$	$CuI \cdot PPh_3$	−107	92	0	0
5	$CH_3(CH_2)_5CHBrCH_3$	$CuI \cdot PPh_3$	−107	33	—	—
6	$C_6H_{11}Cl$	$CuI \cdot PPh_3$	−107	99	—	—
7	$C_6H_{11}Cl$	$CuI \cdot PPh_3$	−107	90	—	—
8	PhCl	$CuI \cdot PPh_3$	−107	93	—	—
9	PhBr	$CuI \cdot PPh_3$	−107	80	—	—
10	PhF	$CuI \cdot PPh_3$	−107	38	—	—
11	$CH_3(CH_2)_7Cl$	$CuCN \cdot 2\,LiBr$	−78	61	19	8
12	$CH_3(CH_2)_7Cl$	$CuCN \cdot 2\,LiBr$	−107	93	0	4

as metal vaporization. It was believed that an extension of the investigations on zerovalent active copper could include a further equivalent reduction of Cu(0) to Cu(−1) [$4s^1 3d^{10} + e^- = 4s^2 3d^{10}$] to yield a closed shell anion. This work initially involved the two-equivalent reduction of $CuI{\cdot}PPh_3$ with lithium naphthalenide in THF at −108°C. The resulting copper anion solutions are homogeneous and, like the zerovalent copper, readily undergo oxidative addition to carbon–halogen bonds. However, the copper anion is a two-equivalent electron reagent, and one equivalent of halide reacts with one equivalent of copper anion. Reaction of 2 mmol of 1-bromooctane with 2 mmol of copper anion at −108°C, followed by warming to −35°C, results in the total consumption of the bromide. Similar results were obtained with 1-chlorooctane. The resulting organocopper species can be cross-coupled with benzoyl chloride to form ketones in high yields, as shown in Table 5.11. The copper anion solution displayed a dramatic increase in reactivity with halides otherwise not reactive with zerovalent active copper, undergoing oxidative addition to alkyl and aryl chlorides and to some extent even aryl fluorides. The significant increase in reactivity and stoichiometry provides evidence of a new copper species, either a copper monoanion or possibly a polyatomic anionic copper complex. The resulting

Table 5.12 1,4-Conjugate addition reactions of anion-based organocopper reagents with 2-cyclohexen-1-one.

Entry	Halide	Product	Yield (%)
1	$CH_3(CH_2)_7Cl$	3-$(CH_2)_7CH_3$-cyclohexanone	73
2	$CH_3(CH_2)_7Br$	3-$(CH_2)_7CH_3$-cyclohexanone	81
3	PhBr	3-Ph-cyclohexanone	42
4	$Br(CH_2)_6Cl$	3-$(CH_2)_6Cl$-cyclohexanone	30
5	$Br(CH_2)_3CN$	3-$(CH_2)_3CN$-cyclohexanone	18

organocopper species derived from copper anion also undergo 1,4-conjugate additions with 2-cyclohexen-1-one in moderate yields, as shown in Table 5.12. However, the copper anion is less suitable for 1,4-conjugate additions when compared with the thienyl-based organocopper species.

While both the thienyl-based and CuCN-based forms of active copper allow the straightforward production of allyl organocopper compounds, the types of cuprates generated possess reactivities comparable to those of the lower-order cyanocuprates. As an aside, Lipshutz et al. have shown that higher-order allyl cyanocuprates possess remarkable reactivities and undergo substitution with alkyl chlorides and epoxides at −78°C [21]. Furthermore, in order to add an allyl moiety in a 1,4-manner across an α,β-unsaturated enone, Lipshutz developed a less reactive "non-ate" allyl copper reagent [22]. A straightforward preparation of both higher-order and "non-ate" allyl copper reagents directly from allyl chloride utilizing a formal copper anion solution is now possible. Table 5.13 shows the results of producing higher-order allyl cyanocuprates

Table 5.13 Formation and reactions of copper anion-based higher-order allyl cyanocuprates.

Entry	Allyl chloride	Electrophile	Product	Yield (%)
1	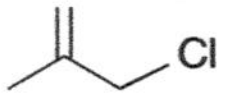 Cl	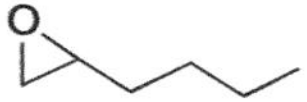 O	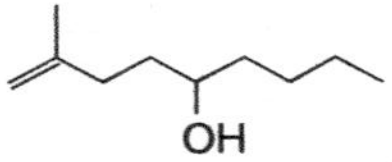 OH	98
2	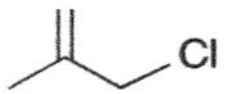 Cl	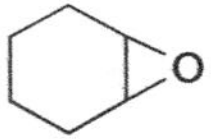 O	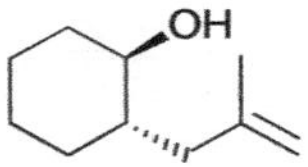 OH	98
3	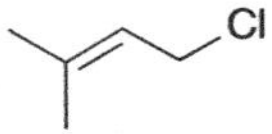 Cl	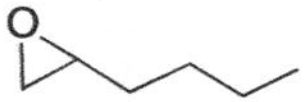 O	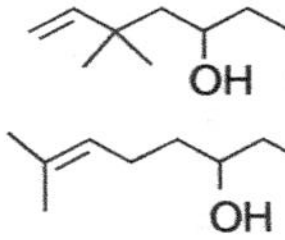 OH (46%) OH (54%)	97
4	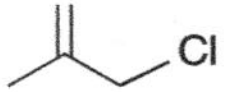 Cl	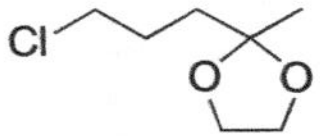 Cl, O, O	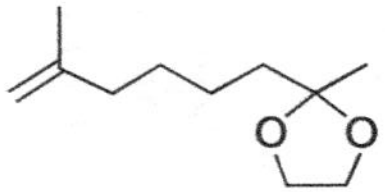 O, O	82
5	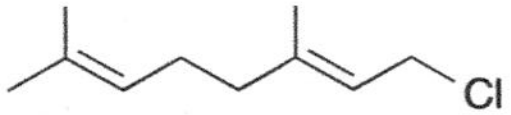 Cl	PhO$(CH_2)_4$Cl	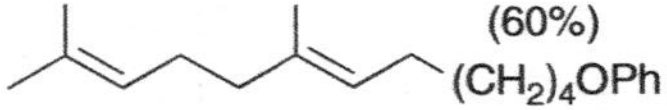 $(CH_2)_4$OPh (60%)	81
6	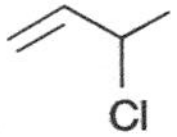 Cl	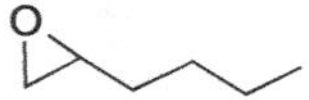 O	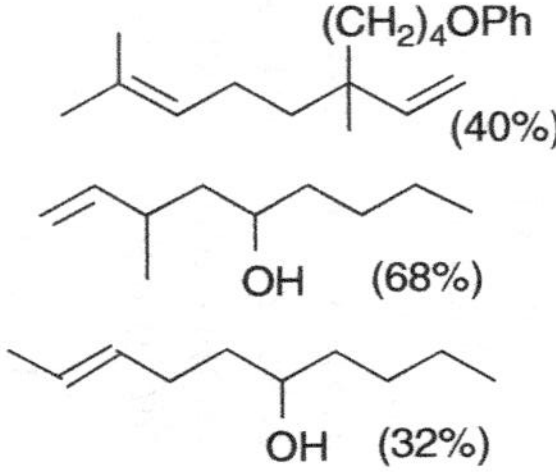 $(CH_2)_4$OPh (40%) OH (68%) OH (32%)	86

with the copper anion. The reaction between the allyl chloride and the copper anion is similar to those previously discussed, but in order to form the higher-order cuprate, methyllithium is added to act as a nontransferable "dummy" ligand. The resulting copper solution shows reactivity comparable to that of the higher order allylcyanocuprates. The regioselectivity of the asymmetrical allyl chlorides proved a disappointment, as substantial yields of products from α and γ attack were observed. Since higher-order allylcyanocuprates produced via transmetallation of allyl stannanes have shown regioselective addition to various electrophiles [23], the nature of these higher-order allyl cyanocuprates must be associated with their method of preparation.

In order to facilitate 1,4-conjugate addition to α,β-unsaturated enones, the production of a less reactive "non-ate" organocopper reagent was sought, since otherwise the 1,2-addition product was formed. Since the cyano ligand is known to remain fixed in organocuprates produced from cuprous cyanide salts [24], the THF-soluble CuCN·LiCl copper(I) complex was chosen as the copper anion precursor. The results with various allyl moieties in their transfer to α,β-unsaturated enones are shown in Table 5.14. As with many less reactive organocopper reagents, such as entry 2 of Table 5.14, the use of TMSCl is essential in these reactions [25]. The use of Lewis acids such as $BF_3 \cdot Et_2O$ did not enhance the conjugate product yields to the same extent as with TMSCl.

5.7 Typical Experimental Details of Copper Chemistry

Active Copper from CuI and K

An example of a typical reduction for copper is as follows: Into a 50-ml two-necked flask is placed 0.3522 g (9.006 mmol) of potassium, 1.7075 g (8.996 mmol) of cuprous iodide, 0.1204 g (0.9393 mmol) of naphthalene, and 10.0 ml of DME. This is stirred vigorously until the reduction is complete (≈8 h). The activated copper appears as a gray-black slurry which settles out of the clear, colorless solution. The cuprous iodide used was purchased from Cerac, Inc., Milwaukee, Wisconsin, and was 99% pure.

The reduction is complete when the solution is clear and colorless. There should be no hint of the green naphthalenide radical anion visible in the supernatant solution. This slurry does not flash or show other indications of alkali metal when syringed onto the surface of water. If the reduction is incomplete or after a partial reaction of the activated copper, the aforementioned aqueous quench will cause the precipitation of white cuprous iodide (decomposition of the soluble copper complexes occurs).

Table 5.14 1,4-Conjugate additions with copper anion-based allyl organocopper reagents.

Entry	Allyl chloride	Enone	Product	Yield (%)
1	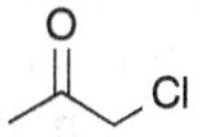	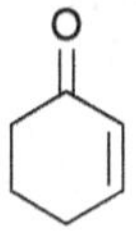	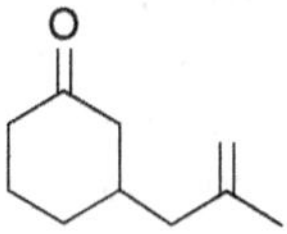	91
2	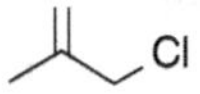	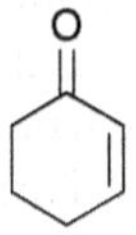	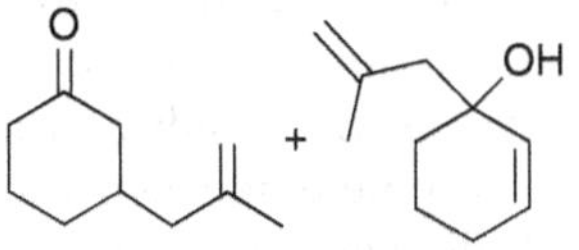 (12%) + (31%)	43
3	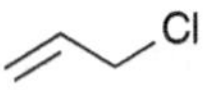	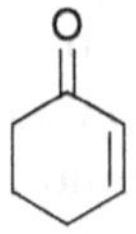	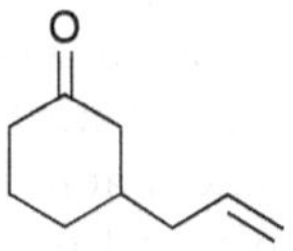	72
4	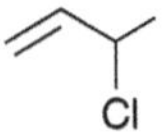	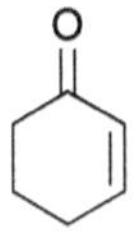	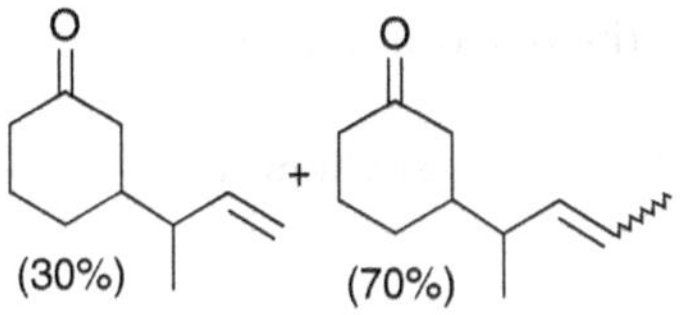 (30%) + (70%)	24
5	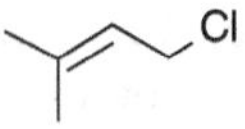	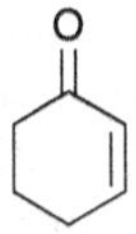	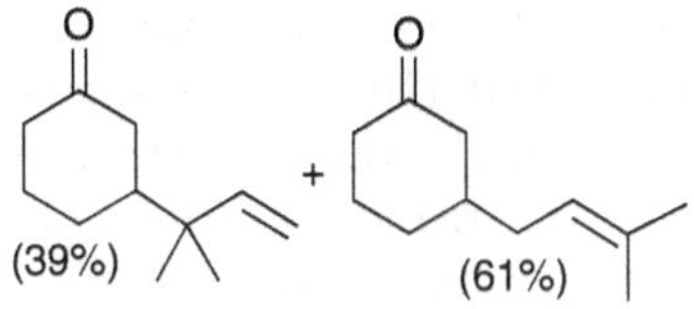 (39%) + (61%)	98
6	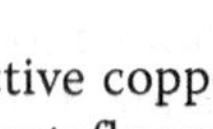	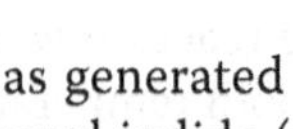	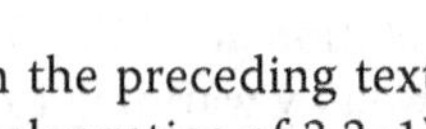	39

Reaction of K-Generated Copper with Pentafluorophenyl Iodide

The active copper as generated in the preceding text was allowed to react with pentafluorophenyl iodide (molar ratios of 3.3 : 1) at room temperature in DME for 30 min. The supernatant solution was then anaerobically transferred to another flask, where the solvent and excess aryl iodide were removed under vacuum to leave a tan solid containing pentafluorophenyl copper. This tan solid was difficult to purify; however, its chemistry matches that published for pure pentafluorophenyl copper. Yields were in the range of 70–80%.

Preparation of Phosphine-Based Copper

Typically Li(0) (21 mmol) and naphthalene (22 mmol) in 20 ml of glyme or THF are stirred for 2 h under argon. Then a solution of CuI·P(Et)$_3$ (20 mmol) in 5 ml of glyme or THF is syringed into the lithium naphthalenide and stirred for 10 min at 0°C. For alkyl halide homocoupling, 10 mmol of the alkyl halide is syringed into 20 mmol of the zerovalent copper which has been precooled to 0°C due to the exothermic nature of these reactions. For the cross-coupled reactions, 5 mmol of each alkyl halide is mixed in a small amount of solvent, and the resulting mixture is syringed into the copper. The reactions are complete between 1 and 10 min. Formation of the stable aryl, alkenyl, and alkynyl copper compounds is accomplished by mixing 10 mmol of the organic halide with 20 mmol of the precooled activated copper. Maximum yield is usually achieved between 10 and 60 min. To cross-couple, add the second reagent and stir for 1 h at 0°C, an additional hour at room temperature, and a final hour at reflux. To homocouple the arylcopper compound, heat the reaction vessel to reflux for several hours or bubble oxygen into the reaction for 1 h.

Phosphine-Based Copper Chemistry

Typical Reaction with Acid Chlorides to Form Ketones

Lithium (70.8 mg, 10.2 mmol) and naphthalene (1.588 g, 12.39 mmol) in freshly distilled THF (10 ml) were stirred under argon until the Li was consumed (~2 h). CuI (1.751 g, 9.194 mmol) and PPh$_3$ (2.919 g, 11.13 mmol) in THF (15 ml) were stirred for 30 min, giving a thick white slurry which was transferred via cannula to the dark-green solution of lithium naphthalenide at 0°C. (Later experiments showed that slightly better results were obtained if the lithium naphthalenide solution was added to the CuI/PPh$_3$ mixture.) The resultant reddish-black solution of active copper was stirred for 20 min at 0°C. Ethyl 4-bromobutyrate (0.3663 g, 1.888 mmol) and GC standard *n*-decane (0.1566 g, 1.101 mmol) were added neat via syringe to the active copper solution at −35°C. The solution was allowed to stir for 10–15 min at −35°C, followed by addition of benzoyl chloride (0.7120 g, 5.065 mmol) neat to the organocopper solution at −35°C. The reaction was allowed to stir 1 h 30 min at −35°C, followed by warming to room temperature for 30 min. GC analysis showed the reaction to be essentially complete after stirring at −35°C. The reaction was then worked up by pouring into saturated aqueous ammonium chloride, extracting with diethyl ether, and drying over anhydrous sodium sulfate. For compounds not sensitive to base, the ether layer was also washed with 5% aqueous sodium hydroxide solution. Silica gel chromatography (hexanes followed by a mixture of hexanes–ethyl acetate) and further purification by preparative thin-layer chromatography (TLC)

(2 mm plate) provided 4-carboethoxy-1-phenyl-1-butanone in 81% isolated yield (93% after quantitation using the isolated product for the preparation of GC standards).

Typical 1,4-Addition Reaction with 2-Cyclohexene-1-One

Lithium (71.2 mg, 10.3 mmol) and naphthalene (1.592 g, 12.42 mmol) in freshly distilled THF (10 ml) were stirred under argon until the Li was consumed (~2 h). A solution of $CuIPBu_3$ (3.666 g, 9.333 mmol) and PBu_3 (2.89 g, 14.3 mmol) in THF (5 ml) was added via cannula to the dark-green lithium naphthalenide solution at 0°C and the resultant reddish-black active copper solution was stirred for 20 min. 1-Bromooctane (0.9032 g, 4.677 mmol) and GC internal standard *n*-decane (0.1725 g, 1.212 mmol) in THF (5 ml) were added rapidly via cannula to the active copper solution at −78°C. The organocopper formation was typically complete within 20 min at −78°C. 2-Cyclohexen-1-one (0.1875 g, 1.950 mmol) in THF (10 ml) was added slowly dropwise over 20 min to the organocopper species at −78°C. The reaction was allowed to react at −78, −50, −30°C, and room temperature for 1 h each. The reaction was then worked up by pouring into saturated aqueous ammonium chloride, extracting with diethyl ether, and drying over anhydrous sodium sulfate. Silica gel chromatography (hexanes followed by mixtures of hexanes–ethyl acetate) and further purification by preparative TLC provided 3-*n*-octylcyclohexanone (93% GC yield after quantitation using the isolated product for the preparation of GC standards).

Typical Procedure for Intermolecular Epoxide-Opening Reaction

Lithium (69.0 mg, 9.9 mmol) and naphthalene (1.418 g, 11.1 mmol) in freshly distilled THF (10 ml) were stirred at ambient temperature for 3 h under argon. Then a solution of $CuIPBu_3$ (3.592 g, 9.2 mmol) and PBu_3 (2.790 g, 13.8 mmol) in THF (7 ml) was transferred via a cannula into the lithium naphthalenide at 0°C. The reaction mixture was stirred at 0°C for 1 h and was then cooled to −78°C. A solution of GC internal standard (*n*-decane) and *t*-butyl-4-bromobutyrate (808.0 mg, 3.6 mmol) in THF (7 ml) was rapidly added to the activated copper at −78°C. After stirring at −78°C for 1 h, 1,2-epoxybutane (117.0 mg, 1.6 mmol) in THF (5 ml) was then added at −78°C. The reaction mixture was stirred at −78°C for 1 h and was gradually warmed to −10°C. Aliquots were withdrawn and quenched with saturated NH_4Cl solution at −78, −50, −20, and −10°C. The organic solution of each aliquot was dried over $MgSO_4$ and was subjected to GC analysis (12′ × ⅛″ stainless steel column packed with OV-17 (3%) on 100/120 Chromosorb G-NAW). The maximum yield was achieved at around −20°C. The organocopper and coupled product decomposed when the reaction temperature was raised above −20°C. The reaction was cooled to −60°C and was then quenched with a saturated NH_4Cl aqueous solution (20 ml). The reaction mixture was allowed to warm to room temperature and was extracted twice with diethyl ether (75 ml each). The combined ether layers were washed with H_2O and dried over anhydrous $MgSO_4$. Solvent was removed under reduced pressure, and the resulting residue was chromatographed on silica gel to give a colorless liquid. The coupled products were usually purified by preparative TLC or preparative GC.

Typical Procedure for Intramolecular Epoxide-Opening Reaction

Lithium (70.5 mg, 10.2 mmol) and naphthalene (1.437 g, 11.2 mmol) in freshly distilled THF (10 ml) were stirred at ambient temperature for 2 h under argon. The dark-green preformed lithium naphthalenide solution was then transferred via a cannula into a 0°C solution of CuI/PPh_3, which was prepared *in situ* by stirring CuI (1.769 g, 9.3 mmol) and PPh_3 (4.870 g, 18.6 mmol) in freshly distilled THF (10 ml) at room temperature for 1 h. After stirring at 0°C for 0.5 h, the reaction mixture was then cooled to −45°C (dry ice–CH_3CN bath). A solution of GC internal standard (*n*-undecane) and 5-bromo-1,2-epoxypentane (362.0 mg, 2.2 mmol) in THF (10 ml) was rapidly added to the activated copper at −45°C. The reaction mixture was stirred at −45°C for an additional 5 min and was then allowed to warm to −23°C (dry ice–CCl_4 bath). The reaction mixture was stirred at −23°C for 3 h and was gradually warmed to room temperature. Aliquots were withdrawn and quenched with saturated NH_4Cl solution at −45, −23, 0°C, and room temperature. The organic solution of each aliquot was dried over $MgSO_4$ and was subject to GC analysis (12′ × ⅛″ stainless steel column packed with OV-17 (3%) on 100/120 Chromosorb G-NAW). The maximum yield was usually achieved around 0°C. The reaction was quenched by adding saturated NH_4Cl solution (15 ml) at room temperature. After stirring at room temperature for 20 min, the reaction mixture was then transferred into a separatory funnel and was extracted twice with diethyl ether (100 ml each). The combined ether layers were dried over anhydrous $MgSO_4$ and were carefully concentrated under reduced pressure at 5°C. The resulting residue was filtered and washed with hexanes/Et_2O (1 : 1, v/v). The combined washings were concentrated and chromatographed on silica gel to give a mixture of cyclopentanol and cyclobutylmethanol (37 : 1) as a colorless liquid. The product was further purified by preparative GC (12′ × ¼″ stainless steel column packed with SP-2100 (10%) on 100/120 Supelcoport).

Lithium 2-Thienylcyanocuprate-Based Copper and Chemistry

Preparation of Thienyl-Based Activated Copper

Li(0) (4.2 mmol) and naphthalene (4.6 mmol) in freshly distilled THF (5 ml) were stirred at room temperature under an argon atmosphere for 2 h. To this dark-green preformed lithium naphthalenide solution (−78°C), lithium 2-thienylcyanocuprate (0.25 M, 16 ml, 4.0 mmol) was added via a syringe. The solution of activated copper was used after it was stirred at −78°C for 0.5 h. For the epoxide-opening reaction cases, the lithium 2-thienylcyanocuprate was usually concentrated *in vacuo* to ~1/3 of original volume prior to use.

Reaction of Organocopper Reagent with Acid Chlorides

To the activated copper solution prepared in the preceding text, organic halide (0.13 mmol) was added via a syringe at −78°C. After stirring at −78°C for 10 min, the reaction mixture was allowed to warm to −35°C. The arylcopper reagent was usually warmed to 0°C prior to the addition of acid chloride. Acid chloride (4.0 mmol) was then added via a syringe at −35°C. After stirring at −35°C for 30 min, the reaction mixture was quenched with aqueous NH_4Cl solution. The reaction mixture was extracted with ether (3 × 70 ml), washed with H_2O, and concentrated under reduced pressure. The resulting mixture was isolated by flash-column chromatography on silica gel.

Epoxide Opening of Organocopper Reagent with 1,2-Epoxybutane

The organocopper reagent was generated as described in the preceding text. Epoxide (4 mmol) was added via a syringe at −35°C. The reaction mixture was gradually warmed to room temperature. For the aryl cases, the reaction mixture was heated at 40°C for 1 h. Workup and isolation were performed as described in the preceding text.

Copper Cyanide-Based Active Copper and Chemistry

Preparation of Active Copper and Reaction with Organic Halides to Yield Organocopper Reagents

A representative procedure for the formation of a functionalized organocopper reagent: Li (8.46 mmol) and naphthalene (10.1 mmol) in anhydrous THF (15 ml) were stirred under argon until the Li was consumed (~2 h). The flask was then cooled to −100°C. CuCN (8.0 mmol) and LiBr (17.27 mmol) in THF (5 ml) were stirred under argon until the Cu(I) salt was solubilized. The CuCN·2LiBr solution was cooled to −40°C and transferred into the LiNap with a cannula. The solution was stirred for 5 min. The active copper solution was warmed to −35°C and charged with ethyl 4-bromobutyrate (1.95 mmol). For aryl halides the solution was warmed to 0°C, immediately charged with the aryl halide, and allowed to mix for 1 h. The solution was stirred for 10 min and was ready to use for acid chloride couplings or conjugate additions.

Cross-Coupling of Benzoyl Chloride with Organocopper Reagents Derived from CuCN·2LiBr-Based Active Copper

To the organocopper species prepared in the preceding text was added benzoyl chloride (3 equiv based on organocopper) neat via syringe at −35°C. The solution was allowed to stir for 30 min, quenched with saturated NH_4Cl

(5 ml), and worked up with standard flash silica gel chromatographic techniques. Based on 1 equiv of CuCN, alkyl halides were allowed to react for 10 min at −35°C. Aryl halides were added at 0°C and allowed to react for 1 h. The reaction was warmed to room temperature and worked up by standard methods.

Conjugate Additions with Organocopper Reagents Derived from CuCN·2LiBr-Based Active Copper

The halide (0.25–0.30 equiv based on 1 equiv of CuCN) was transferred to the active copper solution at −35°C. After 15 min, the flask was cooled to −78°C. A two- to threefold excess of TMSCl, in respect to the equiv of 1,4-adduct, was injected neat into the flask (a sixfold excess of TMSCl was used for entry 6, Table 5.9). The 1,4-adduct was dissolved in THF (10 ml) in a separate vial and delivered dropwise to the stirring organocopper solution. After 1.5 h at −78°C, the flask was gradually warmed to room temperature. The enone, 2-cyclohexen-1-one, was injected neat at −90°C, warmed to −78°C, and stirred for 1 h. 3-(3-Methyl-2-butenyl)cyclohexanone was the sole product isolated upon workup by standard methods.

Reaction of Allyl Organocopper Reagents Derived from CuCN·2LiBr with Benzoyl Chloride

To a solution of active copper at −100°C was added the allyl chloride (0.25 equiv), which was previously cooled to −78°C in a vial admixed with THF (1 ml). The PhCOCl (3 equiv based on organocopper) was added neat via syringe at −100°C and allowed to react for 15 min at −78°C. The reaction was then worked up by standard methods.

Preparation of Copper Anions and Some Resulting Chemistry

Preparation of Cu(−1)Li(+)

A typical procedure for reduction involves weighing lithium ribbon (4.13 mmol) and naphthalene (4.55 mmol) in a two-necked round bottom flask in an argon dry box. After being brought to a double manifold providing argon and vacuum, LiNp is formed in 8 ml of THF for 2 h. Meanwhile, CuI (2 mmol) and PPh_3 (2.09 mmol) are weighed in a second two-necked flask and the complex is formed in 12 ml of THF under argon. The flask is cooled in a Et_2O/liq N_2 bath (∼ −105 to −110°C) for 10 min. The LiNp solution is cooled to −35°C and transferred with a cannula to the CuI·PPh_3 flask. The organic halide to be reacted was dissolved in THF and added dropwise to the copper anion solution at −35°C. The organocopper formation was typically complete after 10 min at −35°C as determined by a 1-ml reaction quench in saturated NH_4Cl and subsequent GC analysis.

References

1 (a) Ullman, F. *Ann.* 1904, **332**, 38. (b) Fanta, P.E. *Chem. Rev.* 1964, **64**, 613. (c) Bacon, R.G.; Hill, H.A. *Proc. Chem. Soc.* 1962, 113.
2 (a) Rieke, R.D.; Wehmeyer, R.M.; Wu, T.C.; Ebert, G.W. *Tetrahedron* 1989, **45**, 443. (b) Ebert, G.W.; Rieke, R.D. *J. Org. Chem.* 1988, **53**, 4482. (c) Wehmeyer, R.M.; Rieke, R.D. *Tetrahedron Lett.* 1988, **29**, 4513. (d) Wu, T.C.; Rieke, R.D. *Tetrahedron Lett.* 1988, **29**, 6753. (e) Wu, T.C.; Wehmeyer, R.M.; Rieke, R.D. *J. Org. Chem.* 1987, **52**, 5057. (f) Wehmeyer, R.M.; Rieke, R.D. *J. Org. Chem.* 1987, **52**, 5056. (g) Ebert, G.W.; Rieke, R.D. *J. Org. Chem.* 1984, **49**, 5280. (h) Rieke, R.D.; Rhyne, L.D. *J. Org. Chem.* 1979, **44**, 3445. (i) Rieke, R.D.; Kavaliunas, A.V.; Rhyne, L.D.; Frazier, D.J. *J. Am. Chem. Soc.* 1979, **101**, 246. (j) Rieke, R.D. *CRC Crit. Rev. Surf. Chem.*. 1991, **1**, 131. (k) Rieke, R.D. *Science* 1989, **246**, 1260.
3 Rieke, R.D.; Rhyne, L.D. *J. Org. Chem.* 1979, **44**, 3445.
4 (a) Rieke, R.D. *CRC Crit. Rev. Surf. Chem.* 1991, **1**, 131. (b) Rieke, R.D. *Science* 1989, **246**, 1260.
5 Ebert, G.W.; Rieke, R.D. *J. Org. Chem.* 1984, **49**, 5280.
6 Suzuki, Y.; Suzuki, T.; Kawagishi, T.; Noyori, R.; *Tetrahedron Lett.* 1980, **21**, 1247.
7 (a) Corey, E.J.; Boaz, N.W. *Tetrahedron Lett.* 1985, **26**, 6019. (b) Alexakis, B.; Berlan, J.; Bescae, Y.; *Tetrahedron Lett.* 1986, **27**, 1047.
8 Wu, T.C.; Rieke, R.D.; *J. Org. Chem.* 1988, **53**, 2381.
9 (a) Ginah, F.O.; Donovan, T.A.; Suchan, S.D.; Pfening, D.R.; Ebert, G.W. *J. Org. Chem.* 1990, **55**, 584. (b) Ebert, G.W.; Klein, W.R. *J. Org. Chem.* 1991, **56**, 4744. (c) Ebert, G.W.; Cheasty, J.W.; Tehrani, S.S.; Aoud, E. *Organometallics* 1992, **11**, 1560. (d) Ebert, G.W.; Pfening, D.R.; Suchan, S.D.; Donovan, T.A. *Tetrahedron Lett.* 1993, **34**, 2279.
10 (a) Smith, J.G. *Synthesis* 1984, 629. (b) Stork, G.; Cohen, J.F. *J. Am. Chem. Soc.* 1974, **96**, 5270. (c) Stork, G.; Cama, L.D.; Coulson, D.R. *J. Am. Chem. Soc.* 1974, **96**, 5268.
11 Baldwin, J.E. *J. Chem. Soc. Chem. Commun.* 1976, 734.
12 Parham, W.F.; Jones, L.D.; Sayed, Y.A. *J. Org. Chem.* 1976, **41**, 1184.
13 Parham, W.E.; Jones, L.D.; Sayed, Y.A. *J. Org. Chem.* 1975, **40**, 2394.
14 (a) Boatman, R.J.; Whitlock, B.J.; Whitlock Jr. H.W. *J. Am. Chem. Soc.* 1978, **100**, 2935. (b) Boatman, R.J.; Whitlock, B.J.; Whitlock Jr. H.W. *J. Am. Chem. Soc.* 1977, **99**, 4822. (c) Brewer, P.D.; Tagat, J.; Hergruetor, C.A.; Helquiist, P. *Tetrahedron Lett.* 1977, **18**, 4573.
15 (a) Rieke, R.D.; Klein, W.R.; Wu, T.C. *J. Org. Chem.* 1993, **58**, 2492. (b) Klein, W.R.; Rieke, R.D. *Synth. Commun.* 1992, **18**, 2635. (c) Rieke, R.D.; Wu, T.C.; Stinn, D.E.; Wehmeyer, R.M. *Synth. Commun.* 1989, **19**, 1833.
16 (a) Oppolzer, W.; Stevenson, T.; Godel, T. *Helv. Chim. Acta* 1985, **68**, 212. (b) Oppolzer, W.; Loher, H.J. *Helv. Chim. Acta* 1981, **64**, 2808.

17 (a) Stack, D.E.; Dawson, B.T.; Rieke, R.D. *J. Am. Chem. Soc.* 1992, **114**, 5110. (b) Stack, D.E.; Dawson, B.T.; Rieke, R.D. *J. Am. Chem. Soc.* 1991, **113**, 4672.

18 (a) Whitesides, G.M.; Fischer, W.F.; SanFilippo, J.; Basche, C.M.; House, H.O. *J. Am. Chem. Soc.* 1969, **91**, 4871. (b) Lipshutz, B.H.; Whitney, S.; Kozolowski, J.A.; Breneman, C.M. *Tetrahedron Lett.* 1986, **27**, 4273.

19 Stack, D.E.; Rieke, R.D. *Tetrahedron Lett.* 1992, **33**, 6575.

20 (a) Stack, D.E.; Klein, W.R.; Rieke, R.D.; *Tetrahedron Lett.* 1993, **34**, 3063–3066. (b) Rieke, R.D.; Dawson, B.T.; Stack, D.E.; Stinn, D.E. *Synth. Commun.* 1990, **20**, 2711.

21 (a) Lipshutz, B.H.; Crow, R.; Dimock, S.H.; Ellsworth, E.L. *J. Am. Chem. Soc.* 1990, **112**, 4063. (b) Lipshutz, B.H.; Elworthy, T.R. *J. Org. Chem.* 1990, **55**, 6095. (c) Lipshutz, B.H.; Ellsworth, E.L.; Dimock, S.H. *J. Org. Chem.* 1989, **54**, 4977. (d) Knochel, P. *J. Am. Chem. Soc.* 1989, **111**, 6474.

22 Lipshutz, B.H.; Ellsworth, E.L.; Dimock, S.H.; Smith, R.A.J. *J. Am. Chem. Soc.* 1990, **112**, 4404.

23 (a) Lipshutz, B.H.; Crow, R.; Dimock, S.H.; Ellsworth, E.L. *J. Am. Chem. Soc.* 1990, **112**, 4063. (b) Lipshutz, B.H.; Elworthy, T.R. *J. Org. Chem.* 1990, **55**, 1695. (c) Lipshutz, B.H.; Ellsworth, E.L.; Dimock, S.H. *J. Org. Chem.* 1989, **54**, 4977.

24 Bertz, S.H. *J. Am. Chem. Soc.* 1990, **112**, 4031.

25 (a) Alexakis, A.; Berlan, J.; Besace, Y. *Tetrahedron Lett.* 1986, **27**, 1047–1050. (b) Corey, E.J.; Boaz, N.W. *Tetrahedron Lett.* 1985, **26**, 6019.

6

Indium

6.1 Background and Synthesis of Rieke Indium

Indium was discovered in Germany in 1863. However, it was not until 1924 that gram quantities of the metal became available. The use of indium in organic chemistry has been very limited. Indium oxide does have extensive use in the area of photovoltaic devices. It serves as a transparent coating on glass which is electrically conducting.

In 1974, we reported a general procedure for the preparation of highly reactive indium metal powder [1]. The procedure involves the reduction of anhydrous indium trichloride with alkali metals in hydrocarbon solvents under an argon atmosphere. A typical combination is potassium metal and xylene as the solvent. The process works well for most any combination with the requirement that the boiling point of the solvent should exceed the melting point of the alkali metal.

6.2 Preparation of Organoindium Compounds

Little work has been reported concerning the direct synthesis of alkylindium halides. Gynane and coworkers [2] reported the direct reaction of indium metal with alkyl halides. The reaction times were, however, on the order of a day or more, and the product was a mixture of R_2InX and $RInX_2$. We have found that our reactive indium will react with methyl and ethyl iodide in 2 h at 80°C in xylene to give nearly quantitative yields of a single product, the dialkylindium iodide. The overall reaction scheme is summarized as follows:

$$InCl_3 + 3M \rightarrow In^* + 3MCl \ \left(M = K \text{ or } Na\right)$$
$$2In^* + 2RI \rightarrow R_2InI + InI$$

Chemical Synthesis Using Highly Reactive Metals, First Edition. Reuben D. Rieke.

Under our reaction conditions, the indium monoiodide formed does not react further with additional alkyl iodide as has been reported [3, 4]. Additional reactions were attempted with indium monoiodide and alkyl iodides in the presence of alkali salts under our reaction conditions, and no reaction was observed.

Further proof of the high reactivity of the indium powder comes from its reaction with iodine. The standard method for the preparation of indium monoiodide involves the heating of stoichiometric amounts of indium metal and iodine at 400°C for 24h [4]. We find that iodine reacts with the activated indium powder in refluxing xylene quantitatively in 30min, and the indium monoiodide formed is easily separated and purified.

For over 70years, one of the most convenient methods used for the preparation of triaryl and trialkyl indium compounds has been the direct reaction of indium metal with diorganomercury compounds [5, 6]. The method is, however, quite time consuming due to the general unreactive nature of indium metal. In some instances, reaction times of a week or more are required to obtain modest yields [5, 7]. Of greater concern, the reaction generates mercury. We reported a new procedure for the rapid preparation of triaryl and trialkyl indium compounds in essentially quantitative yields [9].

The highly reactive indium metal is prepared by reducing anhydrous indium halides in hydrocarbon solvents with an alkali metal under an argon atmosphere [1]. The reduction yields a light black material which has not been totally characterized but from all indications is indium metal in a very finely divided state. The role of the alkali salt, generated in the reduction, on the activity of the metal is also unclear at this time. In any event, the resulting indium metal exhibits a high degree of reactivity toward diorganomercury compounds yielding the corresponding triaryl and trialkyl indium compounds. The general reaction scheme is outlined as follows. The reactions of R_2Hg ($R = CH_3$, C_6H_5, *p*-tolyl) with the activated indium in hydrocarbon solvents have been studied, and the results are summarized in Table 6.1:

$$InCl_3 + 3M \rightarrow In^*(\text{activated}) + 3MCl \quad \left(M = K \text{ or } Na\right)$$

$$2In^* + 3R_2Hg \rightarrow 2R_3In + 3Hg$$

The high degree of reactivity can be seen by comparing the literature preparation of these indium compounds. For example, Dennis prepared $(CH_3)_3In$ from indium metal and $(CH_3)_2Hg$ by heating at 100°C for 8days, and no yield was reported. However, when $(CH_3)_2Hg$ was allowed to react with the activated indium metal, the reaction reached 100% yields after only 3h at 100°C. Similarly, the preparation of the triaryl indium compounds is greatly facilitated by using the activated indium metal, and the yields are essentially quantitative.

Table 6.1 Reactions of activated indium and diorganomercury.

Expt. no.	Source of In* $InCl_3$ + M	Solvent[a]	Time to[b] generate In* (h)	R_2Hg	Reaction time (h)	Temp.	%
1	K	B	22	CH_3	1.5	Reflux	63[c]
2	K	X	6	CH_3	1.5	110°	76[c]
3	K	X	6	CH_3	3.0	100°	100[d]
4	K	X	6	C_6H_5	1.5	Reflux	92[c]
5	K	X	6	C_6H_5	2.5	130°	100[d]
6	Na	X	5	C_6H_5	2.5	Reflux	100[d]
7	K	X	5	*p*CH_3-C_6H_4	1.5	Reflux	100[c]
8	K	X	6	*p*-CH_3-C_6H_4	2.5	130°	100[d]
9	Na	X	5	*p*CH_3-C_6H_4	1.5	Reflux	85[c]

[a] B = benzene, X = xylene.
[b] Activated In was generated under reflux.
[c] % was based on the amount of Hg collected.
[d] The presence of R_2Hg was tested by mercury test paper which was obtained from Macherey-Nagel GmbH & Co., D-516 Düren, Germany. Limit of sensitivity: 25 mg Hg/l.

The presence of the alkali metal salts causes no problem in isolating the final product. Eisch has reported that the addition of KCl in the reaction of $InCl_3$ and Et_3Al will increase the yield of Et_3In [8]. The KCl forms complexes with Et_2InCl and excess Et_3Al but not with Et_3In, allowing for a much more convenient separation of the desired trialkyl indium compounds. We also have found that under our reaction conditions, the alkali metal salts do not form complexes with the organoindium compounds. As the alkali metal salts are very insoluble in xylene while the desired triorganoindium compounds are soluble, they can be separated from each other by simple filtration of the reaction mixture.

The Direct Synthesis of Diphenylindium Iodide and Ditolyindium Iodide from Activated Indium and Aryl Iodides

Only two methods for the preparation of phenylindium halides have been reported: (i) transmetallation of Grignard reagents with indium trihalides [10] and (ii) the reaction of triphenylindium with Br_2 or I_2 [11]. No work had appeared in the literature concerning the direct preparation of diphenylindium halides from indium metal and aryl halides. The only related work reported was the preparation of pentafluorophenylindium compounds from indium metal by Deacon and Parrott [12]. We reported the first successful direct route to prepare diphenylindium iodide and ditolylindium iodide from activated indium metal and the corresponding aryl iodide [13].

Results and Discussion

The activated indium powder was prepared as previously reported by the reduction of anhydrous indium trichloride with potassium in refluxing xylene under an argon atmosphere. The reduction of the indium salt yields a finely divided black powder. Particle size analysis [14] of the black powder indicated a size distribution of 10–0.2 μm with the average particle size being 4.0 μm. Scanning electron microscope photographs [15] indicated that the particles are conglomerates of even smaller crystals of 0.5 μm or less. Thus the activated indium metal clearly has a high surface area. As to the origin of the high reactivity, it is not clear yet whether it is just a result of high surface area or if the alkali salt generated in the reduction is in any way enhancing the reactivity. The finely divided metal shows little tendency to form larger conglomerates of indium metal during the reaction. One possible explanation is the specific adsorption of chloride ions on the surface of the metal. This could have two results. One might be to lower the work function of the metal, thereby increasing the reactivity of the metal. The second expected effect would be the retardation of conglomeration due to the effective negative charge on the surfaces of the small crystals due to the adsorbed chloride ions.

This black indium powder was found to react rapidly with aryl iodides. For example, the activated indium reacted with neat iodobenzenes at 150°C, and in 2 h over 94% yield of diphenylindium iodide was obtained. For comparison, the same reaction was run with commercial indium powder (atomized, 325 mesh) and yielded only 30% diphenylindium iodide after 25 h. It is interesting to note that considerable conglomeration of the indium metal particles occurred with the commercial indium powder. It is felt that this conglomeration is, to a large extent, responsible for the low reactivity of the commercial indium metal. Attempts to prevent the conglomeration by the addition of potassium chloride failed.

Recently Deacon and Parrott [12] reported the direct reaction of indium metal with iodopentafluorobenzene. The reaction yield was <40%, and the product composition was dependent upon the stoichiometric ratios of reactants; trispentafluorophenylindium was prepared with an excess of indium metal but was mixed with bispentafluorophenylindium when the stoichiometric amount was used. We have found that our indium will react with neat iodobenzenes or iodotoluene at 150°C for 2 h to give a very high yield, 94% and 93%, respectively, and only a single product. The diphenylindium iodide or ditolylindium iodide is formed regardless of the stoichiometric ratios of reactants. The reactions of aryl halide with activated indium powder at various

Table 6.2 Reactions of aryl halides with activated indium.

Expt. no.	Reactant	Reactant/In*	Solvent	Reaction temp. (°C)	Reaction time (h)	% Yield[a]
1	C_6H_5I	6.7	Neat	150	2	94[b]
2	p-$CH_3C_6H_4I$	5.8	Neat	150	2	93
3	C_6H_5Br	9.8	Neat	156	26	11
4	C_6H_5I	4.0	Xylene	140	4	64
		4.0	Xylene	140	10	84
5	p-$CH_3C_6H_4I$	3.3	Xylene	140	4	68
		3.3	Xylene	140	10	88
6	C_6H_5I	0.5	Xylene	140	8	54
7	p-$CH_3C_6H_4I$	0.5	Xylene	140	8	58

[a] Yield was based on the smallest amount of reactant, and quenched aliquots were analyzed for benzene and toluene with a Perkin Elmer Model 881 gas chromatograph equipped with a disc integrator model 3200 and with 8′ × 1/8″ stainless steel columns packed with 10% SE30.
[b] Yield was obtained by measuring the concentration of indium in solution EDTA.

stoichiometric ratios have been studied, and the results are summarized in Table 6.2. The overall reaction scheme is outlined as follows:

$$InCl_3 + 3K \rightarrow In^*(\text{activated}) + 3KCl$$

$$2In^* + 2RI \rightarrow R_2InI + InI\left(R = \text{phenyl, tolyl}\right)$$

The activated indium not only reacts with iodobenzenes and iodotoluene but also reacts with bromobenzene; however, with bromobenzene the yield is low (see Table 6.2).

The xylene solution layer of the reaction mixtures of activated indium with iodobenzenes or iodotoluene is light yellow in color, and the resulting precipitates are dark red. The reaction products, diphenylindium iodide and ditolylindium iodide, are only slightly soluble in xylene at room temperature but more soluble in xylene at high temperatures. The products, diphenylindium iodide and ditolylindium iodide, were isolated by filtering the reaction mixtures in the dry box when the solution was hot and crystallizing from petroleum ether. The products were identified by elementary analysis and IR spectra. The results of elementary analysis on the product from the reaction of activated indium with iodobenzenes showed that the relative ratio of C_6H_5:In:halide was 1.99:1.00:0.92. This indicated an empirical formula of $(C_6H_5)_2InI$ for the product. The IR spectrum of the product was scanned from 1700 to 450 cm^{-1} and compared with that of an authentic sample of $(C_6H_5)_2InI$ which was obtained from the reaction of $(C_6H_5)_3InI$ and I_2. They were exactly identical in this region. In the reaction of activated indium with iodotoluene, the elementary analysis showed that the relative ratio of p-$CH_3C_6H_4$:In:halide was 1.80:1.00:0.90. The IR spectrum of the product $(p\text{-}CH_3C_6H_4)_2InI$ was recorded from 1500 to 460 cm^{-1} and compared with that of an authentic sample in this region.

The dark red residue remaining after filtering the reaction mixtures was charged in the sublimer and heated to 140°C under reduced pressure, 0.05 mm. No sublimation occurred, suggesting that no indium triiodide, which is volatile [16], is present in the residue. The possibility that the product, R_2InI (R = phenyl or tolyl), is a mixture of suitable amounts of R_3In and InI_3 or $RInI_2$ can be eliminated by comparing the IR spectrum with that of R_3In with the IR obtained from the sublimation experiment.

One note of caution is necessary regarding the preparation of the activated indium. Excess potassium or indium trichloride should be avoided as they can catalyze side reactions. Thus the ratio of potassium to indium trichloride should be 3 : 1; under these conditions GC analysis indicated little or no by-products. Thus the use of activated indium allows the ready preparation of diaryl indium iodides directly from the aryl iodide.

6.3 Preparation and Reactions of Indium Reformatsky Reagents

The highly reactive indium powder reported in Sections 6.1 and 6.2 reacts readily with α-haloesters to give an indium Reformatsky-type reagent which will add to ketones and aldehydes to give β-hydroxy esters. The Reformatsky reaction using zinc metal has been used in the preparation of β-hydroxy esters for many years. Recent improvements have made this reaction a very reliable reaction (see Section 3.10).

The activated indium powder was prepared by reducing anhydrous $InCl_3$ (Alfa ultrapure) with freshly cut potassium metal in dry, freshly distilled xylene. The mixture is heated to reflux and stirred with a magnetic stirrer under argon for 4–6 h, yielding a light black powder:

$$InCl_3 + 3K \rightarrow In^* + 3KCl$$

Reaction of this activated indium metal with α-haloesters is rapid. For example, reaction of a mixture of ethyl-α-bromoacetate and cyclohexanone in xylene at 55°C for 2 h gives an almost quantitative yield of the corresponding β-hydroxyester. The activated indium will also react with ethyl-α-chloroacetate. In this case reaction for 7 h at 55°C in xylene gave a 42% yield of the β-hydroxy ester. In contrast, commercial indium metal (325 mesh) was reacted with ethyl-α-bromoacetate for periods of time up to 18 h with a maximum yield of only 17%.

The reaction of the black indium powder with the α-bromoacetate can be readily observed by the rapid disappearance of the black powder and the formation of a brownish-yellow material. The proposed structure for the intermediate indium compounds is $(EtOCOCH_2)_2InBr$. This is based on an analysis of the material which was xylene soluble and showed an In/halide ratio of 1.05 : 1.00. The overall reaction scheme is shown in Equation 6.1.

$$2In^* + 2BrCH_2CO_2Et \longrightarrow BrIn(CH_2CO_2Et)_2 + InBr$$

$$BrIn(CH_2CO_2Et)_2 + R_2C{=}O \longrightarrow BrIn(CH_2CO_2Et)(O{-}CR_2{-}CH_2CO_2Et) \xrightarrow{H^+{-}H_2O} R{-}C(OH)(R){-}CH_2CO_2Et \quad (6.1)$$

When we extracted the brownish-yellow solid with THF, in which it is much more soluble, we obtained different analysis results. In this case it was found that the K : In : halide ratio was 0.85 : 1.00 : 2.02 and it contained 25.6% of In. This fits the structure of the ate complex of $(EtOCOCH_2)_2InBr$ with KCl shown in Equation 6.2.

$$BrIn(CH_2CO_2Et)_2 + KCl \xrightarrow[\text{solvent}]{\text{Polar}} K^+ [Br(Cl)In(CH_2CO_2Et)_2]^- \quad (6.2)$$

Thus it appears that the composition of this Reformatsky reagent is quite solvent dependent, with the ate complex being favored in polar solvents. We examined the effect of the solvent on the yield of the β-hydroxy ester, and the results are summarized in Table 6.3. It is readily apparent that the reactions in polar solvents gave lower yields and more side products. We attribute these results to the formation of the ate complex in polar solvents.

In order to obtain good yields of β-hydroxy esters, several reaction conditions must be carefully controlled. In the preparation of the activated indium, it is very important that the molar ratio of $InCl_3$ to K be exactly 1 : 3. If there is an excess of either $InCl_3$ or K, side reaction products become considerable. The ratio of the carbonyl compound to the α-bromoacetate is also very critical. In order to obtain high yields, a onefold excess of the carbonyl compound is

Table 6.3 Solvent effect on the yield of the reactions with activated indium.[a]

Carbonyl compd	Solvent	Reaction temp (°C)	Reaction time (h)	% Yield[b]
Cyclohexanone	Xylene	55	2	100
Cyclohexanone	Et_2O	34	12	100
Cyclohexanone	THF	66	4.5	39–71[c]
Benzaldehyde	Xylene	55	2	83–100
Benzaldehyde	Et_2O	34	6	81
Benzaldehyde	THF	66	4	67

[a] All reactions were carried out in the ratio of $In^*:BrCH_2CO_2Et:>{=}O = 1.0:0.95:0.95$.
[b] Yield was measured by GC based on $BrCH_2CO_2Et/2 = 100\%$ yield[d].
[c] Unreacted cyclohexanone was not completely recovered.
[d] The factor of ½ is necessary because only one $-CH_2CO_2Et$ group of $BrIn(CH_2CO_2Et)_2$ adds to a ketone or aldehyde.

Table 6.4 Effect of stoichiometric ratio of ethyl bromoacetate to carbonyl compounds on the yield of the reaction with activated indium.[a]

Carbonyl compd	$In^*:BrCH_2CO_2Et:>{=}O$	% Yield
Cyclohexanone	1.0:0.95:0.95	100
Cyclohexanone	2.1:2.0:1.0	18
Benzaldehyde	1.0:0.95:0.95	83–100
Benzaldehyde	2.1:2.0:1.0	54

[a] All reactions were carried out in xylene at 55° for 2 h.

necessary; the excess carbonyl compound can be recovered later. The reason for the necessity of the excess carbonyl compound is not readily obvious. Some of the results with varying ratios of bromoacetate to ketone are summarized in Table 6.4.

Finally, the results of reactions of a variety of carbonyl compounds with the activated indium and ethyl-α-bromoacetate are summarized in Table 6.5. In general, the yields of β-hydroxy esters are good with ketones and also with benzaldehyde in xylene and diethyl ether. However, alkyl aldehydes give relatively low yields. In conclusion, highly activated indium has been prepared and some interesting new results have been reported. However, it is clear that much work remains to be done to fully explore the utility of this new material.

Table 6.5 Summary of Reformatsky reaction of carbonyl compounds with activated indium and ethyl bromoacetate.[a]

Carbonyl compd	Product	Registry no.	Solvent	Reaction temp (°C)	% Yield[b]
Cyclohexanone	1-hydroxycyclohexyl with OH and CH_2CO_2Et	5326-50-1	Xylene	55	100
Cyclopentanone	1-hydroxycyclopentyl with OH and CH_2CO_2Et	3197-76-0	Xylene	55	80
Benzaldehyde	$C_6H_5CH(OH)CH_2CO_2Et$	5764-85-2	Xylene	55	83–100
p-Methylacetophenone	$CH_3C_6H_4C(OH)(CH_3)CH_2CO_2Et$	55319-45-4	Et_2O	34	59

[a] All reactions were carried out for 2 h.

[b] Based upon GLC analysis using internal standard, $BrCH_2CO_2Et/2 = 100\%$[c].

[c] The factor of ½ is necessary because only one $-CH_2CO_2Et$ group of $BrIn(CH_2CO_2Et)_2$ adds to a ketone or aldehyde.

6.4 Experimental Details for Preparation and Reactions of Activated Indium

Preparation of Active Indium and Reaction with Alkyl Iodides

The activated indium powder was prepared in the following manner. Anhydrous $InCl_3$ (Alfa ultrapure, 20 mmol) was placed in a flask with a side arm equipped with a septum cap in a dry box. Freshly distilled xylene (25 ml) was added, followed by adding freshly cut potassium (2.34 g, 60 mmol). The mixture was heated to reflux and stirred with a magnetic stirrer under argon. The reduction was generally complete in 4–6 h, yielding a light black powder. The alkyl iodide (30 mmol, R = Me, Et) in 50% excess and internal standard were added, and the mixture was kept at 80°C. Aliquots were periodically quenched in 10% HCl solution, and the percentage of reaction was determined by measuring the disappearance of alkyl iodide. The reactions of methyl iodide and ethyl iodide with activated indium reached 100% and 97% after 2 h, respectively.

Dimethylindium iodide was obtained by either crystallization from xylene or sublimation under reduced pressure. The IR spectrum showed C—In stretching frequencies at 552 s and 478 w/cm (lit. [18], 548 s, 480 w/cm), and mass spectrum showed molecular ion *m*/*e* 272. Diethylindium iodide was also obtained by sublimation, mp 171–174°C (lit. [18], 171–173°C). The IR spectrum showed C—In stretching frequencies at 450 s and 505 s/cm (lit. [17], 455, 506 cm^{-1}), and mass spectrum showed molecular ion *m*/*e* 300.

Reaction of Active Indium with Iodine

Sublimed iodine (1.27 g, 5 mmol) was added to activated indium (prepared from 2.21 g of $InCl_3$ and 1.17 g of potassium) in xylene (25 ml). After refluxing for 30 min, the reaction was complete. The red indium monoiodide was collected by filtration, washed with water to remove the alkali salts, and dried under reduced pressure. The product analyzed correctly for InI (Calcd.: In, 47.50; Found: In 46.80, 47.91). The indium was determined by EDTA titration using 4-(2-pyridylazo) resorcinol as indicator or atomic absorption spectroscopy.

All air-sensitive compounds were manipulated in an inert atmosphere box (Kewaunee Model 2C1982) equipped with a recirculation system to remove moisture and oxygen.

Anhydrous indium trichloride (ultrapure) was obtained from Alfa. Dimethylmercury, diphenylmercury, and ditolylmercury were obtained from Eastman and were used without further purification. Potassium (purified) and sodium were obtained from Baker and cleaned under hexane just prior to use.

Xylene (Baker) was distilled under argon over $NaAlH_4$ through 60 cm glass helix packed column. Benzene (Baker) was distilled over $LiAlH_4$.

The following procedure for carrying out these reactions is representative of the general technique. $InCl_3$ (4.41 g, 20 mmol) was placed in a 100 ml round bottomed flask with a side arm equipped with a septum cap in a dry box, and then freshly distilled solvent (25 ml) was added and followed by adding freshly cut potassium (2.34 g, 60 mmol) or sodium (1.38 g, 60 mmol) into the flask under argon. The flask was connected with a water condenser, and then the mixture was heated to reflux and stirred with a magnetic stirrer. The reduction was generally complete in 4–6 h in xylene, yielding a light black powder.

Triphenylindium

In a typical experiment 20 mmol of activated indium metal in 25 ml xylene, which was prepared by following the above procedure, was allowed to react with 7.10 g of diphenylmercury under an argon atmosphere. The mixture was heated to reflux for 2.5 h, and then the solution was allowed to cool to room temperature, and 0.5 ml of solution was withdrawn and hydrolyzed by concentrated HNO_3. This sample was tested for the presence of R_2Hg and was found to contain none. The solution was then warmed up again and filtered in the dry box when it was hot. The filtrate was allowed to cool, and the mass of needles which separated was collected and washed with dry petroleum ether and dried under vacuum, mp 206–208°C (lit. [6] 208°C). Anal. Calcd. for $(C_6H_5)_3In$: C_6H_5: 66.83, In: 33.17; found for C_6H_5: 67.75, In: 33.98 [14].

Tritolylindium

The preparation of this compound is similar to that of triphenylindium. The reaction temperature was 130°C. The product could not be purified by sublimation at high temperature due to thermal decomposition and was purified by crystallization. (Anal. Calcd. for $(CH_3{-}C_6H_4)_3In$, *p*-$CH_3C_6H_4$: 70.42, In: 29.58; found: *p*-$CH_3C_6H_4$: 69.09, In: 30.44 [14].)

Trimethylindium

The following was found to be the best procedure. The xylene is stripped off after the activated indium was generated, and then the light black powder was allowed to react with $(CH_3)_2Hg$ at 100°C for 3 h, followed by sublimation of the product under vacuum. The product is conveniently trapped in a U-tube equipped with a ground glass stopcock and cooled in liquid nitrogen (mp 88°C) (lit. [5] 89–89.5°C).

The Reaction of Activated Indium with Iodobenzene

Activated indium (20 mmol) was prepared by the reduction of anhydrous indium trichloride (20 mmol, 4.41 g) with freshly cut potassium (60 mmol, 2.34 g) in 25 ml of xylene by refluxing 4.5 h. The reaction of In* with iodobenzene was carried out either in xylene or without xylene; we prefer the latter way. Thus xylene was stripped off under reduced pressure, leaving the indium metal. Iodobenzene (15 ml) was added to the light black metal and heated to 150°C with stirring under an argon atmosphere for 2 h. A dark red mixture formed, unreacted iodobenzene was stripped off under reduced pressure, and 50 ml of freshly distilled xylene was added to the reaction mixture. A 2 ml aliquot of the reaction mixture was withdrawn with stirring and hydrolyzed with 2 N HCl solution, after which an external standard was added. The organic layer was subject to GC. The yield of the reaction was measured by GC analysis. The rest of the solution was warmed up, and the reaction mixture was filtered in a dry box when the solution was hot. The light yellow filtrate was put under reduced pressure to strip off some xylene until some crystals formed, and then 50 ml of petroleum ether was added, and a large amount of product immediately precipitated out. The light yellow crystals of diphenylindium iodide were filtered in the dry box and dried under vacuum.

The Reaction of Activated Indium with Iodotoluene

This reaction is similar to that of activated indium with iodobenzenes. Light beige crystals of ditolylindium iodide were obtained.

The Reaction of Triphenylindium with Iodine

Triphenylindium was prepared by the reaction of diphenylmercury with activated indium [9]. Then sublimed iodine (0.95 g, 3.74 mmol) was allowed to react with a 50 ml xylene solution of triphenylindium (3.74 mmol). The mixture was stirred at room temperature; the purple color disappeared within seconds. Then the xylene was stripped off until about 20 ml remained and 50 ml of petroleum ether was added. A large amount of white precipitate immediately formed; this was filtered in the dry box and dried under vacuum. White crystals of diphenylindium iodide were obtained.

Materials

Cyclohexanone, cyclopentanone, benzaldehyde, *p*-methylacetophenone, ethyl bromoacetate, and ethyl chloroacetate were obtained from commercial sources. They were used without further purification. Xylene (ortho, 99%) was obtained from Aldrich and distilled over $NaAlH_4$. Diethyl ether (Fisher anhydrous) and THF (MCB) were distilled over $LiAlH_4$ under argon. Commercial indium (325 mesh, 99.9%) and anhydrous indium trichloride were obtained from Alfa. Potassium (Baker purified) was cleaned under heptane prior to use. Activated indium metal was prepared by reducing indium trichloride with potassium in xylene by refluxing 4–6 h under an argon atmosphere.

Indium Reformatsky Reaction

The following procedure for the conversion of cyclohexanone to ethyl (1-hydroxycyclohexyl) acetate is representative. Activated indium (10 mmol) was prepared from the reduction of $InCl_3$ (2.21 g, 10 mmol) and K (1.17 g, 30 g-atoms) in 25 ml of xylene in a 100 ml round-bottom flask with a side arm equipped with a septum cap under an argon atmosphere. After refluxing for 4.5 h, the mixtures were cooled down to 2°C with ice water, 1.0 g of biphenyl was added as an internal standard, and then two loaded syringes with cyclohexanone (0.93 g, 9.5 mmol) and ethyl bromoacetate (1.5 g, 9.5 mmol) were inserted into the flask via a septum cap. Both components were added simultaneously and stirred at 2°C for 10 min; then the mixtures were heated to 55°C. One 2 ml sample was withdrawn periodically and hydrolyzed with 2 N HCl solution. The organic layer was subjected to GLC. GLC analyses were carried out with an HP Model 5750 research gas chromatograph equipped with 6 ft × 0.125 in. stainless steel columns packed with 10% SE-30 on Chromosorb W. The product was identified by comparing the GLC retention time with that of the authentic sample obtained from the regular zinc Reformatsky reaction.

References

1 Chao, L.; Rieke, R.D. *J. Organomet. Chem.* 1974, **67**, C64.
2 Gynane, M.J.S.; Waterworth, L.G.; Worrall, I.J. *J. Organomet. Chem.* 1972, **40**, C9.
3 Poland, J.S.; Tuck, D.G. *J. Organomet. Chem.* 1972, **42**, 315.
4 Gynane, M.J.S.; Waterworth, L.G.; Worrall, I.J. *J. Organomet. Chem.* 1972, **43**, 257.
5 Dennis, L.M.; Work, R.W.; Rochow, E.G. *J. Am. Chem. Soc.* 1934, **56**, 1047.
6 Gilman, H.; Jones, R.G. *J. Am. Chem. Soc.* 1940, **62**, 2353.
7 Visser, H.D.; Oliver, J.P. *J. Organomet. Chem.* 1972, **40**, 7.
8 Eisch, J. *J. Am. Chem. Soc.* 1962, **84**, 3605.
9 Chao, L.; Rieke, R.D. *Synth. React. Inorg. Met.-Org. Chem.* 1974, **4**, 373.
10 Pohlmann, J.L.W.; Brinckmann, F.E. *Z. Naturforsch.* 1965, **20B**, 5.
11 Schumb, W.C.; Crane, H. I. *J. Am. Chem. Soc.* 1938, **60**, 306.
12 Deacon, G.B.; Parrott, J.C. *Aust. J. Chem.* 1971, **24**, 1771.

13 Chao, L.; Rieke, R.D. *Synth. React. Inorg. Met.-Org. Chem.* 1975, **5**, 165.
14 SediGraph 5000 was used for particle size analysis.
15 Scanning Electronic Microscope, Model ETEC U-1 was used for surface study.
16 Carty, A.J.; Tuck, D.G. *J. Chem. Soc. A* 1966, 1081.
17 Maeda, T.; Tada, H.; Yasuda, K.; Okawara, R. *J. Organomet. Chem.* 1971, **27**, 13.
18 Clark, H.C.; Pickard, A.L. *J. Organomet. Chem.* 1967, **8**, 427.

7

Nickel

7.1 Preparation of Rieke Nickel, Characterization of Active Nickel Powder, and Some Chemistry

Nickel was first discovered in 1751 by Axel Cronstedt in Sweden. Its use in chemistry lay dormant until well into the twentieth century. Most of its early chemistry involved its use as a hydrogenation catalyst. Later, complexes of nickel resulted in it being useful in many chemical transformations. However, any chemistry from alkyl or aryl nickel reagents resulted from transmetallation chemistry, from Grignard reagents or organolithium reagents. Accordingly, little if any functionality was present in these organonickel compounds. The general procedure for preparing highly reactive metals we developed proved to be quite applicable for the preparation of highly reactive nickel [1–16]. The reduction of nickel salts with potassium proved to be a significant advance in the ability of nickel powders to undergo direct oxidative addition with alkyl and aryl halides. However, the most dramatic increase in reactivity was found by the reduction of $NiBr_2$ or NiI_2 with lithium and an electron carrier such as naphthalene. These black metal powders showed exceptional reactivity with C_6F_5X (X = Br, I) and yield solvated complexes of the type $Ni(C_6F_5)_2$ or $Ni(C_6F_5)X$. These complexes can be readily transformed into a wide range of valuable precursors of other new organometallic compounds. Klabunde and coworkers have reported the preparation of some of these compounds by using the metal vaporization technique [17–27]. In the following sections of this chapter, we will report a large number of new chemical transformations utilizing these new nickel powders.

Preparation of Rieke Nickel Slurries

Finely divided, extremely reactive black nickel powders are readily obtained by reducing in THF or glyme anhydrous $NiBr_2$ or NiI_2 with lithium in the presence of a small amount of naphthalene as an electron carrier. Reduction times vary depending on the metal halide but are usually complete in 20 h or less.

Chemical Synthesis Using Highly Reactive Metals, First Edition. Reuben D. Rieke.

Occasionally, if coating of the piece of lithium occurs, reduction times may be somewhat longer. Even though some of the metal halides are hygroscopic, we find that they can be weighed in air, if this is done quickly, and still yield satisfactory results. Glyme and THF are freshly distilled from sodium–potassium alloy under argon prior to use. In the rare situations where difficulties have been encountered in reductions, invariably the problem has been found to be contaminated solvents.

We have observed that during a reduction, a portion of the lithium is involved in reaction with the solvent [15]. To ensure complete reduction of the metal halide, we have found that a slight excess of lithium over stoichiometry is needed. The ratio of lithium to the metal halide is given in Table 7.1. The use of the particular ratio in a given reduction may or may not result in the formation of a slight amount of excess lithium naphthalide. This variability is possibly caused by small amounts of contaminants introduced in a given reaction.

Reductions involving lithium are easy to judge complete since the piece of lithium is usually pink (covered with dilithium naphthalide), floats, and is thus readily visible during a reduction. If an excess of lithium has been added, the excess can easily be removed from the reaction vessel, or, if it is converted to lithium naphthalide, this can be removed by allowing the slurry to settle, syringing off the olive green solution above the slurry, and adding fresh solvent. This can be repeated until a colorless solution exists with the black powder. This also results in the removal of the electron carrier naphthalene. During a reduction, the mixtures of partially reduced halides display some unusual properties. For example, a mixture may undergo changes from a very thick slurry to a very free-moving slurry during the course of a reduction. The reverse sequence of changes has also been observed. Very often in the early stages of reduction of most of these salts, a greasy material is observed that smears on the glass and appears to be insoluble in the solvent. Yet in 1 or 2 h, this substance disappears, and a free-moving slurry is obtained. The highly reactive nickel powders are clearly seen to be complex materials from the bulk elemental analysis shown in Table 7.2:

Surface Analysis

Surface (Table 7.3) as well as bulk analysis (Table 7.2) shows that the metal powders produced by this reduction technique are quite complex materials containing considerable amounts of carbon, hydrogen, and oxygen. The ESCA results are summarized in Table 7.3.

Table 7.1 Molar ratio of lithium to metal halide for reduction in glyme.

Metal halide	Lithium	Metal halide	Lithium
Ni (II)	2.3	Pt (IV)	5.1

Table 7.2 Bulk elemental analysis of active nickel powder.

Ni powder	C	H	Ni	I	Li	O
	4.45	1.36	64.16	1.42	5.35	28.68

Table 7.3 Relative atomic concentrations in percent as a function of sputtering depth[a]—activated Ni.

Depth	C	O	Ni	Cl	S	Li
As received	17.2	42.1	0.8	0.1	0.2	39.6
30 Å	19.4	40.4	1.3	0.3	0.0	38.7
100 Å	23.3	39.0	2.7	0.2	0.0	34.7
200 Å	25.4	35.6	4.4	0.2	0.1	34.0

[a] The surface material was removed by Ar^+ bombardment. The sputtering rates are determined from known thickness of Ta_2O_5 films. Since the sputtering rate of this material relative to Ta_2O_5 is unknown, the absolute depths are approximate. The relative atomic percents are calculated by using experimental sensitivity factors and peak areas for each of the elements shown. The relative atomic concentrations show reproducibility to ±0.2% and are correct to within ±5% of the absolute concentrations.

BET surface area measurements were carried out on the activated Ni powder and were found to have a specific surface area of 32.7 m^2/g. Bulk analysis was performed on Ni powders, and the results are shown in Table 7.2.

Discussion

The binding energies of the O 1s and the kinetic energies of the O KVV and its shape indicate that most of the Li exists as LiOH. The electron energies for the Ni lines show that the Ni is mostly Ni(0). A small amount of oxide NiO is seen on the high-resolution spectrum of the Ni 2p line; this oxide layer is completely removed after sputtering to a depth of 30 Å. This indicates that most of the O is bound to the Li. By sputtering to a depth of 200 Å, one sees an increase of the Ni/Li ratio while the Li/O ratio remains somewhat constant. Some reduction of the LiOH to Li_2O is observed in the O 1s spectrum after sputtering. A small amount of carbonate is seen in the C 1s spectrum that is most likely bound to the Li.

The BET analysis on the activated Ni clearly shows the high surface area of these materials. The bulk analysis data indicate the very complex nature of these materials. All samples contained considerable carbon and hydrogen. It is not clear at this point whether this is due to trapped solvent molecules or degradation products from the solvent. Klabunde has demonstrated that codeposition of transition metals with ethereal and hydrocarbon solvents leads to cleavage of C—H as well as C—C bonds [24, 25]. These reactions result in the incorporation of carbon and hydrogen in the final metal powders. At this point, it is not clear whether these reactions are occurring in our preparation of finely divided metals. As 2.3–2.6 equiv of lithium is used to reduce Ni(II), it is highly likely that some solvent cleavage to alkoxide ions is occurring. Much of the high oxygen content is likely to originate from the reaction of the metal powders with spurious oxygen encountered during all the manipulations. There is, of course, one primary concern with these attempts to do surface analysis on these very highly reactive metal powders. Do the results have any bearing on the freshly reduced metal powders before they are manipulated in any way? The answer is most likely that certain data will be valid. The bulk of the metal is in the zerovalent state, and we do not have a simple physical mixture of M(II) salts and some reducing agent. The metal powders are very complex materials containing carbon and hydrogen. The surfaces seem to be covered with LiOH and some carbonaneous material. However, the surface composition is the one aspect that is most likely affected by the various manipulations.

The origin of the high reactivity of these metals is thus still open to much speculation. Part of this reactivity certainly comes from the high surface area and small particle size. Also, before any manipulations, this surface area must be relatively free of passivation oxide coatings. It is also possible that the adsorbed hydroxide ions, alkoxide anions, and possibly halide anions may be adsorbed on the surfaces and reduce the work function of the metal. This would enhance the metal's ability to transfer an electron to the organic substrate in the initial step of these oxidative addition reactions.

Reactions of Slurries

The black nickel powders prepared by lithium reduction of a metal halide are not pyrophoric when exposed to air, yet they display remarkable reactivity toward organic molecules. These powders are most conveniently used as prepared, that is, they are not separated from the lithium salts, naphthalene, nor any other products formed in the reduction.

In some previous studies, Klabunde has shown that a variety of transition metal atoms when cocondensed with C_6F_5X (X = I, Br) undergo rapid oxidative addition to yield C_6F_5MX complexes [18, 19, 23, 24, 26, 27]. Klabunde demonstrated that these complexes can be converted to a variety of new compounds depending on the metal and halogen. It was also demonstrated that C_6H_5PdX can be prepared but is only stable at low temperatures [23]. We felt it would be of value and interest to see if transition metal powders prepared by the simple and convenient reduction method were reactive enough to undergo oxidative addition to C_6F_5X and other aryl halides.

Summary

Lithium reduction of transition metal halides is not only readily effected with inexpensive apparatus and a simple procedure but also yields exceptionally reactive metal powders. These powders react with C_6F_5X to yield ether-coordinated species of the type $M(C_6F_5)_2$ or $M(C_6F_5)X$, where X = I and Br. The coordinated ether is exceptionally labile and is readily, and in most cases instantly, displaced by a variety of ligands to yield organometallic compounds in high yield based on two reactions, that is, a reaction of a metal powder with C_6F_5X followed by the addition of another ligand to displace the coordinated ether of the initially formed organometallic species. The displacement of the coordinated ether allows preparation of compounds with ligands which would in some cases not survive a reaction involving a typical arylating agent.

The addition of a stabilizing ligand to a metal slurry prior to the addition of an organic halide allows preparation and isolation of organometallic compounds with groups other than C_6F_5. This route may allow the preparation of entirely new classes of organometallic compounds.

Surface analysis indicates that these powders are finely divided and very complex materials. Carbon as well as hydrogen is found in these metals. The metal surfaces appear to be coated with LiOH and some carbonaneous material, leaving the question of origin of high reactivity still somewhat in doubt.

The reaction of nickel metal powder with C_6F_5I results in the solvated $Ni(C_6F_5)_2$ and NiI_2. The glyme coordinated to $Ni(C_6F_5)_2$ is very labile and can be displaced with other ligands. Addition of $P(C_2H_5)_3$ to a mixture of solvated $Ni(C_6F_5)_2$ and NiI_2 results in an immediate reaction and formation of the known compound $Ni(C_6F_5)_2[P(C_2H_5)_3]_2$ in 69% yield. The ether displacement is also readily affected with $(C_6H_5)_2PH$ and yields, after recrystallization from toluene, $Ni(C_6F_5)_2[(C_6H_5)_2PH]_2C_6H_5CH_3$. Noteworthy in this preparation is that this compound is probably not readily accessible by standard arylating methods since the reaction of a nickel halide with $(C_6H_5)_2PH$ results in $Ni[(C_6H_5)_2P]_2[(C_6H_5)_2PH]_2$ [28].

Amines can also displace the coordinated glyme molecule from $Ni(C_6F_5)_2(C_4H_{10}O_2)$. Thus the addition of pyridine to the orange-brown mixture of solvated $Ni(C_6F_5)_2$ and NiI_2 results in an immediate color change and the formation of $Ni(C_6F_5)_2(C_5H_5N)_2$.

Experimental Procedures

Preparation of a Typical Nickel Slurry

A 50 ml two-necked flask equipped with a magnetic stirrer and a condenser topped with an argon inlet was charged with 2.285 g (0.01048 mol) of $NiBr_2$ (quickly weighed in air), 0.152 g (0.0233 mol) of freshly cut lithium, and 0.156 g (0.00121 mol) of naphthalene. Glyme (18 ml) was syringed onto this mixture and stirring started. In 1 h, the mixture was essentially black with a pink and in part a black piece of lithium. After 18 h, reduction was complete; a black slurry existed in a colorless solution.

Preparation of $Ni(C_6F_5)_2[P(C_2H_5)_3]_2$

A nickel metal slurry was prepared in the typical manner and then allowed to settle. About half of the liquid above the slurry was removed via syringe, and then 1 equiv of C_6F_5I per nickel was added dropwise. A yellow color formed in the reaction mixture. The mixture was stirred at about 40°C for 9 h, and then the deep orange mixture cooled to about 0°C. Two equivalents of $P(C_2H_5)_3$ per nickel were slowly syringed in; the mixture initially became greenish black and then solidified into a crystalline mass. The product was extracted with CH_2Cl_2, and the greenish brown solution filtered. Removal of solvent under reduced pressure resulted in a dark brown solid. $NiI_2[P(C_2H_5)_3]_2$ was removed from this mixture with several small portions of 2-propanol, leaving a light yellow crystalline powder. Recrystallization from CH_2Cl_2 by slow solvent evaporation resulted in large golden yellow crystals of $Ni(C_6F_5)_2[P(C_2H_5)_3]_2$ in 69% yield: mp 206–208°C (lit. [22] mp 213–214°C).

Preparation of $Ni(C_6F_5)_2(C_5H_5N)_2$

A nickel slurry was prepared in the usual manner and allowed to settle. About half of the beige solution above the slurry was removed, and 1 equiv of C_6F_5I was slowly added in. The mixture became slightly yellow and was then stirred at about 50°C for 24 h, after which time it became orange brown. The mixture was allowed to cool to room temperature, and then 1 equiv of pyridine per nickel was added to the mixture. The color became yellow green. The mixture was stirred for about 5 min and then allowed to stand for 1 h. Essentially the entire contents solidified into a crystalline mass. The volatiles were removed under vacuum, and the product was extracted with CH_2Cl_2. The greenish-yellow solution was filtered, and the solvent removed under reduced pressure to yield a brown crystalline solid. Recrystallization from CH_2Cl_2 by slow solvent evaporation resulted in straw yellow, needlelike crystals of $Ni(C_6F_5)_2(C_5H_5N)_2$ in 40% yield: mp > 280°C, discolors at about 250°C. Anal. Calcd for $C_{22}H_{10}F_{10}N_2Ni$: C, 47.95; H, 1.83; F, 34.48; N, 5.08. Found: C, 47.81; H, 1.83; F, 34.26; N, 5.07.

In a similar reaction employing 2 equiv of pyridine per nickel, essentially the same yield of $Ni(C_6F_5)_2(C_5H_5N)_2$ was obtained. The light green crystals, presumable $NiI_2(C_5H_5N)_4$, which were also obtained were readily washed out with water prior to recrystallizing $Ni(C_6F_5)_2(C_5H_5N)_2$ from CH_2Cl_2.

Preparation of $Ni(C_6F_5)_2[(C_6H_5)_2PH]_2{\cdot}C_6H_5CH_3$

A nickel slurry was prepared in the usual manner and allowed to settle. The pale beige solution above the slurry was removed, and 1 equiv of C_6F_5I was syringed in. The mixture was stirred at about 60°C for 19 h, after which time a deep orange-brown solution existed. The solution was allowed to cool, and 3 equiv of $(C_6H_5)_2PH$ was slowly syringed in. Initially the mixture became yellow and then deep green in a mildly exothermic reaction. The mixture was stirred for 0.5 h, and then the volatiles were removed under vacuum. The product was extracted with CH_2Cl_2 and the black solution filtered. Solvent evaporation under reduced pressure yielded a black tarry crystalline mass. The mixture was washed with several portions of methanol, leaving a dirty crystalline mass. This was dissolved in CH_2Cl_2. Slow solvent evaporation from the golden yellow solution resulted in yellow crystals in 38% yield. Recrystallization from toluene affords the compound containing a toluene of crystallization: mp 161–162°C; crystals are completely black before melting; IR (Nujol) P–H 2360 cm^{-1}; ^1H NMR ($CDCl_3$) δ 7.28 (aromatic m, 25H), 2.32 (aliphatic s, 3H). Anal. Calcd for $C_{43}H_{30}F_{10}NiP_2$: C, 60.24; H, 3.53; F, 22.16. Found: C, 59.74; H, 3.70; F, 21.71.

7.2 Preparation of 3-Aryl-2-hydroxy-1-propane by Nickel-Mediated Addition of Benzylic Halides to 1,2-Diketones

Acyloins (**1a**) [29, 30] and bis-silylated derivatives of their enediols (**1b**) [31] have been demonstrated to be useful acyl anion equivalents (RCO^-) and have been employed in an indirect but facile way for the preparation of ketones (**3**) (Equation 7.1). The hydroxy ketones (**2**), which are the key intermediates in this method, have been prepared by the reaction of acyloins (**1a**) with organic halides in the presence of sodium hydroxide in dimethyl sulfoxide [29, 30] according to the Hein's procedure [32] or using sodium hydride in 1,2-dimethoxyethane (DME) (glyme) [33]. They also have been prepared by the alkylation of lithium enediolates of acyloins prepared from the corresponding **1b** and methyllithium [31]. These reactions proceed with C-alkylation; however, the reaction conditions employed are quite basic which would limit substrate choice.

$$\underset{\mathbf{1a}}{R^1CH(OH)-C(=O)R^1} \text{ or } \underset{\mathbf{1b}}{R^1C(OSi(CH_3)_3)=C(OSi(CH_3)_3)R^1} \xrightarrow{R^2-X/\text{base}} \underset{\mathbf{2}}{R^1-C(R^2)(OH)-C(=O)-R^1} \xrightarrow[\text{or } 1.\ H_2O_2/HO^{\ominus}]{1.\ NaBH_4;\ 2.\ Pb(OAc)_4\ [\text{or } NaJO_4]} \underset{\mathbf{3}}{R^1-C(=O)-R^2} + R^1-CH{=}O \text{ or } R^1-COOH \quad (7.1)$$

An alternative method has been reported for the preparation of **2** by the reaction of 1,2-diketones with organometallic compounds of lithium [34], magnesium [35–37], cadmium [38], and aluminum [39]. The organometallic compounds of these main group metals are generally quite nucleophilic and often have the limitation of not being compatible with functional groups such as halogen, cyano, and alkoxy-carbonyl groups. π-Allylnickel bromide complexes have been shown to add to benzyl and 2,3-butanedione [40]; however, this approach is somewhat limited by the fact that the complexes are prepared using the highly toxic nickel tetracarbonyl.

We reported that smooth oxidative addition of organic halides such as aryl, benzyl, and allyl halides to metallic nickel proceeded to afford organonickel halides under mild conditions, which yielded homocoupled products [11, 41] or ketones by the reaction with acid chlorides [42] or alkyl oxalyl chlorides [43]. We describe here a new method for the preparation of 3-aryl-2-hydroxy-1-propanones (**4**) in good yield by the Grignard-type addition of benzyl halides to 1,2-diketones mediated by metallic nickel under neutral conditions [44].

$$NiJ_2 \xrightarrow[\text{Glyme, rt, 12 h}]{\text{Li(2.3 equiv) / naphthalene (0.1 equiv) /}} Ni$$

$$Ar{-}CH_2{-}X + R^1{-}C(=O){-}C(=O){-}R^2 \xrightarrow{\text{1. Ni/glyme, 85°C; 2. } H_2O} Ar{-}CH_2{-}C(R^1)(OH){-}C(=O){-}R^2 \quad \mathbf{4}$$

$R^1, R^2 = C_6H_5$ or CH_3 (7.2)

Metallic nickel was easily prepared in glyme by the reduction of nickel iodide (1 equiv) with lithium (2.3 equiv) in the presence of naphthalene (0.1 equiv) as an electron carrier (Equation 7.2). After stirring these reagents at room temperature for 12 h under an atmosphere of argon, the lithium metal was completely consumed, and the finely divided metallic nickel appeared as a black powder which settled to leave a clear colorless solution after standing.

The reaction with benzyl was carried out by adding benzyl halides to a mixture of benzyl halides and the nickel in refluxing glyme (85°C). With 1-phenyl-1,2-propanedione or 2,3-butanedione, a mixture of 1,2-diketone and benzyl chloride was added to the nickel at 85°C because the former procedure mainly gave the homocoupled product [45] 1,2-diphenylethane, and the yields of the expected adducts were poor. In the case of 1-phenyl-1,2-propanedione, two adducts, 3,4-diphenyl-3-hydroxy-2-butanone (**4b**, 34%) and 1,3-diphenyl-2-hydroxy-2-methyl-1-propanone (**4c**, 15%), were formed. The carbonyl group next to the phenyl group was more reactive than that next to the methyl group, and the major adduct was **4b**. As is shown in Table 7.4, a wide range of

Table 7.4 3-Aryl-2-hydroxy-1-propanones (4) prepared.[a]

Benzyl halide		1,2-Diketone						1 R (neat or K Br) v (cm^{-1})	
Ar	X	R^1	R^2	Product	Yield[b] (%)	mp (°C) or bp (°C)/Torr	Molecular formula[c] or lit. data	OH	C=O
C_6H_5–	Cl	C_6H_5–	C_6H_5–	**4a**	78	118–118.5°	115–116° [29]	3480	1670
C_6H_5–	Cl	C_6H_5–	CH_3	**4b**	34	135°/0.37	148°/0.50 [46]	3450	1710
C_6H_5–	Cl	CH_3	C_6H_5–	**4c**	15	Oil	$C_{16}H_{16}O_2$ (240.3)	3450	1670
C_6H_5–	Cl	CH_3	CH_3	**4d**	35	114°/8	118–119°/8 [47]	3470	1710
H_3C–C_6H_4–	Cl	C_6H_5–	C_6H_5–	**4e**	75	121–122°	$C_{22}H_{20}O_2$ (316.4)	3395	1650
F_3C–C_6H_4–	Cl	C_6H_5–	C_6H_5–	**4f**	73	110.5–111°	$C_{22}H_{17}F_3O_2$ (370.4)	3400	1650
F–C_6H_4–	Cl	C_6H_5–	C_6H_5–	**4g**	74	132–133°	$C_{21}H_{17}FO_2$ (320.4)	3490	1665
Cl–C_6H_4–	Cl	C_6H_5–	C_6H_5–	**4h**	83	149–150°	150–151° [29]	3500	1665

(Continued)

Table 7.4 (Continued)

Benzyl halide		1,2-Diketone						1 R (neat or K Br) ν (cm^{-1})	
Ar	X	R[1]	R[2]	Product	Yield[b] (%)	mp (°C) or bp (°C)/Torr	Molecular formula[c] or lit. data	OH	C=O
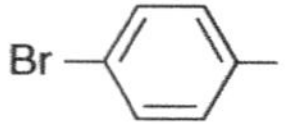 Br–	Br	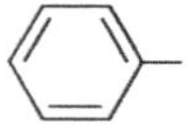	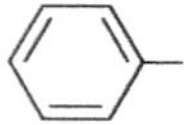	**4i**	52	155–156°	$C_{21}H_{17}BrO_2$ (380.9)	3490	1665
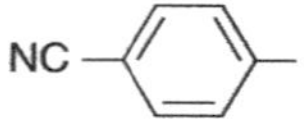 NC–	Br	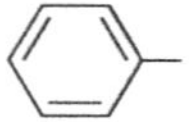	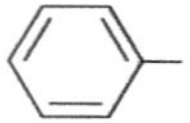	**4j**	66	148–149°	$C_{22}H_{17}NO_2$ (327.4)	3450	1670
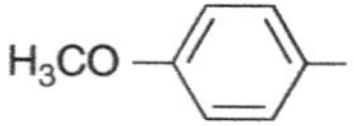 H_3CO–	Cl	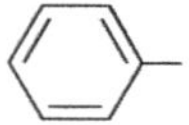	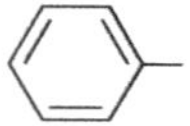	**4k**	73	138–139°	$C_{23}H_{20}O_4$ (360.4)	3440	1720, 1660
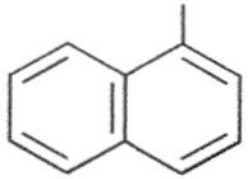	Cl	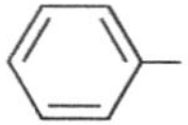	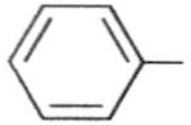	**41**	82	173–173.5°	$C_{25}H_{20}O_2$ (352.4)	3420	1660

[a] Reactions were carried out in glyme under argon at 85°C using the reagents in the ration of benzyl chloride 1.2-diketone metallic nickel = 0.8/1.0/1.0 unless otherwise noted.
[b] Yield of product isolated by silica gel chromatography.
[c] Microanalyses were in good agreement with the calculated values: C ± 0.23; H ± 0.27.

functional groups including halogen, cyano, and methoxycarbonyl groups are tolerated under the reaction conditions employed. Further advantages of the present method are that the preparation of metallic nickel is quite easy, no special apparatus is required, and the subsequent addition reaction can be carried out in a one-pot flask under neutral conditions.

The present Grignard-type addition may be reasonably explained by smooth oxidative addition of benzyl halides to nickel in the metallic state to give benzylnickel halide intermediates, and the insertion of 1,2-diketone into the carbon–nickel bond would afford the corresponding hydroxy ketones after hydrolysis.

2-Hydroxy-1,2,3-triphenyl-1-propanone (4a: Ar = R^1 = R^2 = C_6H_5): Typical Procedure

Benzil (1.76 g, 8.38 mmol) is added to the nickel [48] (8.38 mmol) in glyme, and the mixture is heated to reflux. Benzyl chloride (0.853 g, 6.74 mmol) in glyme (10 ml) is added dropwise for 1 h. Additional heating is continued for 15 min, and the brown-colored reaction mixture is poured into a separatory funnel containing 3% hydrochloric acid (100 ml) and is extracted with chloroform (2 × 100 ml). The chloroform solution is washed with water (200 ml), and the aqueous phase is extracted with chloroform (150 ml). The combined extracts are dried with anhydrous sodium sulfate and concentrated. The crude solid is chromatographed on silica gel eluting with chloroform to give **4a**; yield: 1.59 g (78%).

7.3 Preparation of 3-Arylpropanenitriles by Nickel-Mediated Reaction of Benzylic Halides with Haloacetonitriles

The alkylation of α-cyano anions [49, 50] and metal cyanides [49] with organic halides has been among the most popular methods for the preparation of nitriles. Some years ago, Brown et al. [51, 52] showed that the reaction of trialkylboranes with chloroacetonitrile gave the corresponding nitriles in good yields. Their results suggest that the alkylation of haloacetonitriles with organometallic compounds would provide an alternative route for the preparation of nitriles by carbon–carbon bond formation. However, no related examples have been reported.

We found that oxidative addition of aryl, benzyl, and allyl halides to metallic nickel proceeded smoothly to generate the corresponding organonickel intermediates under mild conditions, which afforded homocoupled products [11, 41, 45] and ketones by the reaction with acid chlorides [42] or alkyl oxalyl chlorides [43]. We describe here a simple method for the preparation of 3-aryl-propanenitriles (**3**) by the reaction of benzylic halides (**1**) with haloacetonitriles (**2**) mediated by metallic nickel [44]; see Equation 7.3.

$$NiJ_2 \xrightarrow[-LiJ]{\text{Li(2.3 equiv) / naphthalene (0.1 equiv) / Glyme, rt}} Ni$$

$$\underset{\mathbf{1}}{Ar{-}CH_2{-}X^1} + \underset{\mathbf{2}}{X^2{-}CH_2{-}CN} \xrightarrow[-NiX^1X^2]{\text{Ni/glyme, 85°C}} \underset{\mathbf{3}}{Ar{-}CH_2{-}CH_2{-}CN} \quad (7.3)$$

Metallic nickel was easily prepared by stirring a mixture of nickel iodide and lithium metal with a catalytic amount of naphthalene as an electron carrier at room temperature for 12 h in DME (glyme) (Equation 7.3). The reaction of benzylic halides with haloacetonitriles was carried out by adding a mixture of these reagents to the metallic nickel in glyme. The coupling reaction of benzyl chloride with bromoacetonitrile in the presence of nickel proceeded at 65°C to give 3-phenylpropanenitrile (**3a**) in 21% yield, and improved results (57%) were obtained under refluxing glyme (85°C). The use of iodo- and chloroacetonitriles as substrates also worked well at 85°C in the present system, and **3a** was formed in 57 and 52% yields, respectively. The preparation of a variety of 3-arylpropanenitriles (**3**) was carried out at 85°C using bromoacetonitrile; the results are summarized in Table 7.5.

Compound **3** could be generally prepared by the benzylation of sodioacetonitrile [60, 61], by the condensation of carbonyl compounds with α-cyano anions [62, 63] followed by a catalytic hydrogenation of 3-arylcinnamonitriles formed, [54] or by the cyanoethylation of aromatic substrates with 3-chloropropanenitrile or acrylonitrile/hydrogen chloride [53, 64]. However, the first method afforded a mixture of mono- and dibenzylated products. The second method required two reaction steps although the catalytic hydrogenation occurred quantitatively. The last method proceeded with poor regioselectivity when substituted aromatic compounds were employed as substrates.

Our method overcomes these difficulties, and the yields are in general compatible with those reported. Furthermore, the preparation of metallic nickel and the following reaction involves a very simple one-pot procedure, and the reaction conditions employed are compatible with a variety of substituents including halogen, cyano, and alkoxycarbonyl groups.

Preparation of Metallic Nickel

A 50 ml two-necked flask is equipped with a magnetic stirrer, a rubber septum, and a reflux condenser topped with an argon inlet and outlet to an oil pump. Lithium metal is cut under mineral oil. One piece of lithium with a shining metal surface is rinsed in hexane and transferred into a glass tube with a stopcock and a rubber septum which has been filled with argon. The glass tube is evacuated to evaporate the hexane, filled with argon, and weighed. Nickel iodide (3.84 g, 12.3 mmol), lithium (0.196 g, 28.2 mmol), and naphthalene (0.157 g, 1.23 mmol) are placed in the flask through the side neck. The flask is evacuated and filled with argon two or three times. The use of a glove box or bag is not required if contact of the lithium with air is kept to a minimum. Then glyme (25 ml, distilled prior to use from sodium–potassium alloy) is added through the septum with a syringe, and the mixture is stirred for 12 h. During the reduction the surface of lithium is observed to be pink colored. After the lithium metal has been completely consumed, the stirring is stopped. Metallic nickel which has adhered to the walls of the flask is scraped off with the stirrer and a magnet. The nickel is precipitated as bulky black powders in a clear colorless solution after standing. The septum on the side neck was replaced with an addition funnel, and a mixture of reagents in glyme is added to the nickel.

Typical Procedure for 3-Phenylpropanenitrile (3a)

A mixture of benzyl chloride (1.24 g, 9.82 mmol) and bromoacetonitrile (1.18 g, 9.80 mmol) in glyme (10 ml) is added dropwise to the nickel (12.3 mmol) in refluxing glyme for 30 min. After additional heating is continued for 15 min, the mixture is cooled and poured into a separatory funnel containing 3% hydrochloric acid (100 ml) and is extracted with chloroform (2 × 100 ml). The aqueous phase is extracted with chloroform (1 × 150 ml), and the combined extracts are washed with water (200 ml), dried with anhydrous sodium sulfate, and concentrated. The residual oil is chromatographed on silica gel, eluting with chloroform to give **3a**; yield: 0.732 g (57%) (Table 7.5).

7.4 Reformatsky-Type Additions of Haloacetonitriles to Aldehydes Mediated by Metallic Nickel

The Reformatsky-type reaction of α-halonitriles and the related addition reaction of α-cyano anions to carbonyl compounds have been known as a useful method for the preparation of β-hydroxynitriles. However, the promoting reagents in these reactions were limited to main group metals such as zinc [65], magnesium [66], or lithium [63] and to inorganic bases [67, 68], and the substrates used were generally ketones and aromatic aldehydes [69]. Transition metals in the metallic state have not been acceptable reagents except for a few examples [70] due to their low reactivity toward organic halides. We reported

Table 7.5 3-Arylpropanenitriles (3) prepared by the reaction of benzyl halides (1) with bromoacetonitrile (2; X^2 = nickel).[a]

Benzyl Ar	Halide 1 X^1	Product	Yield[b] (%)	mp (°C) or bp (°C) Torr	Molecular formula[c] or lit. data
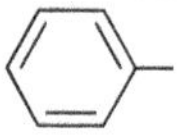	Cl	**3a**	57	129°/16	125–126°/15 [53]
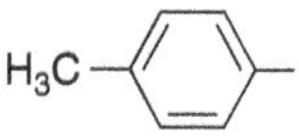 H_3C	Cl	**3b**	60	141°/16	137–141°/15 [53]
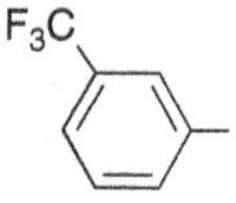 F_3C	Cl	**3c**	58	86°/0.42	$C_{10}H_8F_3N$ (199.2)
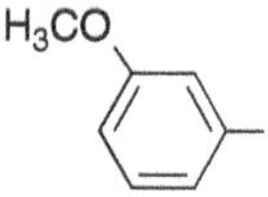 H_3CO	Cl	**3d**	60	104°/0.43	110–120°/0.5 [54]
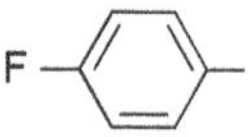 F	Cl	**3e**	46	79°/0.65	73–76°/0.05 [55]
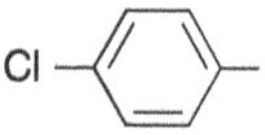 Cl	Cl	**3f**	61	163°/18	153–156°/15 [53]
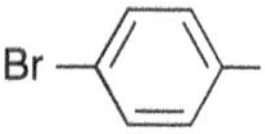 Br	Br	**3g**	52	114°/0.32	116°/0.05 [56]
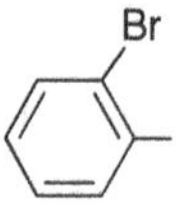 Br	Br	**3h**	43	98°/0.42	C_9H_8BrN (210.1)
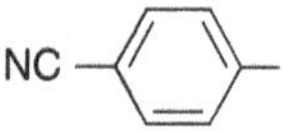 NC	Br	**3i**	43	76.5–77°	85–87° [57]
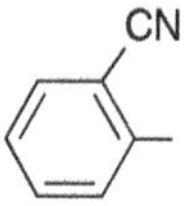 CN	Br	**3j**	44	123°/0.49	110°/0.2 [58]
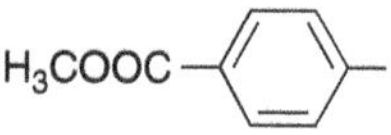 H_3COOC	Cl	**3k**	55	60.5–61°	$C_{11}H_{11}NO_2$ (189.2)
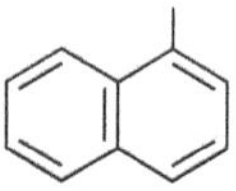	Cl	**3l**	20	139°/0.20	$C_{13}H_{11}N$ (181.2)
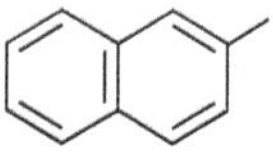	Br	**3m**	53	78–78.5°	$C_{13}H_{11}N$ (181.2)
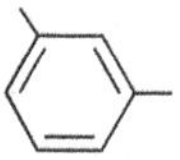	2Br	**3n**	36	164°/0.34	165–169°/0.1 [59]

[a] Reactions were carried out in glyme under argon at 85°C using the reagents in the ratio of benzylic halide/bromoacetonitrile/metallic nickel = 0.8/0.8/1.0 unless otherwise noted.
[b] Isolated by silica gel chromatography upon elution with chloroform.
[c] Microanalyses were in good agreement with the calculated values: C ± 0.23; H ± 0.24; N ± 0.18.

that haloacetonitriles react with aryl and alkyl aldehydes in the presence of metallic nickel at 85°C in DME (glyme) to give β-hydroxynitriles in good yields after hydrolysis [71, 72]; see Equation 7.4.

$$\text{R-CHO} + \text{XCH}_2\text{CN} \xrightarrow[\text{Glyme, 85°C}]{\text{Metallic Ni}} \xrightarrow{\text{H}^+} \text{R-CH(OH)CH}_2\text{CN} \quad (7.4)$$

Metallic nickel was prepared in glyme (25 ml) at room temperature in 12 h by the reduction of nickel iodide (4.71 g, 15.1 mmol) with lithium (0.241 g, 34.7 mmol) in the presence of naphthalene (0.193 g, 1.23 mmol) as an electron carrier. To the nickel in refluxing glyme, a mixture of benzaldehyde (1.23 g, 11.6 mmol) and bromoacetonitrile (1.81 g, 15.1 mmol) was added dropwise for 30 min. Additional heating was continued for 10 min, and the reaction mixture was cooled and poured into separatory funnels containing 3% hydrochloric acid (100 ml) and was extracted with chloroform. The extracts were washed with water, dried over anhydrous sodium sulfate, and concentrated. Crude oil was submitted to silica gel chromatography, eluted with hexane/EtOAc (3/1) to give 2-hydroxy-2-phenylpropanenitrile (1.44 g, 85%): bp 115–117°C/45 mmHg (lit. [3] bp 154–155°C/1 mmHg); see Table 7.6.

In a similar manner, chloro- and iodoacetonitriles reacted with benzaldehyde to give the nitrile in 67% and 77% yields, respectively. When bromoacetonitrile was added to a mixture of benzaldehyde and nickel, reduction of the aldehyde by the nickel was observed; benzyl alcohol was formed in 15% yield along with β-hydroxynitrile (60%). The use of bromoacetate failed to yield satisfactory results. Also, the addition of bromoacetonitrile to ketones gave poor yields under similar conditions. The addition reaction to aldehydes is reasonably explained by the smooth oxidative addition of haloacetonitriles to metallic nickel to give cyanomethylnickel halides [73], which would add to aldehydes to yield β-hydroxynitriles after hydrolysis. The advantages of the present method are that the self-condensation of aldehydes is minimized because of lower basicity of cyanomethylnickel halides compared with the corresponding zinc reagents and that the gently exothermic reaction proceeds smoothly at 85°C. Moreover, the selective addition to aldehydes and not to ketones may prove to be a great synthetic advantage in certain cases. The results are summarized in Table 7.6.

7.5 Preparation of Symmetrical 1,3-Diarylpropan-2-ones from Benzylic Halides and Alkyl Oxalyl Chlorides

Carbonylation with group VIII transition metal complexes has been shown to be a powerful tool in organic synthesis because it is easy to undergo oxidation addition of organic halides to transition metals and insertion of carbon monoxide to organotransition metals formed [74]. Metal carbonyls such as diiron

Table 7.6 The preparation of β-hydroxynitriles by the reaction of bromoacetonitriles with aldehydes mediated by the metallic nickel.[a]

Aldehyde	Product[b]	Bp (°C/mmHg)	IR (cm^{-1})		NMR (δ)		Yield[c] (%)
			VOH	VCN	$C\underline{H}_2CN$	$C\underline{H}OH$	
C_6H_5CHO	$C_6H_5CH(OH)CH_2CN$	115–117/0.45[d]	3430	2240	2.63	4.90	84
Br–C_6H_4–CHO	Br–C_6H_4–$CH(OH)CH_2CN$	147–148/0.49	3430	2245	2.68	4.93	76
NC–C_6H_4–CHO	NC–C_6H_4–$CH(OH)CH_2CN$	165/0.24[e]	3530	2240 2220	2.78	5.10	81
$C_6H_5C{\equiv}CCHO$	$C_6H_5C{\equiv}CCH(OH)CH_2CN$	142/0.83	3410	2240	2.82	4.88	63
C_3H_7CHO	$C_3H_7CH(OH)CH_2CN$	114/11[f]	3420	2240	2.52	3.93	57
$C_5H_{11}CHO$	$C_5H_{11}CH(OH)CH_2CN$	83/0.34	3430	2240	2.50	3.91	59
$C_3H_7CH{=}CHCHO$	$C_3H_7CH{=}CHCH(OH)CH_2CN$	93/0.92	3430	2240	2.57	4.42	76
$C_6H_5CH{=}CHCHO$	$C_6H_5CH{=}CHCH(OH)CH_2CN$	143/0.41	3420	2240	2.58	4.53	77
$C_4H_9CH{=}CHCH{=}CHCHO$	$C_4H_9CH{=}CHCH{=}CHCH(OH)CH_2CN$	107/0.15	3420	2240	2.58	4.45	59

[a] The reaction was carried out by adding a mixture of aldehyde and bromoacetonitrile to metallic nickel in refluxing glyme (85°C).
[b] All compounds gave satisfactory results on microanalyses (C, H, and N: ± 0.29%).
[c] Isolated by silica gel chromatography.
[d] Lit. 63 bp 115–117°C/0.45 mmHg.
[e] mp 84.5–85.5°C.
[f] Lit. 67 bp 57°C/0.2 mmHg.

nonacarbonyl [75–77], triiron dodecacarbonyl [77, 78], nickel carbonyl [79, 80], and dicobalt octacarbonyl [81] have been employed for the preparation of symmetrical ketones from aryl, benzyl, and arylmercuric halides [82] in a stoichiometric fashion [83]. Disodium tetracarbonylferrate [77, 84] or potassium hexacyanodinickelate-carbon monoxide [85] have also been used for dibenzyl ketone synthesis.

In spite of the usefulness of these complexes, it is generally not possible to cause the satisfactory reaction with transition metals in the metallic state [86] under mild conditions due to their poor reactivity. We have reported that activated metallic nickel, prepared by the reduction of nickel halide with lithium, underwent oxidative addition of benzylic halides to give homocoupled products [45]. We reported that carbonylation of the oxidative adducts of benzylic halides to the nickel proceeded smoothly to afford symmetrical 1,3-diarylpropan-2-ones in moderate yields, in which the carbonyl groups of alkyl oxalyl chlorides served as a source of carbon monoxide [43]; see Equation 7.5.

$$2\ ArCH_2X + XCOCO_2R \xrightarrow[\text{Glyme, 85°C}]{\text{Metallic Ni}} ArCH_2COCH_2Ar + 1/2\ RO_2CCO_2R \quad (7.5)$$

X = Cl, Br

Metallic nickel was prepared in glyme (25 ml) by the reduction of nickel iodide (3.13 g, 10 mmol) with lithium (0.160 g, 23 mmol) using naphthalene (0.128 g, 1 mmol) as an electron carrier. After stirring these reagents under an atmosphere of argon at room temperature for 12 h, the finely divided metal appeared as black powders which settled in a clear colorless solution. To the nickel in refluxing glyme (85°C), a mixture of benzyl chloride (1.01 g, 8.0 mmol) and ethyl oxalyl chloride (1.09 g, 8.0 mmol) in glyme (10 ml) was added dropwise for 30 min. Additional heating was continued for 15 min, and the red-brown reaction mixture was poured into a separatory funnel containing 3% hydrochloric acid solution (100 ml) and was extracted with chloroform. The aqueous phase was extracted with chloroform, and the combined extracts were washed with water, dried over anhydrous sodium sulfate, and concentrated. 1,3-Diphenylpropan-2-one (0.403 g, 48%) was isolated by silica gel chromatography upon elution with chloroform: bp 125–127°C/0.8 Torr (lit. [78] bp 105–110°C/0.4 Torr). The other products, diethyl oxalate (39%) and 1,2-diphenylethane (18%), were formed by the homocoupling reaction of benzyl chloride by metallic nickel [45].

The reaction of benzyl chloride with ethyl oxalyl chloride proceeded at room temperature or at 60°C, and the yields of 1,3-diphenylpropan-2-one were 11% and 45%, respectively. Thus, the carbonylation was carried out at 85°C, and the results are summarized in Table 7.7. The stoichiometric ratio of benzylic halide/alkyl oxalyl chloride was 2 : 1, and the most satisfactory results were obtained with a ratio of 1 : 1.

Table 7.7 Preparation of symmetrical 1,3-diarylpropan-2-ones by the reaction of benzylic halides with alkyl oxalyl chlorides in the presence of metallic nickel.[a]

Benzyl halide (1)	Alkyl oxalyl chloride (2)	Ratio of (1)/(2)	Product (3)[b]	Yield[c] (%)
$C_6H_5CH_2Cl$	$ClCOCO_2C_2H_5$	2.5/1	$C_6H_5CH_2COCH_2C_6H_5$	35[d]
$C_6H_5CH_2Cl$	$ClCOCO_2C_2H_5$	1/1	$C_6H_6CH_2COCH_2C_6H_5$	48
$C_6H_5CH_2Cl$	$ClCOCO_2C_2H_5$	1/2.5	$C_6H_5CH_2COCH_2C_6H_5$	47
$C_6H_5CH_2Cl$	$ClCOCO_2CH_3$	1/1	$C_6H_5CH_2COCH_2C_6H_5$	46
4-$CH_3C_6H_4CH_2Cl$	$ClCOCO_2C_2H_5$	1/1	4-$CH_3C_6H_4CH_2COCH_2C_6H_4CH_3$-4[e]	46
3-$CH_3OC_6H_4CH_2Cl$	$ClCOCO_2C_2H_5$	1/1	3-$CH_3OC_6H_4CH_2COCH_2C_6H_4OCH_3$-3[f]	43
4-$ClC_6H_4CH_2Cl$	$ClCOCO_2C_2H_5$	1/1	4-$ClC_6H_4CH_2COCH_2C_6H_4Cl$-4[g]	54
4-$BrC_6H_4CH_2Br$	$ClCOCO_2C_2H_5$	1/1	4-$BrC_6H_4CH_2COCH_2C_6H_4Br$-4[h]	46
4-$NCC_6H_4CH_2Br$	$ClCOCO_2C_2H_5$	1/1	4-$NCC_6H_4CH_2COCH_2C_6H_4CN$-4[i]	46
4-$CH_3O_2CC_6H_4CH_2Cl$	$C1COCO_2C_2H_5$	1/1	4-$CH_3O_2CC_6H_4CH_2COCH_2C_6H_4CO_2CH_3$-4[j]	45
3-$F_3CC_6H_4CH_2Cl$	$ClCOCO_2C_2H_5$	1/1	3-$F_3CC_6H_4CH_2COCH_2C_6H_4CF_3$-3[k]	49
1-$C_{10}H_7CH_2Cl$[l]	$ClCOCO_2C_2H_5$	1/1	1-$C_{10}H_7CH_2COCH_2C_{10}H_7$-1[m]	59

[a] Reaction was carried out in glyme at 85°C.
[b] Carbonyl stretching frequencies were observed in the region of 1700–1710 cm^{-1} and NMR signals of methylene protons appeared at 63.64–4.07 ppm.
[c] Isolated yield based on benzylic halide used unless otherwise noted.
[d] Isolated yield based on ethyl oxalyl chloride used.
[e] mp 54–55°C (lit. 54–55°C).
[f] mp 135.5–136.5°C (semicarbazone) [(lit. mp 136–136.5°C (semicarbazone)].
[g] mp 96–97°C (lit. mp 95–96°C).
[h] mp 119–119.5°C (lit. mp 116–118°C).
[i] mp 149–150°C, m/e = 260 (M^+) (lit. [87]) mp 79°C.
[j] mp 140–141°C, m/e = 326.1136 calcd for $C_{19}H_{18}O_5$ 326.1153.
[k] mp 53–54°C, bp 127°C/0.47 Torr, m/e = 346.0769 calcd for $C_{17}H_{12}F_6O$ 346.0792.
[l] 1-(Chloromethyl)naphthalene.
[m] mp 107–107.5°C (lit. mp 108–109°C).

One possible mechanistic sequence for the present reaction is shown in Scheme 7.1. The carbon monoxide insertion into the carbon–metal σ bond of alkyltransition metal complexes is well known [88]. Thus, the oxidative addition of benzyl halide to metallic nickel gives benzylnickel (II) halide **4**, and the insertion of carbon monoxide, which is formed by decarbonylation of alkyl oxalyl chloride into the benzyl–nickel bond of complex **4**, would afford arylacetyl (II) complex **5**. The metathesis of complexes **4** and **5** seems to give (arylacetyl)benzylnickel complex **6**, which undergoes reductive elimination to yield 1,3-diarylpropan-2-one, **3**. The formation of 1,2-diarylethane may be explained by the reductive elimination of bisbenzylnickel complex **7** formed by metathesis of benzylnickel complex **4** [89]. It is also possible that the reaction of benzyl halide with complex **4** or **5** gives homocoupled product or ketone, respectively.

In the dibenzyl ketone synthesis using benzyl halides and transition metal complexes [75–79, 84, 85], readily available benzyl chlorides were not as reactive and generally gave poor results (up to 51% yield) [79]. In spite of the effectiveness of benzyl bromides and iodides, the compatibility with functional groups was not examined. In conclusion, these results show that yields are comparable or greater than those using complexes of iron and nickel and that functional groups such as chloro, bromo, cyano, and carbomethoxy groups are compatible with the reaction conditions employed.

7.6 Nickel-Mediated Coupling of Benzylic Halides and Acyl Halides to Yield Benzyl Ketones

The coupling reaction of acyl halides with organometallic reagents such as organomagnesium, zinc, cadmium [90], and copper [91] compounds is an extremely useful tool for the synthesis of ketones. Palladium–phosphine complexes have been employed as catalysts for the reaction with organozinc [92], tin [93], and mercury [94] compounds utilizing easy oxidative addition of acyl halides to palladium. We have reported that the oxidative addition of benzyl halides to metallic nickel proceeded smoothly to give homocoupled products [45]. We also reported that the cross-coupling reaction of benzyl halides with acyl halides can be mediated by metallic nickel to afford benzyl ketones [42]; see Equation 7.6.

$$ArCH_2X^2 + RCOX^3 + Ni \xrightarrow[\text{Glyme}]{\text{Reflux}} ArCH_2COR + NiX^2X^3 \qquad (7.6)$$

Metallic nickel was prepared in glyme (25 ml) by the reduction of nickel iodide (2.97 g, 9.50 mmol) with lithium (0.152 g, 21.9 mmol) using naphthalene (0.122 g, 0.95 mmol) as an electron carrier. After stirring these reagents under an atmosphere of argon at room temperature for 12 h, the finely divided nickel

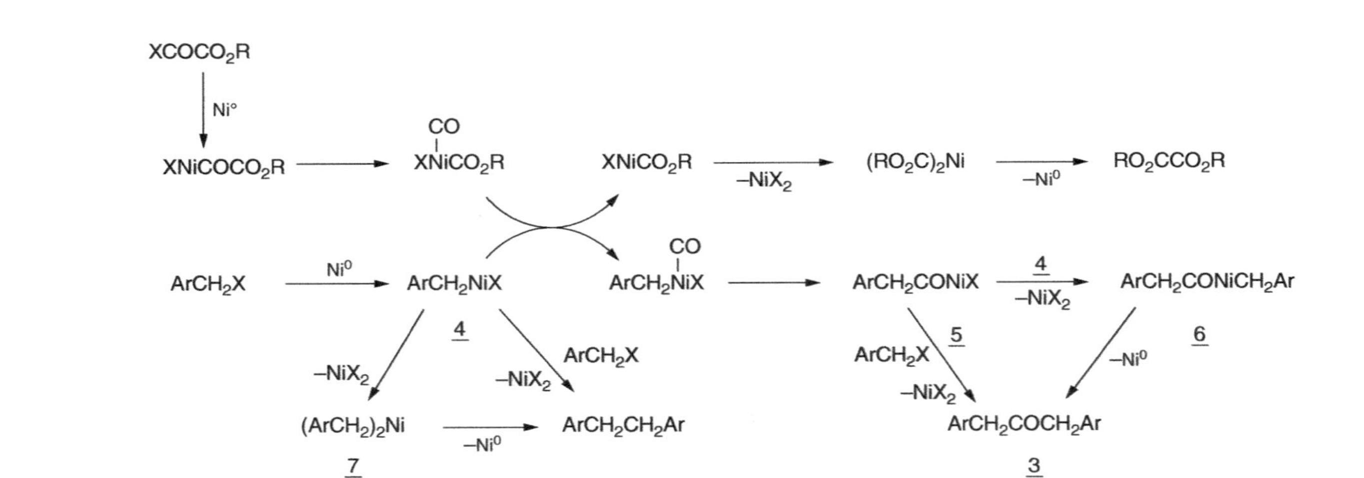

Scheme 7.1 Postulated mechanisms.

appeared as a black powder which settled in a clear colorless solution. To the nickel powder in refluxing glyme, a mixture of benzyl chloride (0.916 g, 7.24 mmol) and benzoyl chloride (1.07 g, 7.61 mmol) in glyme (10 ml) was added dropwise for 30 min. Additional heating was continued for 15 min, and the red-brown reaction mixture was poured into a separatory funnel containing 3% hydrochloric acid solution (100 ml) and was extracted with chloroform. The aqueous phase was extracted with chloroform, and the combined extracts were washed with water, dried over anhydrous sodium sulfate, and concentrated. Crude oil was chromatographed on silica gel and eluted with chloroform to give benzyl phenyl ketone (1.04 g, 73%); mp 55–56°C (lit. [95] mp 56°C). Bibenzyl formed by the homocoupling reaction of benzyl chloride [45] was also isolated in 14% yield.

The reductive coupling reaction of benzyl chloride with benzoyl chloride proceeded even at room temperature; however, improved results were obtained under refluxing glyme (at 85°C). The choice of nickel halide that was reduced was important. Metallic nickel prepared from nickel iodide, bromide, and chloride gave benzyl phenyl ketone in 73, 42, and 11% yields, respectively. Thus, the reaction of benzyl halides with acyl halides using metallic nickel derived from nickel iodide was carried out under refluxing glyme, and the results are summarized in Table 7.8.

The reaction may be reasonably explained by the smooth oxidative addition of halides to nickel to afford benzylnickel and acylnickel halides. The metathesis of these complexes could give the acylbenzylnickel complex, which upon reductive elimination would yield the benzyl ketone.

7.7 Nickel-Assisted Room Temperature Generation and Diels–Alder Chemistry of *o*-Xylylene Intermediates

The Diels–Alder reaction of various dienophiles with the highly reactive diene *o*-xylylene has been utilized extensively in the construction of various polycyclic ring systems. Although calculations show that the *o*-xylylene species may actually be best represented as a biradicaloid **1** [97], its reactivity parallels that of a highly reactive diene **2** (see Figure 7.1). The high reactivity presumably results from attainment of aromaticity in the Diels–Alder cycloadduct.

A variety of methods have been employed to generate the *o*-xylylene intermediate [98]. Among them are ring openings of benzocyclobutenes **3** [99] and various

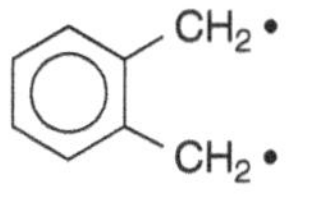

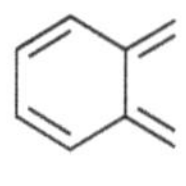

Figure 7.1 Proposed intermediates.

Table 7.8 Coupling reaction of benzyl halides with acyl halides mediated by metallic nickel.

Benzyl halide	Acyl halide	Conditions	Product[a]	Yield (%)[b]
$C_6H_5CH_2Cl$	C_6H_5COCl	rt, 12 h	$C_6H_5CH_2COC_6H_5$	54
$C_6H_5CH_2Cl$	C_6H_5COCl	50°C, 3 h	$C_6H_5CH_2COC_6H_5$	61
$C_6H_5CH_2Cl$	C_6H_5COCl	85°C, 15 min	$C_6H_5CH_2COC_6H_5$	73
$C_6H_5CH_2Cl$	CH_3COCl	85°C, 15 min	$C_6H_5CH_2COCH_3$[c]	68
$C_6H_5CH_2Cl$	(H)–COCl (cyclohexyl)	85°C, 15 min	$C_6H_5CH_2CO$–(H) (cyclohexyl)[d]	55
$C_6H_5CH_2Cl$	$C_6H_5(H)C{=}C(H)COCl$	85°C, 15 min	$C_6H_5CH_2CO(H)C{=}C(H)C_6H_5$[e]	64
$C_6H_5CH_2Cl$	2-furyl–COCl	85°C, 15 min	$C_6H_5CH_2CO$–2-furyl[f]	39
$C_6H_5CH_2Br$	$CH_3OCOCH_2CH_2COCl$	85°C, 15 min	$C_6H_5CH_2COCH_2CH_2CO_2CH_3$[g]	55
p-$ClC_6H_4CH_2Cl$	C_6H_5COCl	85°C, 15 min	p-$ClC_6H_4CH_2COC_6H_5$[h]	82
p-$NCC_6H_4CH_2Br$	CH_3COCl	85°C, 15 min	p-$NCC_6H_4CH_2COCH_3$[i]	85
m-$BrCH_2C_6H_4CH_2Br$	CH_3COCl	85°C, 15 min	m-$CH_3COCH_2C_6H_4CH_2COCH_3$[j]	62
2-naphthyl–CH_2Br	CH_3COCl	85°C, 15 min	2-naphthyl–CH_2COCH_3[k]	62

[a] Stretching frequencies of keto-carbonyl groups were observed in the region of 1655–1720 cm^{-1} and NMR signals of methylene protons adjacent to aromatic rings appeared at δ 3.67–4.27 ppm.
[b] Yield isolated by silica gel chromatography or recrystallization.
[c] bp 95–96°C/11 mmHg (lit. bp 101/14 mmHg).
[d] bp 102–103°C/0.6 mmHg (lit. bp 138–139°C/5 mmHg).
[e] mp 72–73°C (lit. mp 68–70°C).
[f] bp 104°C/0.6 mmHg (lit. bp 161–163°C/10 mmHg).
[g] bp 110–111°C/0.45 mmHg (lit. bp 150°C/0.01 mmHg) [96].
[h] mp 135–136°C (lit. mp 136.5°C).
[i] mp 79–79.5°C (lit. mp 78–80°C).
[j] bp 121°C/0.4 mmHg (lit. bp 146°C/3 mmHg).

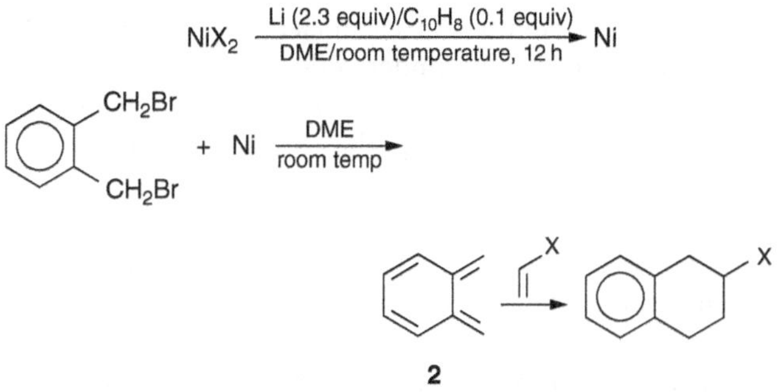

NiX_2 —Li (2.3 equiv)/$C_{10}H_8$ (0.1 equiv), DME/room temperature, 12 h→ Ni

o-$C_6H_4(CH_2Br)_2$ + Ni —DME, room temp→ **2** —X-substituted olefin→ 2-X-tetralin

Scheme 7.2 Proposed reaction scheme.

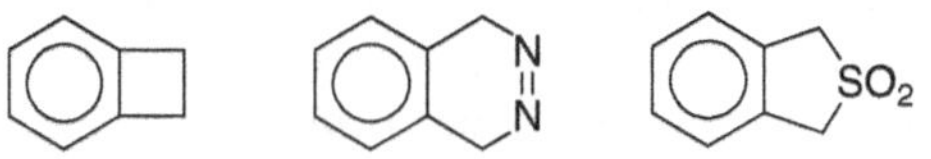

3 **4** **5** (SO_2)

Figure 7.2 Precursors.

"extrusion" reactions involving loss of a small molecule from a cyclic system (see Scheme 7.2). For example, losses of N_2 from **4** [100] or SO_2 from **5** [101] have been shown to lead to the *o*-xylylene intermediate **2** (see Figure 7.2).

One of the more fundamental methods for generation of the *o*-xylylene intermediate is 1,4-elimination from *o*-xylene derivatives. *o*-Xylene di-, tri-, or tetrabromides undergo intramolecular 1,4-dehalogenation by means of sodium iodide [102], lithium [103], zinc [104], copper [105], iron [106], and chromium [107]. Hoffman degradation of (*o*-methylbenzyl)trimethylammonium hydroxides [108], as well as the more recently developed fluoride ion-induced elimination from [*o*-[α-(trimethylsilyl)alkyl]benzyl]trimethylammonium halides [109], have also been effectively utilized in the generation of **2**.

In the course of our investigations concerning the scope of reactions of activated nickel with various organic substrates, we discovered that 1,4-dihalide eliminations from α,α′-dibromo-*o*-xylene derivatives were induced at room temperature with activated nickel, giving *o*-xylylene intermediates. Electron-withdrawing or electron-donating groups can be present in the starting *o*-xylene dihalides. The resultant *o*-xylylene intermediates were trapped with a variety of electron-deficient olefins, giving the Diels–Alder cycloadducts in moderate to good yield [110].

Results and Discussion

Previous communications [111] have shown the high reactivity and utility of activated nickel powder prepared by the lithium metal reduction of nickel iodide with a catalytic amount of naphthalene as an electron carrier in DME. The active nickel produced reacts readily under mild conditions with α,α′-dibromo- and α,α′-dichloro-*o*-xylene in the presence of electron-deficient olefins, yielding substituted 1,2,3,4-tetrahydronaphthalene (tetralin) derivatives in moderate to good yields (Scheme 7.2). A plausible explanation for this process would invoke the *o*-xylylene intermediate **2**. Insertion of Ni into one of the carbon–halogen bonds, followed by rapid elimination of $NiBr_2$, would produce **2**. Whether the *o*-xylylene produced exists free in solution or is coordinated by Ni is uncertain.

A variety of electron-deficient olefins were shown to be effective dienophiles, giving the expected cycloaddition products (Table 7.9). Reaction of the *o*-xylylene species **2** with maleic anhydride yielded only the *cis* product, while *cis* olefins (e.g., dimethyl or diethyl maleate) yielded a mixture of *cis* and *trans* cycloaddition products. *Trans* esters (diethyl fumarate and methyl crotonate) yielded *trans* adducts; however, fumaronitrile gave a mixture of *cis* and *trans* adducts. In a separate experiment, it was discovered that upon exposure to the activated nickel, dimethyl maleate was isomerized to dimethyl fumarate. The exact mechanism of the isomerization is at this point unexplained. Noncyclic *cis* olefins appear to give mixtures of *cis* and *trans* cycloadducts, while *trans* olefins may yield *trans* products or mixtures of *cis* and *trans* products. In cases where mixtures are produced, the *trans* isomer is the major adduct isomer with the ratio of *cis*/*trans* isomers apparently being affected by the nature of the olefin. Thus, although the cycloaddition reaction itself is still not proven to be a "concerted" rather than a stepwise process, the mixture of *cis* and *trans* cycloadducts could arise from isomerization of the olefin prior to the cycloaddition process.

2-Cyclohexene-1-one, methyl vinyl ketone, phenylacetylene, diphenylacetylene, benzaldehyde, perfluoro-1-heptene, and cyclohexene were found to not be effective dienophiles for the reaction. Only polymeric *o*-xylylene products were seen in the reaction mixture. In addition, no cycloadduct was produced by using *p*-benzoquinone as the dienophile. Only hydroquinone and α,α′-diiodo-*o*-xylene were recovered. The hydroquinone presumably results from the reduction of benzoquinone by nickel. Since sodium iodide is known to react with dibromo-*o*-xylene to give diiodo-*o*-xylene [112], the diiodo-*o*-xylene could result from the reaction of unconsumed dibromo-*o*-xylene with lithium iodide which is present in the reaction flask. These results are analogous to those reported in a similar reaction by Scheffer [113], where the use of zinc metal and ultrasound gave only hydroquinone and a quantitative yield of the unreacted dibromo-*o*-xylene.

Various attempts were made to determine the nature of the reactive *o*-xylylene species. In previous results [111a], α,α′-dibromo-*m*-xylene reacted with metallic nickel to give a presumed *m*-xylene bis(nickel bromide) species **6**, which was effectively trapped with acetyl chloride to give the diketone product **7** in 62% yield (Scheme 7.3). Attempted acetyl chloride trapping of an *o*-xylene

Table 7.9 Nickel-mediated cycloadditions of *o*-xylylene with dienophiles.[a]

Starting dihalides	Dienophiles (molar equiv)[c]	Adducts (% isolated yield)[b]	Starting dihalides	Dienophiles (molar equiv)[c]	Adducts (% isolated yield)[b]
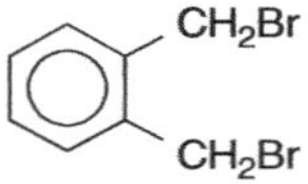 CH_2Br, CH_2Br	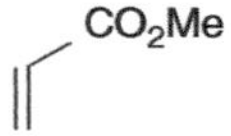 CO_2Me (2)	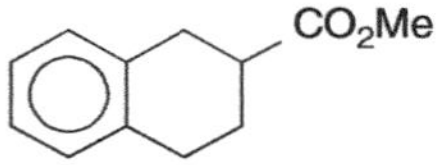 CO_2Me **20** (67)		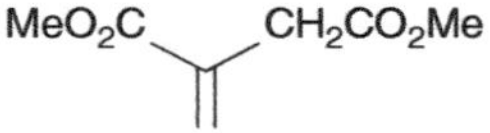 MeO_2C, CH_2CO_2Me (2)	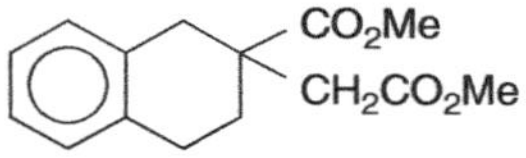 CO_2Me, CH_2CO_2Me **29** (42)
	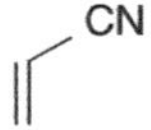 CN (2)	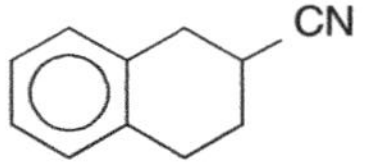 CN **21** (67)		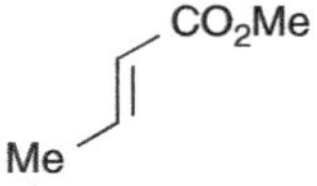 CO_2Me, Me (2)	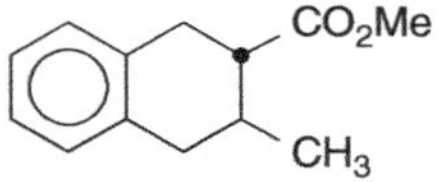 CO_2Me, CH_3 **30** (14)[d]
	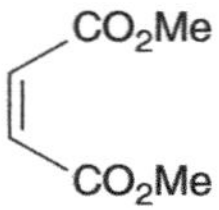 CO_2Me, CO_2Me (2)	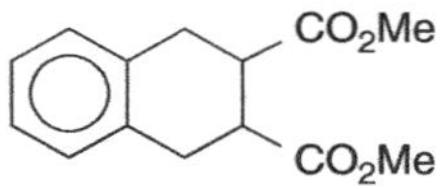 CO_2Me, CO_2Me **22,23** (47)		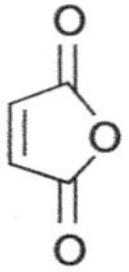 O, O, O (2)	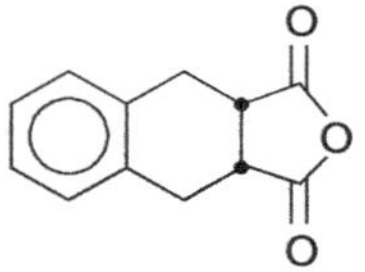 O, O, O **31** (7)
	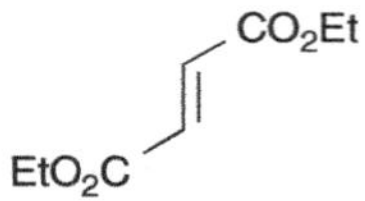 CO_2Et, EtO_2C (2)	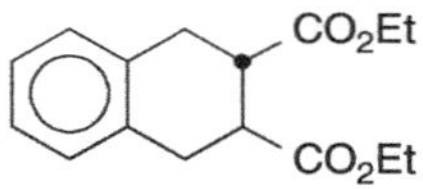 CO_2Et, CO_2Et **24** (76)			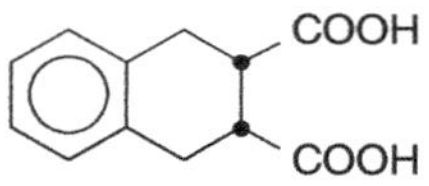 COOH, COOH **32** (61)

(*Continued*)

Table 7.9 (Continued)

Starting dihalides	Dienophiles (molar equiv)[c]	Adducts (% isolated yield)[b]	Starting dihalides	Dienophiles (molar equiv)[c]	Adducts (% isolated yield)[b]
	CO_2Et, CO_2Et (2)	CO_2Et, CO_2Et **24,25** (90) (*trans/cis* = 1/1)[e]	CH_2Cl, CH_2Cl	CN (2)	CN **33** (61)
	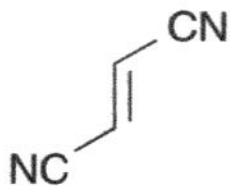 NC, CN (2)	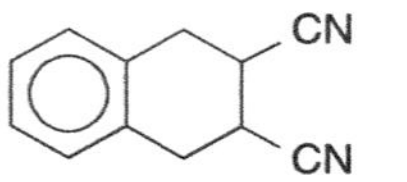 CN, CN **26,27** (54) (*trans/cis* = 2.1)[e]		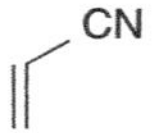 CN (1)	**33** (63)
	 CO_2Me–C≡C–CO_2Me (2)	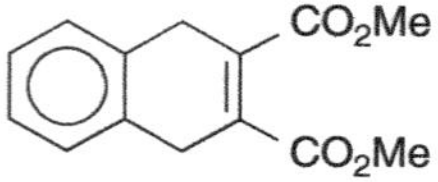 CO_2Me, CO_2Me **28** (63)			

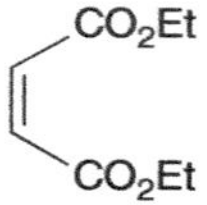

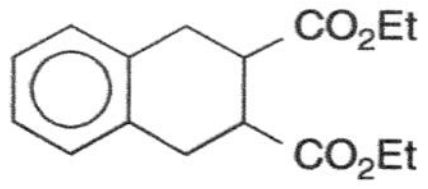

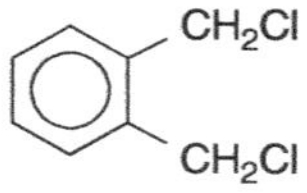

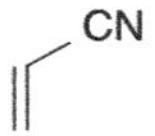

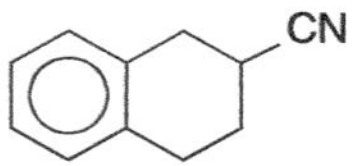

[a] All reactions were carried out at room temperature.
[b] Isolated by silica gel chromatography.
[c] Molar equivalents of dienophile are relative to starting dibromide.
[d] A small amount (<5%) of 1,2-bis(2-(iodomethyl)phenyl)ethane was also isolated.
[e] *trans/cis* ratio determined by GLPC analysis and by ^{1}H NMR integration of crude products.

bis(nickel bromide) species **8** in the reaction of metallic nickel with α,α′-dibromo-*o*-xylene failed to yield any diketone product **9**. These results suggest a different mechanism may be involved in the reactions of the *o*- and *m*-dibromoxylene isomers with metallic nickel.

In conjunction with the aforementioned trapping experiment, several attempts were made to isolate a nickel complex such as **10** analogous to the PT complex **11** [114] or such as **12** similar to the Fe complex **13** [115] by using 1,5-cyclooctadiene, 2,2′-bipyridine, triethylphosphine, or triphenylphosphine as stabilizing ligands; see Figure 7.3.

All attempts produced only *o*-xylylene polymer [116] and the ligand-coordinated nickel halide salts. Therefore, although the existence of an *o*-xylylene nickel complex in the reaction medium has not been disproven, such a complex if present probably has limited stability which precludes its isolation at room temperature.

In addition to the *o*-xylylene intermediate **2**, isoindene **14** and 2,3-dihydronaphthalene **15** were generated by the 1,4-eliminations from 1,3-dibromoindan and 1,4-dibromotetralin, respectively, using activated nickel (Figure 7.4). The reactive intermediates **14** and **15** gave cycloadducts analogous to the reactions using **2** but with one or two carbon bridges, respectively. Both species

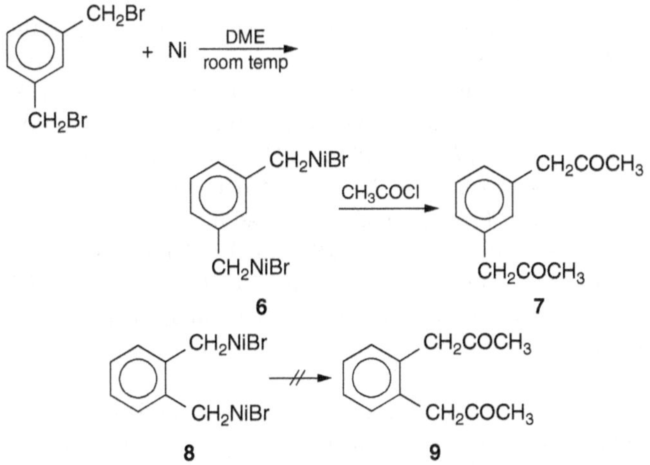

Scheme 7.3 Reaction with acetyl chloride.

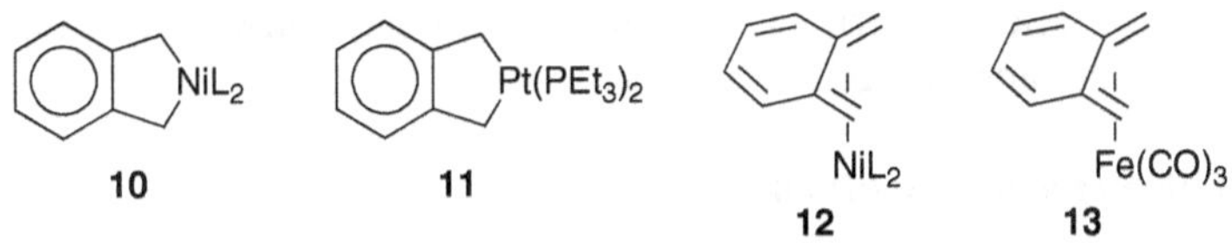

Figure 7.3 Proposed organometallic intermediates.

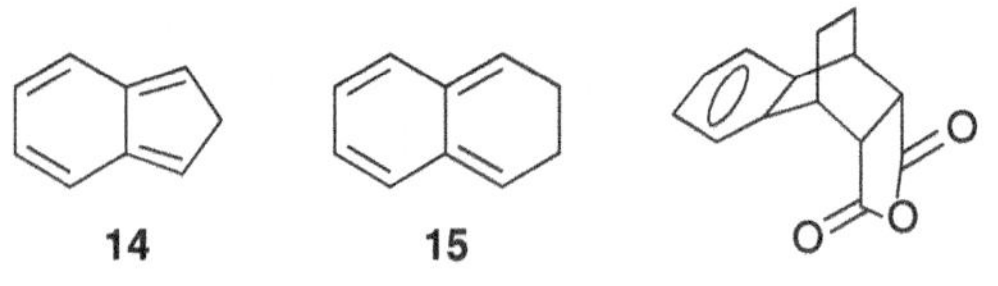

Figure 7.4 Intermediates and product.

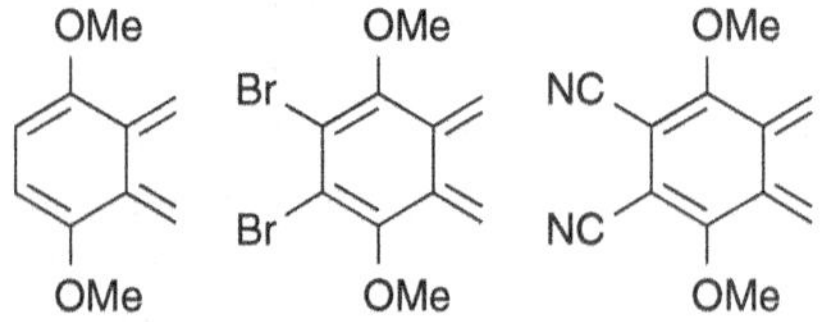

Figure 7.5 Reaction intermediates.

reacted with fumaronitrile to give the bridged *trans* cycloadducts in moderate yields with no detectable amounts of either the *cis–endo* or *cis–exo* cycloadducts. The formation of exclusively the *trans* adducts is somewhat surprising in light of the results of cycloaddition using the parent *o*-xylylene intermediate **2** with fumaronitrile, which gave the mixture of *cis* and *trans* cycloadducts. The reaction of the 2,3-dihydronaphthalene species with maleic anhydride gave the known *endo* adduct **16** [117] in 46% yield with no detectable amount of the *exo* isomer. This result is similar in nature to the results reported by Warrener [118] in which generation of an isoindene intermediate and its trapping with *N*-methylmaleimide produced only the *endo* cycloadduct. This stereoselectivity is in good agreement with the "*endo* rule" for the Diels–Alder reactions in which secondary orbital overlap directs formation of the *endo* adduct.

The substituted *o*-xylylene intermediates **18** and **19** demonstrate that electron-deficient substituents can be tolerated. The use of intermediates **17**, **18**, and **19** provide a convenient entry into the synthesis of 1,4-naphthoquinones with electron-withdrawing groups in the 2,3,6,7-positions as well as just in the 6,7-positions (Figure 7.5).

In summary, the nickel-assisted generation of *o*-xylylene and substituted *o*-xylylenes represents a mild and convenient method for generation of this useful synthetic intermediate. While not all dienophiles are effective in trapping this intermediate, an extensive list has been shown to provide modest to good yields of the Diels–Alder products (Table 7.10). Finally, it has been demonstrated that the *o*-xylylene intermediates can tolerate electron-withdrawing groups such as nitriles and halogens.

Typical Preparation of Activated Nickel

A 50 ml three-necked flask was equipped with a rubber septum, an addition funnel topped with rubber septum, a condenser topped with an argon inlet,

Table 7.10 Nickel-mediated cycloadditions of substituted dibromides with dienophiles.[a]

Starting dihalides	Dienophiles (molar equiv)[b]	Adducts (% isolated yield)[c]
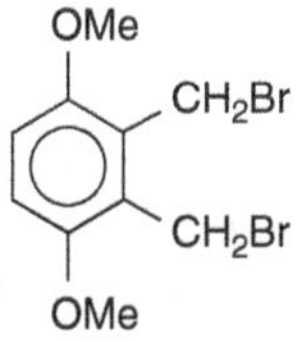 OMe, CH_2Br, CH_2Br, OMe	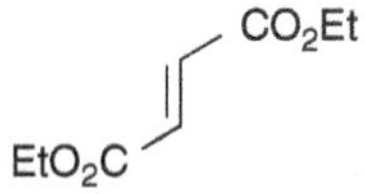 CO_2Et, EtO_2C (1.5)	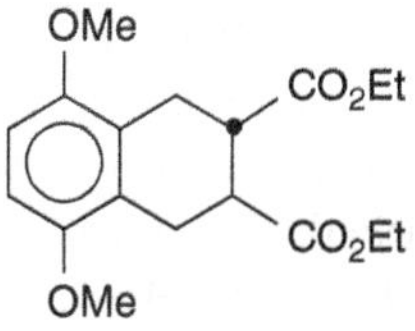 OMe, CO_2Et, CO_2Et, OMe **34** (59)
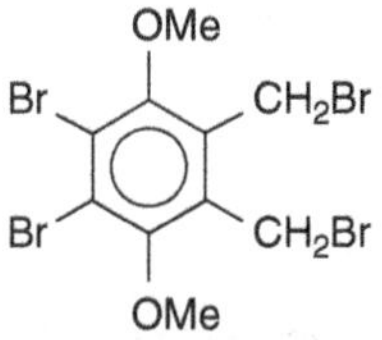 OMe, Br, CH_2Br, Br, CH_2Br, OMe	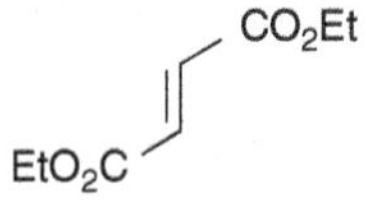 CO_2Et, EtO_2C (1.5)	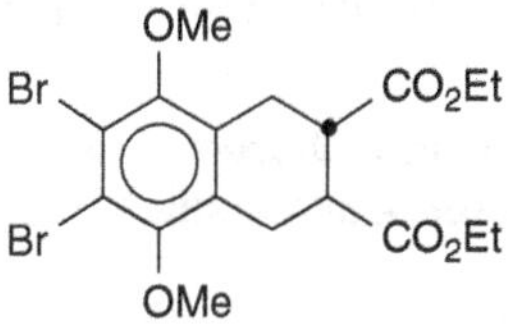 OMe, Br, CO_2Et, Br, CO_2Et, OMe **35** (41)
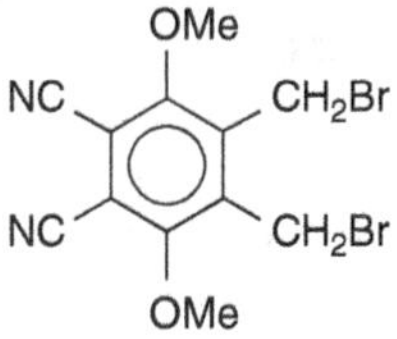 OMe, NC, CH_2Br, NC, CH_2Br, OMe	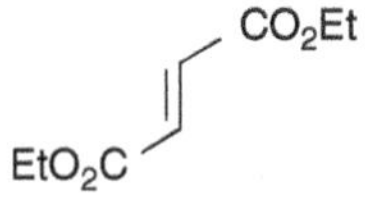 CO_2Et, EtO_2C (1.5)	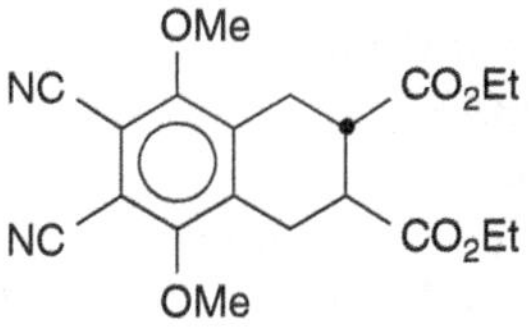 OMe, NC, CO_2Et, NC, CO_2Et, OMe **36** (61)
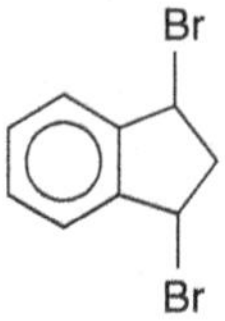 Br, Br	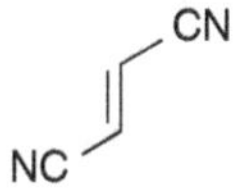 CN, NC (2)	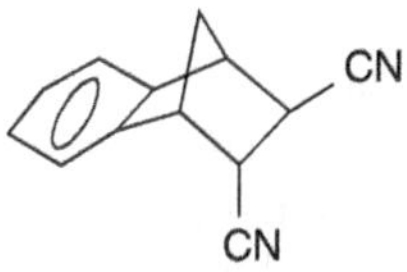 CN, CN **37** (66)
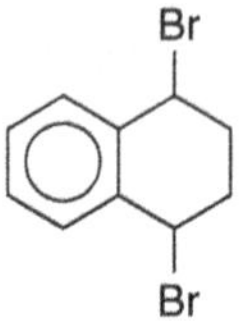 Br, Br	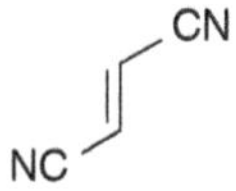 CN, NC (2)	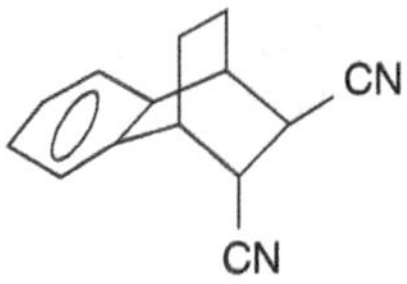 CN, CN **38** (61)
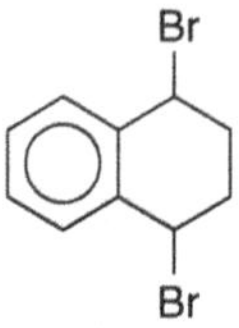 Br, Br	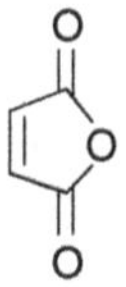 O, O, O (2)	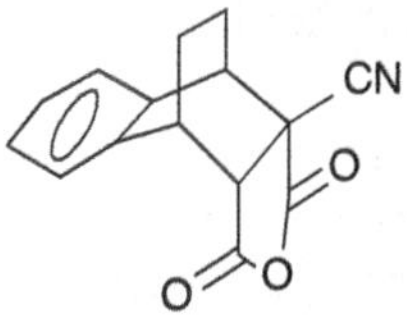 CN, O, O, O **16** (46)

[a] All reactions were carried out at room temperature.
[b] Molar equivalents of dienophile are relative to starting dibromide.
[c] Isolated by silica gel chromatography.

and a Teflon-coated magnetic stir bar. The flask was charged with nickel iodide (5.595 g, 17.90 mmol), freshly cut lithium (0.2866 g, 41.30 mmol), and naphthalene (0.2302 g, 1.796 mmol) in the inert atmosphere dry box. The flask was transferred to the manifold system and an argon inlet fitted. Freshly distilled DME was then added to the flask via syringe, and the mixture stirred vigorously at room temperature. Within the first hour, the Li was pink colored within the reddish mixture. The reduction was complete within 12 h with all the Li consumed. The black Ni metal settled from a clear solution within 20–30 min. The clear solution above the slurry was removed via syringe and distilled DME (20 ml) added. The slurry was stirred for 1 min and Ni allowed to settle. This procedure was repeated twice to remove the naphthalene.

Reaction of α,α′-Dibromo-*o*-xylene with Diethyl Fumarate in the Presence of Metallic Nickel

To nickel powder prepared as described in the preceding text from nickel iodide (5.595 g, 17.90 mmol) in a three-necked flask fitted with an addition funnel was added diethyl fumarate (4.930 g, 28.64 mmol) neat via syringe. α,α′-Dibromo-*o*-xylene (3.751 g, 14.21 mmol) dissolved in DME (20 ml) was then added slowly dropwise via an addition funnel over 1 h at room temperature with stirring to the nickel plus diethyl fumarate. The reaction mixture began to warm slightly and turn greenish within 5–10 min after beginning the addition of the α,α′-dibromo-*o*-xylene. Thin layer chromatography of reaction quenches showed the dibromide to be essentially consumed within 7 h. The greenish reaction mixture was allowed to stir a total of 24 h at room temperature and then was poured into 3% HCl (100 ml). The DME/HCl solution was then extracted with three portions (50 ml each) of dichloromethane. The dichloromethane extracts were washed with a solution of sodium bisulfite to remove residual iodine and then were dried over anhydrous sodium sulfate. Removal of the dichloromethane gave an oil which was chromatographed on silica gel to give *trans*-2,3-dicarbethoxy-1,2,3,4-tetrahydronaphthalene **24** [105] (2.998 g, 76%) as an oil: bp 132–133°C (0.23 mmHg).

7.8 Active Nickel-Mediated Dehalogenative Coupling of Aryl and Benzylic Halides

Considerable interest has centered on the use of zerovalent transition metals and their complexes as selective and mild reagents in organic synthesis [74]. However, some of the useful complexes coordinated with ligands such as phosphine or carbon monoxide are unstable and not easy to prepare or are toxic. Other studies using the metal atom vaporization techniques have also been successful in employing transition metals in synthesis [119]. In many cases, however, this approach does not lend itself to being completely general.

There are few reports of oxidative addition of metallic transition metals under mild conditions [120]; two reports involving group 8 elements have appeared. Fischer and Bürger reported the preparation of the π-allylpalladium complex by the reaction of a palladium sponge with allyl bromide [121]. The Grignard-type reaction addition of allyl halides to aldehydes has been carried out by reacting allylic halides with cobalt or nickel metal prepared by reduction of cobalt or nickel halides with manganese/iron alloy thiourea [122].

In a series of our studies on the chemistry of activated metals, we have shown that the transition metal powders prepared by the reduction of metal halides with alkali metal in an ethereal or a hydrocarbon solvent have high reactivities in their metallic state toward organic and inorganic substrates [123]. We reported results showing that activated nickel powder can induce the dehalogenative coupling of iodo- or bromobenzene [15]. We reported the application of activated metallic nickel as a reagent for the homocoupling of a variety of halobenzenes [41]. In addition, the mechanistic considerations will be discussed on the basis of the isolation of intermediates, ArNiX and Ar_2Ni species ($Ar = C_6F_5$), as their phosphine complexes.

Results and Discussion

Activated nickel powder employed for the reactions was simply prepared by stirring a 1 : 2.3 mixture of nickel halide and lithium metal (Equation 7.7) under argon with a catalytic amount of naphthalene (10 mmol % based on nickel halide) at room temperature for 12 h in DME (glyme). The resulting black slurry, which slowly settled after the stirring was stopped, was washed with freshly distilled glyme to remove the naphthalene and was used for the reaction with halobenzenes. As the presence of the naphthalene did not disturb the following reaction, the removal process is optional. The dehalogenative coupling reactions were carried out at 80°C by using a 0.7–0.8 : 1 mixture of halobenzene and the nickel powder (Equation 7.8). The course of the reaction was followed, and the yields were determined by GLC.

$$NiX_2 \xrightarrow[\text{Glyme}]{\text{Li (2.3 equiv)/}C_{10}H_8\text{ (0.1 equiv)}} [Ni] \tag{7.7}$$

$$ArX' \xrightarrow[\text{Glyme (Me}_2\text{SO or DMF)}]{[Ni]} \tfrac{1}{2}ArAr \tag{7.8}$$

The results of the homocoupling reaction of iodobenzenes by activated nickel powder are summarized in Table 7.11. Metallic nickel worked well for the dehalogenative coupling of unsubstituted and 4-substituted iodobenzenes. For instance, the reaction of 4-iodomethoxybenzene reached completion in 2 h at 80°C to afford 4,4′-dimethoxybiphenyl in 85% yield (via GLC) along with the reduction product anisole (15%). The usual workup of the reaction mixture gave the biphenyl in 68% isolated yield. Previous workers obtained 4,4′-dimeth-

Table 7.11 Reaction of iodobenzenes with activated nickel powder.

Iodobenzene	NiX_2	Time (h)	Product (%)[a] Coupling	Reduction	% conversion[b]
C_6H_5I	NiI_2	2	83	15	100
4-$CH_3OC_6H_4I$	NiI_2	2	85 (68)c	15	100
4-ClC_6H_4I	NiI_2	4	75	22	100
4-ClC_6H_4I	$NiBr_2$	4	77	25	100
4-ClC_6H_4I	$NiCl_2$	5	74	23	96
4-NCC_6H_4I	NiI_2	2	85[c]	<1	100
C_6F_5I	NiI_2	2	100	0	100
2-$CH_3OC_6H_4I$	NiI_2	4	<1	85	56
2-$O_2NC_6H_4I$	NiI_2	6	<1	<1	<1

[a] The yields were determined by GLC based on the iodobenzene consumed unless otherwise noted.
[b] Conversion = 100 × [(millimoles of the iodobenzene consumed)/(millimoles of the iodobenzene charged)].
[c] Isolated yield.

oxybiphenyl in 85% yield using copper powder; however, drastic conditions (225°C) were required to promote the reaction [124].

In the case of 4-iodochlorobenzene, oxidative addition occurred exclusively at the carbon–iodine bond, and the chlorine–carbon bond was unreactive toward the nickel powder. In this particular reaction, the effect of varying the halide of the nickel salt reduced was examined. There was essentially no difference in the yield of the reaction in the series $NiCl_2$, $NiBr_2$, or NiI_2. This effect was not studied in detail for the other iodobenzene reactions but most probably would have been the same.

Ortho substituents on the aryl groups seem to inhibit the coupling. For example, 2-iodomethoxybenzene failed to couple and gave exclusively anisole in moderate yield. 2-Iodonitrobenzene did not react at all under similar conditions. These facts are significantly different from those of copper-promoted reactions, in which electron-withdrawing substituents such as a nitro group in the *ortho* position enhance the reactivity [125].

Table 7.12 summarizes the results of the reaction of bromobenzenes with activated nickel. Although the reactivity of bromobenzenes toward nickel was relatively lower than that of iodobenzenes, they reacted at 85°C to give biphenyls in moderate to good yields, together with the corresponding reduction products. The reaction of 4-bromochlorobenzene with nickel powder from nickel bromide in glyme afforded chlorobenzene as the major product. Dimethyl sulfoxide and dimethylformamide were tried; these solvents did

Table 7.12 Reaction of bromobenzenes with activated nickel powder.

Bromobenzene	NiX_2	Solvent	Time (h)	Product (%)[a]		% conversion[b]
				Coupling	Reduction	
4-ClC_6H_4Br	$NiBr_2$	Glyme	20	36	55	100
4-ClC_6H_4Br	$NiBr_2$	Me_2SO	15	41	52	95
4-ClC_6H_4Br	$NiBr_2$	DMF	4	<1	89	100
4-ClC_6H_4Br	NiI_2	Glyme	20	61	32	100
4-$CH_3OC_6H_4Br$	$NiBr_2$	Glyme	30	0	100	43
4-$CH_3COC_6H_4Br$	$NiBr_2$	Glyme	18	57	43	75
4-$CH_3COC_6H_4Br$	NiI_2	Glyme	8	46[c]	[d]	100
4-NCC_6H_4Br	NiI_2	Glyme	21	71[c]	[d]	100
4-NCC_6H_4Br	$NiBr_2$	DMF	15	55	<1	100
C_6F_5Br	$NiBr_2$	Glyme	8	37	32	100
C_6F_5Br	NiI_2	Glyme	4	49	22	100

[a] The yields were determined by GLC based on the bromobenzene consumed unless otherwise noted.
[b] Conversion = 100 × [(millimoles of the bromobenzene consumed)/(millimoles of the bromobenzene charged)].
[c] Isolated yield.
[d] Not determined.

not greatly improve the yield of the coupled product. However, the yield of 4,4′-dichlorobiphenyl could be improved to 61% by using nickel powder obtained from nickel iodide. This observation suggests that the reaction was facilitated by the halogen–halogen exchange reaction which occurred between the substrate and iodide present.

Electron-donating substituents seem to lower the reactivity. For example, 4-bromomethoxybenzene failed to couple and gave only anisole in 43% yield. Substituents such as carbonyl and cyano groups were compatible with the reaction conditions employed, and the corresponding biphenyls were obtained in good yields.

As for mechanistic considerations of homogeneous zerovalent nickel-induced biaryl synthesis, several possibilities have been discussed. Semmelhack et al. [126] suggested that the coupling of aryl halides using bis(1,5-cyclooctadiene)-nickel(0) in DMF occurs as in Scheme 7.4.

(L = solvent or COD)

$$ArX + NiL_2 \rightarrow ArNi^{II}XL_2$$

$$ArNi^{II}XL_2 + ArX \rightarrow Ar_2Ni^{IV}X_2 + 2L$$

$$Ar_2Ni^{IV}X_2 \rightarrow ArAr + NiX_2$$

Scheme 7.4 Semmelhack et al. [126] proposed scheme for aryl halide coupling.

$$Ni^{I}X + ArX \rightarrow ArNi^{III}X_2$$
$$ArNi^{III}X_2 + ArNi^{II}X \rightarrow Ni^{II}X_2 + Ar_2Ni^{III}X$$
$$Ar_2Ni^{III}X \rightarrow ArAr + Ni^{I}X$$

Scheme 7.5 Proposed coupling scheme of Tsou and Kochi [127].

This scheme involves a second oxidative addition of the initially formed arylnickel(II) halide complex, yielding the diaryl nickel (IV) species, which yields the biaryl product by reductive elimination.

On the basis of detailed studies, Tsou and Kochi [127] concluded that the nickel(I) and arylnickel(III) species are the reactive intermediates in a radical chain process (Scheme 7.5). However, these studies were carried out in hydrocarbon solvents, and the conclusion, as pointed out by Semmelhack et al. [128], is not completely consistent with the product studies performed in DMF.

(L = solvent)

$$[Ni] + ArX + L_n \rightarrow ArNi^{II}XL_n$$
$$2ArNi^{II}XL_n \rightarrow Ar_2Ni^{II}L_n + Ni^{II}X_2 + L_n$$
$$ArNi^{II}XL_n \rightarrow ArAr + [Ni] + L_n$$

Scheme 7.6 Aryl coupling scheme proposed by Rieke and Kavaliunas [123].

The following experiments were made concerning the mechanism of the coupling reaction mediated by metallic nickel. The reaction of iodopentafluorobenzene with nickel powder prepared from nickel iodide was investigated in detail. Iodopentafluorobenzene was found to react with nickel at room temperature; however, decafluorobiphenyl could not be detected in the reaction mixture. Triphenylphosphine (2 equiv) was then added to the mixture to trap the arylnickel intermediate as the phosphine complex, and a usual workup afforded bis(triphenylphosphine) bis(pentafluorophenyl)nickel(II) in 45% yield. As the diarylnickel(II) species seemed to be derived by disproportionation from an initially formed oxidative adduct, (pentafluorophenyl)nickel(II) iodide, the trapping of the arylnickel(II) iodide was tried. In the presence of 2 equiv of triethylphosphine, the reaction of iodopentafluorobenzene with nickel was carried out at room temperature, and bis(triethylphosphine)(pentafluorophenyl)-nickel(II) iodide was isolated in 15% yield.

These results strongly suggest that the initial oxidative addition of aryl halide yields the arylnickel(II) halide species, which most likely is coordinated with glyme. Klabunde has shown that the arylnickel(II) halide species are not stable unless they are coordinated to some ligand [129]. The arylnickel(II) halide then disproportionates to diarylnickel(II) and nickel halide. The final step is the formation of biaryl by reductive elimination of diarylnickel(II) under the reaction conditions employed (Scheme 7.6).

It is well known that thermal decomposition of arylnickel(II) halide complexes such as bis(triphenylphosphine)phenylnickel(II) bromide [130] or bis(triethylphosphine)(4-fluorophenyl)nickel(II) chloride [131] gave the corresponding biphenyls by successive disproportionation and reductive elimination reactions [132]. Scheme 7.6 is also consistent with these facts.

In conclusion, oxidative addition of aryl halides to metallic nickel proceeded smoothly under mild conditions, and the corresponding biaryls were obtained in good yields. The present reaction is superior to the Ullmann synthesis in the scope of the reaction conditions [125] and to the coupling reaction of Grignard reagents catalyzed by transition metals [133]. Finally, because of the simple and easy procedure for the preparation of nickel powder, it is far more convenient to work with than zerovalent nickel complexes such as tetrakis(triphenylphosphine)nickel(0) or bis(1,5-cyclooctadiene)nickel(0) [128].

Transition metals in a low oxidation state have been used extensively as reagents for the reductive homocoupling of benzylic halides. For example, titanium [134], vanadium [135], chromium [136], and tungsten [137] have been employed as reagents for this purpose and were prepared *in situ* by reducing the appropriate metal halide with lithium aluminum hydride. Nickel(I) [138], cobalt(I) [139], and vanadium(II) [140] complexes or metal carbonyls of nickel [79, 141], cobalt [142], iron [76, 78, 143], molybdenum [144], and tungsten [137] as well as metallic iron [145–147] have been found to be useful as low-valent coupling reagents of benzylic mono- and polyhalides.

However, low-valent reagents originating from transition metal halides/lithium aluminum hydride have the limitation of not being compatible with functional groups such as cyano and nitro groups. Chlorotris(triphenylphosphine) cobalt(I), dicyclopentadienylvanadium(II), and metallic iron worked well for the coupling of benzyl halides; nevertheless, the examples reported were limited to benzyl halides which did not contain functional groups sensitive to reduction. Although potassium hexacyanodinickelate(I) was applied to 4-(bromomethyl)benzonitrile, the yield of the corresponding homocoupled product was low. Metal carbonyls were generally tried with benzylic polyhalides, and the applicability to benzylic monohalides and the compatibility with functional groups were not made clear. We report here the detailed applications of metallic nickel as a mild and selective reagent for the reductive coupling reaction of a variety of benzylic halides as well as polyhalides. The mechanistic aspects are discussed on the basis of the trapping of benzylnickel complexes with electron-deficient olefins [45].

Results and Discussion

Metallic nickel was easily prepared by stirring a mixture of nickel halide and lithium metal (2.3 equiv) with naphthalene (0.1 equiv) as an electron carrier at room temperature for 12 h in DME (glyme) (Equation 7.9). The resulting black powders which slowly settled in a colorless solution after stirring was stopped were used for the reductive homocoupling reaction of benzylic halides (Equation 7.10).

$$NiX_2 \xrightarrow[\text{Glyme/room temperature, 12 h}]{\text{Li (2.3 equiv)/}C_{10}H_8\text{ (0.1 equiv)}} Ni \tag{7.9}$$

$$R\text{-}C_6H_4\text{-}CH_2Y \xrightarrow[\text{Glyme/room temperature or 70°C}]{\text{Ni}} R\text{-}C_6H_4\text{-}CH_2CH_2\text{-}C_6H_4\text{-}R + R\text{-}C_6H_4\text{-}H$$

X, Y = I, Br, Cl
R = H, 4-H_3C, 3-CH_3O, 3-F_3C, 4-Cl, 4-Br, 4-O_2N, 4-NC, 4-CH_3OCO

(7.10)

The results of the reaction of benzyl and substituted benzyl halides with metallic nickel powders are summarized in Table 7.13. Benzyl chloride reacted at room temperature with metallic nickel prepared from nickel chloride to give a mixture of the coupled product bibenzyl (40%) and the reduction product toluene (60%). However, the coupled product was found to be formed mainly when the reaction was run at 70°C and when the nickel powders used were prepared from nickel iodide. Under these conditions, a yield of 86% bibenzyl was attained. The fact that iodide ion present in the system facilitates the homocoupling reaction may be ascribed to the chlorine–iodine exchange during the reaction [148]. Higher reaction temperatures such as 70°C may also accelerate the exchange reaction.

Similarly, substituted benzyl halides reacted with metallic nickel to afford the corresponding bibenzyls in good to high yields, and the use of nickel iodide generally gave more satisfactory results [148]. The aromatic carbon–halogen bond was unreactive toward the nickel in the reaction of 4-chloro- and 4-bromobenzly halides. When substituents such as methoxy, nitro, cyano, and alkoxycarbonyl groups were used and compatible reaction conditions were employed, the corresponding bibenzyls were obtained in good yields. Benzyl halides having the hydroxycarbonyl group failed to couple with metallic nickel and were reduced to give the corresponding carboxylic acids. For example, α-bromophenylacetic acid was converted to phenylacetic acid in 80% yield. 4-(Bromomethyl)benzoic acid afforded 4-methylbenzoic acid (54%) together with 4-(iodomethyl)benzoic acid (20%). The latter seems to be formed by the halogen–halogen exchange reaction catalyzed by nickel between the substrate and lithium iodide present in the system.

Benzylic monohalides such as 1-(chloromethyl)-naphthalene, 2-(bromomethyl)naphthalene, chlorodiphenylmethane, and 9-bromofluorene also underwent a reductive homocoupling reaction at room temperature to afford the corresponding ethane derivatives in 55–76% yields (Table 7.14). The corresponding reduction products were formed as side products in all cases.

Table 7.13 Reaction of benzyl halides with metallic nickel powders.

			Product[c] (%)	
Benzyl halide	**NiX_2[a]**	**Conditions[b]**	**Coupling**	**Reduction**
$C_6H_5CH_2Cl$	$NiCl_2$	rt, 1 h	40	60
$C_6H_5CH_2Cl$	$NiCl_2$	70°C, 0.5 h	65	33
$C_6H_5CH_2Cl$	NiI_2	70°C, 0.5 h	86	10
$C_6H_5CH_2Br$	$NiBr_2$	rt, 1 h	77	23
$4\text{-}CH_3C_6H_4CH_2Cl$	$NiCl_2$	rt, 2 h	76	20
$4\text{-}CH_3C_6H_4CH_2Cl$	NiI_2	70°C, 0.5 h	96	4
$3\text{-}CH_3OC_6H_4CH_2Cl$	NiI_2	rt, 6 h	69[c]	[e]
$3\text{-}F_3CC_6H_4CH_2Cl$	NiI_2	rt, 6 h	75[d]	[e]
$4\text{-}ClC_6H_4CH_2Cl$	$NiCl_2$	70°C, 0.5 h	72	20
$4\text{-}ClC_6H_4CH_2Cl$	NiI_2	rt, 3 h	72[d]	[e]
$4\text{-}ClC_6H_4CH_2Cl$	NiI_2	70°C, 0.5 h	85	13
$4\text{-}BrC_6H_4CH_2Br$	NiI_2	rt, 6 h	70[d]	[e]
$4\text{-}O_2NC_6H_4CH_2Cl$	NiI_2	rt, 1 h	78[d]	[e]
$4\text{-}NCC_6H_4CH_2Br$	$NiBr_2$	rt, 1 h	68[d]	[e]
$4\text{-}NCC_6H_4CH_2Br$	NiI_2	rt, 1 h	75[d]	[e]
$4\text{-}CH_3O_2CC_6H_4CH_2Cl$	NiI_2	rt, 1 h	73[d]	14[d]
$4\text{-}HO_2CC_6H_4CH_2Br$	NiI_2	85°C, 0.5 h	[f]	54[d,g]
$C_6H_5CHBrCO_2H$	NiI_2	rt, 9 h	[f]	80[d]

[a] Nickel halide used for the preparation of nickel powders.
[b] rt = room temperature.
[c] Yields were determined by GLPC based on benzyl halides used unless otherwise noted.
[d] Isolated yield.
[e] Not determined.
[f] Not detected.
[g] 4-(Iodomethyl)benzoic acid (20%) was also isolated.

On the other hand, benzylic di- and trihalides afforded substituted ethenes. For example, dichlorodiphenylmethane gave tetraphenylethene in quantitative yield at room temperature, and α,α-dibromotoluene gave predominantly the *trans* isomer of stilbene (*cis*/*trans* = 27/73). At 80°C after 1 h, α,α,α-trichlorotoluene afforded a mixture of *cis*- and *trans*-1,2-dichloro-1,2-diphenylethene with predominantly the *cis* isomer (*cis*/*trans* = 74/26) and a trace amount of diphenylacetylene.

There are two likely mechanistic pathways for the reductive coupling of benzylic polyhalides, either via a carbene or carbene–nickel complex intermediate

Table 7.14 Reaction of benzylic mono- and polyhalides with metallic nickel.[a]

Halide	Conditions[b]	Product	Yield[c] (%)
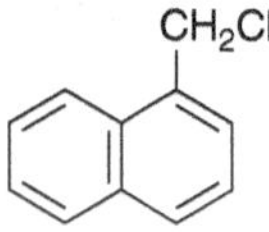 CH_2Cl	rt, 5 h	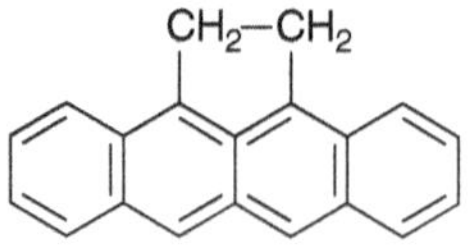 $CH_2{-}CH_2$	55
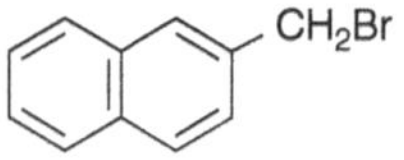 CH_2Br	rt, 2 h	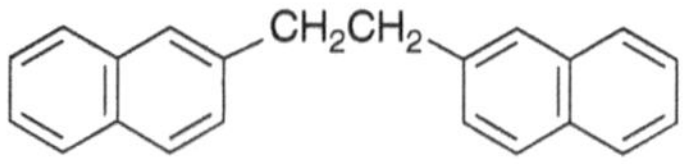 CH_2CH_2	67
$(C_6H_5)_2CHCl$	rt, 9 h	$(C_6H_5)_2CHCH(C_6H_5)_2$	76
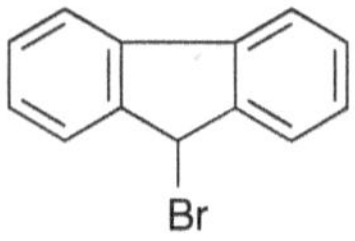 Br	rt, 1 h	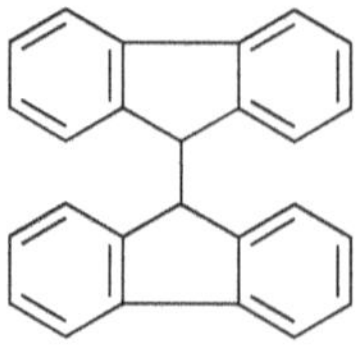	75
$(C_6H_5)_2CCl_2$	rt, 2 h	$(C_6H_5)_2C{=}C(C_6H_5)_2$	99
$C_6H_5CHBr_2$	rt, 1 h	$(C_6H_5)CH{=}CH(C_6H_5)$[d]	65
$C_6H_5CCl_3$	85°C, 1 h	$(C_6H_5)CCl{=}CCl(C_6H_5)$[e]	71
$C_6H_5CHBrCHBrC_6H_6$ (meso)	rt, 5 h	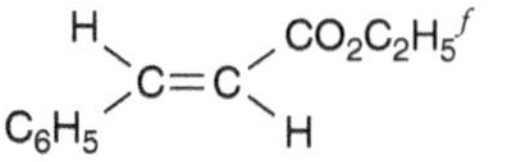 H, $CO_2C_2H_5$[f], C=C, C_6H_5, H	89
$4\text{-}O_2NC_6H_4CHBrCHBrCO_2C_2H_5$ (erythro)	rt, 1 h	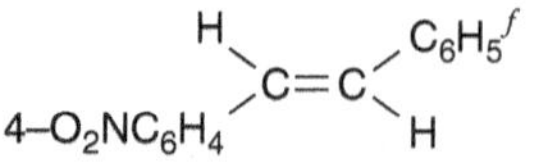 H, C_6H_5[f], C=C, $4\text{-}O_2NC_6H_4$, H	76
$C_6H_5CCl_2CCl_2H_5$	85°C, 1 h	$(C_6H_5)CCl{=}CCl(C_6H_5)$[g]	97

[a] Metallic nickel prepared by reducing nickel iodide was used.
[b] rt = room temperature.
[c] Isolated yield.
[d] Ratio of *cis/trans* determined by GLPC was 27/73.
[e] Ratio of *cis/trans* determined by ^1H NMR was 74/26.
[f] Trace amount of *cis*-stilbene was detected by GLPC.
[g] Ratio of *cis/trans* determined by ^1H NMR was 70/30.

or via a step-by-step dehalogenation. Coffey [143] suggested a carbene mechanism for the iron pentacarbonyl-promoted reaction. Later, carbenoid and carbene-transition metal complex intermediates were suggested on the basis of the trapping of cyclopropane derivatives in the reaction of $Ni(COD)_2$, $Fe(CO)_5$,

$Co_2(CO)_8$ [149], or $W(CO)_6$ [137] with *gem*-dihalides. On the other hand, it was suggested that the reaction with $Co(CO)_8$ [142] or metallic iron [150] proceeds via a step-by-step mechanism.

The reaction of α,α-dibromotoluene or dichlorodiphenylmethane was carried out in the presence of an excess of cyclohexene or an electron-deficient olefin such as diethyl fumarate in order to try to trap the divalent intermediate. However, the formation of cycloadducts could not be detected. The coupling reaction of α,α,α-trichlorotoluene was then carried out under milder conditions (0°C for 1 h followed by room temperature for 12 h) and was found to afford 1,2-diphenyl-1,1,2,2-tetrachloroethane in 50% yield. The results suggest that the reaction of benzylic polyhalides proceeds via a stepwise manner such as shown in Equation 7.11.

$$PhRCX_2 \xrightarrow{Ni} PhRC(X)\text{—}C(X)RPh \xrightarrow{Ni} PhRC{=}CRPh$$

$$R = Ph, H, X;\ X = Cl, Br \qquad (7.11)$$

Dehalogenation of vicinal polyhalides with nickel was then investigated to establish that the second step in Equation 7.11 proceeds as shown. 1,2-Dibromo-1,2-diphenylethane (meso) reacted with metallic nickel at room temperature to yield stilbene (*trans*) in 89% yield and a trace amount of *cis*-stilbene. Similarly, ethyl 2,3-dibromo-3-(4-nitrophenyl)propanoate (erythro) underwent debromination to give ethyl 4-nitrocinnamate (*trans*) in 76% yield. Dechlorination of 1,2-diphenyl-1,1,2,2-tetrachloroethane at 85°C afforded 1,2-dichloro-1,2-diphenylethene (97%, *cis*/*trans* = 70/30).

These results show that the second step of Equation 7.11 proceeds smoothly with a variety of vicinal dihalides. Thus, it may be concluded that the reaction of benzylic polyhalides with nickel proceeds via a step-by-step dehalogenation to give ethene derivatives.

Additional mechanistic aspects of the homocoupling of benzylic monohalides need to be addressed. Although the oxidative addition of organic halides to transition metal complexes has been well known [74], few examples are reported on the corresponding reaction of transition metals in the metallic state under mild conditions [121, 122, 151, 152]. Previously we reported that the homocoupling reaction of aromatic halides mediated by metallic nickel proceeds via ArNiX followed by Ar_2Ni intermediates, which were isolated as the phosphine complexes ($Ar = C_6F_5$) [41, 45]. The benzyl halide reaction seems to occur in a similar way. However, the possibility of a radical mechanism cannot be excluded. Both radical [145] and organoiron [150] intermediates were proposed in the homocoupling reaction of benzylic halides using metallic iron in water.

Organonickel compounds which contain carbon–metal σ bonds are known to add to active unsaturated bonds [130, 138, 153–157]. For example, Hashimoto et al. [138] reported that benzylnickel(II) complex prepared from benzyl bromide and potassium hexacyanodinickelate was added to acrylonitrile or ethyl acrylate. The addition reaction of benzyl chloride to electron-deficient olefins mediated by metallic nickel was carried out to help establish the presence of benzylnickel intermediates.

The reaction of benzyl chloride with metallic nickel in the presence of methyl acrylate was carried out at 85°C, and the expected addition product methyl 4-phenylbutanoate was formed in 17% yield (Equation 7.12). The reaction with acrylonitrile gave 4-phenylbutanenitrile in 14% yield together with *cis*- and *trans*-4-phenyl-2-butenenitriles, 4-cyano-6-phenylhexanenitrile, and 2-benzyl-4-phenylbutanenitrile (Equation 7.13). The results suggest the presence of a benzylnickel(II) chloride complex (I), which could have been formed by the oxidative addition of benzyl chloride to the metallic nickel (Scheme 7.7). The complex (I) would then be expected to add to the electron-deficient olefins, affording the addition product (III) via intermediate complex (IV). The formation of *cis*- and *trans*-4-phenyl-2-butenenitrile (V) is reasonably explained by the reductive elimination of nickel hydride from intermediate (IV), which is analogous to the substitution reaction of olefins with alkylpalladium compound [158] and to the addition–elimination reaction of bis(triphenylphosphine) phenylnickel(II) bromide with methyl acrylate to yield methyl cinnamate [130]. Furthermore, intermediate (IV) seems to add another molecule of acrylonitrile to give the 1:2 adduct 4-cyano-6-phenylhexanenitrile (VI). 2-Benzyl-4-phenylbutanenitrile (VIII) would be formed by the metathesis of complex IV and the benzylnickel chloride (I).

$$PhCH_2Cl + H_2C{=}CHCO_2CH_3 \xrightarrow[\text{Glyme, reflux}]{Ni} \underset{17\%}{PhCH_2CH_2CH_2CO_2CH_3}\ (4) \tag{7.12}$$

$$PhCH_2Cl + H_2C{=}CHCN \xrightarrow[\text{Glyme, reflux}]{Ni} \underset{14\%}{PhCH_2CH_2CH_2CN} + \underset{3\%}{(H)(PhCH_2)C{=}C(H)(CN)} + \underset{6\%}{(H)(PhCH_2)C{=}C(CN)(H)} + \underset{4\%}{PhCH_2CH_2CH(CN)CH_2CH_2CN} + \underset{15\%}{PhCH_2CH_2CH(CH_2Ph)CN}\ (5) \tag{7.13}$$

Thus, the homocoupling reaction might be reasonably explained by the smooth oxidative addition of benzyl halide to nickel in the metallic state, which

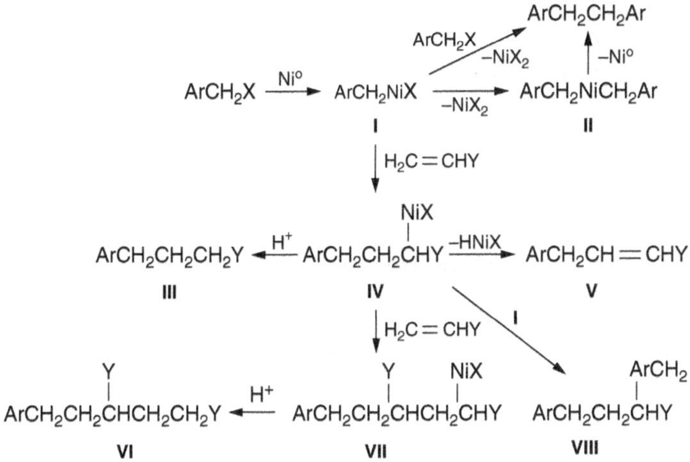

$$ArCH_2X \xrightarrow{Ni^0} \underset{\mathbf{I}}{ArCH_2NiX} \xrightarrow[-NiX_2]{} \underset{\mathbf{II}}{ArCH_2NiCH_2Ar} \xrightarrow{-Ni^0} ArCH_2CH_2Ar$$

$$\mathbf{I} + ArCH_2X \xrightarrow{-NiX_2} ArCH_2CH_2Ar$$

$$\mathbf{I} \xrightarrow{H_2C=CHY} \underset{\mathbf{IV}}{ArCH_2CH_2CH(NiX)Y}$$

$$\underset{\mathbf{III}}{ArCH_2CH_2CH_2Y} \xleftarrow{H^+} \mathbf{IV} \xrightarrow{-HNiX} \underset{\mathbf{V}}{ArCH_2CH=CHY}$$

$$\mathbf{IV} \xrightarrow{H_2C=CHY} \underset{\mathbf{VII}}{ArCH_2CH_2CH(Y)CH_2CH(NiX)Y} \xrightarrow{H^+} \underset{\mathbf{VI}}{ArCH_2CH_2CH(Y)CH_2CH_2Y}$$

$$\mathbf{IV} \xrightarrow{\mathbf{I}} \underset{\mathbf{VIII}}{ArCH_2CH_2CH(CH_2Ar)Y}$$

[a] X = halogen; Y = CO_2CH_3 or CN.

Scheme 7.7 Proposed reaction mechanism.[a]

gives benzylnickel(II) halide complex (I). The metathesis of the complex (I) would afford dibenzylnickel(II) (complex (II), which could undergo reductive elimination to yield coupled product, 1,2-diarylethane. The formation of 1,2-diarylethane could also possibly result from the reaction of benzylnickel halide (I) with benzyl halide.

In conclusion, a very convenient method is presented for the reductive coupling of benzylic mono-, di- and trihalides. The procedure involves the use of highly reactive nickel in the metallic state simply prepared by the lithium reduction of anhydrous nickel halides. The yields are good to high, and the system is compatible with a wide variety of substituents on the aromatic ring including nitro, cyano, alkoxycarbonyl, methoxy, bromo, and chloro groups. Because of the ease of preparation of the highly reactive nickel powders and of the mild conditions, this procedure should prove to be a convenient and general method for carrying out reductive coupling of benzylic halides.

Preparation of Activated Nickel Powder

A 50 ml two-necked flask was equipped with a magnetic stirrer, a rubber septum, and a condenser topped with an argon inlet. The flask was charged with nickel iodide (3.82 g, 12.22 mmol), freshly cut lithium (0.195 g, 28.1 mmol), naphthalene (0.16 g, 1.25 mmol), and glyme (30 ml), and the mixture was stirred vigorously at room temperature for 12 h. The nickel powder precipitated as a black slurry in a colorless clear solution after the stirring was stopped. The top of the glyme was removed by syringe, and freshly distilled glyme (20 ml) was added and was stirred for 5 min. This procedure was repeated two times to remove the naphthalene.

In those cases where the coupled product was easily separable from naphthalene, the removal procedure of naphthalene was optional because the naphthalene did not disturb the coupling reaction at all. Glyme could be replaced with other solvents such as DMF or Me_2SO, after the evaporation of the glyme under reduced pressure.

Reaction of 4-Iodomethoxybenzene with Activated Nickel Powder

To the nickel powder (12.22 mmol) was added 4-iodomethoxybenzene (1.88 g, 8.02 mmol), and the mixture was stirred at 85°C for 2 h. The reaction mixture changed to reddish brown, and most of the nickel powder was consumed. GLC analysis of the reaction mixture showed that 4,4′-dimethoxybiphenyl and anisole were formed in 85% and 15% yields, respectively. The reaction mixture was poured into ether (100 ml) and was filtered. The filtrate was washed with water (50 ml) and dried over anhydrous magnesium sulfate. The solution was concentrated to 10 ml, and ethanol (10 ml) was added to precipitate the product (0.69 g, 78%). Recrystallization from ether–ethanol (1 : 1) gave 4,4′-dimethoxybiphenyl (0.58 g, 68%) as colorless flakes: mp 176–177°C (mp 176–178°C).

In the case of less soluble biphenyls such as 4,4′-dicyano- or 4,4′-diacetylbiphenyl, chloroform was used for extraction. The reaction mixture was poured into chloroform (150 ml) and washed with 3% HCl solution followed by water (100 ml).

4,4′-Dicyanobiphenyl: mp 236–237°C (sublimation) (mp 237–238°C)
4,4′-Diacetylbiphenyl: mp 190–191°C (glyme) (mp 192–193°C)

Trapping of Bis(pentafluorophenyl)nickel(II) Species with Triphenylphosphine

Activated nickel was prepared by stirring a mixture of nickel iodide (2.54 g, 8.14 mmol), lithium (0.130 g, 18.7 mmol), and naphthalene (0.104 g, 0.81 mmol) at room temperature for 12 h. To the nickel powder was added iodopentafluorobenzene (1.85 g, 6.29 mmol), and the mixture was stirred for 24 h at the same temperature. To the formed reddish-brown reaction mixture was added triphenylphosphine (4.27 g, 16.28 mmol), and this was stirred overnight at room temperature. The glyme was removed under reduced pressure, the residue was dissolved in benzene (30 ml), and the solution was filtered under argon. The benzene solution was concentrated to 3 ml, and methanol (20 ml) was added. The yellow powder which precipitated was separated by filtration. Recrystallization from benzene–methanol gave the yellow crystalline solid bis(triphenylphosphine)bis(pentafluorophenyl)nickel(II): 1.31 g (45%); mp 203–204°C dec (lit. [22] mp 201–204°C dec); IR spectrum is consistent with the reported values [22].

Trapping of (Pentafluorophenyl)nickel(II) Iodide Species with Triethylphosphine

To nickel powder in glyme (30 ml) prepared from nickel iodide (3.95 g, 12.7 mmol) in a similar manner was added triethylphosphine (2.99 g, 25.3 mmol) at room temperature. After the mixture was stirred for 1 h, iodopentafluorobenzene (3.01 g, 10.3 mmol) was added, and the mixture stirred at room temperature for 24 h. The reaction mixture was evaporated under reduced pressure, benzene (20 ml) was added to the residue, and the mixture was filtered under argon. The filtrate was evaporated, and the residue was recrystallized from methanol to give the brown crystalline solid bis(triethylphosphine)(pentafluorophenyl)nickel(II) iodide: 0.91 g (15%); mp 149–149.5°C (lit. [150] mp 151–152°C). Anal. Calcd for $C_{18}H_{30}F_5INiP_2$: C, 37.10; H, 5.19. Found: C, 36.96; H, 5.17.

Preparation of Metallic Nickel Powders and Their Reaction with 4-Nitrobenzyl Chloride

A 50 ml two-necked flask was equipped with a magnetic stirrer, a rubber septum, and a condenser topped with an argon inlet. The flask was charged with nickel iodide (4.53 g, 14.5 mmol), freshly cut lithium (0.232 g, 33.4 mmol), naphthalene (0.186 g, 1.45 mmol), and glyme (30 ml), and the mixture was stirred vigorously at room temperature for 12 h. The nickel powders were precipitated as bulky black slurries in colorless clear solution after the stirring was stopped. Lithium metal was completely consumed.

To the nickel powders 4-nitrobenzyl chloride (1.97 g, 11.5 mmol) in glyme (10 ml) was added and stirred at room temperature for 1 h. The reddish-brown colored reaction mixture formed was poured into 3% hydrochloric acid solution (100 ml) and was extracted with chloroform twice (150 ml). The extracts were washed with water (50 ml), and the aqueous phase was extracted with chloroform. The chloroform solution was dried over anhydrous sodium sulfate. After evaporation of the solvent, the residue was washed with pentane twice (15 ml) to give pale yellow needles (mp 176–179°C). Recrystallization from ethanol gave 1,2-bis(4-nitrophenyl)ethane (1.22 g, 78%): mp 182–183°C (lit. [78] mp 183–184°C).

The reaction of 4-nitrobenzyl chloride (3.37 g, 19.6 mmol) with nickel prepared from nickel iodide (3.86 g, 12.4 mmol) proceeded smoothly at room temperature for 2 h and gave 1,2-bis(4-nitrophenyl)ethane in 76% yield.

All other coupling reactions were carried out using the reagents in the ratio of metallic nickel/one halogen for coupling = 1/0.8 to optimize the yield and to shorten the reaction time.

References

1 (a) Rieke, R.D.; Kavaliunas, A.V.; Rhyne, L.D.; Fraser, D.J.J. *J. Am. Chem. Soc.* 1979, **101**, 246.(b) Kavaliunas, A.V.; Taylor, A.; Rieke, R.D. *Organometallics.* 1983, **2**, 377.

2 (a) Rieke, R.D. Hudnall, P.M. *J. Am. Chem. Soc.* 1972, **94**, 7178. (b) Rieke, R.D.; Hudnall, P.M.; Uhm, S. *J. Chem. Soc., Chem. Commun.* 1973, 269.

3 Rieke, R.D.; Bales, S.E. *J. Chem. Soc., Chem. Commun.* 1973, 789.

4 Rieke, R.D.; Bales, S.E. *J. Am. Chem. Soc.* 1974, **96**, 1775.

5 Rieke, R.D.; Chao, L. *Synth. React. Inorg. Met.-Org. Chem.* 1974, **4**, 101.

6 Rieke, R.D.; Ofele, K.; Fischer, E.O. *J. Organomet. Chem.* 1974, **76**, C19.

7 Rieke, R.D. *Top. Curr. Chem.* 1975, **59**, 1.

8 Rieke, R.D. *Acc. Chem. Res.* 1977, **10**, 301 and references therein.

9 Rieke, R.D.; Wolf, W.J.; Kujundzic, N.; Kavaliunas, A.V. *J. Am. Chem. Soc.* 1977, **99**, 4159.

10 Uhm, S. Preparation of Activated Zinc Metal and its Reactions with Organic Halides in the Presence of Various Lewis Bases, Ph.D. thesis, University of North Carolina-Chapel Hill, 1974.

11 Rieke, R.D.; Kavaliunas, A.V.; Rhyne, L.D.; Frazier, D.J.J. *J. Am. Chem. Soc.* 1979, **101**, 246.

12 Rieke, R.D.; Kavaliunas, A.V. *J. Org. Chem.* 1979, **44**, 3069.

13 Rieke, R.D.; Rhyne, L.D. *J. Org. Chem.* 1979, **44**, 3445.

14 Rieke, R.D.; Bales, S.E.; Hudnall, P.M.; Poindexter, G.S. *Org. Synth.* 1979, **59**, 85.

15 Kavaliunas, A.V.; Rieke, R.D. *J. Am. Chem. Soc.* 1980, **102**, 5944.

16 Kavaliunas, A.V.; Rieke, R.D. *Angew. Chem. Int. Ed. Engl.* 1982, 2196.

17 Shchukavev, S.A.; Tolmacheva, T.A.; Pazukhina, Y.L. *Russ. J. Inorg. Chem.* (Engl. Transl.) 1964, **9**, 1354.

18 Anderson, B.B.; Behrens, C.L.; Radonovich, L.I.; Klabunde, K.J. *J. Am. Chem. Soc.* 1976, **98**, 5390.

19 Klabunde, K.J.; Anderson, B.B.; Neuenschwander, K. *Inorg. Chem.* 1980, **19**, 3719.

20 Heck, R.F. *J. Am. Chem. Soc.* 1968, **90**, 5518, 5526, 5531, 5535, 5538.

21 Herzog, S.; Taube, R. *Z. Chem.* 1962, **2**, 208.

22 Phillips, J.R.; Rosevear, D.T.; Stone, F.G.A. *J. Organomet. Chem.* 1964, **2**, 455.

23 Klabunde, K.J.; Low, J.Y.F. *J. Am. Chem. Soc.* 1974, **96**, 7674.

24 Klabunde, K.J.; Murdock, T.O. *J. Org. Chem.* 1979, **44**, 3901.

25 Davis, S.C.; Klabunde, K.J. *J. Am. Chem. Soc.* 1978, **100**, 5973.

26 Gastinger, R.G.; Anderson, B.B.; Klabunde, K.J. *J. Am. Chem. Soc.* 1980, **102**, 4959.

27 Klabunde, K.J.; Low, J.Y.F.; Efner, H.F. *J. Am. Chem. Soc.* 1974, **96**, 1984.

28 Issleib, K.; Wenschuh, E.Z. *Anorg. Allg. Chem.* 1960, **305**, 15.

29 Ueno, Y.; Okawara, M. *Synthesis.* 1975, 268.

30 Hase, T. *Synthesis.* 1980, 36.
31 Wakamatsu, T.; Akasaka, K.; Ban, Y. *Tetrahedron Lett.* 1974, **15**, 3879 and 3883.
32 Hein, H.-G. *Liebigs Ann. Chem.* 1970, **735**, 56.
33 van De Sande, J.H.; Kopecky, K.R. *Can. J. Chem.* 1969, **47**, 163.
34 Chambers, R.D.; Clark, M. *J. Chem. Soc. Perkin Trans* **1**. 1972, 2469.
35 Sharp, D.B.; Miller, E.L. *J. Am. Chem. Soc.* 1952, **74**, 5643.
36 Baddar, F.G.; Fahmy, A.F.M.; Aly, N.F. *J. Indian Chem. Soc.* 1973, **50**, 586.
37 Trost, B.M.; Keinan, E. *J. Org. Chem.* 1980, **45**, 2741.
38 Jones, P.R.; Sherman, P.D., Jr.; Schwarzenberg, K. *J. Organomet. Chem.* 1967, **10**, 521.
39 Cognacq, J.-C.; Skowronski, R.; Chodkiewicz, W.; Cadiot, P. *Bull. Soc. Chim. Fr.* 1967, 880.
40 Hegedus, L.S.; Wagner, S.D.; Waterman, E.L.; Siirala-Hansen, K. *J. Org. Chem.* 1975, **40**, 593.
41 Matsumoto, H.; Inaba, S.; Rieke, R.D. *J. Org. Chem.* 1983, **48**, 840.
42 Inaba, S.; Rieke, R.D. *Tetrahedron Lett.* 1983, **24**, 2451.
43 Inaba, S.; Rieke, R.D. *Chem. Lett.* 1984, 25.
44 Inaba, S.; Rieke, R.D. *Synthesis.* 1984, 844.
45 Inaba, S.; Matsumoto, H.; Rieke, R.D. *Tetrahedron Lett.* 1982, **23**, 4215.
46 Sullivan, H.R.; Beck, J.R.; Pohland, A. *J. Org. Chem.* 1963, **28**, 2381.
47 Temnikova, T.I.; Martynov, V.F. *J. Gen. Chem.* 1945, USSR 15, 499. C.A. **1946**, 40, 4695.
48 Inaba, S.; Rieke, R.D. *Synthesis.* 1984, 842.
49 Harrison, I.T.; Harrison, S. *Compendium of Organic Synthetic Methods*, Vol. **1**, John Wiley & Sons, New York, 1971, p. 457; see also Vol. 2, **1974**, p. 185; Vol. 3, **1977**; p. 296; Vol. 4, **1980**, p. 297.
50 Cope, A.C.; Holmes, H.L.; House, H.O. *Org. React.* 1957, **9**, 107.
51 Brown, H.C.; Nambu, H.; Rogic, M.M. *J. Am. Chem. Soc.* 1969, **91**, 6854.
52 Nambu, H.; Brown, H.C. *J. Am. Chem. Soc.* 1970, **92**, 5790.
53 Grebenyuk, A.D.; Tsukervanik, I.P. *Zh. Obshch. Khim.* 1955, **25**, 286; C.A. **1956**, *50*, 1639.
54 McFarland, J.W.; Howes, H.L., Jr. *J. Med. Chem.* 1972, **15**, 365.
55 McFarland, J.W.; Howes, H.L. Conover, L.H. Lynch, J.E.; Austin, W.C.; Morgan, D.H. *J. Med. Chem.* 1970, **13**, 113.
56 Parham, W.E.; Jones, L.D. *J. Org. Chem.* 1976, **41**, 1187.
57 Nicolson, J.A.; Jarboe, C.H.; Haddon, W.; McLafferty, F.W. *Experientia.* 1968, **24**, 251.
58 Snyder, H.R.; Poos, G.I. *J. Am. Chem. Soc.* 1949, **71**, 1395.
59 Ziegler, K.; Luttringhaus, A. *Liebigs Ann. Chem.* 1934, **511**, 1.
60 Ziegler, K.; Ohlinger, H. *Liebigs Ann. Chem.* 1932, **495**, 84.
61 Bergstrom, F.W.; Agostinho, R. *J. Am. Chem. Soc.* 1945, **67**, 2152.
62 Wadsworth, W.S.; Emmons, W.D. *J. Am. Chem. Soc.* 1961, **83**, 1733.
63 Kaiser, E.M.; Hauser, C.R. *J. Org. Chem.* 1968, **33**, 3402.

64 Chatterjee, A.; Hazra, B.G. *Tetrahedron Lett.* 1969, **10**, 73.
65 Gaudemar, M. *Organomet. Chem. Rev. A.* 1972, **8**, 183.
66 Kaiser, E.M.; Hauser, C.R. *J. Am. Chem. Soc.* 1967, **89**, 4566.
67 Birkofer, L.; Ritter, A.; Widen, H. *Chem. Ber.* 1962, **95**, 971.
68 Zupancic, B.; Mokalj, M. *Synthesis.* 1981, 913.
69 Very few reactions are common to both our studies and the cited work. Reference 63 reports a 78% yield of reaction of benzaldehyde and bromoacetonitrile while Reference 66 reports a 32% yield. Our study includes more varied and sensitive aldehydes.
70 (a) Cu: Fanta, P.E. *Synthesis,* 1974, 9. (b) Fe: Nozaki, H.; Noyori, R. *Tetrahedron.* 1966, **22**, 2163 and references cited therein. (c) Co, Ni: Agnes, G.; Chiusoli, G.P.; Marraccini, A. *J. Organomet. Chem.* 1973, **49**, 239. (d) Ni: Inaba, S.; Matsumoto, H.; Rieke, R.D. *Tetrahedron Lett.* 1982, **23**, 4215. (e) Inaba, S.; Rieke, R.D. ibid. 1983, **24**, 2451. (f) PD: Fischer, E.O.; Bürger, G.; *Z. Naturforsh. B.* 1961, **16**, 702.
71 The only β-halonitriles we have studied are the β-haloacetonitriles.
72 Inaba, S.; Rieke, R.D. *Tetrahedron Lett.* 1985, **26**, 155.
73 Nickel carbonyl-promoted addition of haloacetonitriles to benzaldehyde via similar intermediates was mentioned briefly as the unpublished results by Yoshisato, E.; Abe, T.; Murai, S.; Sonoda, N.; Tsutsumi, S. in M. Ryang, *Organomet. Chem. Rev. A.* 1970, **5**, 67.
74 Tsuji, J. *Organic Synthesis by Means of Transition Metal Complexes*; Springer-Verlag, Berlin/Heidelberg/New York, 1975.
75 Nesmeyanov, A.N.; Zol'nikova, G.P.; Babakhina, G.M.; Kritskaya, I.I.; Yakobson, G.G.; *Obshch, Zh. Khim.* 1973, **43**, 2007.
76 Nakanishi, S.; Oda, T.; Ueda, T.; Otsuji, Y. *Chem. Lett.* 1978, 1309.
77 Rhee, I.; Ryang, M.; Tsutsumi, S. *J. Organomet. Chem.* 1967, **9**, 361.
78 Rhee, I.; Mizuta, N.; Ryang, M.; Tsutsumi, S. *Bull. Chem. Soc. Jpn.* 1968, **41**, 1417.
79 Yoshisato, E.; Tsutsumi, S. *J. Org. Chem.* 1968, **33**, 869.
80 Hirato, Y.; Ryang, M.; Tsutsumi, S. *Tetrahedron Lett.* 1971, **12**, 1531.
81 Seyferth, D.; Spohn, R.J. *J. Am. Chem. Soc.* 1968, **90**, 540; **1969**, *91*, 3037.
82 S-(2-Pyridyl) aromatic thiolates have been used as substrates for the symmetrical ketone synthesis by the reaction with bis(1,5-cyclooctadiene) nickel in DMF; Goto, T.; Onaka, M.; Mukaiyama, T. *Chem. Lett.* 1980, 51.
83 (a) For symmetrical ketone synthesis from carbon monoxide and organometallic compounds catalyzed by transition metals, see Heck, R.F. *J. Am. Chem. Soc.* 1968, **90**, 5546. (b) Seyferth, D.; Spohn, R.J. *J. Am. Chem. Soc.* 1969, **91**, 6192. (c) Tanaka, M. *Tetrahedron Lett.* 1979, **20**, 2601.
84 Kimura, Y.; Tomita, Y.; Nakanishi, S.; Otsuji, Y. *Chem. Lett.* 1979, 321.
85 (a) Hashimoto, I.; Ryang, M.; Tsutsumi, S. *Tetrahedron Lett.* 1969, **10**, 3291. (b) Hashimoto, I.; Tsuruta, N.; Ryang, M.; Tsutsumi, S. *J. Org. Chem.* 1970, **35**, 3748.

86 As for the application of metallic transition metals as reagents to organic synthesis, copper-promoted Ullmann synthesis is typical; however, drastic conditions in the range of 150–280°C are required. Fischer and Bürger reported the preparation of π-allylpalladium bromide by the reaction of palladium sponge with allyl bromide. The Grignard-type addition of allyl halides to aldehydes has been carried out using cobalt or nickel metal prepared by the reduction of cobalt or nickel halides with manganese/iron alloy thiourea. Metallic iron powder has been used for the homocoupling reaction of benzylic halides in water.

87 Okubo, M.; Goto, R. *Nippon Kagaku Zasshi,* 1961, **82**, 261.

88 Heck, R.F. *Organotransition Metal Chemistry. A Mechanistic Approach*; Academic Press, Inc., New York, 1974, Chap. 9.

89 Intermediates, ArNix and Ar_2Ni ($Ar = C_6F_5$), were isolated as their phosphine complexes in the homocoupling reaction of aromatic halides mediated by metallic nickel [99]; see also Matsumoto, H.; Inaba, S.; Rieke, R.D. *J. Org. Chem.* 1983, **48**, 840; see also Kavaliunas, A.V.; Taylor, A.; Rieke, R.D. *Organometallics* 1983, **2**, 377.

90 Shirley, D.A. *Org. Reactions.* 1954, **8**, 28.

91 Posner, G.H. ibid. 1975, **22**, 253.

92 Sato, T.; Naruse, K.; Enokiya, M.; Fujisawa, T. *Chem. Lett.* 1981, 1135.

93 (a) Kosugi, M.; Shimizu, Y.; Migita, T. *Chem. Lett.* 1977, 1423; (b) Milstein, D.; Stille, J.K. *J. Am. Chem. Soc.* 1978, **100**, 3636; (c) idem., *J. Org. Chem.* 1979, **44**, 1613.

94 Takagi, K.; Okamoto, T.; Sakakibara, Y.; Ohno, A.; Oka, S.; Hayama, N. *Chem. Lett.* 1975, 951.

95 Fischer, A.; Grigor, B.A.; Packer, J.; Vaughan, J. *J. Am. Chem. Soc.* 1961, **83**, 4208.

96 Spectral data were consistent with reported values; Wegmann, H.; Schulz, G.; Steglich, W. *Liebigs Ann. Chem.* 1980, 1736.

97 Ichikawa, H.; Ebisawa, Y.; Honda, T.; Kametani, T. *Tetrahedron.* 1985, **41**, 3643.

98 These topics are covered in two review articles: (a) McCullough, J.J. *Acc. Chem. Res.* 1980, **13**, 270. (b) Oppolzer, W. *Synthesis.* 1978, **11**, 793.

99 (a) Kametani, T.; Kato, Y.; Honda, T.; Fukimoto, K. *J. Chem. Soc. Perkin Trans.* **1**. 1975, 2001. (b) Fleming, I.; Gianni, F.L.; Mah, T. *Tetrahedron Lett.* 1976, 881.

100 Flynn, C.R.; Michl, J. *J. Am. Chem. Soc.* 1974, **96**, 3280.

101 (a) Cava, M.P.; Deana, A.A. *J. Am. Chem. Soc.* 1959, **81**, 4266. (b) Sisido, K.; Noyori, R.; Nozaki, H. *J. Am. Chem. Soc.* 1962, **84**, 3562. (c) Oppolzer, W.; Robers, D.A.; Bird, T.G. *Helv. Chim. Acta.* 1979, **62**, 2017.

102 (a) Cava, M.P.; Napier, D.R. *J. Am. Chem. Soc.* 1957, **79**, 1701. (b) Cava, M.P.; Deana, A.A.; Muth, K. *J. Am. Chem. Soc.* 1959, **81**, 6458.

103 Avram, M.; Dinulescu, I.G.; Dinu, D.; Matescu, G.; Nenitzescu, C. D. *Tetrahedron.* 1963, **19**, 309.

104 (a) Alder, K.; Fremery, M. *Tetrahedron.* 1961, **14**, 190.(b) Boudjouk, P.; Han, B.H. *J. Org. Chem.* 1982, **47**, 751.

105 Ito, Y.; Yonezawa, K.; Saegusa, T. *J. Org. Chem.* 1974, **39**, 2769.

106 (a) Sisido, K.; Kusano, N.; Noyori, R.; Nozaki, Y.; Simosaka, M.; Nozaki, H. *J. Polym. Sci. Ser. A.* 1963, **1**, 2101. (b) Nozaki, H.; Noyori, R. *Tetrahedron.* 1966, **22**, 2163.

107 Stephan, D.; Gorgues, A.; Le Coq, A. *Tetrahedron Lett.* 1984, **25**, 5649.

108 Errede, L.A. *J. Am. Chem. Soc.* 1961, **83**, 949.

109 Ito, Y.; Nakatsuka, M.; Saegusa, T. *J. Am. Chem. Soc.* 1982, **104**, 7609.

110 Inaba, S.; Wehmeyer, R.M.; Forkner, M.W.; Rieke, R.D. *J. Org. Chem.* 1988, **53**, 339.

111 (a) Inaba, S.; Rieke, R.D. *J. Org. Chem.* 1985, **50**, 1373. (b) Matsumoto, H.; Inaba, S.; Rieke, R.D. *J. Org. Chem.* 1983, **48**, 840. (c) Inaba, S.; Matsumoto, H.; Rieke, R.D. *J. Org. Chem.* 1984, **49**, 2093.

112 Kerdesky, F.A.J.; Ardecky, R.J.; Lakshmikanthan, M.V.; Cava, M.J. *J. Am. Chem. Soc.* 1981, **103**, 1992.

113 Askari, S.; Lee, S.; Perkins, R.R.; Scheffer, J.R. *Can. J. Chem.* 1985, **63**, 3526.

114 Chappell, S.D.; Cole-Hamilton, D.J. *J. Chem. Soc. Chem. Commun.* 1980, 238.

115 Roth, W.R.; Meier, J.D. *Tetrahedron Lett.* 1967, **8**, 2053.

116 Errede, L.A. *J. Polym. Sci.* 1961, **49**, 253.

117 (a) Schenck, G.O.; Kuhls, J.; Mannsfield, S.; Krauch, C.H. *Chem. Ber.* 1963, **96**, 813. Mp 196–197°C. (b) Takeda, K.; Kitahonoki, K.; Sugiura, M.; Takano, Y. *Chem. Ber.* 1962, **95**, 2344. Mp 196.5–197.5°C.

118 Warrener, R.N.; Russell, R.A.; Lee, T.S. *Tetrahedron Lett.* 1977, **18**, 49. Similar results were reported by Alder and Fremery [104a] using maleic anhydride as the dienophile.

119 Klabunde, K.J. *Chemistry of Free Atoms and Particles*; Academic Press: New York, 1980.

120 (a) The oxidative addition of aryl halide to palladium black is postulated in the catalytic arylation of olefins with aryl halides. For example: Mizoroki, T.; Mori, K.; Ozaki, A. *Bull. Chem. Soc. Jpn.* 1971, **44**, 581. (b) Mori, K.; Mizoroki, T.; Ozaki, A. *Ibid.* 1973, **46**, 1505. (c) Heck, R.F.; Nolley, J.P., Jr. *J. Org. Chem.* 1972, **37**, 2320. (d) Metallic copper is employed in the Ullmann synthesis of biaryls; however, drastic conditions in the range of 150–280°C are required (see Ref. 125a).

121 Fischer, E.O.; Bürger, G. *Z. Naturforsh., B: Anorg. Chem. Org. Chem.* 1961, **16B**, 702.

122 Agnès, G.; Chiusoli, G.P.; Marraccini, A. *J. Organomet. Chem.* 1973, **49**, 239.

123 (a) Rieke, R.D.; Ofele, K.; Fischer, E.O. *J. Organomet. Chem.* 1974, **76**, C19. (b) Rieke, R.D.; Wolf, W.J.; Kujundzic, N.; Kavaliunas, A.V. *J. Am. Chem. Soc.* 1977, **99**, 4159. (c) Rieke, R.D.; Kavaliunas, A.V. *J. Org. Chem.* 1979, **44**, 3069. (d) Rieke, R.D.; Rhyne, L.D. Ibid. 1979, **44**, 3445. (e) Rieke, R.D.; Kavaliunas, A.V.; Rhyne, L.D. *J. Am. Chem. Soc.* 1979, **101**, 246.

124 Mascarelli, L.; Long, B.; Ravera, A. *Gazz. Chim. Ital.* 1938, **68**, 33; *Chem. Abstr.* **1938**, *32*, 4565.

125 (a) Fanta, P.E. *Synthesis*. 1974, 9; *Chem. Rev.* **1974**, *64*, 613. (b) Goshaev, M.; Otroshchenko, O.S.; Sadykov, A.S. *Russ. Chem. Rev. (Engl. Transl.)* 1972, **41**, 12.
126 Semmelhack, M.F.; Helquist, P.M.; Jones, L.D. *J. Am. Chem. Soc.* 1971, **93**, 5908.
127 Tsou, T.T.; Kochi, J.K. *J. Am. Chem. Soc.* 1979, **101**, 7547.
128 Semmelhack, M.F.; Helquist, P.; Jones, L.D.; Keller, L.; Mendelson, L.; Ryono, L.S.; Smith, J.G.; Stauffer, R.D. *J. Am. Chem. Soc.* 1981, **103**, 6460.
129 Klabunde, K.J.; Anderson, B.B.; Bader, M.; Radonovich, L.J. *J. Am. Chem. Soc.* 1978, **100**, 1313.
130 Otsuka, S.; Nakamura, A.; Yoshida, T.; Naruto, M.; Ataka, K. *J. Am. Chem. Soc.* 1973, **95**, 3180.
131 Parshall, G.W. *J. Am. Chem. Soc.* 1974, **96**, 2360.
132 Nakamura, A.; Otsuka, S. *Tetrahedron Lett.* 1974, **15**, 463.
133 Kharasch, M.S.; Reinmuth, O. *Grignard Reactions of Non-metallic Substances*; Constable and Co., Ltd.: London, 1954.
134 Olay, G.A.; Prakash, G.K.S. *Synthesis*. 1976, 607.
135 Ho, T.-L.; Olah, G.A. *Synthesis*. 1977, 170.
136 Okude, Y.; Hiyama, T.; Nozaki, H. *Tetrahedron Lett.* 1977, 3829.
137 Fujiwara, Y.; Ishikawa, R.; Teranishi, S. *Bull. Chem. Soc. Jpn.* 1978, **51**, 589.
138 Hashimoto, I.; Tsuruta, N.; Ryang, M.; Tsutsumi, S. *J. Org. Chem.* 1970, **35**, 3748.
139 Yamada, Y.; Momose, D. *Chem. Lett.* 1981, 1277.
140 Cooper, T.A. *J. Am. Chem. Soc.* 1973, **95**, 4158.
141 Kunieda, T.; Tamura, T.; Takizawa, T. *J. Chem. Soc. Chem. Commun.* 1972, 885.
142 Seyferth, D.; Millar, M.D. *J. Organomet. Chem.* 1972, **38**, 373.
143 Coffey, C.E. *J. Org. Chem.* **1961**, 83, 1623.
144 Alper, H.; Roches, D.D. *J. Org. Chem.* 1976, **41**, 806.
145 Ogata, Y.; Oda, R. *Bull. Inst. Phys. Chem. Res. Tokyo*. 1942, **21**, 616; *Chem. Abstr.* **1949**, *43*, 2194.
146. Nozaki, H.; Noyori, R. *Tetrahedron*. 1966, **22**, 2163 and references cited therein.
147 Buu-Hoi, Ng. Ph.; Hoan, Ng. *J. Org. Chem.* 1949, **14**, 1023.
148 (a) A similar effect was observed in the homocoupling reaction of aromatic bromides mediated by metallic nickel; see Ref. 41. The remarkable effect by iodide ion was also reported in the homocoupling reaction of aryl halides catalyzed by nickel complexes: Zembayashi, M.; Tamao, K.; Toshida, J.; Kumada, M. *Tetrahedron Lett.* 1977, **18**, 4089. (b) Takagi, K.; Hayama, N.; Inokawa, S. *Chem. Lett.* 1979, 917.
149 Furukawa, J.; Matsumura, A.; Matsuoka, Y.; Kiji, J. *Bull. Chem. Soc. Jpn.* 1976, **49**, 829.
150 Ogata, Y.; Nakamura, H. *J. Org. Chem.* 1956, **21**, 1170.
151 (a) Metallic copper is employed in the Ullmann synthesis of biaryls, however drastic conditions in the range of 150–280°C are required: Fanta, P.E.

Synthesis. 1974, 9. (b) Goshaev, M.; Otroshchenko, O.S.; Sadykov, A.S. *Russ. Chem. Rev. (Engl. Transl.)* 1972, **41**, 1046.
152 (a) The oxidative addition of aryl halide to palladium black is postulated in the catalytic arylation of olefin with aryl halide. For example: Mizoroki, T.; Mori, K.; Ozaki, A. *Bull. Chem. Soc. Jpn.* 1971, **44**, 581. (b) Mori, K.; Mizoroki, T.; Ozai, A. *Ibid.* 1973, **46**, 1505. (c) Heck, R.F.; Nolley, Jr., J.P. *J. Org. Chem.* 1972, **37**, 2320.
153 Sawa, Y.; Hashimoto, I.; Ryang, M.; Tsutsumi, S. *J. Org. Chem.* 1968, **33**, 2159.
154 Yoshisato, E.; Ryang, M.; Tsutsumi, S. *J. Org. Chem.* 1969, **34**, 1500.
155 Corey, E.J.; Hegedus, L.S. *J. Am. Chem. Soc.* 1969, **91**, 4926.
156 Hashimoto, I.; Ryang, M.; Tsutsumi, S. *Tetrahedron Lett.* 1970, 4567.
157 Ryang, M.; Toyoda, Y.; Murai, S.; Sonoda, N.; Tsutsumi, S. *J. Org. Chem.* 1973, **38**, 62.
158 Heck, R.F. *Organotransition Metal Chemistry: A Mechanistic Approach*; Academic Press: New York, 1974.

8

Manganese

8.1 Preparation of Rieke Manganese

Manganese metal was first isolated in 1774 by Johann Gahn in Sweden. The free metal was produced by reducing MnO_2 with charcoal. The use of manganese in organic chemistry had to wait until 1937 when Gilman and Bailie reported the preparation of diphenylmanganese and phenylmanganese iodide [1]. The next advance occurred in the 1980s and 1990s primarily by Normant and Cahiez [2, 3]. Much of this work involved the metathesis of organolithium or magnesium reagents with manganese halides. A number of cross-coupling reactions with these reagents were also reported. In the 1990s, a significant advance was reported when we announced the preparation of highly reactive manganese metal using the Rieke method [4, 5]. The highly reactive metal was prepared by reducing manganese halides with lithium and a catalytic amount of naphthalene as an electron carrier. This extremely divided black powder was partially soluble in THF at room temperature (rt). This finely divided metal was found to be one of the most reactive metals toward oxidative addition generated by the Rieke method. In fact, based on a number of reactions involving the most unreactive halides, it can be clearly stated that the manganese metal is the most reactive metal to date. This metal undergoes direct oxidative addition to primary, secondary, and tertiary halides, aryl iodides and bromides, vinyl halides, and even 3-bromothiophene. Moreover, this metal undergoes oxidative addition to a number of carbon–oxygen bonds. The formation of these organomanganese reagents and their cross-coupling reactions will be discussed in the following chapters. One significant feature of these reagents is that for many of the cross-coupling reactions, no catalyst is required.

In a typical preparation, lithium (9.68 mmol), naphthalene (1.48 mmol), and anhydrous manganese chloride (4.71 mmol) under argon were stirred in freshly distilled THF (10 ml) for 3 h at rt.

As a representative procedure, to a slurry of Rieke manganese (10.0 mmol) in THF (10 ml) under argon was added 1-bromododecane (0.80 mmol) at 0°C,

Chemical Synthesis Using Highly Reactive Metals, First Edition. Reuben D. Rieke.

and the mixture was allowed to warm to rt over 1 h. 1,2-Dibromoethane (0.30 mmol) was added neat to the reaction mixture at 0°C and stirred for 20 min. Benzoyl chloride (0.40 mmol) was added neat, and the mixture was stirred at rt for 30 min. The reaction mixture was quenched with 3 M HCl (10 ml) and extracted with ether (2 × 10 ml), and the combined organic layers were sequentially washed with saturated $NaHCO_3$ and NaCl solutions, dried over $MgSO_4$, and concentrated. Flash chromatography (ethyl acetate–hexanes) afforded 1-phenyldodecanone in 68% isolated yield. After the oxidative addition of 0.8 equiv of organohalide to 1.0 equiv of active manganese was totally complete, 0.3 equiv of 1,2-dibromoethane was added to consume the remaining manganese metal. We did not observe the formation of homocoupling product of benzoyl chloride under these reaction conditions.

8.2 Direct Formation of Aryl-, Alkyl-, and Vinylmanganese Halides via Oxidative Addition of the Active Metal to the Corresponding Halide

As we mentioned before, since Rieke manganese is partially soluble in THF, the supernatant containing the organomanganese bromide could not be removed from the manganese by simple settling. In order to achieve a successful cross-coupling reaction, 1,2-dibromoethane was used to remove the excess manganese from the reaction mixture because Rieke manganese will react readily with benzoyl chloride, resulting in the homocoupling product. Primary alkyl bromides readily underwent oxidative insertion in THF to form the organomanganese reagents and coupled with benzoyl chloride to give moderate yields of the product ketones (Table 8.1, entries 1–3). Cyclohexylmanganese bromide reacted with benzoyl chloride at 0°C to rt in 1 h to give the coupling product (Table 8.1, entry 5). However, *sec*-octylmanganese bromide gave a low yield (Table 8.1, entry 4) because *sec*-octylmanganese bromide easily formed alkenes via β-hydride elimination. 1-Bromoadamantane also gave a ketone in low yield due to the difficulty of the formation of the tertiary organomanganese reagent (Table 8.1, entry 6). In contrast to alkylmanganese halides, the reaction of Rieke manganese with 3-bromothiophene afforded 3-substituted thiophene derivatives in good yield (Table 8.1, entry 7). The Gronowitz method, the metal–halogen exchange reaction of 3-bromothiophene with *n*-butyllithium, is the only methodology so far found for the preparation of 3-thienyl organometallics from 3-bromothiophene [6]. No report pertaining to the direct formation of 3-thienyl organometallics via oxidative addition of metals to 3-bromothiophene was reported.

Most benzylic lithium reagents have been prepared via bond cleavage reactions [7] and/or transmetallation reactions [8, 9]. Unfortunately, these

Table 8.1 Formation and coupling reactions of organomanganese bromides.

Entry	Organic bromide	Electrophile	Product[a]	Yield (%)[b]
1	 Br, 5	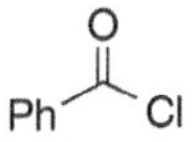 O, Ph, Cl	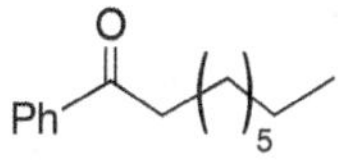 O, Ph, 5	49
2	 Br, 7	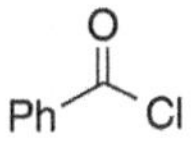 O, Ph, Cl	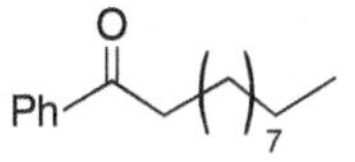 O, Ph, 7	50
3	 Br, 9	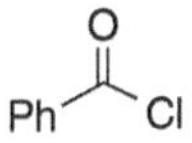 O, Ph, Cl	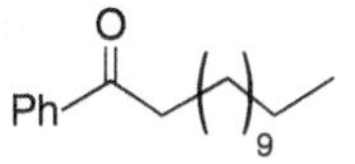 O, Ph, 9	68
4	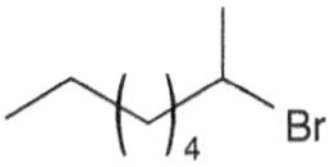 Br, 4	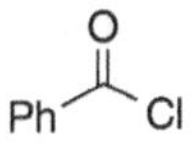 O, Ph, Cl	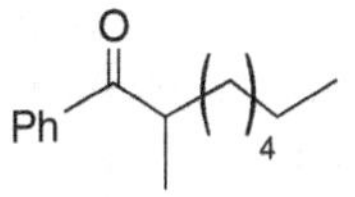 O, Ph, 4	34
5	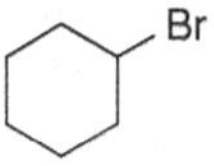 Br	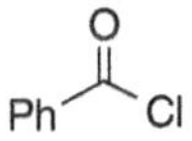 O, Ph, Cl	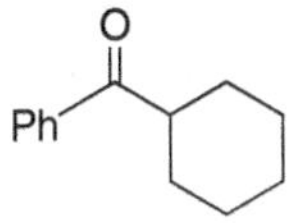 O, Ph	68
6[c]	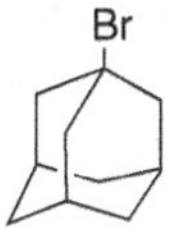 Br	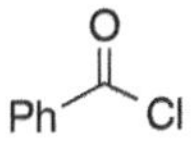 O, Ph, Cl	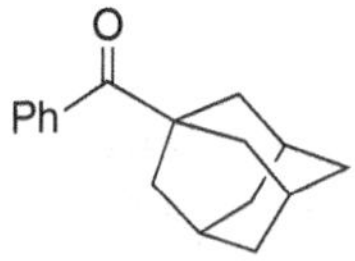 O, Ph	29
7[c]	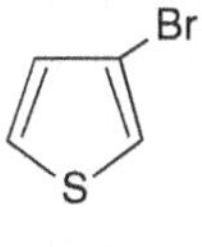 Br, S	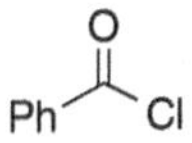 O, Ph, Cl	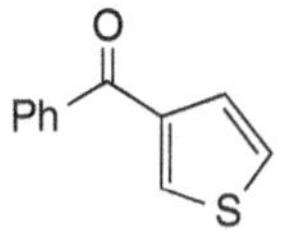 O, Ph, S	87

[a] ^{1}H, ^{13}C NMR, and FTIR were consistent with literature.
[b] Isolated yields (based on electrophile).
[c] MnI_2 was used.

methods are often accompanied by the formation of complex mixtures and homocoupling products even at low temperature [7].

Other metals which undergo oxidative addition to benzylic halides are zinc [10] and cadmium [11]. Recently, an interesting new synthetic method for benzylic zinc reagent has been reported using triorganozincate [12].

We developed an alternative synthetic route for the direct formation of non-functionalized and functionalized benzylic manganese halides from highly reactive manganese and benzylic halides [13].

Treatment of the highly active manganese (Mn*) [13, 14] with benzyl halides (bromide and chloride) gave high yields of the corresponding benzylic manganese halides. The resulting benzylic manganese halides reacted readily with an appropriate electrophile to give the corresponding cross-coupled product. Significantly, the majority of these reactions were carried out without a transition metal catalyst. The environmental and economic advantages are obvious [15].

The oxidative addition of Mn* to benzyl halides was completed at rt in 20 min in THF (Scheme 8.1). Small amounts (3–9%) of the homocoupling product of benzyl halide were observed. This problem was improved by using more highly active manganese [16] prepared from manganese iodide. Trace amounts (<1%) of homocoupling product were formed in this case.

The cross-coupling reaction of the benzylic manganese halides with acid chlorides was carried out at rt in 2 h in THF. It is worthy to note that the coupling reaction was performed in the absence of any transition metal catalyst (except Table 8.2, entry 14). An excess of acid chloride was employed in this reaction to avoid the further reaction of the remaining benzylic manganese halides with the ketone formed. Both aryl and alkyl acid chlorides gave excellent yields (Table 8.2). As shown in Table 8.2, some functionalized benzylic manganese halides (Table 8.2, entries 4–9 and 14) have been obtained as well as nonfunctionalized ones (Table 8.2, entries 1 and 2). From these results, it can be inferred that the present conditions tolerate a wide range of functional groups attached to the benzyl halides. Of special interest is entry 8 in Table 8.2. Preparing organometallics with molecules containing a trifluoromethyl group can be problematic. However, **1g** was readily converted to the corresponding organomanganese reagent, and subsequent cross-coupling proceeded in excellent yield. The oxidative addition will tolerate an electron-withdrawing group such as a carbomethoxy group. In contrast to all the rest of the entries in Table 8.2, the manganese derivative of **1k** does not undergo cross-coupling in the absence of a catalyst. However, in the presence of CuI, product **2m** readily forms. The reason for this is not clear but may be the reduced nucleophilicity caused by the carbomethoxy group (Table 8.2, entry 14). The *p*-cyano derivative only leads to homocoupling under a wide variety of solvents and reaction conditions. Interestingly, treatment of Mn* with α,α′-dichloro-*m*-xylene and the consecutive coupling reaction with benzoyl chloride gave a symmetrical biaryl compound in 79% yield (Table 8.2, entry 10). The reason for the homocoupling is not clear at this moment. Presumably, a competitive reaction

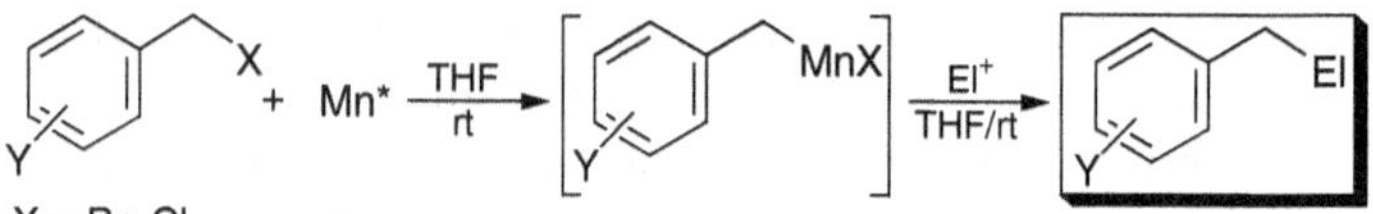

X = Br, Cl
Y = H, Br, Cl, F, OMe, CF_3

Scheme 8.1 Preparation and coupling reaction of benzylic manganese halides.

Table 8.2 Coupling reaction with acid chlorides.[a]

Entry	Halide	Electrophile[b]	Product[c]	Yield (%)[d]
1	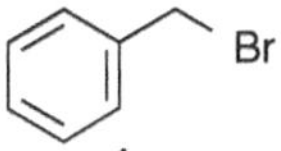 1a	I	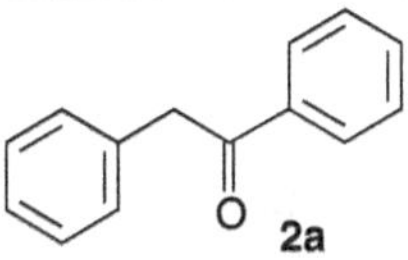 2a	82
2	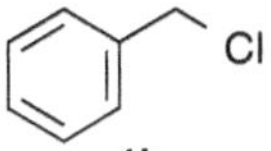 1b	I	2a	91
3	1b	II	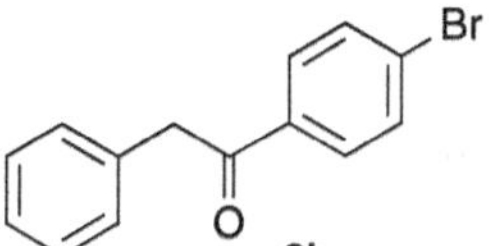 2b	85
4	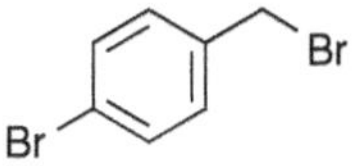 1c	I	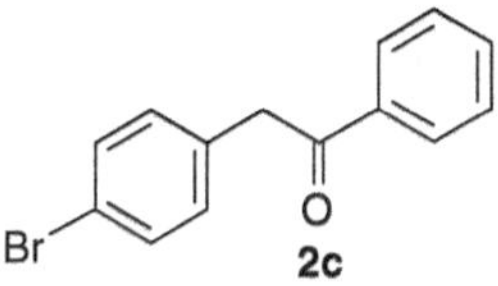 2c	84
5	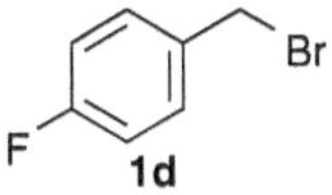 1d	I	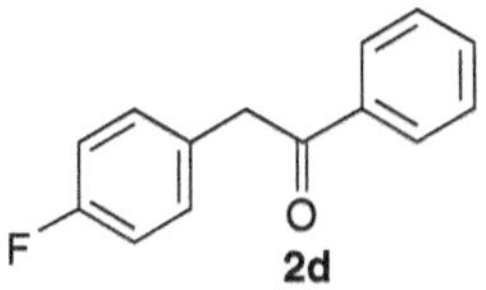 2d	74
6	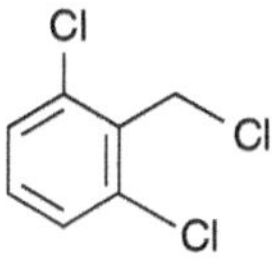 1e	I	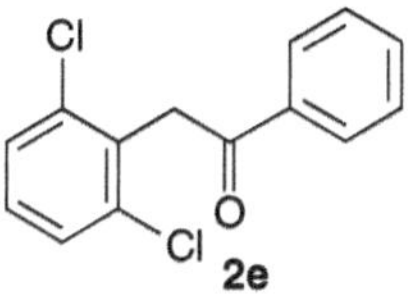 2e	75
7	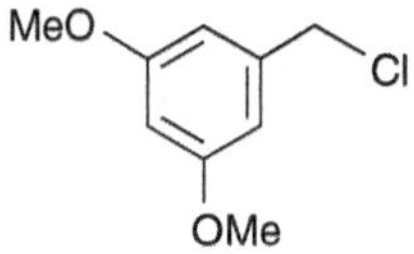 1f	I	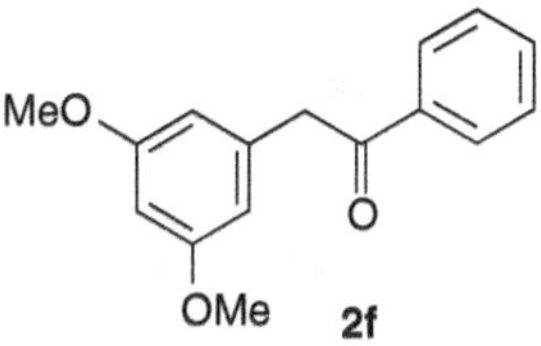 2f	86
8	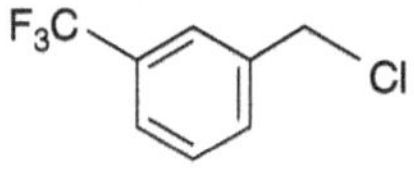 1g	I	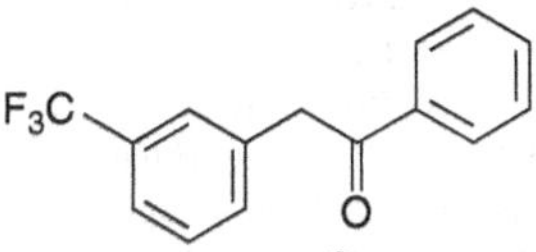 2g	71

(*Continued*)

Table 8.2 (Continued)

Entry	Halide	Electrophile[b]	Product[c]	Yield (%)[d]
9	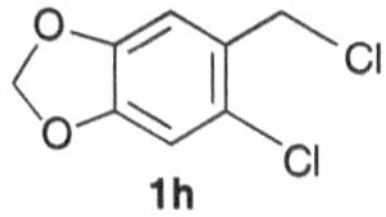 1h	I	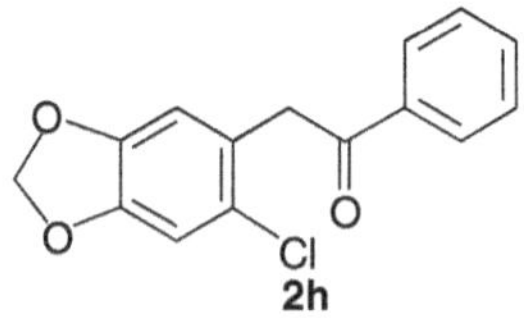 2h	89
10	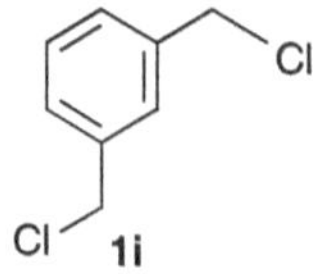 1i	I	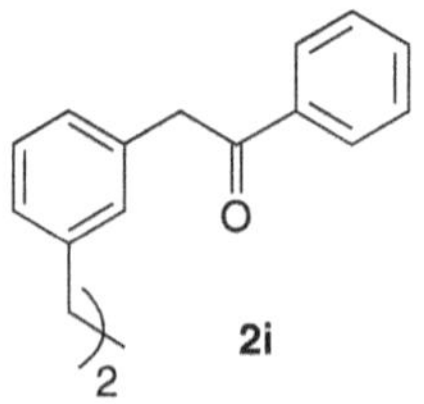 2i	79
11	1a	III	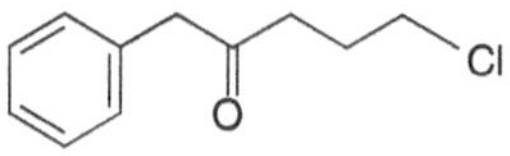 2j	78
12	1b	IV	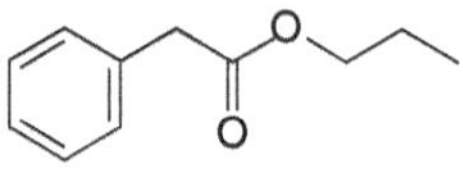 2k	69
13	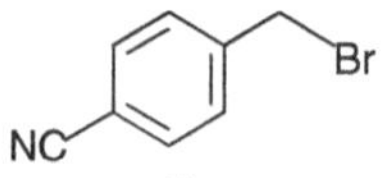 1j	—	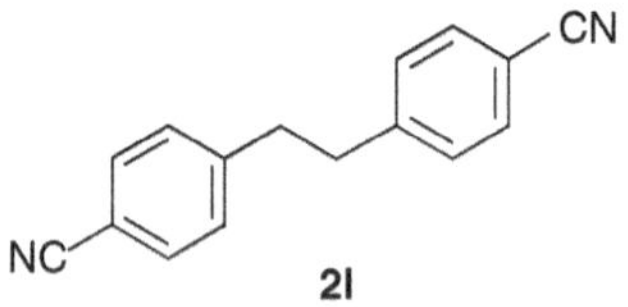 2l	82[e]
14	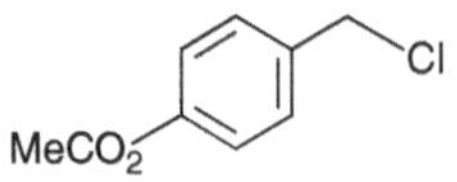 1k	I	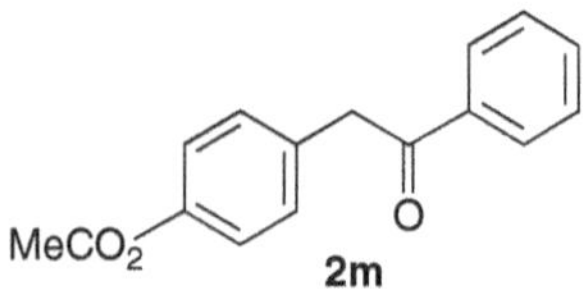 2m	75[f]

[a] Oxidative addition reaction and coupling reaction were carried out at room temperature in THF.
[b] Electrophile; **I**, benzoyl chloride; **II**, *p*-bromobenzoyl chloride; **III**, 4-chlorobutyryl chloride; **IV**, ethyl chloroformate.
[c] All products were fully characterized by ^{1}H, ^{13}C NMR, and HRMS (or EIMS).
[d] Isolated yield (based on benzyl halides).
[e] For recent example of homocoupling product (see Ref. [6]).
[f] 5 mol% CuI was used as a catalyst.

occurred during the following cross-coupling reaction with benzoyl chloride. Unfortunately, in the reaction of **1j**, the homocoupling product **2l** was obtained during the oxidative addition. The benzylic manganese halides were found to add to several other electrophiles including aldehydes, ketones, and di-*tert*-butyl azodicarboxylate (DBAD). The results are summarized in Table 8.3. Addition to aldehydes (Table 8.3, entries 1–4 and 7) gave the corresponding secondary alcohols in good yields (78–93%) even with bulky aldehydes (Table 8.3, entry 8). The reaction tolerated halides or a nitrile group in the aldehydes but not a nitro group (Table 8.3, entry 10). The addition to an alkyl ketone yielded the corresponding tertiary alcohol in good yield (Table 8.3, entry 5). Coupling with acetophenone was successful. However, the yield was uncharacteristically low. DBAD was also employed as an electrophile, and the corresponding coupling product **3i** was obtained in excellent yield (80%).

Benzylic manganese halides were also found to undergo palladium-catalyzed cross-coupling reactions with aryl iodides [10a]. As shown in Scheme 8.2, the corresponding coupling compounds **4a** and **4b** were achieved in moderate to good yields (56 and 78%, respectively) using 5 mol% palladium catalyst ($Pd(PPh_3)_4$) in THF at rt.

In summary, a facile route to benzylic manganese reagents has been developed. The starting benzylic halide can contain a wide variety of substituents, and the subsequent cross-coupling proceeds with good to excellent yields with several electrophiles.

As described in Section 8.1 by Cahiez, both alkylmanganese and arylmanganese reagents have been prepared by the metathesis of organolithium and/or Grignard reagents with manganese halides such as $MnBr_2$, $MnCl_2$, and MnI_2. In these reports, various types of alkylmanganese reagents have been studied to understand the reactivities of organomanganese reagents. In contrast, few examples of using arylmanganese halides were performed.

While our studies were in progress, the direct synthesis of arylmanganese reagents including allyl- and alkenylmanganeses was reported using active manganese–graphite which was prepared from the reduction of the soluble ate-complex $MnBr_2 \cdot nLiBr$ ($n = 1, 2$) with two C_8K in THF [17].

We reported the direct synthesis of arylmanganese bromides via the direct oxidative addition of Rieke manganese (Mn*) to the corresponding aryl bromides under mild reaction conditions. These manganese bromide reagents undergo Cu(I)-catalyzed coupling reaction with acid chlorides and phenyl isocyanate to give the corresponding ketones and amides in moderate to good isolated yields (Tables 8.5 and 8.6).

As has been observed for some of our active metals, the choice of counter ion is critical in determining the ultimate reactivity of the active manganese. The results of these studies are presented in Scheme 8.3 and Table 8.4. The salt giving the highest reactivity is MnI_2, with $MnCl_2$ giving the slowest reactivity. The reactions were run with a 2 : 1 excess of manganese. Higher ratios reacted

Table 8.3 Cross-coupling reaction of benzylic manganese halides.[a]

Entry	Halide	Electrophile	Product[b]	Yield (%)[c]
1	1a	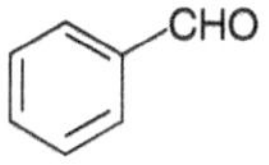 CHO	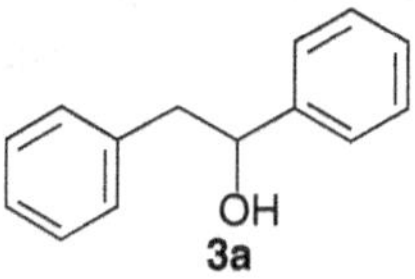 OH **3a**	93
2	1a	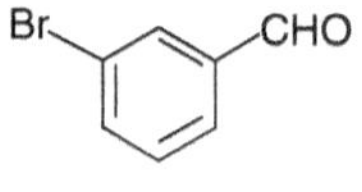 Br CHO	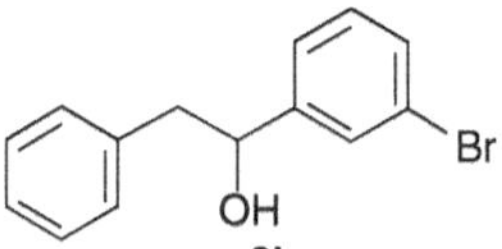 Br OH **3b**	95
3	1a	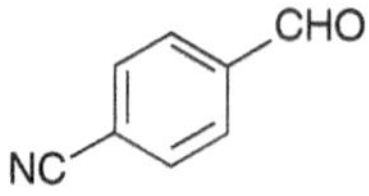 CHO NC	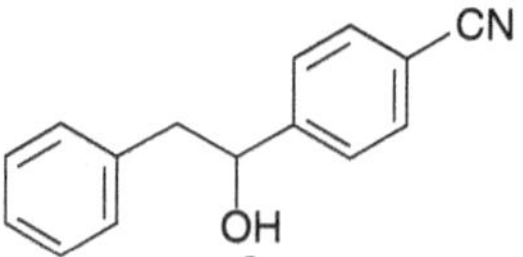 CN OH **3c**	87
4	1a	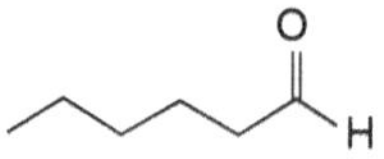 O H	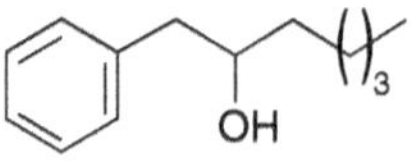 3 OH **3d**	78
5	1a	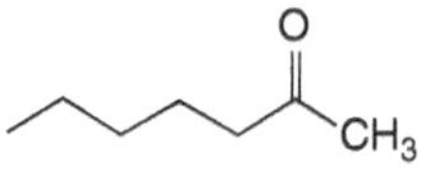 O CH_3	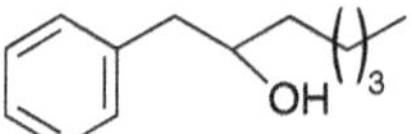 OH 3 **3e**	72
6	1a	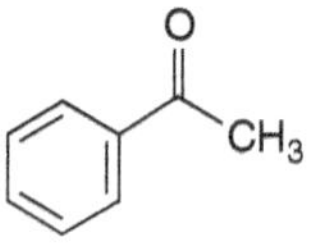 O CH_3	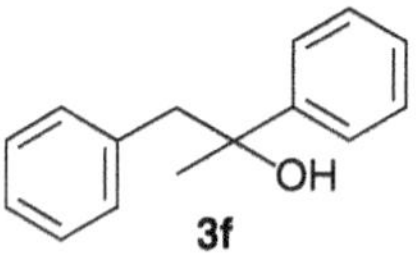 OH **3f**	46
7	1c	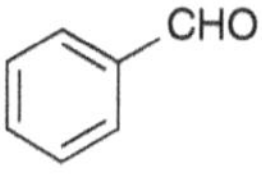 CHO	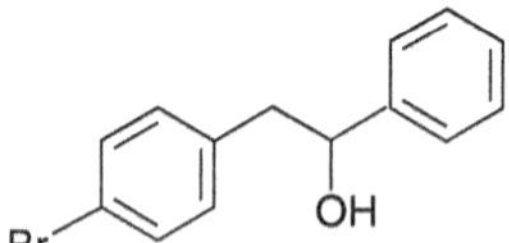 OH Br **3g**	80
8	1c	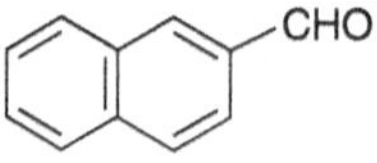 CHO	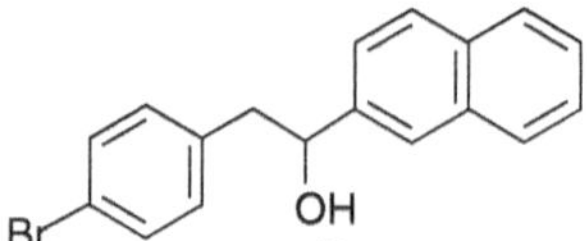 Br OH **3h**	89
9	1a	t-BuO_2C–N=N–CO_2Bu-t	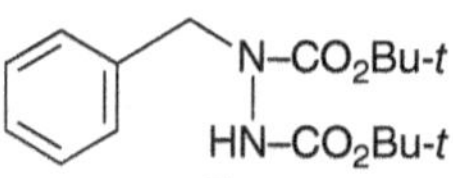 N–CO_2Bu-t HN–CO_2Bu-t **3i**	80
10	1a	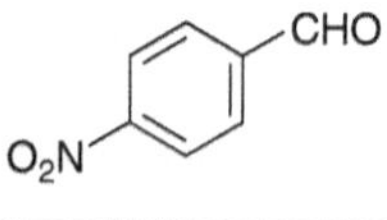 CHO O_2N	—	0[d]

[a] Oxidative addition reaction and coupling reaction were carried out at room temperature in THF.
[b] All products were fully characterized by ^1H, ^{13}C NMR, and HRMS (or EIMS).
[c] Isolated yields (based on electrophile).
[d] According to TLC analysis, no coupling reaction occurred.

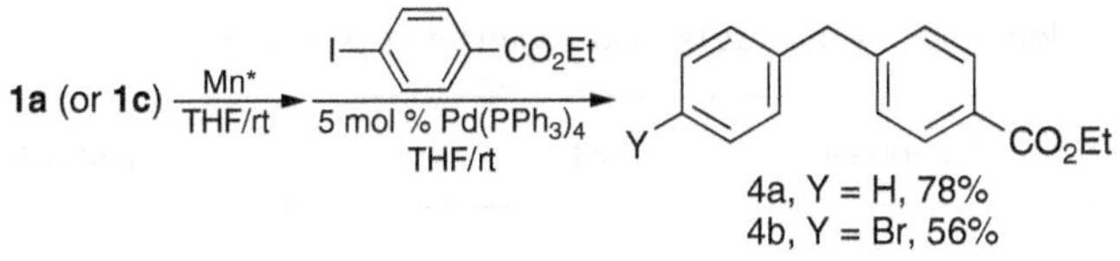

Scheme 8.2 Palladium-catalyzed coupling reaction.

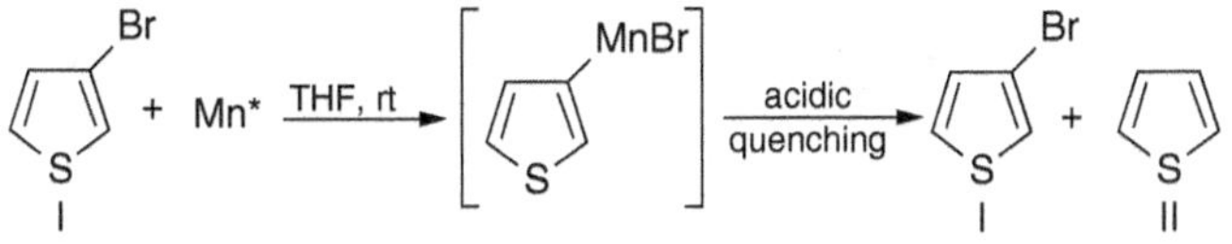

Scheme 8.3 Reaction of manganese with bromothiophene.

Table 8.4 Reactivity of Mn*.

Metal	Equivalent	Ratio[a] II:I						
MnX_2	Mn*:I	0.5 h	1 h	2 h	3 h	8 h	24 h	48 h
MnI_2	2:1	68:32	75:25	80:20	86:14	—	—	—
	3:1	—	96:4	97:3	97:3	—	—	—
$MnBr_2$	2:1	21:79	50:50	53:47	58:42	—	70:30	73:27
	4:1	—	97:3	97:3	—	—	—	—
$MnCl_2$	2:1	—	32:68	39:61	—	47:53	56:44	—
	4:1	—	—	66:34	—	78:22	85:25	93:7

[a] Ratio was determined by GC analysis.

even faster as expected. At this point, we cannot rule out the possibility that halogen exchange is occurring and that this is the primary explanation for the iodide salt being the most reactive. We have not observed any of the possible halide exchanged aryl halides in our GC traces during reaction.

Coupling reactions of the arylmanganese reagents with acid chlorides were performed in the presence of a catalytic amount of Cu(I) catalyst at −10 to −5°C in 6 h. As shown in Table 8.5, the Cu(I)-catalyzed coupling reaction afforded the corresponding ketones in moderate isolated yields. In these reactions, electron-donating groups on the aryl halide gave disappointing yields (Table 8.5, entries 3 and 4). For instance, treatment of 4-phenoxyphenyl manganese bromide and 4-methylphenylmanganese bromide with benzoyl chloride afforded 4-phenoxybenzophenone and 4-methylbenzophenone in 29 and 34% isolated yields, respectively. In contrast, coupling reaction of arylmanganese bromides

Table 8.5 Coupling reaction of arylmanganese bromides with benzoyl chloride.

Entry	Arylbromide	Electrophile	Product	Yield (%)[a]
1	F–C$_6$H$_4$–Br	PhC(O)Cl	F–C$_6$H$_4$–C(O)–Ph	52
2	3,5-F$_2$C$_6$H$_3$–Br	I	3,5-F$_2$C$_6$H$_3$–C(O)–Ph	74
3	PhO–C$_6$H$_4$–Br	I	PhO–C$_6$H$_4$–C(O)–Ph	29
4	H$_3$C–C$_6$H$_4$–Br	I	H$_3$C–C$_6$H$_4$–C(O)–Ph	34
5	3-Cl–C$_6$H$_4$–Br	I	3-Cl–C$_6$H$_4$–C(O)–Ph	79
6	Cl–C$_6$H$_4$–Br	I	Cl–C$_6$H$_4$–C(O)–Ph	83
7	2-Bromonaphthalene	I	2-Naphthyl–C(O)–Ph	36
8	F$_3$C–C$_6$H$_4$–Br	4-Br–C$_6$H$_4$C(O)Cl	F$_3$C–C$_6$H$_4$–C(O)–C$_6$H$_4$–Br	60

[a] Isolated yields (based on electrophile).

which have electron-withdrawing groups with benzoyl chloride resulted in the formation of ketones in good yields (Table 8.5, entries 1, 2, 5, 6, and 8).

We have found that the oxidative addition of Mn* to 2-bromonaphthalene was completed at rt in 1 h. The resulting 2-naphthylmanganese bromide was followed by a coupling reaction with benzoyl chloride to give the 2-naphthophenone in the presence of a Cu catalyst (Table 8.5, entry 7). It was observed that homocoupling of the organomanganese reagent also occurred along with the cross-coupling reaction with the electrophile. 2,2′-Binaphthyl was obtained as well as the ketone 2-naphthophenone. The isolated yield of this reaction was not high. It appears that the homocoupling reaction begins to take place before the cross-coupling reaction is completed.

The coupling reaction with phenyl isocyanate was also examined. The arylmanganese bromide reagents underwent a coupling reaction with phenyl isocyanate without any transition metal catalyst to yield the corresponding amides in moderate yields (Table 8.6).

In conclusion, arylmanganese bromide reagents including functionalized arylmanganese bromides were easily prepared via the direct oxidative addition of Rieke manganese to the corresponding aryl bromides under mild conditions. These resulting organomanganese reagents undergo a coupling reaction with electrophiles to give the C—C bond-forming products.

The following is a representative procedure. To a slurry of Rieke manganese* (10.0 mmol) in THF (10 ml) under argon was added 1-bromo-3,5-difluorobenzene (5.0 mmol) at rt, and the mixture was stirred at rt for 1 h. 1,2-Dibromoethane (6.0 mmol) was added neat to the reaction mixture at 0°C, and the mixture was allowed to warm to rt over 20 min. The resulting 3,5-difluorophenyl manganese bromide reagent was transferred via cannula to a flask containing a THF solution of Cu(I) catalyst (0.04 mmol) at −10°C. Benzoyl chloride (4.0 mmol) was added neat to this mixture at −10°C. The resulting mixture was stirred at −10 to −5°C for 6 h. The reaction mixture was quenched with 3 M HCl (10 ml) solution and extracted with ether (2 × 10 ml), and the combined organic layers were sequentially washed with saturated $NaHCO_3$, $Na_2S_2O_3$, and NaCl solutions, dried over $MgSO_4$, and concentrated. Flash chromatography (ethyl acetate–hexanes) afforded 3,5-difluorobenzophenone in 74% isolated yield.

Table 8.6 Coupling reaction of arylmanganese bromides with phenyl isocyanate.

Entry	Arylbromide	Product	Yield (%)[a]
1	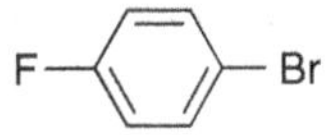 F–C6H4–Br (4-fluorophenyl bromide)	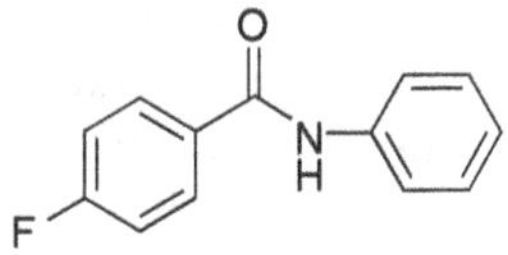 4-F-C6H4–C(=O)–NH–Ph	34
2	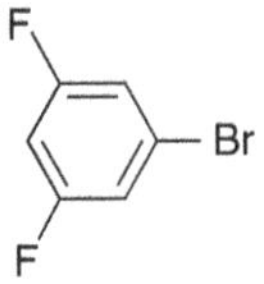 3,5-F2C6H3–Br	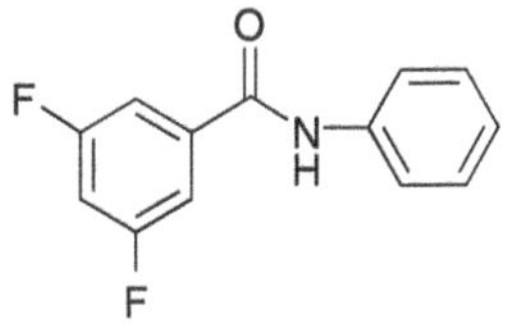 3,5-F2C6H3–C(=O)–NH–Ph	48

[a] Isolated yields (based on electrophile).

8.3 Direct Formation of Organomanganese Tosylates and Mesylates and Some Cross-Coupling Reactions

Until recently, the most widely used approach for the preparation of organomanganese compounds was the metathesis of organolithium or Grignard reagents [18]. This approach allows limited functionality in the organic moiety. Recently, the advent of Rieke manganese* [19] and Mn graphite [20] has allowed the preparation of organomanganese reagents via direct oxidative addition of manganese metal to carbon–halogen bonds. This approach tolerates a much wider spectrum of functional groups in the organic moiety. In fact, the highly reactive manganese (Mn*) prepared via the lithium–naphthalene reduction method represents the most reactive metal toward oxidative addition of all the metals prepared by this method [21]. Considering the high reactivity of the active manganese metal prepared by this approach, we explored the possibility of the oxidative addition to carbon–oxygen bonds. Few reports exist on the oxidative addition of main group or even transition metals to carbon–oxygen bonds [22]. However, the synthetic advantages to such an approach are obvious as the conversion of alcohols into halides can be extremely problematic for many functional group combinations. Of particular interest would be the availability of tosylates and mesylates as precursors to organomanganese reagents (Scheme 8.4).

We reported that Mn* does, in fact, undergo oxidative addition to alkyl and benzylic tosylates and mesylates under mild conditions to yield the corresponding organomanganese tosylates and mesylates in excellent yields. To our knowledge, this is the first example of organomanganese tosylates and mesylates prepared via direct oxidative addition of manganese metal or by any metathesis approach.

To investigate the new protocol, a variety of benzyl sulfonates [23] were examined with active manganese [24]. The oxidative addition of active manganese to benzyl mesylates was easily completed at rt in only 30 min. The resulting benzylic manganese mesylates were cross-coupled with several electrophiles. The coupling reactions were completed under mild conditions and significantly in the absence of any transition metal catalyst. With the preliminary results from nonfunctionalized benzylic mesylates, functionalized benzylic mesylates containing a halogen atom have been investigated to expand this methodology. The results are summarized in Table 8.7.

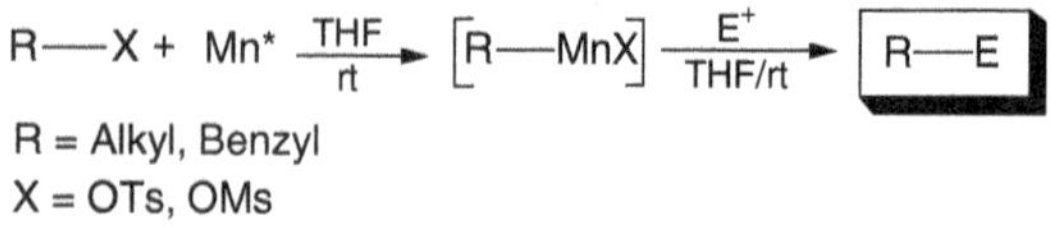

$$\mathrm{R{-}X + Mn^{*}} \xrightarrow[\text{rt}]{\text{THF}} [\mathrm{R{-}MnX}] \xrightarrow[\text{THF/rt}]{\mathrm{E^{+}}} \boxed{\mathrm{R{-}E}}$$

R = Alkyl, Benzyl
X = OTs, OMs

Scheme 8.4 Proposed reaction scheme.

Table 8.7 Cross-coupling reaction of benzylic and functionalized benzylic manganese sulfonates.

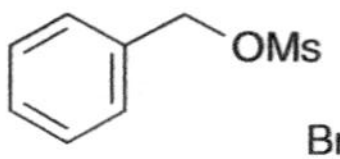 2a	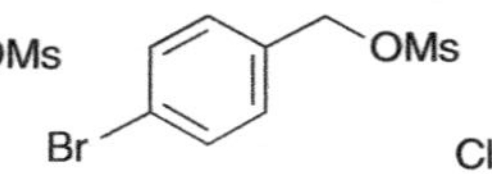 2b	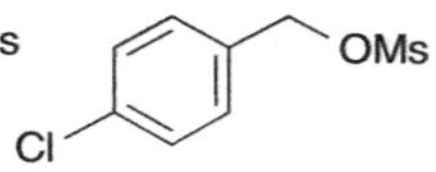 2c	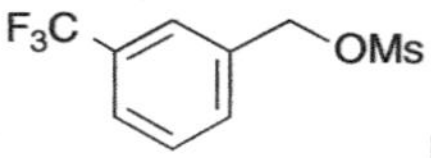 2d	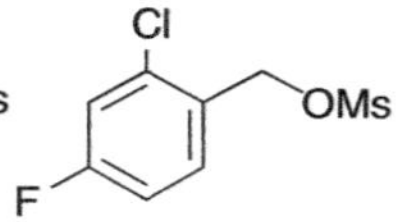 2e	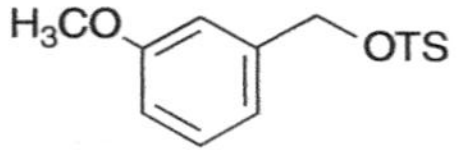 2f
OMs	Br, OMs	Cl, OMs	F_3C, OMs	Cl, F, OMs	H_3CO, OTS

Entry	Sulfonate[a]	E[b]	Product[c]	Yield (%)[d]
1	**2a**	I[e]	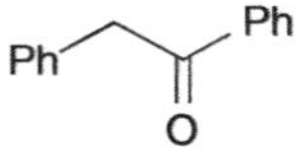 Ph, Ph, O	63[e]
2	**2a**	II	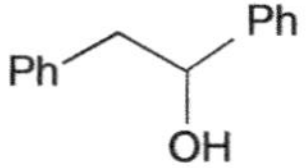 Ph, Ph, OH	72
3	**2a**	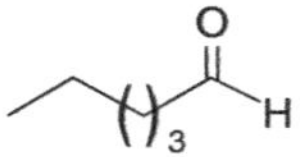 O, H, ( )3	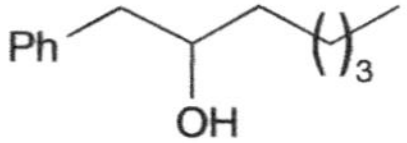 Ph, OH, ( )3	50
4	**2a**	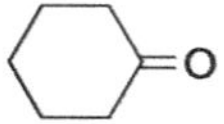 O	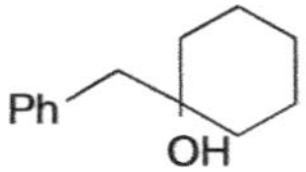 Ph, OH	29
5	**2a**	$PhCOCH_3$	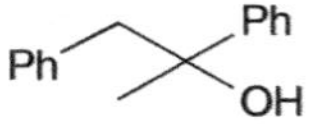 Ph, Ph, OH	36
6	**2b**	I[e]	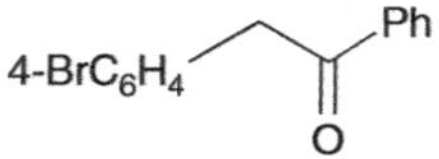 $4\text{-}BrC_6H_4$, Ph, O	50[e]

(*Continued*)

Table 8.7 (Continued)

Entry	Sulfonate[a]	E[b]	Product[c]	Yield (%)[d]
7	**2b**	**II**	$4\text{-}BrC_6H_4CH_2CH(OH)Ph$	90
8	**2b**	**III**	$4\text{-}BrC_6H_4CH_2C(CH_3)(OH)Ph$	80
9	**2c**	**II**	$4\text{-}ClC_6H_4CH_2CH(OH)Ph$	95
10	**2c**	**III**	$4\text{-}ClC_6H_4CH_2C(CH_3)(OH)Ph$	92
11	**2d**	**II**	$3\text{-}CH_3OC_6H_4CH_2CH(OH)Ph$	89
12	**2e**	**II**	$2\text{-}Cl\text{-}4\text{-}FC_6H_3CH_2CH(OH)Ph$	92
13	**2f**	**II**	$3\text{-}CH_3OC_6H_4CH_2CH(OH)Ph$	94

[a] For the preparation, see Ref. [23].
[b] Electrophile; **I**, benzoyl chloride; **II**, benzaldehyde; **III**, acetophenone.
[c] All of the products were fully characterized by ^1H, ^{13}C NMR, FTIR, and HRMS.
[d] Isolated yields (based on electrophile unless otherwise mentioned).
[e] Excess benzoyl chloride was used (yield was based on mesylate).

From these results, it can be inferred that the present conditions tolerate halogen atoms attached to the aryl group of the benzyl mesylates. Of special interest is entry 11 in Table 8.7; clearly this approach tolerates the presence of a trifluoromethyl group.

As shown in Table 8.8, the oxidative addition to alkyl tosylates [25] was also investigated. The oxidative addition was easily completed by simple treatment

Table 8.8 Preparation of organomanganese tosylates and their coupling reaction.

Entry	Tosylate	E[a]	Product[b]	Yield (%)[c]
1	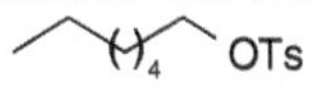 $()_4$ OTs **1a**	I	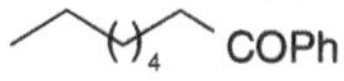 $()_4$ COPh	48
2	1a	II	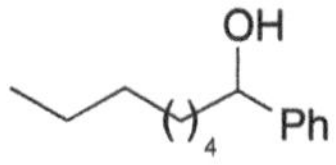 OH, $()_4$, Ph	64
3	 $()_7$ OTs **1b**	I	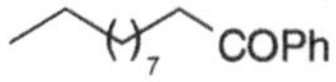 $()_7$ COPh	57
4	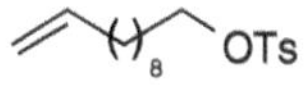 $()_8$ OTs **1c**	I	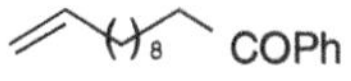 $()_8$ COPh	44
5	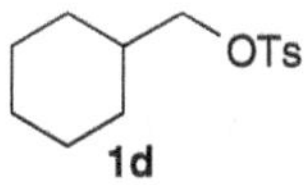 OTs **1d**	I	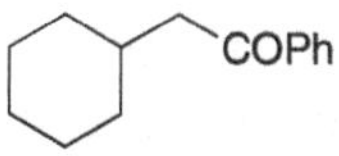 COPh	86
6	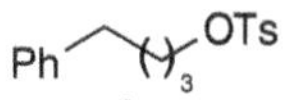 Ph $()_3$ OTs **1e**	I	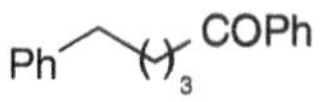 Ph $()_3$ COPh	48
7	1e	II	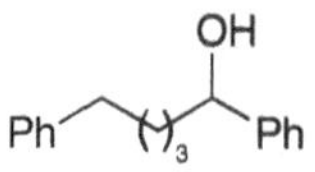 OH, Ph $()_3$ Ph	36
8	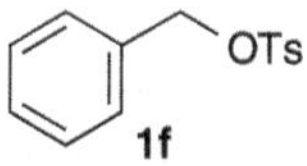 OTs **1f**	I	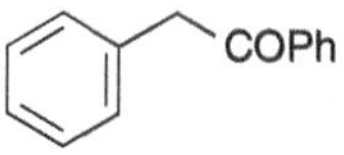 COPh	74[d]
9	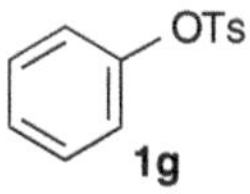 OTs **1g**	—		0[e]

[a] Electrophile; I, benzoyl chloride; II, benzaldehyde.
[b] All of the products were fully characterized by ^{1}H, ^{13}C NMR, FTIR, and HRMS.
[c] Based on electrophile unless otherwise mentioned.
[d] Excess benzoyl chloride was used (yield was based on tosylate).
[e] No oxidative addition occurred.

of Mn* with the alkyl tosylates at rt in THF. Unfortunately, no reaction was found to occur with phenyl tosylate even at refluxing temperatures (Table 8.8, entry 9). The formation of organomanganese tosylates was also confirmed by the subsequent cross-coupling reaction with different electrophiles. The tosylates listed in Table 8.8 reacted with benzoyl chloride to yield the corresponding ketones. The coupling reactions were carried out at rt in THF in the absence of any transition metal catalyst. The results are summarized in Table 8.8. Moderate to excellent yields (44–86%) were obtained. In contrast to all the primary tosylates, a low yield (17%) was obtained from a secondary tosylate. This may be due to the steric hindrance and/or easy elimination of organomanganese tosylates yielding alkenes. This aspect needs further examination. Also, cross-coupling reactions with benzaldehyde gave the corresponding alcohols in moderate yields (36–64%) [26].

In conclusion, a new synthetic route to organomanganese tosylate reagents has been developed. These highly useful reagents can be readily prepared via direct oxidative addition of highly active manganese to the corresponding tosylates under mild conditions. The approach works equally well for primary as well as benzylic tosylates and mesylates. Finally, it should be mentioned that the cross-coupling reaction of the resulting organomanganese reagents is carried out in the absence of any transition metal catalyst.

8.4 Benzylic Manganese Halides, Sulfonates, and Phosphates: Preparation, Coupling Reactions, and Applications in New Reactions

Introduction

Benzylic organometallic reagents have been playing an important role in organic synthesis, especially in homologation of organometallics. In spite of their significance, developments of synthetic methods for the preparation of benzylic metal reagents have been limited.

One of the earliest studies of benzylic metal chemistry concerns benzylic lithium reagents. Most benzylic lithium reagents have been prepared via bond cleavage reactions [27] and/or transmetallation reactions [28]. However, in general, these methods are often accompanied by the formation of complex mixtures and homocoupling products even at low temperature [27]. Similar problems are frequently observed in the preparation of benzylic Grignard reagents. The magnesium anthracene complex involving an electron-transfer mechanism has been extensively used to alleviate these problems [29]. In some cases, more efficient direct synthetic methods utilizing the oxidative addition of a metal to the corresponding benzyl halides have been developed including activated zinc [30] and cadmium [31a]. However, a copper catalyst was needed to complete the cross-coupling reaction of benzylic zinc halides with electrophiles. An interesting new synthetic procedure for preparing benzylic zinc reagent has been reported using triorganozincate and 4-iodobenzyl mesylates [12]. This system appears very useful for the preparation of 4-alkyl-substituted benzylic zinc reagents.

Even though some practical synthetic pathways have been developed using active zinc with benzylic halides [31b, c], more facile approaches would be of value to organic synthesis. We reported an alternative synthetic route for the direct formation of nonfunctionalized and functionalized benzylic manganese halides from the oxidative addition reaction of highly active manganese to benzylic halides [32]. More significantly, benzylic manganese sulfonates and phosphates were prepared by the reaction on Mn* (prepared by the Rieke method) with the corresponding benzylic sulfonates and phosphates. The resulting benzylic manganese reagents were found to undergo a variety of cross-coupling reactions. It was also found that homocoupled products of functionalized benzyl halides could be readily prepared depending on reaction conditions.

Results and Discussion

Preparation and Coupling Reactions of Benzylic Manganese Halides

Highly activated manganese metal (Mn*) can be prepared via the lithium–naphthalene reduction method in freshly distilled THF at rt [33]. Treatment of the highly active manganese (Mn*) with benzyl halides (Br and Cl) gave high yields of the corresponding benzylic manganese halides. The resulting benzylic manganese halides reacted readily with a variety of electrophiles to give the corresponding products. It is worthy to note that most of these coupling reactions were carried out in the absence of a transition metal catalyst. The environmental advantage of carrying out these reactions without a transition metal is significant.

While the oxidative addition of Mn* prepared from manganese bromide and chloride to benzyl halides was completed at rt in 20 min in THF (Scheme 8.5),

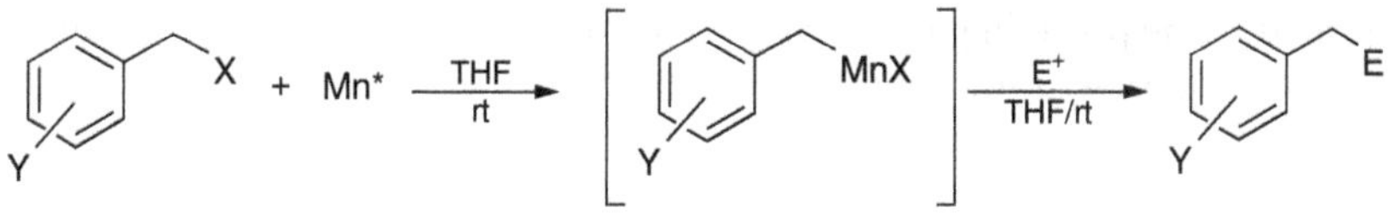

X + Mn* —THF, rt→ [MnX] —E⁺, THF/rt→ E

X = Br, Cl
Y = H, Br, Cl, F, OMe, CF_3, CH_3CO_2

Scheme 8.5 Preparation and coupling reaction of benzylic manganese halides.

small amounts (3–9%) of homocoupling products of benzyl halides were observed. However, the use of manganese iodide [34] to prepare active Mn* alleviated the problem. Less than 1% of homocoupling products were formed in these cases.

The benzylic manganese halides were reacted with acid chlorides to obtain the cross-coupling products, and the results are summarized in Table 8.9. The cross-coupling reactions with acid chlorides were carried out at rt and were completed in 30 min in THF in the absence of any transition metal catalyst (except Table 8.9, entry 13). An excess of acid chloride was employed in these reactions to avoid the addition reaction of the remaining benzylic manganese halides to the ketones formed. Both aryl (I, II) and alkyl (III, IV) acid chlorides gave excellent yields (Table 8.9). Some functionalized benzylic manganese halides (Table 8.9, entries 4–9 and 13) have been prepared as well as nonfunctionalized ones (Table 8.9, entries 1 and 2). Of special interest is entry 8 in Table 8.9. Preparing organometallic reagents with compounds containing a trifluoromethyl group can be problematic. However, **1g** was readily converted to the corresponding organomanganese reagent. Subsequent cross-coupling reactions proceeded in excellent yield. The oxidative addition tolerates an electron-withdrawing group such as a carbomethoxy group (Table 8.9, entry 13). In contrast to the rest of the entries in Table 8.9, the manganese derivative of **1j** does not undergo cross-coupling reactions in the absence of a catalyst. However, product **2l** readily formed in the presence of a catalytic amount of CuI. Interestingly, treatment of Mn* with α,α'-dichloro-*m*-xylene (**1i**) and the consecutive coupling reaction with benzoyl chloride gave a symmetrical biaryl compound **2i** in 79% yield (Table 8.9, entry 10).

The benzylic manganese halides were also found to react with other electrophiles such as aldehydes, ketones, and DBAD. The results are shown in Table 8.10. Addition to aldehydes (Table 8.10, entries 1–4 and 7) gave the corresponding secondary alcohols in good yields (78–93%). The reaction tolerated halides or a nitrile group (Table 8.10, entry 2 or 3) in the aldehyde but not a nitro group. The addition to an alkyl ketone yielded the corresponding tertiary alcohol **3e** in good yield (Table 8.10, entry 5). But coupling with acetophenone afforded the tertiary alcohol in low yield (Table 8.10, entry 6). DBAD was also employed as an electrophile, and the corresponding coupling product **3i** was obtained in excellent yield (80%). It should be noted that the coupling reactions described in the preceding text have been readily accomplished in the absence of any catalyst under mild conditions.

Preparation and Coupling Reactions of Benzylic Manganese Sulfonates and Phosphates

In 1992, Yus et al. reported that allylic and benzylic mesylates reacted with lithium naphthalenide to give the corresponding organolithium reagents, giving cross-coupling products upon treatment with electrophiles [35, 35a].

Table 8.9 Coupling reaction with acid chlorides.[a]

Entry	Halide	Electrophile[b]	Product[c]	Yield (%)[d]
1	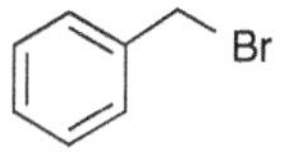 Br **1a**	I	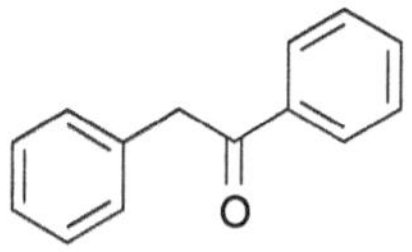 O **2a**	82
2	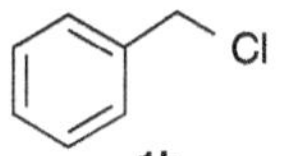 Cl **1b**	I	**2a**	91
3	**1b**	II	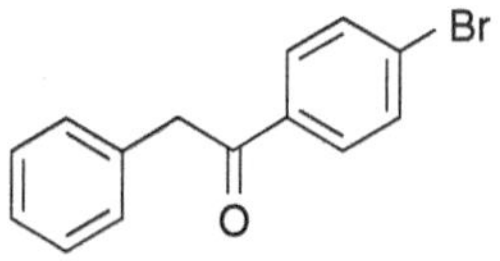 Br O **2b**	85
4	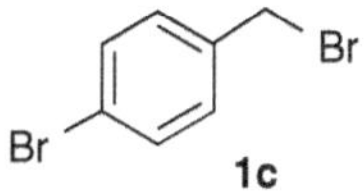 Br Br **1c**	I	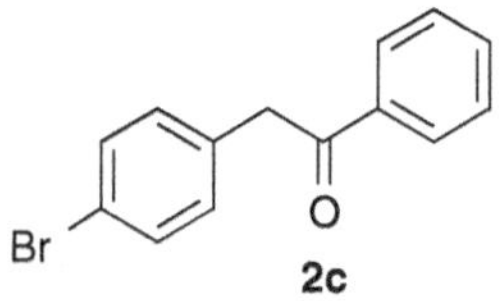 O Br **2c**	84
5	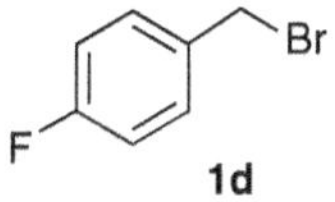 Br F **1d**	I	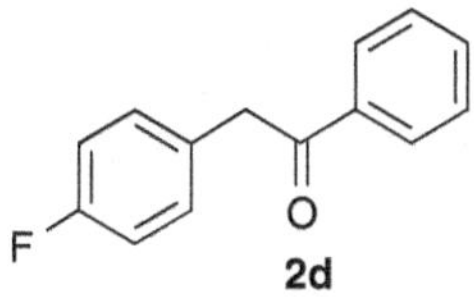 O F **2d**	74
6	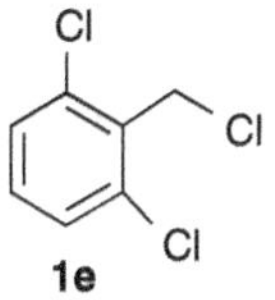 Cl Cl Cl **1e**	I	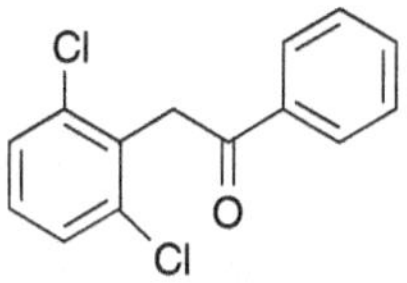 Cl O Cl **2e**	75
7	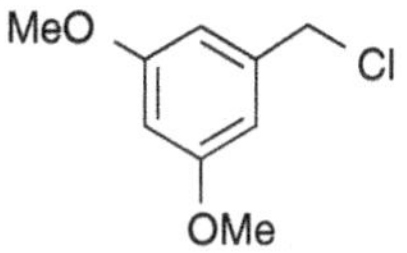 MeO Cl OMe **1f**	I	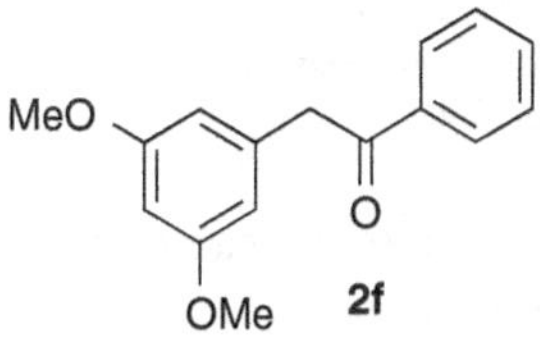 MeO O OMe **2f**	86
8	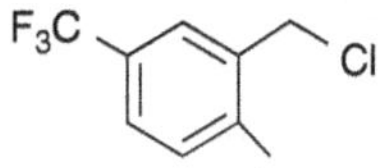 F_3C Cl **1g**	I	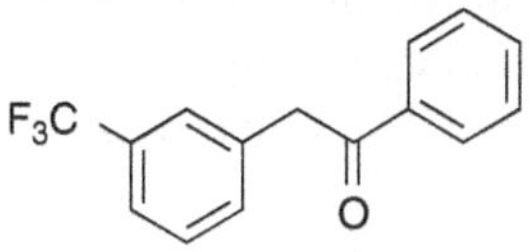 F_3C O **2g**	71

(*Continued*)

Table 8.9 (Continued)

Entry	Halide	Electrophile[b]	Product[c]	Yield (%)[d]
9	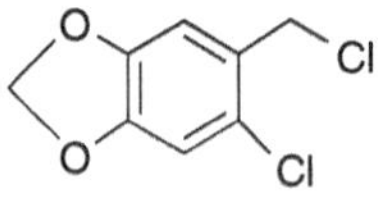 **1h**	**I**	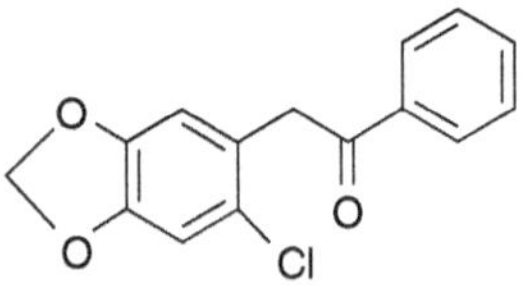 **2h**	89
10	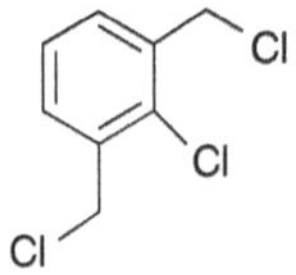 **1i**	**I**	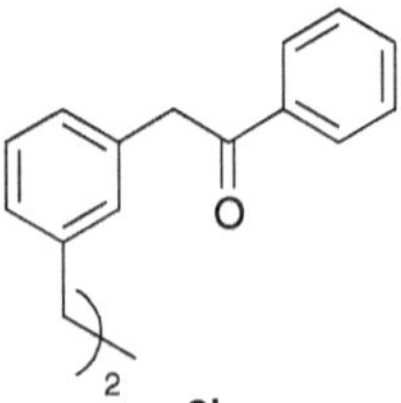 **2i**	79
11	**1a**	**III**	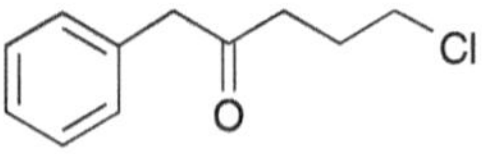 **2j**	78
12	**1b**	**IV**	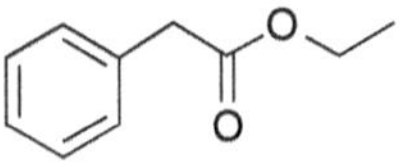 **2k**	69
13	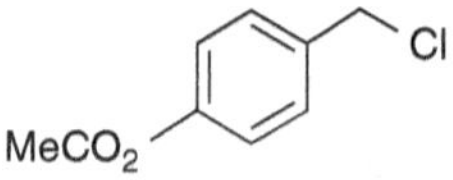 **1j**	**I**	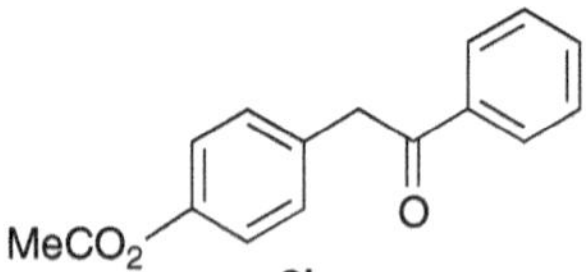 **2l**	75[e]

[a] Oxidative addition reaction and coupling reaction were carried out at room temperature in THF.
[b] Electrophile; **I**, benzoyl chloride; **II**, *p*-bromobenzoyl chloride; **III**, 4-chlorobutyryl chloride; **IV**, ethyl chloroformate.
[c] All products were fully characterized by ^{1}H, ^{13}C NMR, and HRMS (or EMS).
[d] Isolate yield (based on benzyl halides).
[e] 5 mol% CuI was used as a catalyst.

Considering the high reactivity of the active manganese (Mn*) metal prepared by the Rieke method and the exceptional tolerance to a wide range of functionality in the organic moiety, we explored the possibility of the direct oxidative addition to a variety of carbon–oxygen bonds. Our first approach employed benzyl sulfonates. This was expanded to functionalized and

Table 8.10 Cross-coupling reaction of benzylic manganese bromide.[a]

Entry	Halide	Electrophile	Product	Yield (%)[b]
1	**1a**	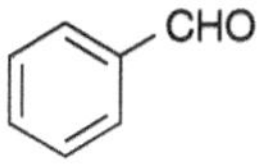 CHO	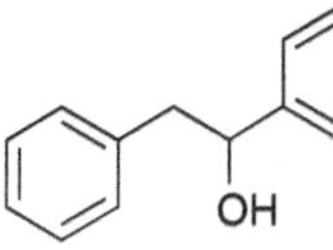 OH **3a**	93
2	**1a**	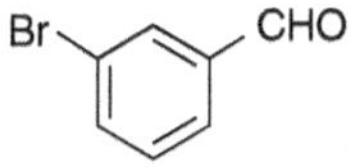 Br CHO	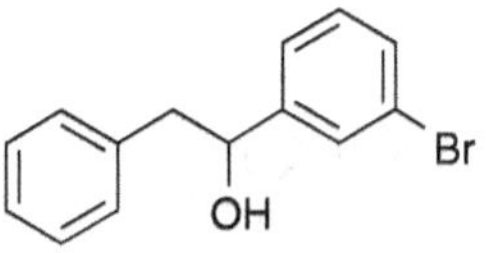 Br OH **3b**	95
3	**1a**	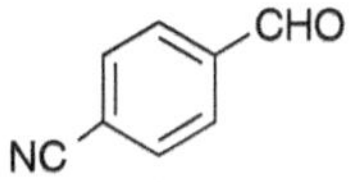 CHO NC	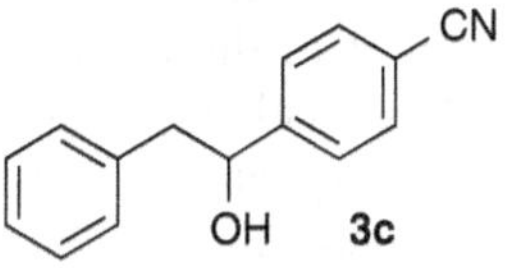 CN OH **3c**	87
4	**1a**	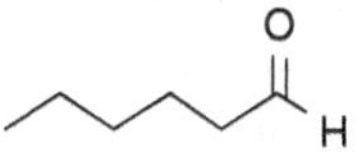 O H	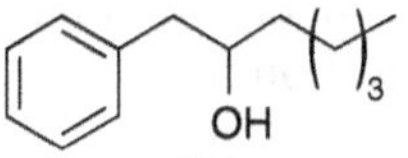 $(\ )_3$ OH **3d**	78
5	**1a**	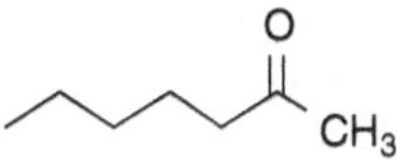 O CH_3	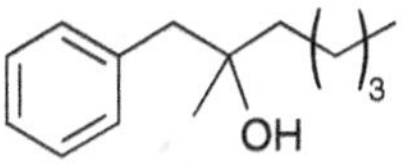 $(\ )_3$ OH **3e**	72
6	**1a**	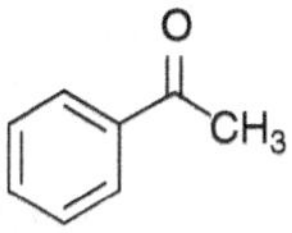 O CH_3	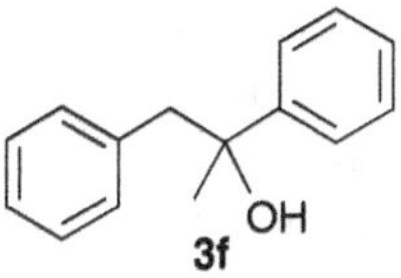 OH **3f**	46
7	**1c**	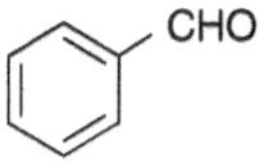 CHO	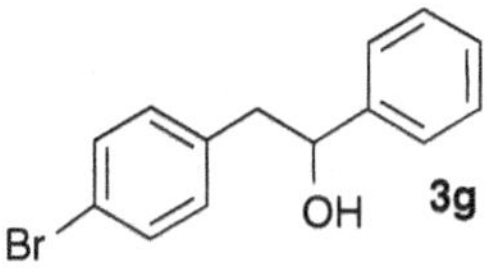 Br OH **3g**	80
8	**1c**	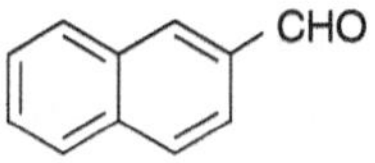 CHO	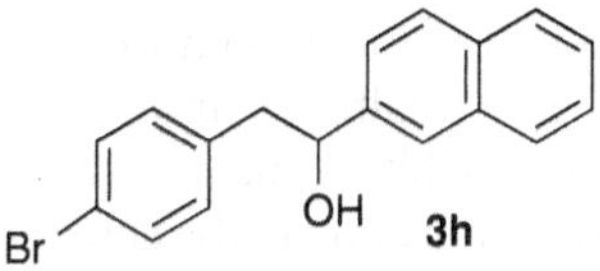 Br OH **3h**	89
9	**1a**	$(t\text{-BuO}_2\text{C-N}\mathbin{=})_2$	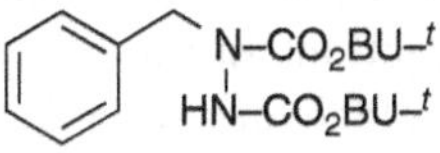 $N\text{–}CO_2BU\text{–}^t$ / $HN\text{–}CO_2BU\text{–}^t$ **3i**	80

[a] Oxidative addition reaction and coupling reaction were carried out at room temperature in THF.

[b] Isolated yields (based on electrophile).

$$\text{R–X} + \text{Mn}^* \xrightarrow[\text{rt}]{\text{THF}} [\text{R–MnX}] \xrightarrow[\text{THF/rt}]{\text{E}^+} \boxed{\text{R–E}}$$

R = Alkyl, Benzyl
X = OTs, OMs, $OP(O)(OEt)_2$, $OP(O)(OPh)_2$

Scheme 8.6 Proposed reaction scheme.

nonfunctionalized benzylic sulfonates and phosphates later. Benzyl sulfonates and phosphates can be easily prepared from the corresponding alcohols using standard literature procedures [36, 37]. Simply stirring the mixture of an alcohol and methane sulfonyl chloride or *p*-toluene sulfonyl chloride in methylene chloride at rt in the presence of trimethylamine followed by a simple workup procedure yielded the sulfonates in over 80% isolated yields. The synthesis of organophosphates was also easily conducted using the corresponding alcohol and commercially available diethyl chlorophosphate in the presence of triethylamine in THF or sodium hydride and diethyl chlorophosphate in THF followed by purification by column chromatography (65% yield). A variety of functionalized benzyl sulfonates and phosphates were prepared by this approach and were used for the preparation of functionalized benzylic manganese reagents. This route provides a significant new approach to benzylic reagents and obviates the need for the corresponding halide, which in many cases is problematic.

Importantly, this study demonstrates that highly active manganese readily undergoes oxidative addition to carbon–oxygen bonds of benzyl sulfonates and phosphates. Also, the subsequent cross-coupling reactions with a variety of electrophiles can be easily carried out. The synthetic advantages of the approach are apparent considering few reports [38] exist on the oxidative addition of main group or even transition metals to carbon–oxygen bonds. Of particular interest is the availability of tosylates, mesylates, and phosphates as precursors to organomanganese reagents (Scheme 8.6).

Table 8.11 contains the summarized results of the coupling reaction of benzylic manganese sulfonates. The oxidative addition of active manganese to benzyl sulfonates was easily completed at rt in only 30 min. According to GC analysis, a small amount (<8%) of homocoupled product, bibenzyl, was formed during the reaction. The resulting benzylic manganese sulfonates were cross-coupled with several electrophiles. The coupling reactions were completed under mild conditions and significantly in the absence of any transition metal catalyst.

Functionalized benzylic mesylates containing a halogen atom were also investigated. Reactions of the halogenated benzyl manganese mesylates with acid chlorides, aldehydes, and ketones yielded the corresponding ketone, secondary alcohol, and tertiary alcohol in good to excellent isolated yields as shown in Table 8.11. As mentioned earlier, it is of interest that the mesylates

Table 8.11 Cross-coupling reaction of benzylic and functionalize benzylic manganese sulfonates.

OMs | Br, OMs | Cl, OMs
4a | 4b | 4c

F_3C, OMs | Cl, F, OMs | MeO, OMs
4d | 4e | 4f

Entry	Sulfonates	Electrophiles[a]	Product[b]	Yield (%)[c]
1	**4a**	I[d]	Ph, Ph, O **2a**	63[d]
2	**4a**	II	Ph, Ph, OH **3a**	72
3	**4a**	O, H, ( )3	Ph, OH, ( )3 **3d**	50
4	**4a**	cyclohexanone (=O)	Ph, OH **5a**	29
5	**4a**	$PHCOCH_3$	Ph, Ph, OH **3f**	36
6	**4b**	I[d]	$4\text{–}BrC_6H_4$, Ph, O **2c**	50[d]
7	**4b**	II	$4\text{–}BrC_6H_4$, Ph, OH **3g**	90

(Continued)

Table 8.11 (Continued)

Entry	Sulfonates	Electrophiles[a]	Product[b]	Yield (%)[c]
8	4b	III	4-BrC_6H_4–$CH_2C(CH_3)(OH)Ph$ 5b	80
9	4c	II	4-ClC_6H_4–$CH_2CH(OH)Ph$ 5c	95
10	4c	III	4-ClC_6H_4–$CH_2C(CH_3)(OH)Ph$ 5d	92
11	4d	II	3-$CF_3C_6H_4$–$CH_2CH(OH)Ph$ 5e	89
12	4e	II	2-Cl-4-F-C_6H_3–$CH_2CH(OH)Ph$ 5f	92
13	4f	II	3-$CH_3OC_6H_4$–$CH_2CH(OH)Ph$ 5g	94

[a] Electrophiles; **I**, benzoyl chloride; **II**, benzaldehyde; **III**, acetophenone.
[b] All of the products were fully characterized by ^{1}H, ^{13}C NMR, FTIR, and HRMS.
[c] Isolate yields (based on electrophiles unless otherwise mentioned).
[d] Excess benzoyl chloride was used (yield was based on mesylate).

containing trifluoromethyl group has been successfully employed for the preparation of benzylic manganese mesylates, and the subsequent coupling reaction proceeded in good yield (89%).

As shown in Table 8.12, the oxidative addition to alkyl tosylates and coupling reactions of organomanganese tosylates were also investigated. The oxidative addition was easily completed by simple treatment of Mn* with the alkyl tosylates at rt in THF.

Unfortunately, no oxidative addition was found to occur with phenyl tosylate even at refluxing temperatures (Table 8.12, entry 9). The formation of organomanganese tosylates was also confirmed by the subsequent cross-coupling reaction with different electrophiles. The tosylates listed in Table 8.12 reacted with benzoyl chloride to afford the corresponding ketones in moderate to excellent yields (44–86%). Once again, the coupling reaction was carried out at

Table 8.12 Preparation of organomanganese tosylates and their coupling reaction.

Entry	Tosylate	E[a]	Product	Yield (%)[b]
1	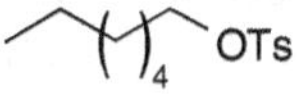 6a	I	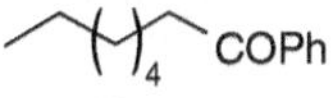 7a	48
2	6a	II	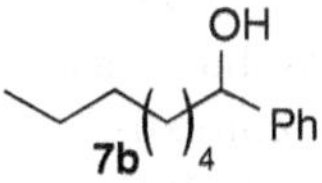 7b	64
3	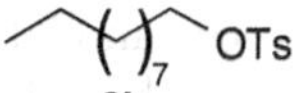 6b	I	 7c	57
4	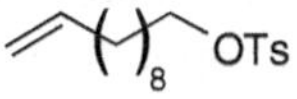 6c	I	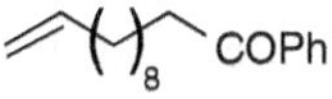 7d	44
5	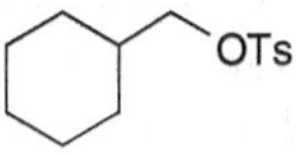 6d	I	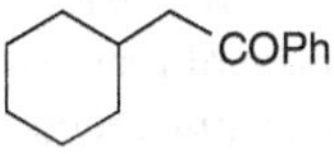 7e	86
6	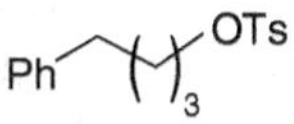 6e	I	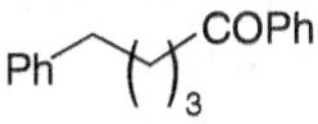 7f	48
7	6e	II	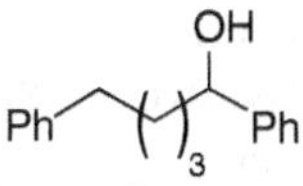 7g	36
8	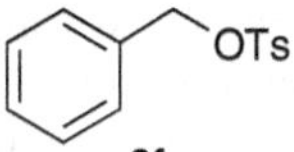 6f	I	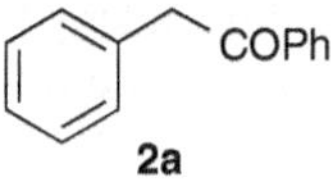 2a	74[c]
9	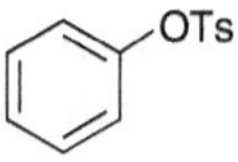 6g		—	0[d]

[a] Electrophile; **I**, benzoyl chloride; **II**, benzaldehyde.
[b] Based on electrophile unless otherwise mentioned.
[c] Excess benzoyl chloride was used (yield was based on tosylate).
[d] No oxidative addition occurred.

rt in THF in the absence of any transition metal catalyst. In contrast to all the primary tosylates, a low yield (17%) was obtained from a secondary tosylate. This may be due to the steric hindrance and/or easy elimination of organomanganese tosylate yielding alkenes. Cross-coupling reaction with benzaldehyde also gave the corresponding alcohols in moderate yields (36–64%).

Nonfunctionalized and functionalized benzyl phosphates were treated with Mn* under the same conditions used for benzyl sulfonates. However, in the case of benzyl phosphates, a longer reaction time was required to complete the oxidative addition compared to the cases of benzyl mesylate and tosylate. The oxidative addition of Mn* to the C—O bond of the phosphate was completed after being stirred at rt overnight (8 h). According to GC and TLC analysis, a trace amount of homocoupling product was observed. The subsequent coupling reactions of the resulting benzylic manganese phosphates with acid chlorides, benzaldehydes, and ketones gave the corresponding products in moderate to good yields at rt in dry THF in the absence of any transition metal catalyst. To facilitate the oxidative addition to the organophosphates, use of a sonicator or increase of reaction temperature was examined. But this resulted in the formation of unexpected side products and was not continued. The results of the coupling reactions of nonfunctionalized as well as functionalized benzylic manganese phosphates with acid chlorides are summarized in Table 8.13. These results indicate that a wide range of functional groups can be tolerated under these conditions. The ethyl groups in the starting phosphate were replaced with phenyl groups with little or no effect on the overall reaction (Table 8.13, entries 2 and 3). In contrast to all the rest of the entries in Table 8.13, the manganese derivatives of **8j** do not undergo cross-coupling without a Cu catalyst. In the presence of CuI, product **2l** was obtained in 39% yield.

The benzylic manganese phosphates were also found to react with other electrophiles including aldehydes and ketones. The results are summarized in Table 8.14. Addition to aldehydes gave the corresponding secondary alcohols in moderate yield (38–76%). However, the addition reaction to acetophenone afforded the tertiary alcohol in low yield (19%). The reaction tolerated a nitrile group in the aldehyde (Table 8.14, entry 8) but not a nitro group. When *p*-cyanobenzyl diethyl phosphate was treated with Mn*, only the homocoupled product was observed under a wide range of reaction conditions.

To expand the range of phosphates for preparation of organomanganese phosphates, alkyl-, phenyl-, and allyl diethyl-phosphates were attempted. Unfortunately, the oxidative addition of active manganese (Mn*) to these phosphates and the subsequent coupling reactions with electrophiles failed to give the corresponding cross-coupled products.

Overall, the isolated yields obtained from using benzyl sulfonates and phosphates were lower than those from using benzyl halides. However, this unusual oxidative addition to the C—O bond of benzyl sulfonates and phosphates under mild conditions and the resulting organomanganese reagents provide a new

Table 8.13 Coupling reactions of benzylic manganese phosphates with acid chlorides.[a]

Entry	Phosphates	Electrophiles	Product[b]	Yield (%)[c]
1	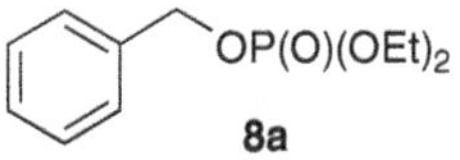 $OP(O)(OEt)_2$ **8a**	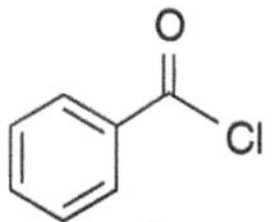 Cl (benzoyl chloride) **I**	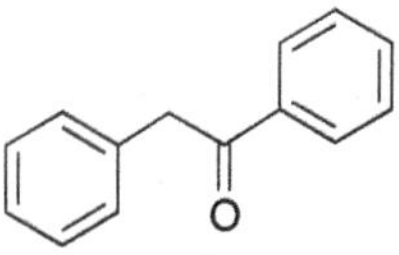 **2a**	52[d]
2	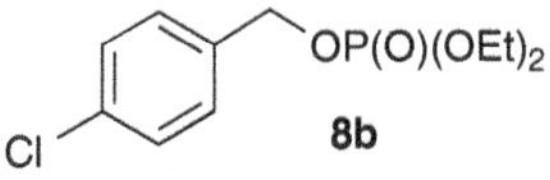 Cl, $OP(O)(OEt)_2$ **8b**	I	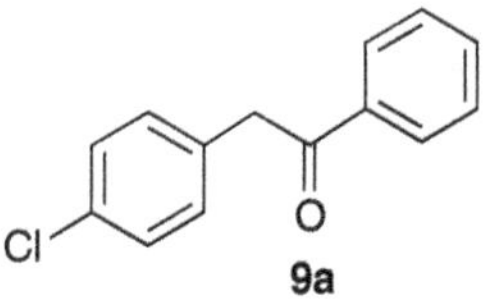 Cl, **9a**	45
3	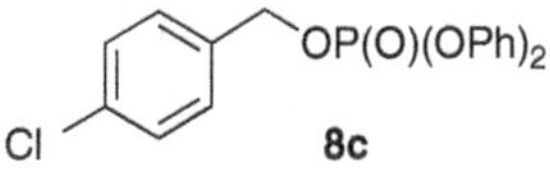 Cl, $OP(O)(OPh)_2$ **8c**	I	9a	47
4	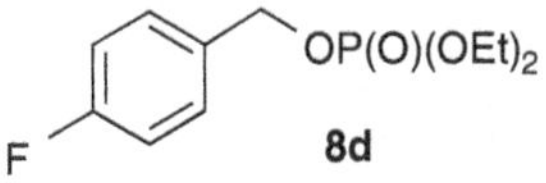 F, $OP(O)(OEt)_2$ **8d**	I	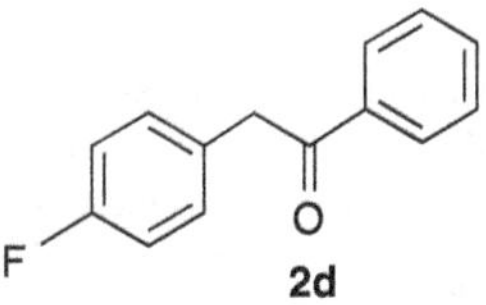 F, **2d**	51
5	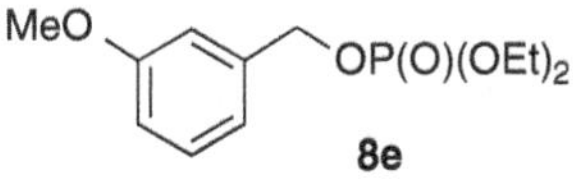 MeO, $OP(O)(OEt)_2$ **8e**	I	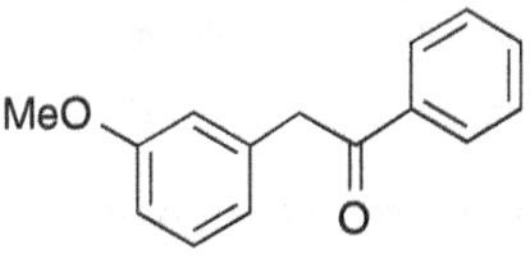 MeO, **9b**	58[d]
6	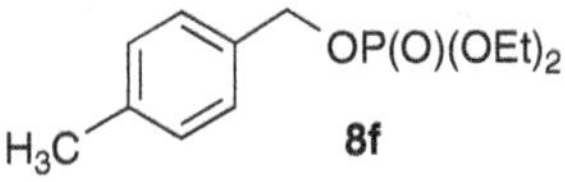 H_3C, $OP(O)(OEt)_2$ **8f**	I	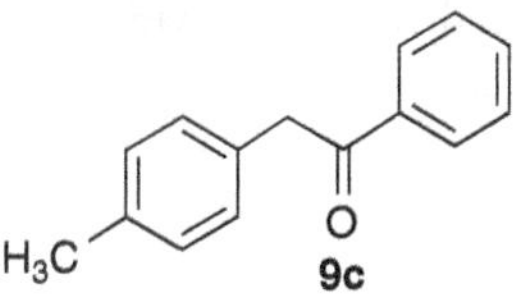 H_3C, **9c**	72
7	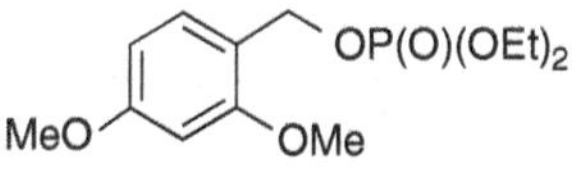 MeO, OMe, $OP(O)(OEt)_2$ **8g**	I	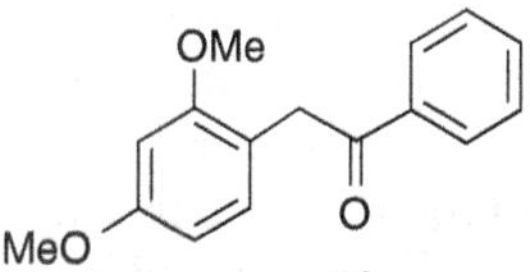 OMe, MeO, **2f**	72
8	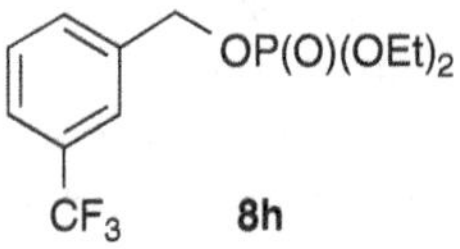 CF_3, $OP(O)(OEt)_2$ **8h**	I	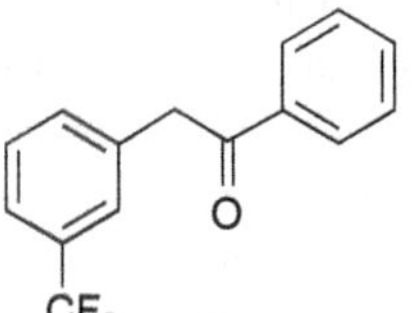 CF_3, **2g**	49

(Continued)

Table 8.13 (Continued)

Entry	Phosphates	Electrophiles	Product[b]	Yield (%)[c]
9	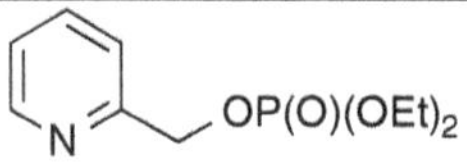 8i	I	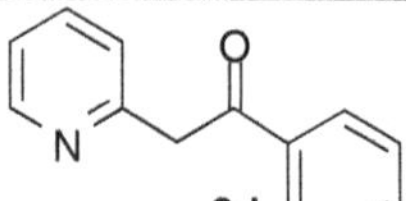 9d	54[d]
10	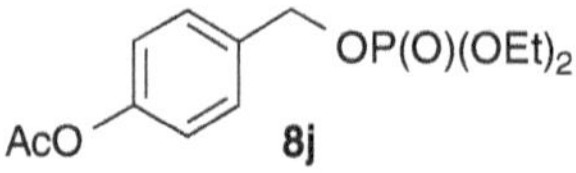 8j	I	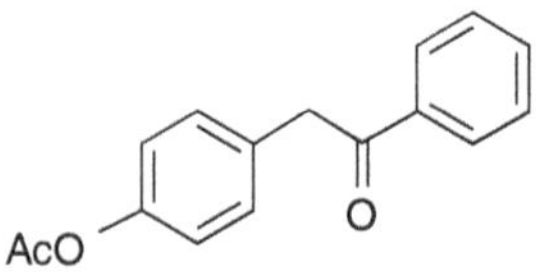 2l	39[e]
11	**8k**	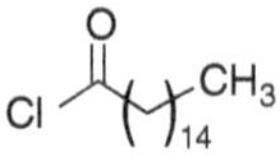 $Cl\text{-}CO\text{-}(CH_2)_{14}\text{-}CH_3$	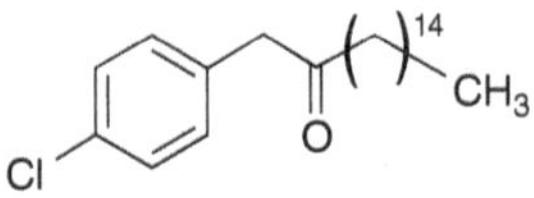 9e	62[f]

[a] Oxidative addition reactions and coupling reactions were carried out at rt in THF.
[b] All products were fully characterized by ^{1}H, ^{13}C NMR, and/or HRMS (EIMS).
[c] Isolated yields (based on electrophiles).
[d] Excess benzoyl chloride was used (yield was based on phosphate).
[e] 5 mol% CuI was used as a catalyst.
[f] The yield after recrystallization over hexanes.

synthetic route to benzylic manganese reagents. Moreover, to our knowledge, these are the first examples of organomanganese sulfonates and phosphates prepared via direct oxidative addition of manganese metal or by any metathesis approach.

Homocoupling Reactions of Functionalized Benzylic Manganese Reagents

We examined the homocoupling of benzyl halides under the influence of active manganese in a wide variety of conditions. As shown in Table 8.15, 1 equiv of active manganese and 2 equiv of benzyl bromide produced 47% of homocoupling products. But the homocoupling yield was decreased to 27% by using 1 equiv of benzyl bromide.

Best results were obtained when the active manganese was added to the neat benzyl bromide. The reaction was completed within 10 min at rt without a catalyst [39]. Several other benzyl halides produced good to excellent yields of homocoupling products, which showed a wide tolerance of functional groups such as nitrile, ester, nitro, chloro, bromo, methoxy, and methyl groups (Table 8.16).

Table 8.14 Addition reactions of benzylic manganese phosphates.[a]

Entry	Phosphates	Electrophiles	Product	Yield (%)[b]
1	**8a**	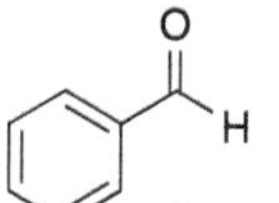 II	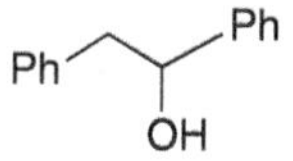 **3a**	73
2	**8a**	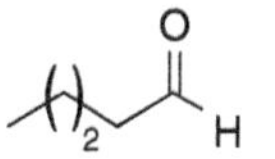	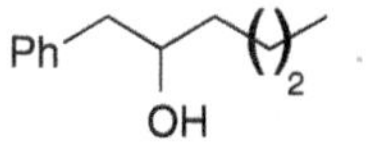 **3d**	49
3	**8b**	II	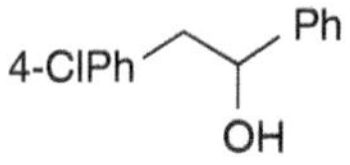 **5c**	38
4	**8d**	II	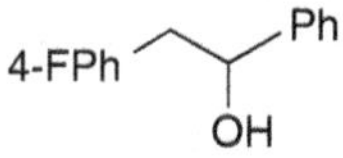 **10a**	44
5	**8f**	II	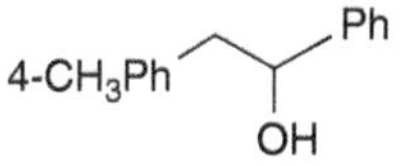 **10b**	52
6	**8a**	Acetophenone	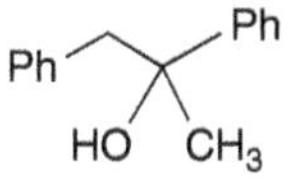 **3f**	19
7	**8a**	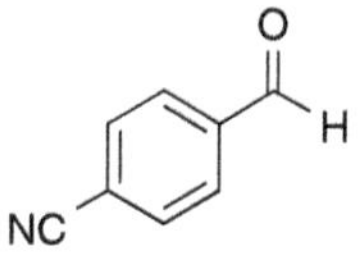	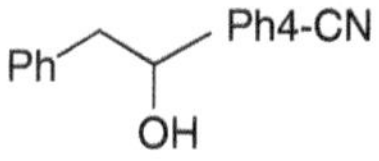 **3c**	76
8	**8h**	II	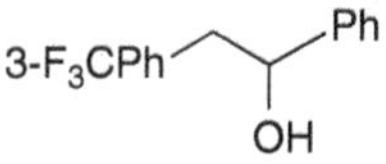 **5e**	46
9	**8a**	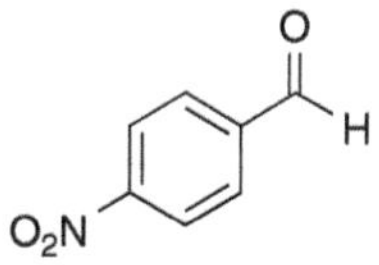	—	0[c]

[a] Oxidative addition reactions and cross-coupling reactions were carried out at room temperature in THF.
[b] Isolated yield (based on electrophile).
[c] According to TLC and/or GC analyses, no coupling reaction occurred.

Table 8.15 Homocoupling reactions of benzyl bromides at different reaction conditions.

Br
MeO_2C
Mn*
THF, Temp, Time
2
MeO_2C

Entry	Mn*:$ArCH_2Br$	Temp.	Time	Yield (%)[a]
1	1.00:1.99	rt	3h	47
2	1.00:0.90	rt	30min	27
3	1.00:1.91	rt	10min	68[b]
4	1.00:1.90	0°C	30min	60
5	1.00:1.80	rt	10min	80[c]

[a] Isolated yields.
[b] The benzyl bromide was slowly added to the active manganese for 30min with 10ml of THF.
[c] THF solution of active manganese was added to the benzyl bromide.

Table 8.16 Homocoupling reactions of benzyl halides.

Entry	R^1	R^2	X	Product	Yield (%)[a]
1	4-CN	H	Br	**11a**	82
2	3-CN	H	Br	**11b**	90
3	2-CN	H	Br	**11c**	84
4	H	CH_3	Br	**11d**	85
5	4-CO_2Me	H	Br	**11e**	80
6	4-NO_2	H	Br	**11f**	34[b]
7	4-Br	H	Br	**11g**	60[c]
8	4-Cl	H	Cl	**11h**	65[c]
9	2,6-*di*Cl	H	Cl	**11i**	77[c]
10	3,5-*di*OMe	H	Cl	**11j**	70[c]
11	2-CH_3	H	Cl	**11k**	59[c]

[a] Isolated yields.
[b] 0.64equiv of $ArCH_2X$ was used.
[c] Yields obtained after recrystallization over hexanes.

Our method is more efficient compared to the previous attempts to prepare functionalized bibenzyl compounds [40] in terms of mild conditions and speed of the reaction. More significantly, no transition metal catalyst (Pd or Ni) was required for completion of the reaction nor was a solvent required.

Palladium-Catalyzed Cross-Coupling Reactions of Benzylic Manganese Reagents

Cross-Coupling reactions of alkyl organometallic compounds with alkyl or aryl halides are difficult to perform because of the slow oxidative addition of alkyl halides to the palladium and slow reductive elimination of dialkyl or alkyl–aryl products. Only limited examples are reported for the cross-coupling of the benzyl organometallic compounds with aryl halides. Betzemeier and Knochel [30a] reported the palladium-catalyzed cross-coupling of benzylzinc bromides with aryl iodides in perfluorinated solvents, and Park et al. [41] prepared diarylmethanes from benzylmagnesium chlorides and aryl bromides with a nickel catalyst. Yoshida et al. [42] prepared functionalized diarylmethanes from cross-coupling reactions of (2-pyridyl)silylmethylstannanes and aryl iodides with a palladium catalyst and additive. The aforementioned methods required additives and long reaction times, and in some cases they did not tolerate functional groups.

Benzylic manganese halides and phosphates were also found to undergo palladium-catalyzed cross-coupling reactions with aryl iodides (sp^3–sp^2 coupling). As shown in Schemes 8.7 and 8.8 and in (Table 8.17), the corresponding cross-coupling compounds were achieved in moderate to good yields using 5 mol% palladium catalyst [$Pd(PPh_3)_4$] in THF within 2 h at rt.

Limitations of This Approach

A number of different types of benzyl substrates, diethyl benzylphosphonate, benzyl methyl sulfide, benzyl phenyl sulfide, benzyl benzoate, benzyl phenyl ether, and benzyloxytrimethylsilane were also reacted with Mn*. Unfortunately, according to TLC and/or gas chromatography analyses, no oxidative addition occurred with these substrates.

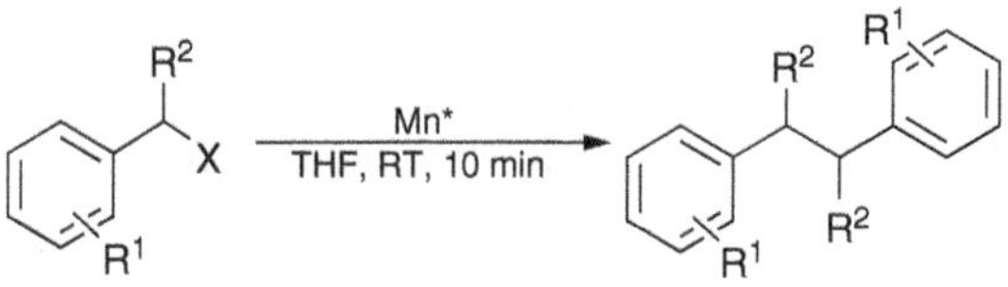

Mn*
THF, RT, 10 min

R^1 = CN, H, CO_2Me, NO_2, Br, Cl, OMe, CH_3

R^2 = H, CH_3

X = Br, Cl

Scheme 8.7 Coupling of benzylic halides.

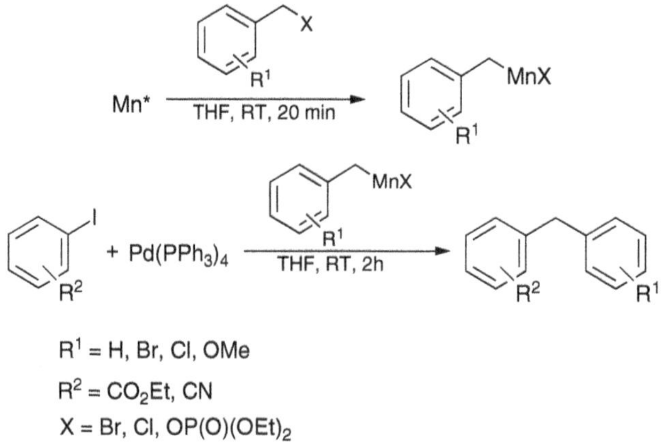

R^1 = H, Br, Cl, OMe
R^2 = CO_2Et, CN
X = Br, Cl, $OP(O)(OEt)_2$

Scheme 8.8 Cross-coupling of benzylic halides.

Table 8.17 Palladium-catalyzed cross-coupling reactions of benzyl manganese halides and phosphate with aryl iodides.

Entry	R^1	R^2	X	Product	Yield (%)[a]
1	H	4-CO_2Et	Br	**12a**	78
2	4-Br	4-CO_2Et	Br	**12b**	56
3	3,5-*di*OMe	3-CN	Cl	**12c**	40
4	4-Cl	4-CO_2Et	Cl	**12d**	64
5	4-Cl	4-CO_2Et	$OP(O)(OEt)_2$	**12d**	40

[a] Isolated yields.

As shown in previous tables containing cross-coupling products, a number of different types of electrophiles were used to complete the cross-coupling reaction of benzylic manganese reagents prepared via direct oxidative addition of Mn* to the corresponding benzylic halides, sulfonates, and phosphates. Apparently, functional group tolerance is a big advantage of this system. However, no coupling products were obtained from the coupling reactions with the following substrates: *p*-nitrobenzaldehyde, epoxides, and ester and alkyl cyanides.

Experimental

Preparation of Highly Active Manganese (Mn*)

To the mixture of lithium (4.4 mmol), naphthalene (0.4 mmol), and manganese halides (2 mmol) was added via syringe freshly distilled THF (10 ml) at rt, and then the resulting mixture was allowed to stir at rt for 1 h. The black slurry was obtained and ready for use. (Note: The number of millimoles of Mn* cited in this paper refers to the theoretical amount possible based on the original amount of anhydrous manganese halide.) No attempt was made to determine whether the Mn*, either wet or as a slurry or predried, was pyrophoric. However, it should be treated as if it is pyrophoric and kept under argon at all times.

Typical Preparation of Benzylic Manganese Halides and Their Coupling Reactions with Benzoyl Chlorides to Give Ketones (2a–2l)

Benzyl halide (9 mmol) in THF (2 ml) was added via a cannula to the slurry of highly active manganese (10 mmol) being stirred in THF (15 ml) at rt. The slurry was stirred at rt for 20 min. The reaction was monitored by gas chromatography. After the completion of the oxidative addition, the mixture was cooled to 0°C. 1,2-Dibromoethane was added to the mixture at this temperature to react with the excess Mn*, and the mixture was stirred for 5 min. The resulting mixture was transferred via cannula to the benzoyl chloride solution in THF at rt. After being stirred for 30 min, the mixture was quenched with 3 M HCl solution and extracted with diethyl ether. The combined organic layers were washed with $NaHCO_3$, $Na_2S_2O_3$, and brine, dried over anhydrous $MgSO_4$, and concentrated using a rotary evaporator. Flash column chromatography (ethyl acetate–hexanes) afforded the corresponding ketones.

Typical Procedure for the Cross-Coupling Reactions of Benzyl Manganese Mesylates

Benzyl mesylate (9 mmol) was added via a syringe to the slurry of active manganese (10 mmol) being stirred in THF (10 ml) at rt. The resulting mixture was stirred at rt for 20 min. The reaction was monitored by gas chromatography. After the completion of the oxidative addition, the mixture was cooled to 0°C. 1,2-Dibromoethane was added to the mixture at this temperature, and the mixture was stirred for 5 min. Benzoyl chloride was added to the resulting mixture at rt. After being stirred for 30 min, the mixture was quenched with 2 M HCl solution and extracted with diethyl ether. The combined organic layers were washed with $NaHCO_3$, $Na_2S_2O_3$, and brine, dried over anhydrous $MgSO_4$, and concentrated using a rotary evaporator. Flash column chromatography (ethyl acetate–hexanes) afforded the corresponding ketones.

Preparation of Alcohols from the Reactions of Benzylic Manganese Halides with Aldehydes and Ketones (3a–3i)

Benzyl halide (9 mmol) in THF (2 ml) was cannulated to the slurry of highly active manganese (10 mmol) being stirred in THF (15 ml) at rt. The slurry was stirred at rt for 20 min. The reaction was monitored by gas chromatography. After the completion of the oxidative addition, the mixture was cooled to 0°C.

1,2-Dibromoethane was added to the mixture at this temperature, and the mixture was stirred for 5 min. To the resulting mixture was added aldehydes (or ketones) at rt. After being stirred for 1 h, the mixture was quenched with 2 M HCl solution and extracted with diethyl ether. The combined organic layers were washed with $NaHCO_3$, $Na_2S_2O_3$, and brine, dried over anhydrous $MgSO_4$, and concentrated using a rotary evaporator. Flash column chromatography (ethyl acetate–hexanes) afforded the corresponding alcohols.

Typical Preparation of Benzylic Manganese Phosphates and Their Cross-Coupling Reactions

Benzyl diethyl phosphate (9 mmol) was added via a syringe to the slurry of highly active manganese (10 mmol) being stirred in THF (12 ml) at rt. The slurry was stirred overnight at rt. The reaction was monitored by TLC and/or GC. After the completion of the oxidative addition, the mixture was cooled to 0°C using an ice bath. 1,2-Dibromoethane was added to the mixture at this temperature and the mixture was stirred for 5 min. To the reaction mixture was added an appropriate electrophile (8 mmol) through a syringe. After being stirred for 1 h, the mixture was quenched with 2 M HCl solution and extracted with diethyl ether. The combined organic layers were washed with $NaHCO_3$, $Na_2S_2O_3$, and brine, dried over anhydrous $MgSO_4$, and concentrated using a rotary evaporator. Flash column chromatography (ethyl acetate–hexanes) afforded the corresponding product.

Homocoupling Reactions of Functionalized Benzyl Halides

The slurry of active manganese (2 mmol) in 15 ml of THF was added to benzyl halides (3.8 mmol), and the reaction mixture was stirred at rt. The reaction progress was monitored by gas chromatography. After being stirred for 10 min, the reaction mixture was quenched with saturated ammonium chloride aqueous solution, and the organic layer was extracted with diethyl ether (2 × 10 ml). The combined organic layers were washed with saturated sodium thiosulfate solution (10 ml), dried over anhydrous magnesium sulfate, and concentrated using a rotary evaporator. Flash column chromatography (hexane–ethyl acetate) or recrystallization afforded the corresponding bibenzyls.

Typical Procedure for the Coupling Reaction of (1f) with Aldehydes and Acid Chlorides (13a–13d, 14–18)

The benzylic manganese reagent was prepared as before. To the resulting benzylic reagent was added the aldehydes (or acid chloride) at rt. After being stirred for 1 h, the mixture was quenched with 2 M HCl solution and extracted with diethyl ether. The combined organic layers were washed with a saturated aqueous solution of sodium thiosulfate (10 ml) and then dried over magnesium sulfate. Removal of solvents and flash column chromatography (hexane–ethyl acetate) afforded the corresponding coupling product.

Typical Preparation of Functionalized Benzylic Manganese Halides and Their Cross-Coupling Reactions with Aryl Iodides under a Palladium Catalyst

Benzyl halide (1.8 mmol) in THF (10 ml) was added via cannula to a slurry of active manganese (2 mmol) being stirred in THF (10 ml) at rt. The reaction mixture was stirred at rt for 20 min. The reaction progress was monitored by gas chromatography. After the completion of the oxidative addition, the mixture was cooled to 0°C using an ice bath. 1,2-Dibromoethane was added to the mixture via syringe and the mixture was allowed to stir for 5 min. The resulting mixture was added via a cannula to a mixture of aryl iodides (1.2 mmol) and 5 mol% of $Pd(PPh_3)_4$ catalyst in THF (10 ml). The mixture was stirred at rt for 2 h. A saturated aqueous solution of ammonium chloride (10 ml) was added, and then the mixture was extracted with diethyl ether (2 × 10 ml). The combined organic layers were washed with a saturated aqueous solution of sodium thiosulfate (10 ml) and then dried over magnesium sulfate. Removal of the solvents and flash column chromatography (hexane–ethyl acetate) afforded the diarylmethane.

8.5 Preparation and Coupling Reactions of Thienylmanganese Halides

Rieke manganese (Mn*) was shown to exhibit superior reactivity toward oxidative addition to organic halides [43]. This discovery prompted us to investigate whether we could apply this finding to 3,4-dibromothiophene and 4-substituted 3-bromothiophene. Due to the possibility of functional group conversion of both bromine atoms of the thiophene ring, 3,4-dibromothiophene appears to be an ideal starting material [44] for the preparation of symmetrical and/or unsymmetrical 3,4-disubstituted thiophene derivatives, which are currently of interest in drug design and for the construction of new polymers (Scheme 8.9) [45].

Acylation–reduction, lithiation–alkylation, thiophene ring construction, and Grignard cross-coupling with bromothiophene are the representative methods for the preparation of alkyl- and arylthiophenes [46].

We reported a facile synthesis of 3-bromo-4-thienylmanganese bromide and 4-substituted 3-thienylmanganese bromides via the direct oxidative addition

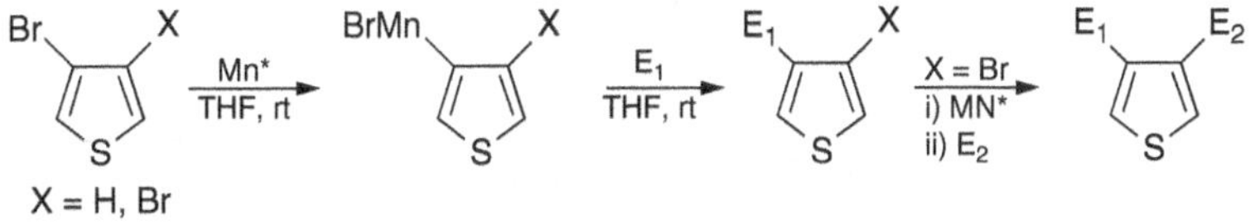

Br X S X = H, Br —Mn*, THF, rt→ BrMn X S —E_1, THF, rt→ E_1 X S —X = Br, i) MN*, ii) E_2→ E_1 E_2 S

Scheme 8.9 Preparation of 3,4-disubstituted thiophenes.

of Rieke manganese to 3,4-dibromothiophene and 4-substituted 3-bromothiophenes, respectively. In addition, the coupling reaction of the resulting thienylmanganese bromides with various electrophiles will be described. Also, the coupling reaction of 3-thienylmanganese bromide will be included.

As shown in Table 8.18, 3-bromo-4-thienylmanganese bromide was formed as a major product from the reaction of 1 equiv of 3,4-dibromothiophene and 2 equiv of Mn* [47]. This reagent easily underwent a cross-coupling reaction with acid chlorides and aryl iodides. According to the high-resolution mass spectra of the products of entries 1, 4, 5, and 6 in Table 8.18, the second bromine atom of the thiophene ring is still retained.

Table 8.18 Coupling reaction of 3-bromo-4-thienylmanganese bromide.

Entry	Electrophile	Product[a]	Yield (%)[b]
1	PhCOCl	3-Br-4-(C(O)Ph)thiophene	80
2	4-BrC_6H_4COCl	3-Br-4-(C(O)C_6H_4-4-Br)thiophene	83
3	3-BrC_6H_4COCl	3-Br-4-(C(O)C_6H_4-3-Br)thiophene	62
4[c]	PhI	3-Br-4-Ph-thiophene	65
5[c]	4-$EtO_2CC_6H_4$	3-Br-4-(C_6H_4-4-CO_2Et)thiophene	86
6[c]	4-$CH_3OC_6H_4$	3-Br-4-(C_6H_4-4-OCH_3)thiophene	69

[a] ^1H, ^{13}C NMR, and HRMS spectra were consistent with literature.
[b] Isolated yields (based on electrophile).
[c] 10 mol% $Pd(PPh_3)_4$ was used as a catalyst (based on electrophile).

The coupling reaction of 3-bromo-4-thienylmanganese bromides with acid chlorides gave the corresponding ketones in good isolated yields (Table 8.18, entries 1–3) without requiring any transition metal catalysts. Coupling reactions with aryl iodides were also examined. Using a catalytic amount of $Pd(PPh_3)_4$ catalyst, the corresponding C—C bond-forming products were obtained in good yields (Table 8.18, entries 4–6). It is noteworthy that the consecutive two procedures, oxidative addition and coupling step, can be completed at rt in a few hours.

The same strategy was applied to 4-substituted 3-bromothiophenes to make unsymmetrical 3,4-disubstituted thiophenes. As shown in Table 8.19, cross-coupling products were obtained in moderate yields (36–66%) under mild conditions. Oxidative additions of Mn* to the 4-substituted bromothiophenes were performed at rt in 5 h, and the successive coupling reactions of thienylmanganese bromides with benzoyl chlorides afforded unsymmetrical 3,4-disubstituted thiophenes at rt in 30 min without a catalyst (Table 8.19, entries 1 and 2). With a Pd(0) catalyst, the C—C bond formation was completed from the reaction of the thienylmanganese bromide with 4-iodobenzene (Table 8.19, entry 3).

3-Substituted thiophene derivatives have attracted much attention in both material [48] and pharmaceutical science [49]. To date, the most widely used methods for the preparation of 3-substituted thiophene derivatives are the coupling reactions of 3-thienyl organometallic reagents with electrophiles. The intermediates used in these reactions are generally obtained via either a

Table 8.19 Coupling reaction of 4-substituted 3-bromothiophene.

Entry	Electrophile 1 (E_1)	Electrophile 2 (E_2)	Product[a]	Yield (%)[b]
1	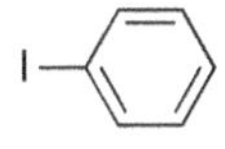 I–C_6H_5	PhCOCl	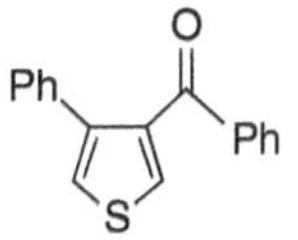 3-Ph-4-(PhCO)thiophene	66
2	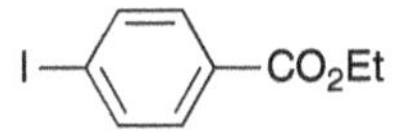 I–C_6H_4–CO_2Et	PhCOCl	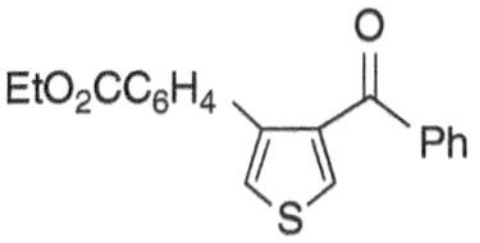 3-($EtO_2CC_6H_4$)-4-(PhCO)thiophene	46
3[c]	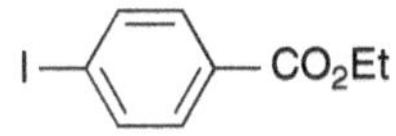 I–C_6H_4–CO_2Et	PhI	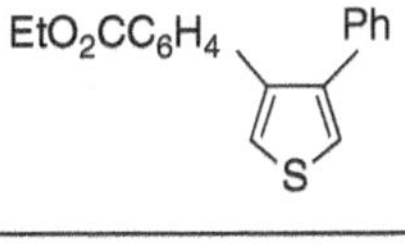 3-($EtO_2CC_6H_4$)-4-Ph-thiophene	36

[a] ^1H, ^{13}C NMR, and HRMS spectra were consistent with literature.
[b] Isolated yields (based on electrophile).
[c] 10 mol% $Pd(PPh_3)_4$ was used as a catalyst (based on electrophile).

metal–halogen exchange reaction of 3-bromothiophene with *n*-butyllithium [50] or the metathesis of the 3-lithiothiophene with different metal halides [51]. However, these reactions suffer from a lack of regiospecificity as well as decomposition of the thiophene ring at rt [50b, 52]. 3-Thienyl organometallic reagents can be prepared by the direct oxidative addition of active metals to 3-iodothiophene [53]. However, 3-iodothiophene is rather difficult to prepare.

3-Thienylmanganese bromide was easily prepared from the reaction of Mn* (2 equiv) and 3-bromothiophene (1 equiv). The oxidative addition of Mn* to 3-bromothiophene was accomplished in THF at rt. The reactions of 3-thienylmanganese bromide with acid chlorides gave the corresponding ketones in moderate to high yields (Table 8.20, entries 1 and 2). In the absence of an additional electrophile, 3-thienylmanganese bromide underwent a homocoupling reaction to give 3,3′-bithienyl with Pd(0) catalyst at rt (Table 8.20, entry 3). Under our reaction conditions, successful cross-coupling products were obtained from Pd(0)-catalyzed reactions with aryl electrophiles in good yields (Table 8.20, entries 4 and 5).

Table 8.20 Coupling reactions of 3-thienylmanganese bromide.

Entry	Electrophile	Product[a]	Yield (%)[b]
1	3-BrC_6H_4COCl	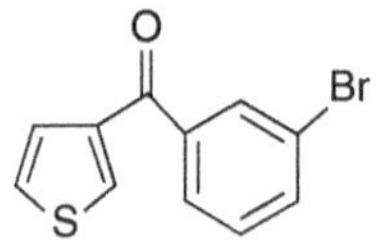 (3-thienyl)(3-bromophenyl) ketone structure; labels: O, Br, S	91
2	4-BrC_6H_4COCl	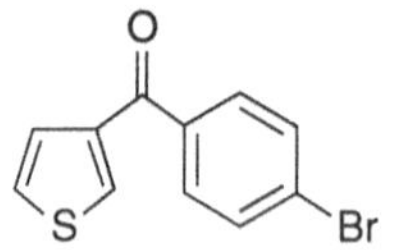 (3-thienyl)(4-bromophenyl) ketone structure; labels: O, S, Br	79
3[c]	–	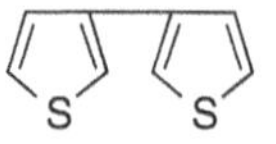 3,3′-bithienyl structure; labels: S, S	58
4[c]	PhI	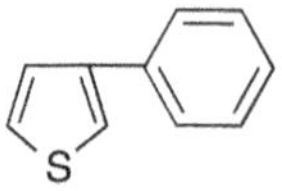 3-phenylthiophene structure; label: S	61
5[c]	4-$IC_6H_4CO_2Et$	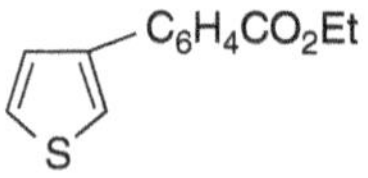 3-thienyl–$C_6H_4CO_2Et$ structure; labels: $C_6H_4CO_2Et$, S	70

[a] ^1H, ^{13}C NMR, and HRMS spectra were consistent with literature.
[b] Isolated yields (based on electrophile).
[c] 10 mol% $Pd(PPh_3)_4$ was used as catalyst (based on electrophile).

In conclusion, thienylmanganese bromide reagents are easily prepared via the direct oxidative addition of Rieke manganese to 3-bromothiophene, 3,4-dibromothiophene, and 4-substituted 3-bromothiophenes. These resulting thienylmanganese bromides undergo coupling reactions with acid chlorides to give the corresponding ketones. The Pd(0)-catalyzed C—C bond formation with aryl iodides afforded the corresponding aryl-substituted thiophenes under mild reaction conditions.

The following is a representative procedure: to a slurry of Rieke manganese [54] (10.0 mmol) in THF (10 ml) under argon was added 3,4-dibromothiophene (5.0 mmol) at rt, and the mixture was stirred at rt for 5 h [47]. 1,2-Dibromoethane [55] (6.0 mmol) was added neat to the reaction mixture at 0°C, and the mixture was allowed to warm to rt over 20 min. The resulting thienylmanganese bromide reagent was transferred via cannula to a flask containing a THF solution of ethyl 4-iodobenzoate (4.0 mmol) and $Pd(PPh_3)_4$ (0.4 mmol) at rt over 30 min. The resulting mixture was stirred at rt for 30 min. The reaction mixture was quenched with 3 M HCl (10 ml) and extracted with ether (2 × 10 ml), and the combined organic layers were sequentially washed with saturated $NaHCO_3$, $Na_2S_2O_3$, and NaCl solutions, dried over $MgSO_4$, and concentrated. Flash chromatography (ethyl acetate–hexanes) afforded ethyl 4-(3-bromo-4-thienyl) benzoate in 70% isolated yield.

8.6 Synthesis of β-Hydroxy Esters Using Active Manganese

The Reformatsky reaction is one of the standard methods for preparing β-hydroxy esters [56]. The reaction typically utilized refluxing benzene or benzene–ether solvents and is characterized by difficulty of initiation and modest yields. Another major problem is the control of the very exothermic reaction. Several attempts have been put forth [57] to improve the reaction conditions including the use of Rieke zinc [58], trimethyl borate/THF, or a continuous-flow procedure requiring refluxing benzene [59, 60]. High-intensity ultrasound-promoted Reformatsky reactions using α-bromoester, zinc dust, and a catalytic amount of iodine produced high yields of β-hydroxy esters [61]. Moreover, a variety of metals or metal salts other than zinc have been studied in order to extend the scope and selectivity [11, 62–67].

However, the study of the reaction using manganese has been limited. In fact, there was only one report [3] where the employed procedure required the presence of a stoichiometric amount of a trapping agent such as acetic anhydride and Lewis acid such as zinc chloride. In the absence of acetic anhydride or Lewis acid, the yields are very low. Therefore, in the course of our studies on the chemistry of active manganese, we attempted to study the reaction in detail using active manganese (Scheme 8.10). Our positive results and improvements

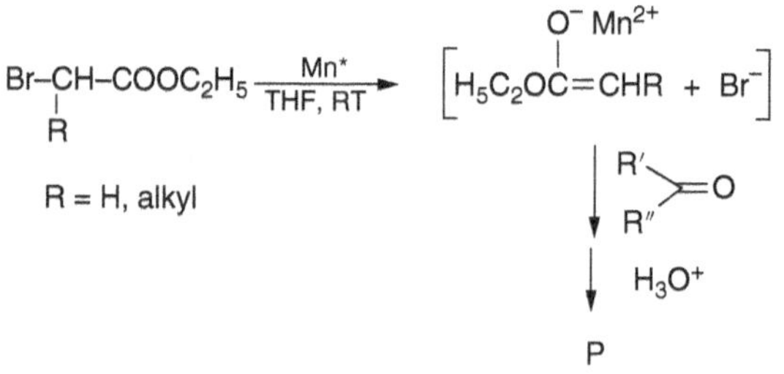

$$Br{-}\underset{R}{CH}{-}COOC_2H_5 \xrightarrow[\text{THF, RT}]{Mn^*} \left[H_5C_2O\overset{O^- Mn^{2+}}{C}{=}CHR + Br^-\right] \xrightarrow{R'R''C{=}O} \xrightarrow{H_3O^+} P$$

R = H, alkyl

Scheme 8.10 Manganese Reformatsky reactions.

on the Reformatsky reaction using active manganese are presented in the succeeding text.

The active manganese was prepared following the procedure presented in the preceding text [68, 69]. The basic procedure involves the reduction of an anhydrous manganese salt with lithium and naphthalene in THF under an argon atmosphere at rt.

Initially, active manganese was simply treated with an α-bromoester followed by addition of the carbonyl compound at rt. But the results were found to be disappointing. Once the reaction started, there was considerable difficulty in moderating the exothermic reaction as expected. Also, the reaction was complicated by the formation of homocoupling products. Lowering the reaction temperature resulted in lower yields. However, we found that the activated manganese would react rapidly with the α-bromoester in the presence of the carbonyl compounds in THF, producing high yields of the corresponding addition product [70]. The reaction was completed in 1–2 h at rt. The results are summarized in Table 8.21. Active manganese (Mn*) prepared from $MnBr_2$ or $MnCl_2$ is extremely reactive toward ketones and aldehydes, yielding pinacols. This precluded its use in this reaction. However, Mn* prepared from MnI_2 is much less reactive, allowing the formation of the Reformatsky reagent in preference to pinacol formation. MnI_2 proved to be ideal for this reaction; see Table 8.22.

The reactivity of the organomanganese reagents was examined with several aldehydes and ketones. The corresponding β-hydroxyester products were obtained in high yields (Table 8.21, entries 1–8). It is noteworthy that this reaction exhibits tolerance of functional groups on the aromatic ring, including cyano and methoxy. Significantly, an α-bromolactone was also found to readily react and undergo a mixed aldol condensation (Table 8.21, entry 7). For all cases, the reaction was conducted in the absence of any Lewis acid, which is required for zinc enolates prepared from α-haloketones. Organomanganese compounds in which the ester group is farther removed from the carbanion center can also be prepared, and the reaction with benzoyl chloride gave a modest yield (Table 8.21, entry 9). This would not be possible for the lithium or

Table 8.21 Reactions of α-bromoester with carbonyl compound using Mn*.

Entry	α-Bromoesters	Carbonyl compound	Product[a]	Yield (%)[b]
1	$BrCH_2COOEt$ **I**	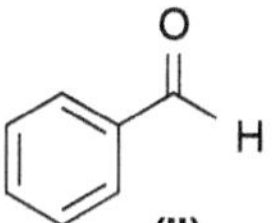 (II)	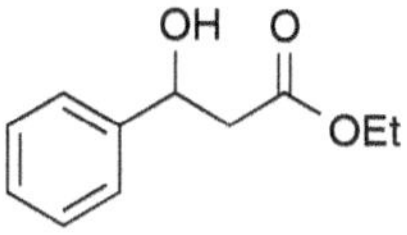 1a	90
2	I	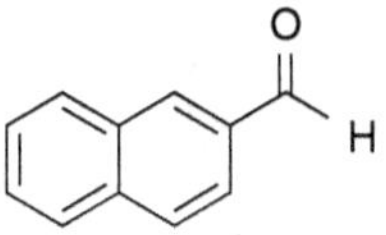	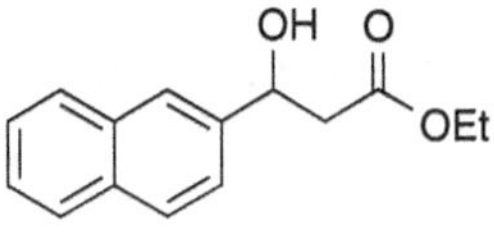 1b	85
3	I	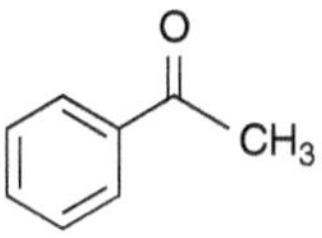	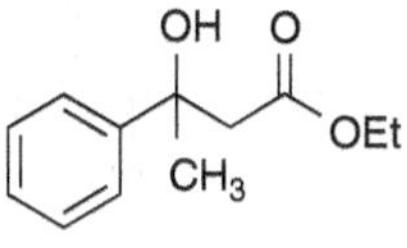 1c	86
4	I	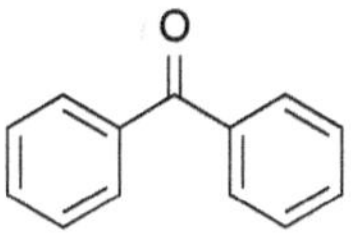	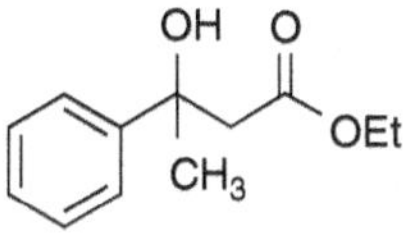 1d	82
5	I	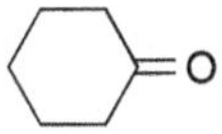	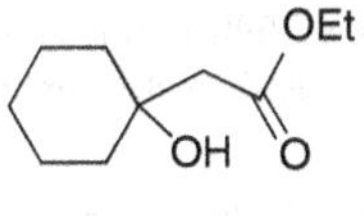 1e	91
6	I	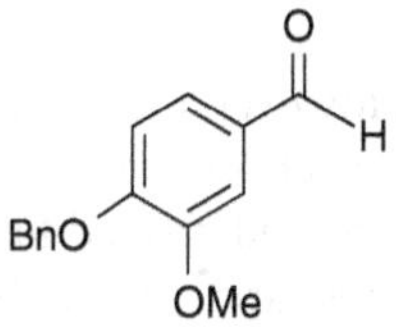	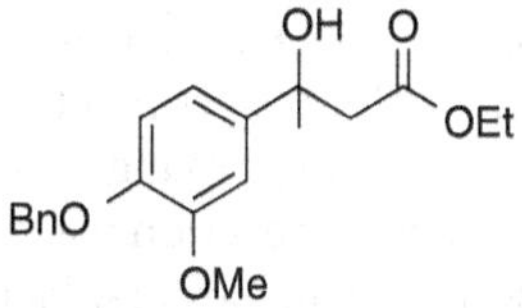 1f	79
7	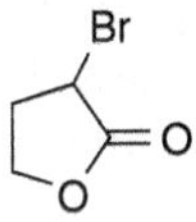	II	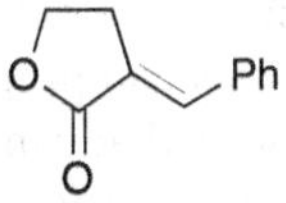 1g	90

(*Continued*)

Table 8.21 (Continued)

Entry	α-Bromoesters	Carbonyl compound	Product[a]	Yield (%)[b]
8	I	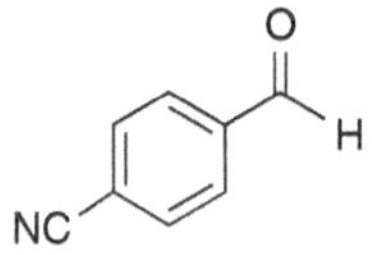 4-NC-C6H4-CHO	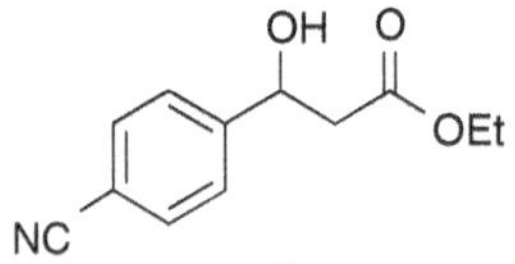 1h	86
9	$C_2H_5O_2C(CH_2)_6Br$	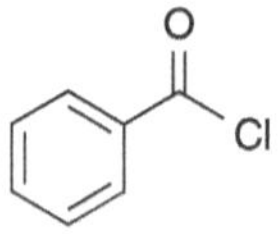 PhCOCl	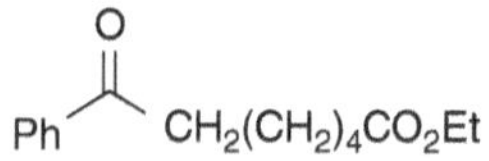 Ph-CO-$CH_2(CH_2)_4CO_2Et$ 1i	68
10	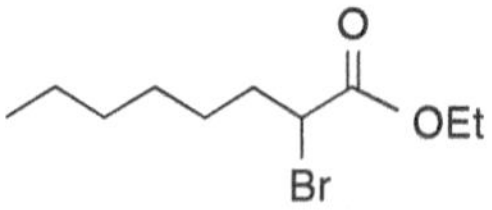 ethyl 2-bromooctanoate (OEt, Br)	II	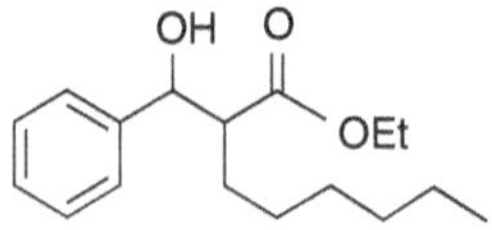 1j	70
11	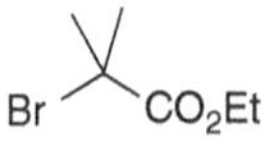 $BrC(CH_3)_2CO_2Et$	II	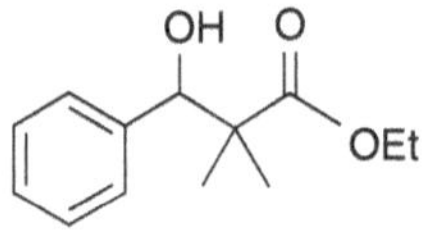 1k	58

[a] All products are fully characterized (^1H, ^{13}C NMR, and MS) and match literatures.
[b] Isolate yield (based on electrophiles).

magnesium reagents. Secondary and tertiary manganese enolates were also tested with aldehydes and were shown to provide the corresponding products readily (Table 8.21, entries 10 and 11). The activated manganese will also react with α-chloroesters in THF. However, the yields are only in the modest range.

In summary, manganese ester enolates were readily prepared for the Reformatsky reaction via direct oxidative addition of highly active Rieke manganese to α-haloesters, lactones, and alkyl halides with remote ester groups. The resulting reagents demonstrated excellent addition reactions with various electrophiles such as aldehydes and ketones. The employed procedure does not require any special preparation or complicated workup process. The wide range of haloesters that can be used in this reaction offers the synthetic chemist an important new tool. The following is a representative procedure: ethyl α-bromoacetate (60 mmol) and benzaldehyde (58 mmol) were weighed separately. To the black slurry of active manganese in THF (30 ml) was added the mixture of benzaldehyde and ethyl α-bromoacetate drop by drop over 25 min.

Table 8.22 Reactivity study of Mn* on halides.

PhCHO → (MnX_2 (1 equiv), Li(2.0 equiv) / Np(0.2 equiv)) → 1,2-diphenyl-1,2-ethanediol (OH, OH), *dl/meso*

MnX_2	Conditions (°C/time)	Yield (%)[a]
$MnCl_2$	rt/30 min	86
$MnBr_2$	rt/overnight	89
MnI_2	rt/24 h	80

[a] Isolate yield (based on aldehyde).

The addition was conducted at 0°C and the entire system was kept under argon. The reaction mixture was stirred at rt for about an hour after the addition was completed. The reaction was then quenched with 2 N hydrochloric acid (20 ml) or saturated ammonium chloride, and the organic layer was separated from the acidic aqueous layer. The aqueous layer was extracted with ether (3 × 20 ml), and the combined organic phase was washed twice with saturated sodium hydrogen carbonate (10 ml) and water (10 ml). The ether extract was dried over anhydrous magnesium sulfate and filtered, and the solvent was evaporated, resulting in a pale yellow liquid. The crude product was subjected to GLC, which showed no by-products. The crude product was purified by vacuum distillation or column chromatography, yielding 90% of the corresponding compound.

8.7 Reductive Coupling of Carbonyl-Containing Compounds and Imines Using Reactive Manganese

One of the most powerful methods for constructing a carbon–carbon bond is the reductive coupling of carbonyl compounds giving olefins and/or 1,2-diols [71]. Of these methods, the pinacol coupling [72], which was described in 1859, is still a useful tool for the synthesis of vicinal diols. The corresponding products of this reaction can be used as intermediates for the preparation of ketones and alkenes [73]. More importantly, this methodology has been applied to the synthesis of biologically active natural compounds [74].

Generally, the reaction is effected by treatment of carbonyl compounds with an appropriate reagent and/or metal complex to give rise to the corresponding coupled product. A number of different types of metal reagents have been used to carry out the pinacol reaction. For instance, the reaction with various low-valent metal complexes of Al [75], Ce [76], Fe [77], Mg [78], Nb [79], Sm [80], Si [81], Ti [82], V [83], Yb [84], Zr [85], and main group organometallic hydrides (Bu_3SnH, Ph_3SnH, Bu_3GeH, and $(TMS)_3SiH$) [86] afforded inter- or intramolecular coupling products of carbonyls.

The coupling products can have two newly formed stereocenters. As a consequence, efficient reaction conditions have been required to control the stereochemistry of the 1,2-diols. Recent efforts have focused on the development of new reagents and reaction systems to improve the reactivity of the reagents and diastereoselectivity of the products. In addition, functional group tolerance has been a challenge for reductive coupling of carbonyl compounds using the reaction systems mentioned in the preceding text. However, many successful results have been reported.

In 1984, Lukehart [87] reported a formal reductive coupling of mangana-β-diketonato complexes to give carbon–carbon bond formation. Manganese complexes also have been used as a good single-electron source in oxidative free-radical cyclizations [88]. On the basis of these findings, we postulated that our active manganese might be a good single-electron donor and, therefore, be utilized for pinacol coupling.

During the course of our investigation on active metals, we have found that treatment of aryl aldehydes and aryl ketones with active manganese (Mn*) prepared via the Rieke method gave the corresponding pinacolic coupling product (Tables 8.23 and 8.24). As shown in Scheme 8.11, the reaction of aldimines with this active manganese also afforded the corresponding coupling products, vicinal diamines. Due to the potential utility of vicinal diamines in

Table 8.23 Study of reactivity depending on halides.

MnX_2 + Li/Np —THF/rt, PhCHO **1a**→ Ph–CH(OH)–CH(OH)–Ph **2a** *dl/meso*

1 equiv 2.1 equiv/0.2 equiv

Np: Naphthalene

MnX_2	Conditions (temp./time)	Yield (%)	*dl*/meso
$MnCl_2$	rt/30 min	86	63:37
$MnBr_2$	rt/overnight	89	68:32
MnI_2	rt/24 h	80	54:46

Table 8.24 Reductive coupling of arylaldehydes.

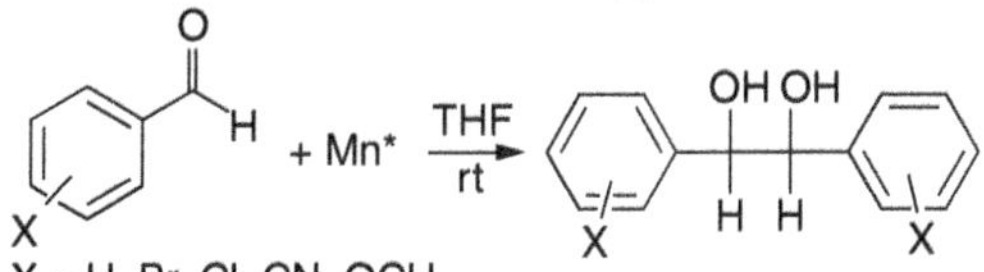

X = H, Br, Cl, CN, OCH_3 (1a–i) + Mn* → (THF, rt) 2a–i

Entry	Carbonyls	Aldehyde : Mn*	Product[a]	Yield (%)[b]	*dl*/meso[c]
1	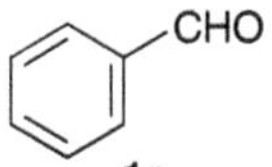 1a	0.8 : 1	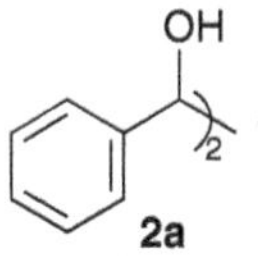 2a	64[c]	63 : 37
		0.8 : 1		62[d]	
		0.5 : 1		86[e]	73 : 27
2	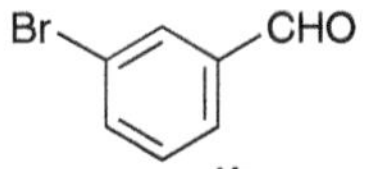 1b	0.5 : 1	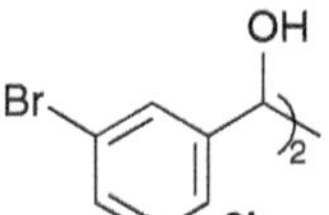 2b	60	93 : 7
3	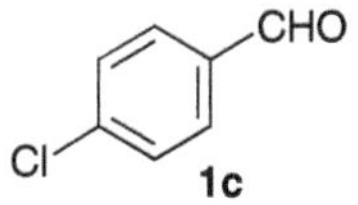 1c	0.8 : 1	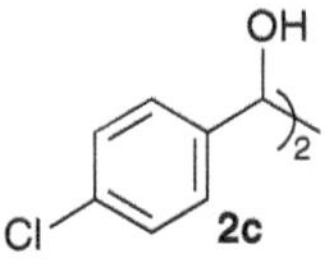 2c	51[f]	58 : 42
		0.5 : 1		81[e]	83 : 17
4	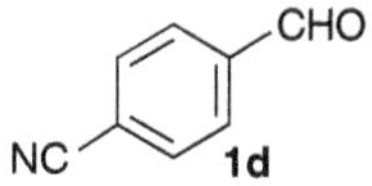 1d	0.5 : 1	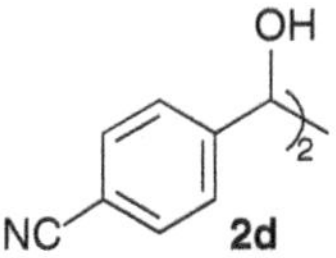 2d	94	63 : 37
5	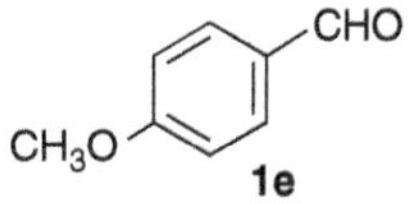 1e	0.5 : 1	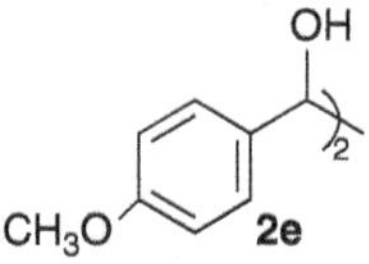 2e	83	66 : 34
6	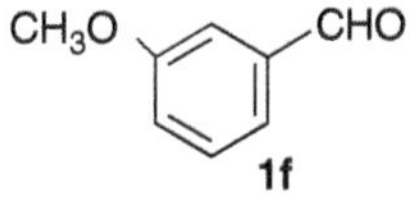 1f	0.5 : 1	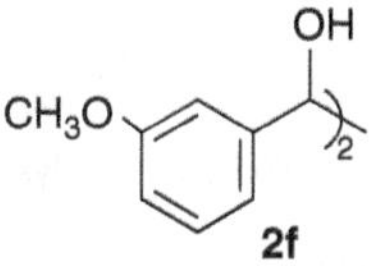 2f	90	64 : 36

(*Continued*)

Table 8.24 (Continued)

Entry	Carbonyls	Aldehyde : Mn*	Product[a]	Yield (%)[b]	*dl*/meso[c]
7	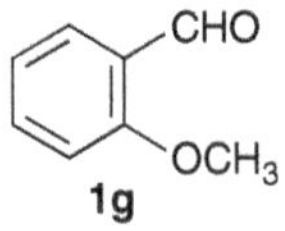 **1g**	0.5 : 1	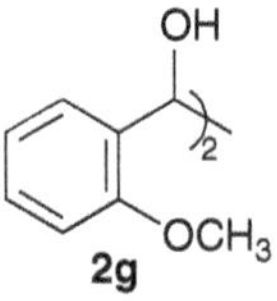 **2g**	90	66 : 34
8	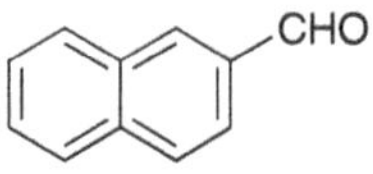 **1h**	0.7 : 1	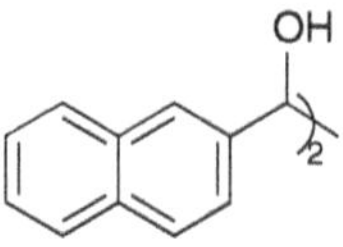 **2h**	72	51 : 49
9	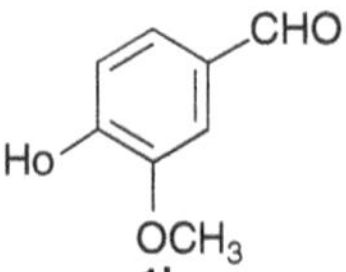 **1i**	0.5 : 1	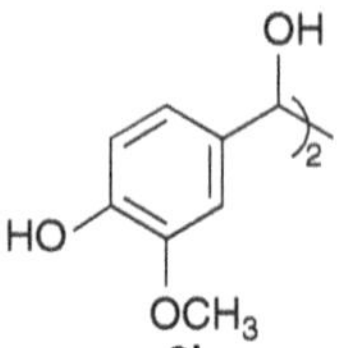 **2i**	0[g]	

[a] All products have been fully characterized by ^{1}H NMR, ^{13}C NMR, and HRFAB-MS.
[b] Isolated yields (based on aldehydes).
[c] Determined by ^{1}H NMR (300 MHz) analysis of the isolated product.
[d] Reaction was carried out at 0°C.
[e] Obtained from recrystallization.
[f] Obtained from flash chromatography.
[g] According to TLC analysis, starting material (aldehyde) was recovered after hydrolysis of the reaction mixture.

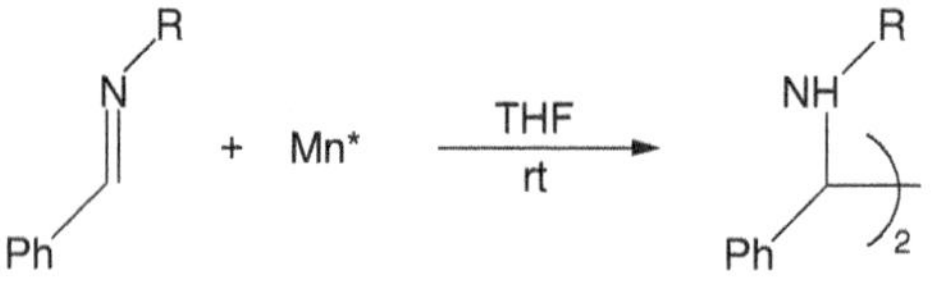

1 equiv + Mn* 2 equiv → THF, rt

13 R = Ph → **14** R = Ph, 62%, *dl*/*meso*[a] 64 : 36

15 R = CH_2Ph → **16** R = CH_2Ph, 56%, *dl*/*meso* 49 : 51

[a] Determined by ^{13}C NMR (75 MHz) analysis

Scheme 8.11 Reductive coupling of aldimines.

organic synthesis, especially in natural products and medicinal reagents [89], synthetic methodologies of vicinal diamines have attracted considerable attention. To date, several methods have been described to perform the reductive dimerization of imines into vicinal diamines [90]. The most widely used method is metal-mediated pinacolic coupling of imines. Several metals have been utilized including samarium [91], sodium [90c, 92], zirconium [93], aluminum, bismuth [94], zinc [95], titanium [96], indium [97], niobium [98], and ytterbium [99].

To date, no report has appeared using manganese metal. We wish to report a mild method for reductive coupling reactions of aryl aldehydes, aryl ketones, and aldimines [100].

Results and Discussion

Reduction of manganese halides (MnI_2, $MnBr_2$, and $MnCl_2$) by Li using naphthalene as an electron carrier in THF affords a slurry of Mn* at rt. The resulting Mn*, however, appears to be partially soluble in THF. Due to this, no washing was performed to remove naphthalene and other salts, and the Mn* was used as a slurry. The preparation of the active metals and the subsequent reaction of the organometallic reagents are conducted under an argon atmosphere.

Reductive Coupling Reactions of Aryl Aldehydes

As described in the preceding text, three different types of manganese halides can be used to prepare active manganese. Therefore, reactivity depending on metal halide was studied first. The results are summarized in Table 8.23, and it was found that there are no significant differences in both yield and stereochemistry of the product (**2a**). Good isolated yields (80–89%) and an almost 1 : 1 mixture of *dl*/meso isomers were obtained from three different attempts described in Table 8.23. Longer reaction times were required to complete the coupling reaction in the case of manganese iodide. While we have not conducted a detailed study of the reaction pathway, we believe that this reaction proceeds via one of the mechanisms presented in reference [72b].

Various aromatic aldehydes were reacted with the active manganese prepared from using manganese chloride. Table 8.24 reports that simple treatment of aryl aldehydes with active manganese gives the corresponding 1,2-diols in moderate to good isolated yields under very mild conditions, rt in THF. The results are shown in Table 8.24.

It is proposed that the reaction proceeds via a single-electron transfer (SET) [72b]. With the active manganese supplying the electrons, the ratio of manganese to aldehyde was found to be very important with respect to overall yield. Entry 1 in Table 8.24 shows that a higher yield is obtained using 2 equiv of Mn* and 1 equiv of aldehyde. This appears to be general for all aldehydes and ketones. All of the reactions with different mole ratios were carried out at rt in

30 min. Reaction temperature had little effect on the yield. Reactions at rt and 0°C in 30 min gave 64 and 62% isolated yields, respectively (Table 8.24, entry 1).

It should be mentioned that this reaction exhibits tolerance of several functional groups on the aromatic ring, including bromine, chlorine, cyano, and methoxy (Table 8.24, entries 2–5). According to the high-resolution FAB-MS of **2b–g** in Table 8.24 and **6** in Table 8.25, the functional groups are still retained in the final products. Unfortunately, according to TLC analysis of the reaction mixture after hydrolysis, the starting material was solely recovered in the coupling reaction of vanillin, which bears a hydroxyl group on the aromatic

Table 8.25 Reductive coupling of arylketones.

Entry	Ketones	Ketone : Mn*	Product[a]	Yield (%)[b]	*dl*/*meso*[c]
1	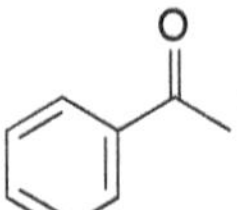 **3**	0.8 : 1	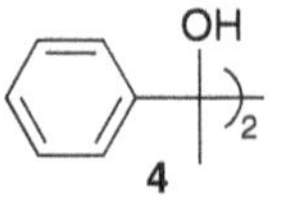 **4**	54	>99 : 1
		0.5 : 1		71	96 : 4
2	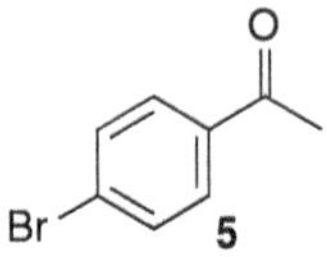 **5**	0.5 : 1	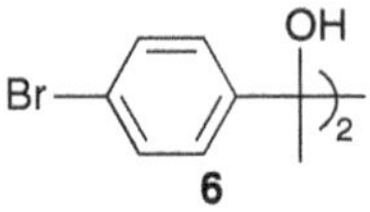 **6**	91	73 : 27
3	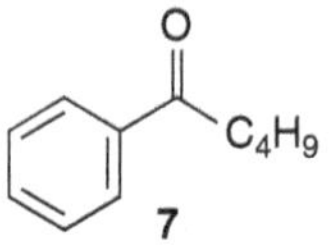 **7**	0.5 : 1	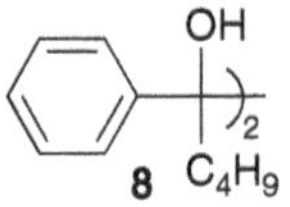 **8**	88	76 : 24
4	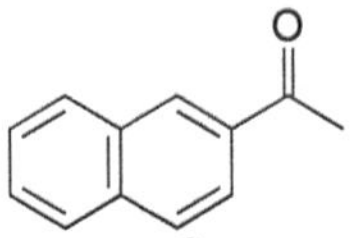 **9**	0.5 : 1	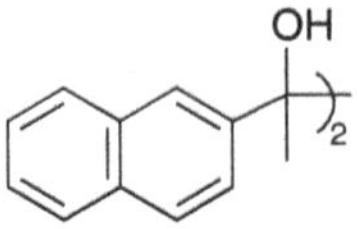 **10**	93	87 : 13
5	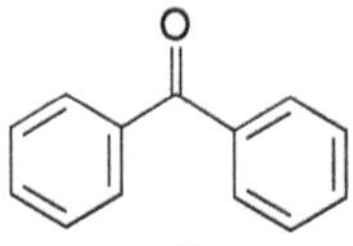 **11**	0.5 : 1	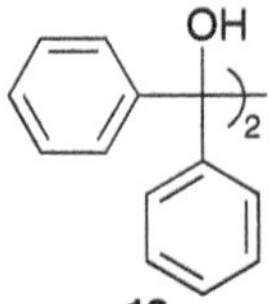 **12**	91	—

[a] All products have been fully characterized by ^{1}H NMR, ^{13}C NMR, and HRFAB-MS.
[b] Isolated yields (based on ketones).
[c] Determined by ^{1}H NMR (300 MHz) or ^{13}C NMR (75 MHz) analysis of the isolated product.

ring (Table 8.24, entry 9). Even when 5 equiv of Mn* was used, the same result was observed. Probably, the acidic hydrogen of the phenolic OH reacts with the Mn*, quenching the reaction.

To investigate the possible steric effects of substituents on the coupling reaction, *o*-, *m*-, and *p*-substituted aldehydes were examined. *o*-, *m*-, and *p*-methoxybenzaldehydes were reacted with Mn* under the same conditions (Table 8.24, entries 5–7). For these molecules, overnight stirring at rt was required to complete the reaction. The results shown in Table 8.24 indicate that no steric hindrance is observed in terms of yield (83–90%). Table 8.24 also shows that high diastereoselectivity is not observed in this reaction system except in the case of 3-bromobenzaldehyde. In the reaction of 3-bromobenzaldehyde, the *dl*-isomer was the major product (*dl*/meso 93 : 7), but the reason for this is not clear. To investigate the change of *dl*/meso ratio depending upon the purification protocols, two different methods, column chromatography and recrystallization, were used to purify the crude reaction mixture obtained from the reaction of 1 equiv of Mn* and 0.5 equiv of benzaldehyde at rt. In both cases, the same *dl*/meso (76 : 24) was observed. This ratio was essentially the same as that of the crude mixture. The ^{1}H NMR spectrum of the crude mixture indicated a *dl*/meso ratio of 74 : 26. A slightly different *dl*/meso ratio (79 : 21) was obtained from the recrystallization (20% ethyl ether–hexanes) of prechromatographically purified product. Interestingly, a significant change in the *dl*/meso ratio was observed in the coupling product of 4-chlorobenzaldehyde. Recrystallization of the prechromatographically purified product (*dl*/meso 63 : 37) was carried out by using 20% ethyl ether–hexanes. The ^{1}H NMR spectrum of the recrystallized product showed a higher *dl*/meso ratio (95 : 5). This is clearly a result of partial separation upon purification.

In addition to the coupling of aryl aldehydes, it should be mentioned that a multicompound mixture was obtained from the reactions using aliphatic aldehydes with Mn*. According to Yanada's work, the radical intermediates from the aliphatic aldehydes are unstable and do not have enough time to react with each other to give diols.

Reductive Coupling Reactions of Aryl Ketones

Table 8.25 represents the reductive coupling reactions of aryl ketones using active manganese. These reactions were also conducted under almost the same conditions used in aryl aldehyde couplings, and the results are summarized in Table 8.25. Isolated yields were good to excellent (71–93%), and much higher stereoselectivity was observed. As observed in the aldehyde couplings, the mole ratio of carbonyl to metal reagent greatly affects the yield of the final product as is shown in entry 1 in Table 8.25. For instance, the reaction of 0.5 equiv of acetophenone (**3**) with 1 equiv of active manganese afforded the 1,2-diol derivative **4** in higher yields (71%) than that (54%) using 0.8 equiv of ketone with 1.0 equiv of Mn*. Interestingly, much higher diastereoselectivity is observed. The reason for this may be steric. The corresponding coupling products (**6**, **8**, and **10**) of alkyl–aryl-substituted ketones (**5**, **7**, and **9**) were obtained in good yields (88–93%) with moderate to good diastereoselectivities under the conditions presented in Table 8.25. Even for a bulky biaryl ketone, the expected coupling product was obtained in high yield. For example, treatment of benzophenone (**11**) with active manganese gave 1,1,2,2-tetraphenylethanediol (**12**) in 91% isolated yield. Once again, attempts to reductively couple alkyl ketones with Mn* gave a multicompound mixture, which was not identified.

Reductive Coupling of Aldimines

It was also found that the active manganese could be used for the coupling reaction of aldimines with Mn*. As shown in Scheme 8.11, mild reaction conditions, rt in THF, were used to carry out the reaction. The imines (**13** and **15**) and active manganese were allowed to stir overnight at rt. After appropriate workup, the corresponding vicinal diamines (**14** and **16**) were afforded in 62 and 56% isolated yields, respectively, with poor diastereoselectivity. ^{1}H NMR and ^{13}C NMR data are consistent with the literature values. Mechanism and optimization studies have not been conducted. However, we believe that this coupling also proceeds via a pinacol-type SET mechanism [72b]. We were not able to detect any of the monoamines resulting from the reduction of the starting aldimines.

Conclusions

In conclusion, we have demonstrated that highly reactive manganese can be used for the reductive dimerization of aryl aldehydes, aryl ketones, and aldimines. With this highly reactive manganese, the corresponding coupling products, 1,2-diols and vicinal diamines, were obtained in good isolated yields. The diastereoselectivity is poor except for aryl ketones. It is well known that pinacol coupling proceeds via a SET mechanism. Accordingly, this work demonstrates that the highly reactive manganese can be used as a good single-electron donor in organic synthesis.

Typical Experimental Procedures

Preparation of Highly Reactive Manganese (Mn*)

Highly reactive manganese was prepared by the reduction of anhydrous manganese halides (chloride, bromide, and iodide) with lithium using naphthalene as an electron carrier. In a typical preparation, lithium (9.68 mmol), naphthalene (1.48 mmol), and anhydrous manganese chloride (4.71 mmol) were stirred in freshly distilled THF (15 ml) for 1–3 h at rt. A black slurry was obtained and ready for use. (Note: The number of millimoles of Mn* cited in here refers to the theoretical amount possible, based on the original amount of anhydrous manganese halide.)

A Typical Procedure for the Preparation of 1,2-Diols (2a–2h) from the Reactions of Aryl Aldehydes with Mn*

To a slurry of Mn* was added aryl aldehyde (0.5–0.8 equiv, based on Mn*) via syringe (or cannula) at rt. The resulting mixture was allowed to stir for 30 min to overnight. The reaction was monitored by TLC. After being stirred, the reaction was quenched with an aqueous solution of 3 M HCl (20 ml) and then extracted with ethyl acetate (3 × 20 ml). The combined organic layers were washed with saturated NaCl solution (2 × 30 ml) and dried over anhydrous $MgSO_4$. Evaporation of solvents and flash chromatography (hexanes–ethyl acetate) and/or recrystallization afforded the desired diols in the indicated yield and diastereoselectivity (Table 8.24).

Typical Procedure for the Preparation of 1,2-Diols from the Reaction of Aryl Ketones with Mn*

The standard procedure used for aryl aldehydes was followed with aryl ketones. Appropriate workup afforded 1,2-diols (**4**, **6**, **8**, **10**, and **12**) in 71–93% yields.

Diol 4 (mixture of *dl*/meso isomers) was obtained from recrystallization (hexanes–ethyl ether) of the crude mixture in 71% yield.

Reductive Coupling Reaction of Aldimines (13 and 15) into Vicinal Diamines (14 and 16)

The standard procedure used for aryl aldehydes was followed using aldimines and Mn* prepared from MnI_2. During the workup step, the combined organic layers were washed with saturated $Na_2S_2O_3$ solution to remove iodine species from the reaction mixture. Flash chromatography (hexanes–ethyl acetate) afforded the corresponding vicinal diamines in 62 and 56% yields, respectively.

Diamine 14 (mixture of dl/meso isomers) Flash chromatography (hexanes–ethyl acetate) of the crude mixture afforded **14** in 62% yield.

Diamine 16 (mixture of dl/meso isomers) Flash chromatography (hexanes–ethyl acetate) of the crude mixture afforded **16** in 56% yield.

8.8 Preparation of Heteroarylmanganese Reagents and Their Cross-Coupling Chemistry

Current interest in the synthesis of natural products containing heteroaromatic moieties as building blocks for pharmaceutical industries, material science, and supramolecular chemistry has resulted in considerable effort in developing new synthetic methods [101].

Among the extensive studies for the preparation of heteroaryl compounds, organometallic reagents are frequently used and mainly prepared from metal–halogen exchange reactions or the metathesis of the corresponding organolithium and organomagnesium derivatives [102]. However, this approach limits the number of functional groups that can be tolerated. The same disadvantage was observed when organozinc reagents were prepared by transmetallation of organolithium and organomagnesium reagents [103]. To alleviate the difficulty, direct preparation of heteroaryl zinc reagents by using Rieke zinc has been reported [104].

In our studies on the utility of active manganese, we examined the range of oxidative addition and cross-coupling reactions with a broad range of heterocyclic halides. Active manganese prepared by the Rieke method has been shown to exhibit superior activity to organic halides under mild reaction conditions. The resulting organomanganese reagents have also shown a great reactivity toward several electrophiles in excellent yields [105]. In spite of the importance of π-deficient heteroarylmanganese halides, few results have been reported about the preparation and application of heteroaromatic manganese reagents [106].

We report the direct formation of heteroarylmanganese halides and the results of several cross-coupling reactions. In addition, the palladium-catalyzed cross-coupling reaction of heteroarylmanganese reagents is also described [107].

In general, heteroarylmanganese halides are easily prepared by the reaction of highly active manganese with various heteroaryl halides under mild reaction conditions. It is of interest that the cross-coupling reaction of these reagents with acid chlorides was accomplished without using any transition metal catalyst, which is required for the reaction of organozinc reagents. The corresponding ketones were obtained in good isolated yield (Scheme 8.12). The results of the cross-coupling reactions are summarized in Tables 8.26 and 8.27.

The oxidative addition of Mn* was completed using 3 equiv of Mn* and 1 equiv of the corresponding halides in a few hours at rt except with 3-bromofuran (Table 8.26, entries 6 and 11). The following coupling reactions were performed in an hour at rt as well. It is noteworthy that the acylation reaction of N-containing π-deficient heteroaryl compounds such as pyridine becomes readily feasible through direct preparation and reaction of pyridinylmanganese reagents (Table 8.26, entries 1, 4, 8, and 9). This is particularly significant considering the fact that Friedel–Crafts acylation or alkylation cannot be done in

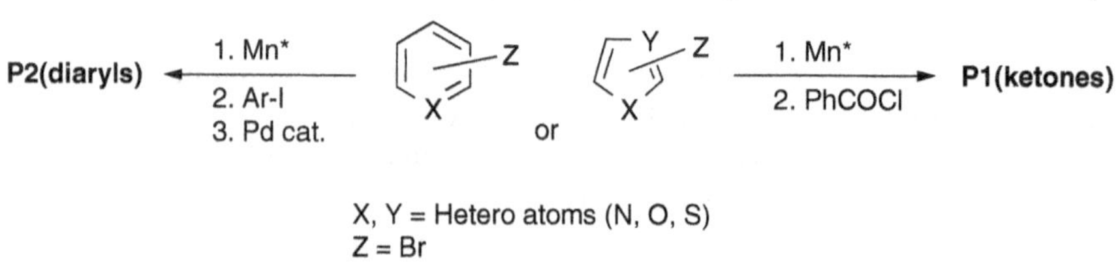

P2(diaryls) ← 1. Mn* 2. Ar-I 3. Pd cat. — [heteroaryl–Z] or [five-membered heteroaryl–Z] — 1. Mn* 2. PhCOCl → P1(ketones)

X, Y = Hetero atoms (N, O, S)
Z = Br

Scheme 8.12 Reaction of heterocyclic compounds with electrophiles.

Table 8.26 Reactions of heteroaryl manganese bromides with benzoyl chloride.

Entry	Substrate	Product[a]	Yield[b] (%)
1	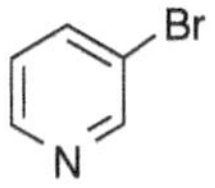	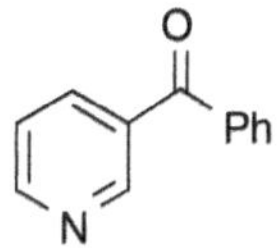 **1a**	61
2	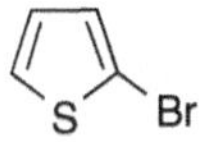	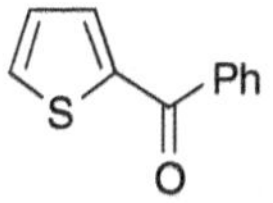 **1b**	92
3	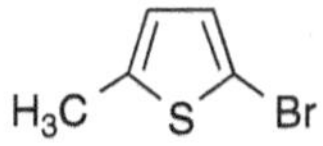	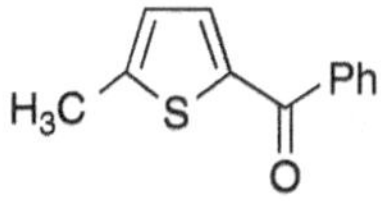 **1c**	90
4	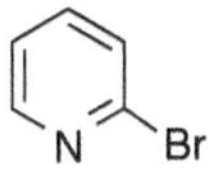	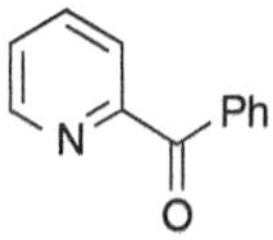 **1d**	33
5	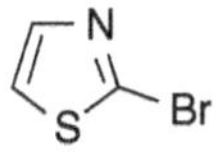	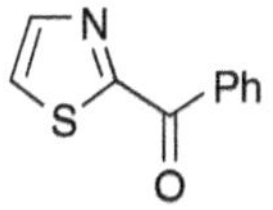 **1e**	55
6[c]	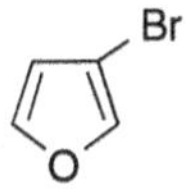	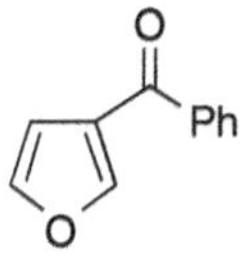 **1f**	60
7	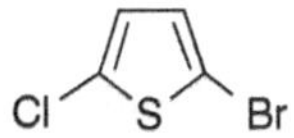	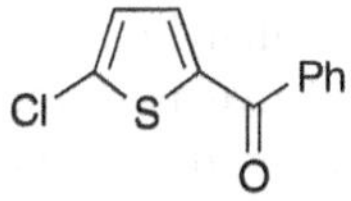 **1g**	88

(Continued)

Table 8.26 (Continued)

Entry	Substrate	Product[a]	Yield[b] (%)
8	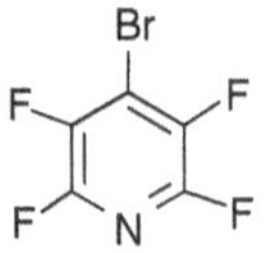	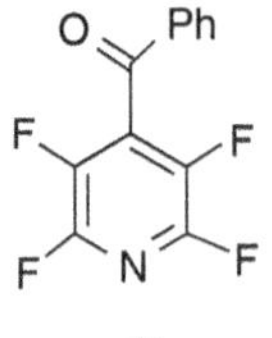 1h	71
9	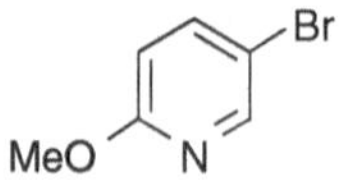	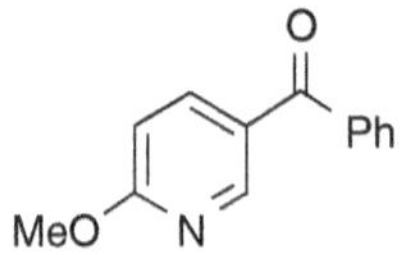 1i	65
10[d]	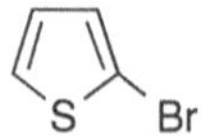	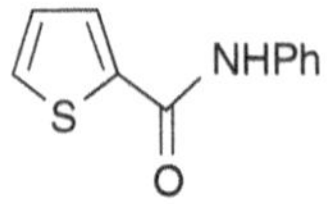 1j	80
11[c]	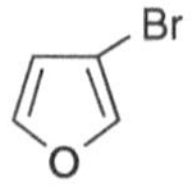	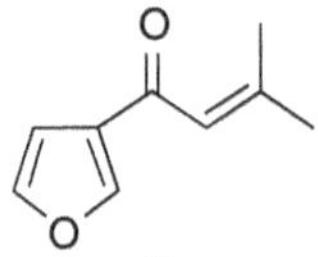 1k	67
12	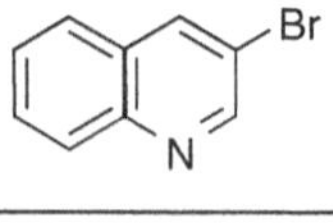	—	0

[a] All products were fully characterized by ^1H, ^{13}C NMR, and HR-MS.
[b] Isolated yield (based on organomanganese reagent).
[c] 4 equiv of Mn* was used for oxidative addition.
[d] Phenyl isocyanate was used as electrophile.

the pyridine ring. Substituted thiophene derivatives were also examined, demonstrating that these thienyl reagents could be prepared readily by the direct oxidative addition of active manganese to thiophene halides with regioselectivity and no decomposition of the thiophene ring at rt. These thiophene derivatives have attracted much attention in both material and pharmaceutical sciences [108] (Table 8.26, entries 2, 3, 7, and 10). In the case of entry 7, selective oxidative addition and coupling reaction with benzoyl chloride was observed. When phenyl isocyanate was used as the electrophile (Table 8.26, entry 10), the reaction gave a secondary amine product. As to heteroaryl

Table 8.27 Cross-coupling reactions of heteroaryl manganese reagents.[a]

Entry	Substrate	Electrophile	Product[b]	Yield[c]
1	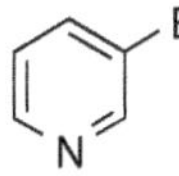 Br, N	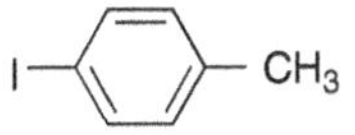 I, CH_3 A	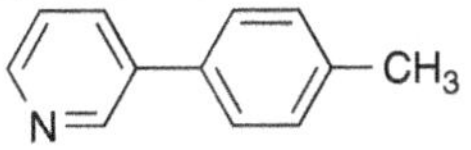 N, CH_3 **2a**	72
2	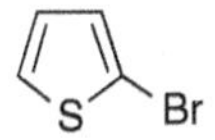 S, Br	A	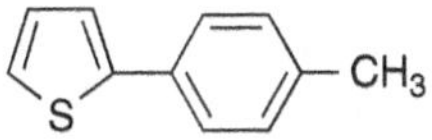 S, CH_3 **2b**	91
3	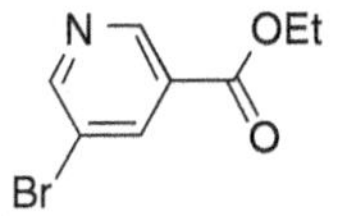 N, OEt, O, Br	A	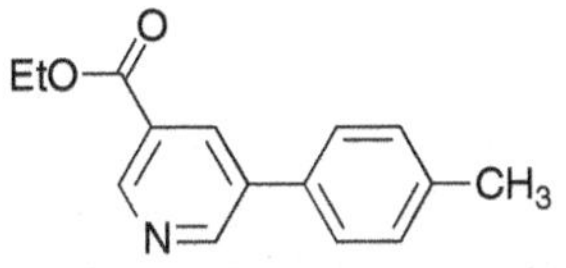 EtO, O, N, CH_3 **2c**	54
4	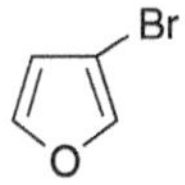 Br, O	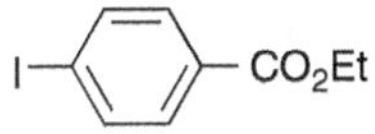 I, CO_2Et	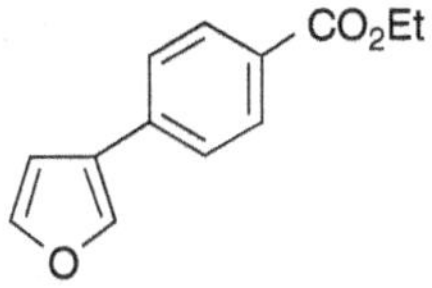 CO_2Et, O **2d**	60
5	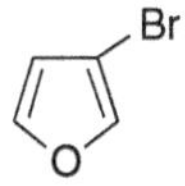 Br, O	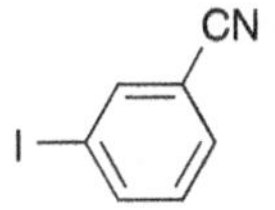 CN, I	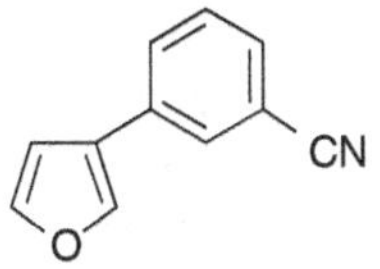 CN, O **2e**	66
6	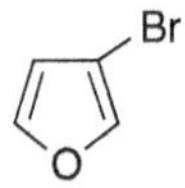 Br, O	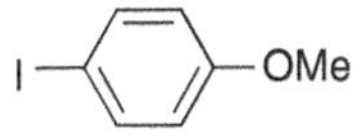 I, OMe	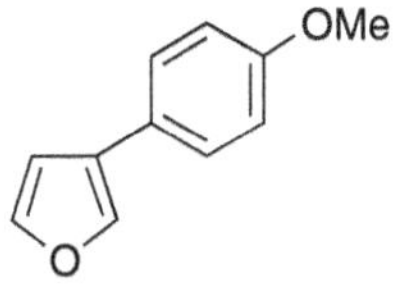 OMe, O **2f**	56
7	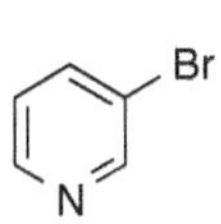 Br, N	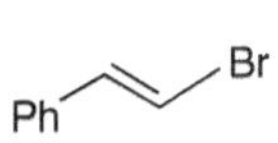 Ph, Br	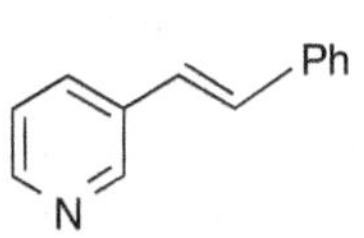 Ph, N **2g**	58

[a] Performed in the presence of 2 mol% $Pd[P(Ph_3)]_4$.
[b] All products were fully characterized by ^1H, ^{13}C NMR, and HR-MS.
[c] Isolated yield (based on electrophile).

five-membered rings, it was interestingly shown that furanic ketones or 3-acylfurans were easily obtained through the acylation of a 3-furyl manganese derivative (Table 8.26, entries 6 and 11). It is well known that such acylations are very difficult because of the low stability of these organometallic reagents, which must be prepared and used at low temperature [109]. However, the organomanganese reagents were readily prepared with highly active manganese and 3-bromofuran at rt. They have been used to prepare the derivative of natural furanic ketones (perilla ketone) [110].

Other possible substrates were also tested for the reactivity studies. However, in the case of 3-bromoquinoline as an example (Table 8.26, entry 12), the reaction gave a high yield of the reduced product, and no coupling product was observed.

Cross-coupling reactions with functionalized aryl halides were also investigated in the presence of a catalytic amount or Pd catalyst. The corresponding C—C bond-forming products were obtained in good yield as shown in Table 8.27. The reactions were also performed readily in a few hours at rt. Interestingly, 3-stilbazole (Table 8.27, entry 7), which is used in a photo-responsive chemical system in plants [111], was readily prepared by the reaction of pyridinylmanganese bromide with β-bromostyrene under the applied conditions. Additionally, furanic manganese bromides and thienyl manganese bromides underwent cross-coupling reactions with electrophiles containing different functional groups in good yields.

In conclusion, organomanganese reagents were easily prepared via direct oxidative addition of highly active manganese to a variety of heteroaryl compounds, and the resulting organomanganese reagents gave the expected cross-coupling products in good yields under mild reaction conditions [112]. It is noteworthy that functional groups such as ester and cyano are tolerated by the highly active manganese.

References

1 (a) Gilman, H.; Bailie, J.C. *J. Org. Chem.* 1937, **2**, 84. (b) Gilman, H.; Kirby, R. *J. Am. Chem. Soc.* 1941, **63**, 2046.

2 For a review, see (a) Normant, J.F.; Cahiez, G. *Modern Synthetic Methods*; John Wiley & Sons, Ltd, Chichester, 1983; Vol. **3**; pp. 172–216, and references cited therein. (b) Cahiez, G.; Chavant, P.Y.; Metais, E. *Tetrahedron Lett.* 1992, **33**, 5245. (c)Cahiez, G.; Laboqe, B. *Tetrahedron Lett.* 1989, **30**, 7369.

3 Cahiez, G.; Chavant, P.Y. *Tetrahedron Lett.* 1989, **30**, 7373.

4 (a) Rieke, R.D.; Hudnall, P.M. *J. Am. Chem. Soc.* 1972, **94**, 7178. (b) Rieke, R.D.; Uhm, S.J.; Hudnall, P.M. *J. Chem. Soc., Chem. Commun.* 1973, 269. (c) Rieke, R.D.; Uhm, S.J. *Synthesis* 1975, 452. (d) Rieke, R.D.; Li, P.J.; Burns, T.P.; Uhm, S.J. *J. Org. Chem.* 1981, **46**, 4342.

5 (a) Rieke, R.D. *Top. Curr. Chem.* 1975, **59**, 1. (b) Rieke, R.D. *Acc. Chem. Res.* 1977, **10**, 301, and references therein. (c) Arnold, R.T.; Kulenovic, S.T. *Synth. Commun.* 1977, **7**, 223.

6 (a) Gronowitz, S.; Hakansson, R. *Arkiv. Kemi.* 1959, **17**, 73–82. (b) Gronowitz, S. *Organic Sulphur Chemistry—Structure, Mechanism, and Synthesis*; Sterling, C.J.M., Ed.; Butterworths: London, 1975; pp. 203–228.

7 (a) Screttas, C.G.; Micha-Screttas, M. *J. Org. Chem.* 1979, **44**, 713. (b) Still, W.C. *J. Am. Chem. Soc.* 1978, **100**, 1481. (c) Parham, W.E.; Jones, L.D.; Sayed, Y.A. *J. Org. Chem.* 1976, **41**, 1184. (d) Gilman, H.; McNinch, H.A. *J. Org. Chem.* 1961, **26**, 3723.

8 (a) Clarembeau, M.; Krief, A. *Tetrahedron Lett.* 1985, **26**, 1093. (b) Seyferth, D.; Suzuki, R.; Murphy, C.J.; Sabet, C.R. *J. Organomet. Chem.* 1964, **2**, 431. (c) Gilman, H.; Rosenberg, S.D. *J. Org. Chem.* 1959, **24**, 3.

9 (a) van den Anker, T.R.; Harvey, S.; Raston, C.L. *J. Organomet. Chem.* 1995, **502**, 35. (b) Bernardon, C. *J. Organomet. Chem.* 1989, **367**, 11. (c) Harvey, S.; Junk, P.C.; Raston, C.L.; Salem, G. *J. Org. Chem.* 1988, **53**(3), 3134. (d) Gallagher, M.J.; Harvey, S.; Raston, C.L.; Sue, R.E. *J. Chem. Soc., Chem. Commun.* 1988, 289. (e) Harvey, S.; Raston, C.L. *J. Chem. Soc. Chem. Commun.* 1988, 652. (f) Itsuno, S.; Darling, G.D.; Stőver, H.D.H.; Frechet, J.M.J. *J. Org. Chem.* 1987, **52**, 4644. (g) Raston, C.L.; Salem, G. *J. Chem. Soc., Chem. Commun.* 1984, 1702.

10 (a) Betzemeier, B.; Knochel, P. *Angew. Chem. Int. Ed. Engl.* 1997, **36**, 2623. (b) Rottländer, M.; Knochel, P. *Synlett* 1997, 1084. (c) Klement, I.; Lennick, K.; Tucker, C.E.; Knochel, P. *Tetrahedron Lett.* 1993, **34**, 4623. (d) Chia, W.-L.; Shiao, M.-J. *Tetrahedron Lett.* 1991, **32**, 2033. (e) Shing, T.-L.; Chia, W.-L.; Shiao, M.-J. Chau, T.-Y. *Synthesis* 1991, 849. (f) Chen, H.G.; Hoechstetter, C.; Knochel, P. *Tetrahedron Lett.* 1989, **30**, 4795. (g) Berk, S.C.; Knochel, P.; Yeh, M.C.P. *J. Org. Chem.* 1988, **53**, 5789.

11 Burkhardt, E.R.; Rieke, R.D. *J. Org. Chem.* 1985, **50**, 416.

12 Harada, T.; Kaneko, T.; Fujiwara, T.; Oku, A. *J. Org. Chem.* 1997, **62**, 8966.

13 (a) An attempt for the preparation of benzylic manganese reagent has been carried out using manganese powder, resulting in the formation of homocoupling product (89%), see: Hiyama, T.; Sawahata, M.; Obayashi, M. *Chem. Lett.* 1983, 1237. (b) Use of manganese for the preparation of allylic manganese reagent, see: Hiyama, T.; Obayashi, M.; Nakamura, A. *Organometallics* 1982, **1**, 1249.

14 For typical procedure for the preparation of Mn* see: Rieke, R.D.; Kim, S.-H.; Wu, X. *J. Org. Chem.* 1997, **62**, 6921.

15 For use of nontoxic manganese in chromium-catalyzed reaction, see: (a) Fürstner, A.; Shi, N. *J. Am. Chem. Soc.* 1996, **118**, 2533. (b) Fürstner, A.; Shi, N. *J. Am. Chem. Soc.* 1996, **118**, 12349.

16 Different reactivity of Mn* depending on manganese halide, see: Kim, S.-H.; Rieke, R.D. *Synth. Commun.* 1998, **28**, 1065.

17 Furstner, A.; Brunner, H. *Tetrahedron Lett.* 1996, **37**, 7009.

18 For a review, see: Normant, J.F.; Cahiez, G. *Modern. Synthetic Methods*; John Wiley & Sons, Ltd: Chichester, 1983; Vol. **3**, pp. 172–216, and references cited therein.

19 (a) Kim, S.-H.; Hanson, M.V.; Rieke, R.D. *Tetrahedron Lett.* 1996, **37**, 2197. (b) Kim, S.H.; Rieke, R.D. *Synth. Commun.* 1997, **28**, 1065. (c) Kim, S.-H.; Rieke, R.D. *Tetrahedron Lett.* 1997, **38**, 993. (d) Rieke, R.D.; Kim, S.-H.; Wu, X. *J. Org. Chem.* 1997, **62**, 6921. (e) Kim, S.-H.; Rieke, R.D.; *J. Org. Chem.* 1998, **63**, 6766. (f) Kim, S.-H.; Rieke, R.D. *J. Org. Chem.* 1998, **63**, 5235.

20 Fürstner, A.; Brunner, H. *Tetrahedron Lett.* 1996, **37**, 7009.

21 (a) Rieke, R.D. *Science* 1989, **24**, 1260. (b) Rieke, R.D. *Crit. Rev. Surf. Chem.* 1991, **1**, 131. (c) Rieke, R.D. *Top. Curr. Chem.* 1975, **59**, 1, and references cited therein.

22 (a) Freijee, F.; Schat, G.; Mierop, R.; Blomberg, C.; Bickelhaupt, F. *Heterocycles* 1977, **7**, 237. (b) Pearson, R.G.; Figdore, P.E. *J. Am. Chem. Soc.* 1980, **102**, 1541, and references cited therein. (c) Alonso, E.; Ramōn, D.J.; Yus, M. *J. Org. Chem.* 1997, **62**, 417. (d) For the reactions of sulfonates with lithium naphthalenide, see: Yus, M. *Chem. Soc. Rev.* 1996, **25**, 155, and references cited therein.

23 For the preparation of mesylates, see: Crossland, R.K.; Servis, K.L. *J. Org. Chem.* 1970, **35**, 3195.

24 For the preparation of benzylic metal reagents and their applications, see: Ref. 19f, and references cited therein.

25 For the preparation of tosylates, see: Kabalka, G.W.; Varma, M.; Varma, R.S. *J. Org. Chem.* 1986, **51**, 2386.

26 The following is a representative procedure: to a slurry of Rieke manganese (10.0 mmol) in THF (10 ml) under argon was added benzyl mesylate (9.0 mmol) at room temperature, and the mixture was stirred at room temperature for 30 min. 1,2-Dibromoethane (2.0 mmol) was added neat to the reaction mixture of 0°C, and the mixture was allowed to warm to room temperature over 5 min. To the resulting organomanganese reagent was added benzaldehyde at room temperature, and then the resulting mixture was stirred at room temperature for 30 min. A typical workup procedure—quenching, washing, and column chromatography—was carried out to purify the product.

27 (a) Screttas, C.G.; Micha-Screttas, M. *J. Org. Chem.* 1979, **44**, 713. (b) Still, W.C. *J. Am. Chem. Soc.* 1978, **100**, 1481. (c) Parham, W.E.; Jones, L.D.; Sayed, Y.A. *J. Org. Chem.* 1976, **41**, 1184. (d) Gilman, H.; McNinch, H. *J. Org. Chem.* 1961, **26**, 3723.

28 (a) Clarembeau, M.; Krief, A. *Tetrahedron Lett.* 1985, **26**, 1093. (b) Seyferth, D.; Suzuki, R.; Murphy, C.J.; Sabet, C.R. *J. Organomet. Chem.* 1964, **2**, 431. (c) Gilman, H.; Rosenberg, S.D. *J. Org. Chem.* 1959, **24**, 2063.

29 (a) van den Anker, T.R.; Harvey, S.; Raston, C.L. *J. Organomet. Chem.* 1995, **502**, 35. (b) Bernardon, C.J. *J. Organomet. Chem.* 1989, **367**, 11. (c) Harvey, S.; Junk, P.C. Raston, C.L.; Salem, G. *J. Org. Chem.* 1988, **53**, 3134. (d) Gallagher, M.J.; Harvey, S.; Raston, C.L.; Sue, R.E. *J. Chem. Soc., Chem. Commun.* 1988,

289. (e) Harvey, S.; Raston, C.L. *J. Chem. Soc., Chem. Commun.* 1988, 652. (f) Itsuno, S.; Darling, G.D.; Stover, H.D.; Frechet, J.M.J. *J. Org. Chem.* 1987, **52**, 4644. (g) Raston, C.L.; Salem, G. *J. Chem. Soc., Chem. Commun.* 1984, 1702.

30 (a) Betzemeier, B.; Knochel, P. *Angew. Chem. Int. Ed. Engl.* 1997, **36**, 2623. (b) Rottlander, M.; Knochel, P. *Synlett* 1997, **1084**. (c) Klement, I.; Lennick, K.; Tucker, C.E.; Knochel, P. *Tetrahedron Lett.* 1993, **34**, 4623. (d) Chia, W.-L.; Shiao, M.-J. *Tetrahedron Lett.* 1991, **32**, 2033. (e) Shing, T.-L.; Shiao, W.-L.; Chau, T.-Y. *Synthesis* 1991, 849. (f) Chen, H.G.; Hoechstetter, C.; Knochel, P. *Tetrahedron Lett.* 1989, **30**, 4795. (g) Berk, S.C.; Knochel, P.; Yeh, M.C.P.; *J. Org. Chem.* 1988, **53**, 5789.

31 (a) Burkhardt, E.R.; Rieke, R.D. *J. Org. Chem.* 1985, **50**, 416. (b) Zhu, L.; Wehmeyer, R.M.; Rieke, R.D. *J. Org. Chem.* 1991, **56**, 1445. (c) Guijarro, A.; Rosenberg, D.M.; Rieke, R.D. *J. Am. Chem. Soc.* 1999, **121**, 4135.

32 (a) Hiyama, T.; Sawahata, M.; Obayashi, M. *Chem. Lett.* 1983, **9**, 1237. (b) Hiyama, T.; Nakamura, A.; Obayashi, M. *Organometallics* 1982, **1**, 1249. (c) Cahiez, G.; Martin, A.; Delacroix, T. *Tetrahedron Lett.* 1999, **40**, 6407. (d) Suh, Y.; Lee, J.; Kim, S.; Rieke, R.D. *J. Organomet. Chem.* 2003, **684**, 20.

33 (a) Kim, S.-H.; Hanson, M.H.; Rieke, R.D. *Tetrahedron Lett.* 1996, **37**, 2197. (b) Kim, S.-H.; Rieke, R.D. *Tetrahedron Lett.* 1997, **38**, 993.

34 Kim, S.-H.; Rieke, R.D. *Synth. Commun.* 1998, **28**, 1065.

35 (a) Guijarro, D.; Mancheno, B.; Yus, M. *Tetrahedron* 1992, **48**, 4593. (b) A review. Yus, M. *Chem. Soc. Rev.* 1996, **25**, 155, and references therein.

36 Crossland, R.K.; Servis, K.L. *J. Org. Chem.* 1970, **35**, 3195.

37 (a) Guijarro, D.; Mancherio, B. *Tetrahedron* 1994, **50**, 8551. (b) Rossi, R.A.; Bunnett, J.F. *J. Org. Chem.* 1973, **38**, 2314.

38 (a) Freijee, F.; Schat, G.; Mierop, R.; Blomberg, C.; Bickelhaupt, F. *Heterocycles* 1977, 7, 237. (b) Pearson, R.G.; Figdore, P.E. *J. Am. Chem. Soc.* 1980, **102**, 1541. (c) Alonso, E.; Ramon, D.J.; Yus, M. *J. Org. Chem.* 1997, **62**, 417. (d) Jubert, C.; Knochel, P. *J. Org. Chem.* 1992, **57**, 5425.

39 Benzyl manganese halides were shown not to react with starting benzyl halides via an S_N2 route.

40 (a) Del Campo, F.J.; Maisonhaute, E.; Compton, R.G.; Marken, F.; Aldaz, A. *J. Electroanal. Chem.* 2001, **506**, 170. (b) Lawless, J.G.; Bartak, D.E.; Hawley, M.D. *J. Am. Chem. Soc.* 1969, **91**, 7121. (c) Higuchi, H.; Otsubo, T.; Ogura, F.; Yamaguchi, H.; Sakata, Y.; Misumi, S. *Bull. Chem. Soc. Jpn.* 1982, **55**, 182. (d) Trahanovsky, W.S.; Ong, C.C. Lawson, J.A. *J. Am. Chem. Soc.* 1968, **90**, 2839. (e) Agrios, K.A.; Srebnik, M. *J. Org. Chem.* 1993, **58**, 6908. (f) Inaba, S.; Matsumoto, H.; Rieke, R.D. *J. Org. Chem.* 1984, **49**, 2093. (g) Lei, A.; Zhang, X. *Org. Lett.* 2002, **4**, 2285.

41 Park, K.; Seo, Y.-S.; Yun, H.-S. *Bull. Korean Chem. Soc.* 1999, **20**, 1345.

42 Yoshida, J.; Itami, K.; Mineno, M.; Kamei, T. *Org. Lett.* 2002, **4**, 3635.

43 (a) Kim, S.H.; Hanson, M.V.; Rieke, R.D. *Tetrahedron Lett.* 1996, **37**, 2197. (b) Kim, S.H.; Rieke, R.D. *Tetrahedron Lett.* 1997, **38**, 993.

44 Gronowitz, S.; Moses, P.; Hornfeldt, A.-B.; Hakansson, R. *Arkiv. Kemi.* 1961, **17**, 165.
45 (a) Daoust, G.; Leclerc, M.J. *Macromolecules* 1991, **24**, 455. (b) Coffey, M.; McKellar, B.R.; Reinhardt, B.A.; Nijakowski, T.; Felda, W.A. *Synth. Commun.* 1996, **26**, 2205, and references cited therein. (c) Leclerc, M.; Daoust, G.; *J. Chem. Soc., Chem. Commun.* 1990, 273.
46 For representatives, see: (a) Sauter, F.; Stanetty, P.; Frohlich, H. *Heterocycles* 1987, **26**, 2657. (b) Shu, P.; Chiang, L.-Y. *J. Chem. Soc., Chem. Commun.* 1981, 920. (c) Chiang, L.-Y.; Shu, P.; Holf, D. *J. Org. Chem.* 1983, **48**, 4713. (d) Reddinger, J.L.; Reynolds, J.R. *J. Org. Chem.* 1996, **61**, 4833. (e) Tamao, K.; Kodama, S.; Nakajima, I.; Kumada, M.; Minato, A.; Suzuki, K. *Tetrahedron* 1982, **38**, 3347, and references cited therein.
47 The formation of bis-organomanganese bromides cannot be ruled out on the basis of gas chromatography monitoring. After acidic quenching of the reaction mixture followed by gas chromatography analysis, less than 5% of thiophene was detected.
48 (a) Jen, K.-Y.; Miller, G.G.; Elsenbaumer, R.L. *J. Chem. Soc., Chem. Commun.* 1986, 1346. (b) Hotta, S.; Rughooputh, S.D.D.V.; Heeger, A.J.; Wudl, F. *Macromolecules* 1987, **20**, 212. (c) Sato, M.; Tanaka, S.; Kaeiyama, K. *J. Chem. Soc., Chem. Commun.* 1986, 873.
49 (a) Hartman, G.D.; Halczenko, W.; Smith, R.L.; Sugrue, M.F.; Mallorga, P.J.; Michelson, S.R.; Randall, W.C.; Schwam, H.; Sondey, J.M. *J. Med. Chem.* 1992, **35**, 3822. (b) Holmes, J.M.; Lee, G.C.M.; Wijono, M.; Weinkam, R.; Wheeler, L.A.; Garst, M.E. *J. Med. Chem.* 1994, **37**, 1646.
50 (a) Gronowitz, S.; Hankansson, R. *Arkiv. Kemi.* 1959, **17**, 73. (b) Gronowitz, S. *Organic Sulphur Chemistry—Structure, Mechanism, and Synthesis*; Sterling, C.J.M., Ed.; Butterworths: London, 1975, pp. 203–228.
51 (a) Zhang, Y.; Hornfeldt, A.-B.; Gronowitz, S. *J. Heterocycl. Chem.* 1995, **32**, 435. (b) Ritter, S.K.; Noftle, R.E. *Chem. Mater.* 1992, **4**, 872. (c) Arnswald, M.; Neumann, W.P. *J. Org. Chem.* 1993, **58**, 7022. (d) Yamamura, K.; Miyake, H.; Nakatsuji, S. *Chem. Lett.* 1992, 1213. (e) Gronowitz, S. *Arkiv. Kemi.* 1958, **12**, 533.
52 Moses, P.; Gronowitz, S. *Arkiv. Kemi.* 1971, **18**, 119.
53 Wu, X.; Rieke, R.D. *J. Org. Chem.* 1995, **60**, 6658. This report from our laboratory has shown that the direct oxidative additions to 3-iodothiophene were completed by using Rieke magnesium (Mg*) and Rieke zinc (Zn*). However, 3-bromothiophene was unreactive toward Mg* and Zn*.
54 In typical preparation, lithium (9.68 mmol), naphthalene (1.48 mmol), and anhydrous manganese iodide (4.71 mmol) under argon were stirred in freshly distilled THF (10 ml) for 1 h at room temperature.
55 1,2-Dibromoethane was used to consume the remaining active manganese in the reaction mixture because Mn* was active to additional electrophile to give a homocoupling product.

56 Shriner, R.L. *Org. React.* 1942, **1**, 1.

57 Frankenfeld, J.W.; Werner, J.J. *J. Org. Chem.* 1969, **34**, 3689, and references cited therein.

58 Rieke, R.D.; Uhm, S.J. *Synthesis* 1975, 452.

59 Rathke, M.W.; Linder, A. *J. Org. Chem.* 1970, **35**, 3966.

60 Ruppert, J.F.; White, J.D. *J. Org. Chem.* 1974, **39**, 269.

61 Ross, N.A.; Bartsch, R.A. *J. Org. Chem.* 2003, **68**, 360.

62 For a review on C8K and metal–graphite combinations, see: Csuk, R.; Glanzer, B.I.; Furstner, A. *Adv. Organomet. Chem.* 1988, **28**, 85.

63 (a) Moriwake, T. *J. Org. Chem.* 1966, **31**, 983. (b) Mladenova, M.; Blagoev, B.; Kurtev, B. *Bull. Soc. Chim. Fr.* 1979, **11**, 77.

64 Inaba, S.I.; Rieke, R.D. *Tetrahedron Lett.* 1985, **26**, 155.

65 Araki, S.; Ito, H.; Butsugan, Y. *Synth. Commun.* 1988, **26**, 155.

66 Imamto, T.; Kusumoto, T.; Tawarayama, Y.; Sugiura, Y.; Mita, T.; Hatanaka, Y.; Yokoyama, M. *J. Org. Chem.* 1984, **49**, 3904.

67 Villieras, J.; Perriot, P.; Bourgain, M.; Normant, J.F. *J. Organomet. Chem.* 1975, **102**, 129.

68 (a) Kim, S.-H.; Hanson, M.V.; Rieke, R.D. *Tetrahedron Lett.* 1996, **37**, 2197. (b) Kim, S.-H.; Rieke, R.D. *Synth. Commun.* 1997, **28**, 1065. (c) Kim, S.-H.; Rieke, R.D. *Tetrahedron Lett.* 1997, **38**, 993.

69 (a) Rieke, R.D.; Kim, S.-H.; Wu, X. *J. Org. Chem.* 1996, **37**, 2197. (b) Kim, S.-H.; Rieke, R.D. *J. Org. Chem.* 1998, **63**, 6766. (c) Kim, S.-H.; Rieke, R.D. *J. Org. Chem.* 1998, **63**, 5235. (d) Suh, Y.; Rieke, R.D. *Tetrahedron Lett.* 2004, **45**, 1807.

70 The following is a representative procedure: ethyl α-bromoacetate (60 mmol) and benzaldehyde (58 mmol) were weighed separately. To the black slurry of active manganese in THF (30 ml) was added the mixture of benzaldehyde and ethyl α-bromoacetate drop by drop over 25 min. The addition was conducted at 0°C, and the entire system was kept under argon. The reaction mixture was stirred at room temperature for about an hour after the addition was completed. The reaction was then quenched with 2 N hydrochloric acid (20 ml) or saturated ammonium chloride, and the organic layer was separated from the acidic aqueous layer. The aqueous layer was extracted with ether (3 × 20 ml), and the combined organic phase was washed twice with saturated sodium hydrogen carbonate (10 ml) and water (10 ml). The ether extract was dried over anhydrous magnesium sulfate and filtered, and the solvent was evaporated, resulting in a pale yellow liquid. The crude product was subjected to GLC, which showed no by-products. The crude product was purified by vacuum distillation or column chromatography yielding 90% of the corresponding compound.

71 For a review, see: (a) House, H.O. *Modern Synthetic Reactions*, 2nd ed.; W.A. Benjamin: New York, 1972; p. 167. (b) McMurry, J.E. *Chem. Rev.* 1989, **89**, 1513. (c) Khan, B.E.; Rieke, R.D. *Chem. Rev.* 1988, **88**, 733.

72 (a) Fittig, R. *Justus Liebigs Ann. Chem.* 1859, **110**, 23. (b) For comprehensive reviews, see: Wirth, T. *Angew. Chem. Int. Ed. Engl.* 1996, **35**, 61. (c) Dushin, R.G. *Comprehensive Organometallic Chemistry II*; Hegedus, L.S., Ed.; Pergamon: Oxford, 1995; Vol. **12**, pp. 1071–1095. (d) Furstner, A. *Angew. Chem. Int. Ed. Engl.* 1993, **32**, 164. (e) Robertson, G.M. *Comprehensive Organic Synthesis*; Trost, B.M., Ed.; Pergamon: New York, 1991; Vol. **3**, pp. 563–611.

73 For example, see: Masamune, S.; Choy, W. *Aldrichim. Acta* 1982, **15**, 47.

74 (a) Chiaro, J.L.; Cabri, W.; Hanessian, S. *Tetrahedron Lett.* 1991, **32**, 1125. (b) Guidot, J.P.; Le Gall, T.; Mioskowsk, C. *Tetrahedron Lett.* 1994, **35**, 6671. (c) Corey, E.J.; Danheiser, R.L.; Chandrasekaran, S.; Siret, P.; Keck, G.E.; Gras, J. *J. Am. Chem. Soc.* 1978, **100**, 8031. (d) Swindell, C.S.; Fan, W.; Klimko, P.G. *Tetrahedron Lett.* 1994, **35**, 4959. (e) Nicolaou, K.C.; Yang, Z.; Liu, J.J.; Ueno, H.; Nantermet, P.G.; Guy, R.K.; Claiborne, C.F.; Renaud, J.; Couladouros, E.A.; Paulvannan, K.; Sorensen, E.J. *Nature* 1994, **367**, 630.

75 Schreibmann, A.A. *Tetrahedron Lett.* 1970, **11**, 4217.

76 Imamoto, T.; Kusumoto, T.; Hatanaka, Y.; Yokoyama, M. *Tetrahedron Lett.* 1982, **23**, 1353.

77 (a) Ito, K.; Nakanishi, S.; Otsuji, Y. *Chem. Lett.* 1980, 1141. (b) Inoue, H.; Suzuki, M.; Fujimoto, N. *J. Org. Chem.* 1979, **44**, 1722. (c) Mukaiyama, T.; Sato, T.; Hana, J. *Chem. Lett.* 1973, 1041.

78 (a) Furstner, A.; Csuk, R.; Rohrer, C.; Weidmann, H. *J. Chem. Soc., Perkin Trans.* **1** 1988, 1729. (b) Csuk, R.; Furstner, A.; Weidmann, H. *J. Chem. Soc., Chem. Commun.* 1986, 1802. (c) Rausch, M.D.; McEwen, W.E.; Kleinberg, J. *Chem. Rev.* 1957, **57**(7), 417. (d) Adams, E.W. *Organic Synthesis*; John Wiley & Sons, Inc.: New York, 1941; Collect. Vol. **I**, p. 459.

79 (a) Kammermeier, B.; Beck, G.; Jacobi, D.; Jendralla, H. *Angew. Chem. Int. Ed. Engl.* 1994, **33**, 685. (b) Szymoniak, J.; Besancon, J.; Moise, C. *Tetrahedron* 1994, **50**, 2841. (c) Szymoniak, J.; Besancon, J.; Moise, C. *Tetrahedron* 1992, **48**, 3867.

80 (a) Nomura, R.; Matsuno, T.; Endo, T. *J. Am. Chem. Soc.* 1996, **118**, 11666. (b) Anies, C.; Pancrazi, A.; Lallemand, J.-Y.; Prange, T. *Tetrahedron Lett.* 1994, **35**, 7771. (c) Chiara, J.L.; Martin-Lomas, M. *Tetrahedron Lett.* 1994, **35**, 2969. (d) Lebrun, A.; Namy, J.-L.; Kagan, H.B. *Tetrahedron Lett.* 1993, **34**, 2311. (e) Arseniyadis, S.; Yashunsky, D.V.; Freitas, R.P.; Dorado, M.M.; Toromanoff, E.; Potier, P. *Tetrahedron Lett.* 1993, **34**, 1137. (f) Umenishi, J.; Masuda, S.; Wakabayashi, S. *Tetrahedron Lett.* 1991, **32**, 5097. (g) Chiara, J.L.; Cabri, W.; Hanessian, S. *Tetrahedron Lett.* 1991, **32**, 1125. (h) Molander, G.A.; Kenny, C. *J. Am. Chem. Soc.* 1989, **111**, 8236. (i) Molander, G.A.; Kenny, C. *J. Org. Chem.* 1988, **53**, 2134. (j) Namy, J.L.; Souppe, J.; Kagan, H.B. *Tetrahedron Lett.* 1983, **24**, 765.

81 Hiyama, T.; Obayashi, M.; Mori, I.; Nozaki, H. *J. Org. Chem.* 1983, **48**, 912.

82 (a) Swindell, C.S.; Fan, W.; Klimko, P.G. *Tetrahedron Lett.* 1994, **35**, 4959. (b) McMurry, J.E.; Siemers, N.O. *Tetrahedron Lett.* 1994, **35**, 4505.

(c) Nicolaou, K.C.; Yang, Z.; Sorenson, E.J.; Nakada, M. *J. Chem. Soc., Chem. Commun.* 1993, 1024. (d) McMurry, J.E.; Siemers, N.O. *Tetrahedron Lett.* 1993, **34**, 7891. (e) Swindell, C.S.; Chandler, M.C.; Heerding, J.M.; Klimko, P.G.; Rahman, L.T.; Raman, J.V.; Venkataraman, H. *Tetrahedron Lett.* 1993, **34**, 7002. (f) Davey, A.E.; Schaeffer, M.J.; Tayler, R.J. *J. Chem. Soc., Perkin Trans.* **1** 1992, 2657. (g) McMurry, J.E.; Dushin, R.G. *J. Am. Chem. Soc.* 1990, **112**, 6942. (h) McMurry, J.E.; Rico, J.G.; Shih, Y.-N. *Tetrahedron Lett.* 1989, **30**, 1173. (i) Pierrot, M.; Pons, J.-M.; Santelli, M. *Tetrahedron Lett.* 1988, **29**, 5925. (j) Handa, Y.; Inanaga, J. *Tetrahedron Lett.* 1987, **28**, 5717. (k) Suzuki, H.; Manabe, H.; Enokiya, R.; Hanazaki, Y. *Chem. Lett.* 1986, 1339. (l) Clerici, A.; Porta, O. *J. Org. Chem.* 1983, **48**, 1690. (m) Clerici, A.; Porta, O. *Tetrahedron Lett.* 1982, **23**, 3517. (n) Clerici, A.; Porta, O. *Tetrahedron Lett.* 1982, **38**, 1293. (o) Mukaiyama, T.; Sato, T.; Hanna, J. *Chem. Lett.* 1973, 1041.

83 (a) Hirao, T.; Hasegawa, T.; Muguruma, Y.; Ikeda, I. *J. Org. Chem.* 1996, **61**, 366. (b) Konradi, A.W.; Kemp, S.J.; Pederson, S.F. *J. Am. Chem. Soc.* 1994, **116**, 1316. (c) Raw, A.S.; Pedersen, S.F. *J. Org. Chem.* 1991, **56**, 830.

84 (a) Hou, Z.; Takamine, K.; Fuziware, Y.; Taniguchi, H. *Chem. Lett.* 1987, 2061. (b) Hou, Z.; Takamine, K.; Aoki, O.; Shirashi, H.; Fujiwara, Y.; Taniguchi, H. *J. Org. Chem.* 1988, **53**, 6077.

85 Askham, F.R.; Carroll, K.M. *J. Org. Chem.* 1993, **58**, 7328.

86 (a) Hays, D.S.; Fu, G.C. *J. Am. Chem. Soc.* 1995, **117**, 7283. (b) For a review, see: Chatgilialoglu, C. *Acc. Chem. Res.* 1992, **25**, 188.

87 Lenhert, P.G.; Lukehart, C.M.; Srinivasan, K. *J. Am. Chem. Soc.* 1984, **106**, 124.

88 Snider, B.B. *Chem. Rev.* 1996, **96**, 339, and references therein.

89 (a) Pasini, A.; Zunino, F. *Angew. Chem. Int. Ed. Engl.* 1987, **26**, 615. (b) Angerer, F.V.; Egginer, G.; Kranzflder, C.; Bernhauer, H.; Schonenerger, H. *J. Med. Chem.* 1982, **25**, 832.

90 (a) Jung, S.H.; Kohn, H. *J. Am. Chem. Soc.* 1985, **107**, 2931. (b) Natsugari, H.; Whittle, R.R.; Weinreb, S.M. *J. Am. Chem. Soc.* 1984, **106**, 7987. (c) Smith, J.G.; Ho, I. *J. Org. Chem.* 1972, **37**, 653. (d) Smith, J.G.; Veach, C.D. *Can. J. Chem.* 1966, **44**, 2497.

91 Enholm, E.J.; Forbes, D.C.; Holub, D.P. *Synth. Commun.* 1990, **20**, 981.

92 Eisch, J.J.; Kaska, D.D.; Peterson, C.J. *J. Org. Chem.* 1966, **31**, 453.

93 Buchwald, S.J.; Watson, B.T.; Wannamaker, M.W.; Dewan, J.C. *J. Am. Chem. Soc.* 1989, **14**, 4486.

94 Baruah, B.; Prajapati, D.; Sandhu, J.S. *Tetrahedron Lett.* 1995, **36**, 6747.

95 (a) Shomo, T.; Kise, N.; Oike, O.; Yashimoto, M.; Okazaki, E.Q. *Tetrahedron Lett.* 1992, **33**, 5559. (b) Khan, N.H.; Zubari, R.H.; Siddiqui, A.A. *Synth. Commun.* 1980, **10**, 363. (c) Smith, J.G.; Boettger, T.J. *Synth. Commun.* 1981, **11**, 61.

96 Mangeney, P.; Tejero, T.; Alexakis, A.; Grosjean, F.; Normant, J. *Synthesis* 1988, 255.

97 Kalyanam, N.; Rao, G.V. *Tetrahedron Lett.* 1993, **34**, 1647.

98 Roskamp, E.J.; Pedersen, S.F. *J. Am. Chem. Soc.* 1987, **109**, 6551.

99 Takaki, K.; Tsubaki, Y.; Tanaka, S.; Beppu, F.; Fujiwara, Y. *Chem. Lett.* 1990, 203.

100 Rieke, R.D.; Kim, S.H. *J. Org. Chem.* 1998, **63**, 5235.

101 (a) According to the MDL Drug Data Report, the most widespread heterocycles in pharmaceutically active compounds are pyridine, imidazole, indole, and pyrimidine derivatives; (b) Queginer, G.; Marsais, F.; Snieckus, V.; Epsztain, J. *Adv. Heterocycl. Chem.* 1991, **52**, 187. (c) Lehn, J.M. *Supramolecular Chemistry*; VCH Verlagsgesellschaft: Weinheim, 1995.

102 (a) Peterson, M.A.; Mitchell, J.R. *J. Org. Chem.* 1997, **62**, 8237. (b) Hannon, M.J.; Mayars, P.C.; Taylor, P.C. *Tetrahedron Lett.* 1998, **39**, 8509. (c) Gomez, I.; Alonso, E.; Ramon, D.J.; Yus, M. *Tetrahedron* 2000, **56**, 4043. (d) Furukawa, N.; Shibutani, T.; Fujihara, H. *Tetrahedron Lett.* 1987, **28**, 5845. (e) Abarbri, M.; Dehmel, F.; Knochel, P. *Tetrahedron Lett.* 1999, **40**, 7449. (f) Trecourt, F.; Breton, G.; Bonnet, V.; Mongin, F.; Marsais, F.; Queginer, G. *Tetrahedron Lett.* 2000, **56**, 1349. (g) Bell, A.S.; Roberts, D.A.; Ruddoct, K.S. *Synthesis* 1987, 843.

103 (a) Knochel, P.; Perea, J.J.A.; Jones, P. *Tetrahedron* 1998, **54**, 8275. (b) Knochel, P.; Millot, A.; Rodrigues, A.L. *Org. React.* 2001, **58**, 417. (c) Dohle, W.; Lindsay, D.M.; Knochel, P. *Org. Lett.* 2001, **3**, 2871.

104 Sakamoto, T.; Kondo, Y.; Murata, N.; Yamanaka, H. *Tetrahedron Lett.* 1992, **33**, 5373.

105 (a) Kim, S.-H.; Hanson, M.V.; Rieke, R.D. *Tetrahedron Lett.* 1996, **37**, 2197. (b) Kim, S.-H.; Rieke, R.D. *Tetrahedron Lett.* 1997, **38**, 993. (c) Kim, S.-H.; Wu, X.; Rieke, R.D. *J. Org. Chem.* 1997, **62**, 6921. (d) Kim, S.-H.; Rieke, R.D. *J. Org. Chem.* 1998, **63**, 6766. (e) Suh, Y.S.; Lee, J.S.; Kim, S.-H.; Rieke, R.D. *J. Organomet. Chem.* 2003, **684**, 20.

106 (a) Chavant, P.Y.; Metais, E.; Cahiez, G. *Tetrahedron Lett.* 1992, **33**, 5235. (b) Martin, A.; Delacroix, T.; Cahiez, G. *Tetrahedron Lett.* 1999, **40**, 6407. (c) Cahiez, G.; Luart, D.; Lecomte, F. *Org. Lett.* 2004, **6**, 4395.

107 Rieke, R.D.; Suh, Y.S.; Kim, S. *Tetrahedron Lett.* 2005, **46**, 5961.

108 (a) Jen, K.-Y., Miller, G.G.; Elsenbaumer, R.L. *J. Chem. Soc., Chem. Commun.* 1986, 1346. (b) Sato, M.; Tanaka, S.; Kareriyama, K. *J. Chem. Soc., Chem. Commun.* 1986, 873. (c) Hartman, G.D.; Halczenko, W.; Smith, R.L.; Sugrue, M.F.; Mallogra, P.J.; Randall, W.C.; Schwam, H.; Sondey, J.M. *J. Med. Chem.* 1992, **35**, 3822. (d) Holmes, J.M.; Lee, G.C.M.; Wijino, M.; Weinkam, R.; Garst, M.E. *J. Med. Chem.* 1994, **37**, 1646.

109 (a) Kojima, Y.; Wakita, S.; Kato, N. *Tetrahedron Lett.* 1979, **20**, 4577. (b) Ennis, D.S.; Gilchrist, T.L. *Tetrahedron Lett.* 1990, **46**, 2623. (c) Buechi, G.; Kovats, E.S.Z.; Enggist, P.; Uhde, G. *J. Org. Chem.* 1968, **33**, 1227.

110 (a) Brown, H.C.; Srebnik, M.; Bakshi, R.K.; Cole, T.E. *J. Am. Chem. Soc.* 1987, **109**, 5420. (b) Bailey, T.R. *Synthesis* 1991, 242.

111 Phytochrome and Photomorphogenesis, Smith, H., Ed.; McGraw H: London, 1975.

112 The following is a typical procedure for palladium-catalyzed cross-coupling reactions of heteroarylmanganese reagents with electrophiles: to a slurry of Rieke manganese (9.0 mmol) in THF (15 ml) under argon was added 2-bromothiophene (3.0 mmol) at room temperature. The mixture was stirred at room temperature of 3 h. 1,2-Dibromoethane (7.0 mmol) was added neat to the reaction mixture at 0°C, and the mixture was allowed to warm to room temperature over 20 min. The resulting thienylmanganese bromide was cannulated via a cannula into a flask containing iodotoluene (2.0 mmol) and $Pd[P(Ph)_3]_4$ (0.06 mmol) at 0°C over 20 min. After cannulation was completed, the reaction mixture was stirred at rt for 1–2 h. The reaction was quenched with 2 M HCl (5 ml) and then extracted with ether (2 × 10 ml). The combined organic layers were sequentially washed with saturated $NaHCO_3$, $Na_2S_2O_3$, and brine solutions and then dried over Na_2SO_4, and concentrated under vacuum. Flash chromatography (ethyl acetate–hexanes) afforded 4-thienyltoluene in 90% isolated yield.

9

Calcium

9.1 Preparation of Rieke Calcium

We have previously reported that highly reactive Rieke magnesium readily reacts with 1,3-dienes to form substituted 2-butene-1,4-diylmagnesium complexes [1]. Reactions of these complexes with organic dihalides yielded 4-, 5-, or 6-membered ring carbocycles. The bis-Grignard reagents can also be reacted with two different electrophiles, yielding polyfunctionalized ketones. Since calcium has a significantly different redox potential as well as a larger ionic radius from that of magnesium, we anticipated that the corresponding calcium metallocycles would behave quite differently with respect to their reactivity and selectivity. We reported a highly reactive form of calcium, which allows the ready preparation of a wide variety of organocalcium reagents. These reagents in turn can be used to carry out a number of useful transformations.

Highly reactive calcium can be readily prepared by the reduction of calcium halides in tetrahydrofuran (THF) solution with preformed lithium biphenylide under an argon atmosphere at room temperature [2]. This colored calcium species seems to be reasonably soluble in THF. However, the reactive calcium complex prepared from preformed lithium naphthalenide was insoluble in THF solution and precipitated out of solution to give a highly reactive black solid. As this lithium naphthalenide generated calcium species was insoluble in most deuterated solvents and reacted with deuterated DMSO and DMF, the exact nature of this black calcium complex has not been determined. Acid hydrolysis of the black material releases naphthalene as well as THF. Accordingly, the most likely structure of the black material is a Ca-naphthalene-THF complex similar in nature to the soluble magnesium-anthracene complex reported [3].

Chemical Synthesis Using Highly Reactive Metals, First Edition. Reuben D. Rieke.

9.2 Oxidative Addition Reactions of Rieke Calcium with Organic Halides and Some Subsequent Reactions

Grignard-Type Reactions with Highly Reactive Calcium

The development of organocalcium chemistry has been surprisingly slow with respect to the extensive studies of organometallic reagents of other light metals [4]. The neglect of organocalcium chemistry is due in part to the lack of a facile method of preparing the organocalcium compounds. Direct oxidative addition to calcium has traditionally been limited by the reduced reactivity of calcium metal with organic substrates. This is presumably due to surface poisoning factors. The organocalcium derivatives RCaX were most readily formed when X = I, and the preparation of RCaX (X = Br, Cl) usually required activated calcium. Few examples have been reported, and overall yields tend to be low [4]. Although simple primary and secondary alkyl iodides reacted with calcium in reasonable yields [5], the tertiary alkyliodocalcium compounds were very difficult to prepare, and most workers reported only trace amounts [6]. In contrast, the highly reactive calcium complexes reported here react readily with all of these substrates to generate excellent yields of the corresponding organocalcium compounds.

Highly reactive calcium was prepared by the lithium biphenylide reduction of calcium salts in THF. Both $CaBr_2$ and CaI_2 generated the reactive calcium species. The organocalcium compounds prepared directly from this calcium complex and organic halides were found to efficiently undergo Grignard-type reactions. Table 9.1 summarizes some examples of 1,2-addition reactions with cyclohexanone utilizing the reactive calcium. Alkyl bromides and alkyl chlorides rapidly reacted with the calcium complex at temperatures as low as −78°C. As shown in Table 9.1, 1-bromooctane and 1-bromo-3-phenoxypropane reacted with the calcium complex at −78°C to form the corresponding alkylbromocalcium reagents, which underwent Grignard-type reactions with cyclohexanone to produce the tertiary alcohols in 79% and 75% yields, respectively. Oxidative addition of alkyl chlorides to this calcium species was also very efficient at low temperature (−78°C). 1-Chlorooctane gave 1-octylcyclohexanol in 83% yield. Similar results were noted for the secondary halides. Bromocyclohexane reacted readily with the calcium species at −78°C, and the resulting organocalcium reagent underwent carbonyl addition to give the alcohol in 75% yield. Significantly, the highly reactive calcium complex reacted rapidly with tertiary bromides at −78°C. For example, the Grignard-type reaction for 1-bromoadamantane utilizing the reactive calcium afforded 1-(1-adamantyl)cyclohexanol in 80% yield. The direct reaction of 1-bromoadamantane with metals is well known to yield mainly reductive cleavage or dimerization [7]. Accordingly, this method represents a significant new approach to the 1-metalloadamantane.

Table 9.1 Grignard-type reactions of organocalcium reagents with cyclohexanone.

Entry	Halide	CaX_2[a]	Product[b]	% yield[c]
1	$Cl(CH_2)_7CH_3$	CaI_2	$1\text{-}(CH_2)_7CH_3\text{-}1\text{-}OH\text{-}c\text{-}C_6H_{10}$	83
2	$Br(CH_2)_7CH_3$	CaI_2	$1\text{-}(CH_2)_7CH_3\text{-}1\text{-}OH\text{-}c\text{-}C_6H_{10}$	79
3	$Br(CH_2)_3OPh$	$CaBr_2$	$1\text{-}(CH_2)_3OPh\text{-}1\text{-}OH\text{-}c\text{-}C_6H_{10}$	75
4	$Br\text{-}c\text{-}C_6H_{11}$	$CaBr_2$	$1\text{-}c\text{-}C_6H_{11}\text{-}1\text{-}OH\text{-}c\text{-}C_6H_{10}$	75
5	Br	$CaBr_2$	OH	80
6	$BrC_6H_4(m\text{-}CH_3)$	CaI_2	$1\text{-}C_6H_4(m\text{-}CH_3)\text{-}1\text{-}OH\text{-}c\text{-}C_6H_{10}$	76
7	$ClC_6H_4(p\text{-}CH_3)$	CaI_2	$1\text{-}C_6H_4(p\text{-}CH_3)\text{-}1\text{-}OH\text{-}c\text{-}C_6H_{10}$	86
8	FPh	CaI_2	$1\text{-}Ph\text{-}1\text{-}OH\text{-}c\text{-}C_6H_{10}$	85
9	$BrC_6H_4(m\text{-}OCH_3)$	$CaBr_2$	$1\text{-}C_6H_4(m\text{-}OCH_3)\text{-}1\text{-}OH\text{-}c\text{-}C_6H_{10}$	79

[a] Both $CaBr_2$ and CaI_2 generate the highly reactive calcium species.
[b] All new substances have satisfactory spectroscopic data including IR, ^1H NMR, ^{13}C NMR, and high-resolution mass spectral data.
[c] Isolated yields.

Reactions of aryl halides with reactive calcium required slightly higher temperatures, up to −30°C for aryl bromides and up to −20°C for aryl chlorides. The aryl-calcium compounds are very stable at room temperature. Reactions of *m*-bromotoluene, *m*-bromoanisole, and *p*-chlorotoluene with the activated calcium complex gave the corresponding aryl-calcium reagents in quantitative yields based on the GC analyses of reaction quenches. As expected, 1,2-addition of these aryl-calcium compounds with ketones gave the alcohols in excellent yields (76%, 79%, and 86%, respectively). Surprisingly, the activated calcium readily reacted with fluorobenzene at room temperature to form the corresponding organometallic compound, which underwent an addition reaction with cyclohexanone to give 1-phenylcyclohexanol in 85% yield. Except for highly reactive magnesium prepared by the reduction of magnesium salts [8], few metals undergo oxidative addition with aryl fluorides to form organometallic compounds [9]. The active calcium species also reacts rapidly with allylic halides; however, the resulting allylcalcium reagent rapidly homocouples with starting allylic halides.

9.3 Preparation and Reaction of Calcium Cuprate Reagents

While a wide spectrum of different metal cuprates are known, calcium cuprates have not yet been reported. Addition of copper(I) salts to the organocalcium compounds resulted in a new complex of vastly different chemical reactivity.

Table 9.2 Cross-coupling reactions of calcium organocuprate reagents with benzoyl chloride.[a]

Entry	Halide	Product[b]	% yield[c]
1	$Cl(CH_2)_7CH_3$	$PhC(O)(CH_2)CH_3$	84
2	$Br(CH_2)_5OPh$	$PhC(O)(CH_2)_5OPh$	76
3	$Br\text{-}c\text{-}C_6H_{11}$	$PhC(O)\text{-}c\text{-}C_6H_{11}$	82
4	$1\text{-}Cl\text{-}4\text{-}CH_3C_6H_4$	$1\text{-}PhC(O)\text{-}4\text{-}CH_3C_6H_4$	86
5	$1\text{-}Br\text{-}4\text{-}OCH_3C_6H_4$	$1\text{-}PhC(O)\text{-}4\text{-}OCH_3C_6H_4$	71

[a] Active calcium was prepared by the lithium biphenylide reduction of $CaBr_2$ in THF. CuCN·2LiBr was used for transmetalation with organocalcium reagents.
[b] Most products were compared with authentic samples. The new substance, 1-phenyl-6-phenoxy-1-hexanone, has satisfactory IR, ^{1}H NMR, ^{13}C NMR, and high-resolution mass spectral data.
[c] Isolated yields.

Presumably, this new complex is a calcium cuprate. Reaction of the organocalcium reagent prepared from an organic halide and the highly reactive calcium, with benzoyl chloride in the absence of a Cu(I) salt, afforded a complex mixture of products. However, in the presence of a Cu(I) salt, high yields of ketone formation were observed. Table 9.2 presents some of the ketone formation reactions of the calcium cuprates with benzoyl chloride.

A soluble copper(I) complex, CuCN·2LiBr [10], was used for the transmetallations with organocalcium reagents to form the assumed copper calcium complexes. Reaction of these calcium cuprate reagents with benzoyl chloride proceeded smoothly at −35°C to yield ketones in excellent yields. As shown in Table 9.2, the primary alkylcalcium cuprates, *n*-octyl- and (5-phenoxypentyl) calcium cuprate, reacted rapidly with benzoyl chloride at −35°C to give 1-phenyl-1-nonanone and 1-phenyl-6-phenoxy-1-hexanone in 84% and 76% yield, respectively. The secondary alkylcalcium cuprate, cyclohexyl calcium cuprate, reacted smoothly with benzoyl chloride to form cyclohexylphenylmethanone in 82% yield. Although the tertiary alkylcalcium cuprate was not investigated, it should undergo this transformation. In the aryl cases, 4-methylphenyl and 4-methoxyphenyl cuprate, for example, also reacted with benzoyl chloride to afford (4-methlyphenyl)phenylmethanone and (4-methoxyphenyl)phenylmethanone in 86% and 71% yield, respectively.

As expected, these calcium cuprate compounds also undergo the conjugate 1,4-addition reactions with α,β unsaturated ketones. Table 9.3 presents some examples of conjugate 1,4-addition reactions utilizing these calcium cuprates. In the first case, *n*-octylcalcium cuprate, generated by transmetallation of the *n*-octanocalcium compound with CuCN•2LiBr, reacted with 2-cyanohexenone to give 3-octylcyclohexanone in moderate yield (42% yield). However,

Table 9.3 Conjugate 1,4-addition reactions of calcium organocuprate reagents with enones.

Entry	Halide	Cu(I) salt	Enone	Product[a]	% yield[b]
1	$Cl(CH_2)_7CH_3$	CuCN·2LiBr	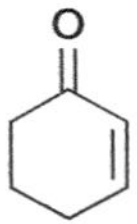 O	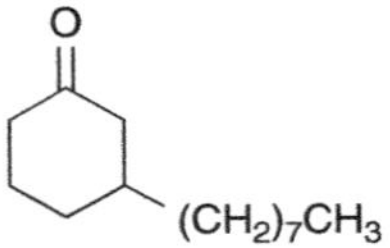 O $(CH_2)_7CH_3$	46
2	$Cl(CH_2)_7CH_3$	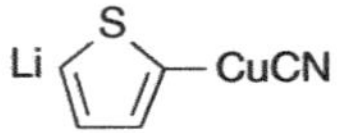 Li S CuCN	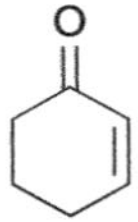 O	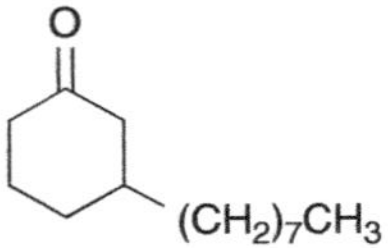 O $(CH_2)_7CH_3$	87
3	$Cl(CH_2)_7CH_3$	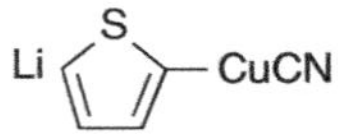 Li S CuCN	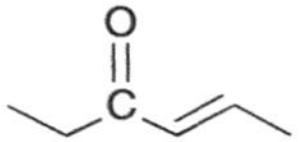 O C	$EtC(O)CH_2CH(CH_3)(CH_2)_7CH_3$	47
4	$Cl(CH_2)_7CH_3$	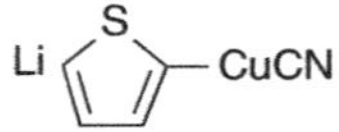 Li S CuCN	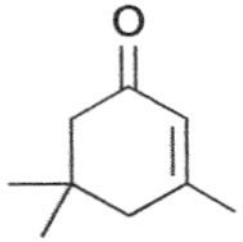 O	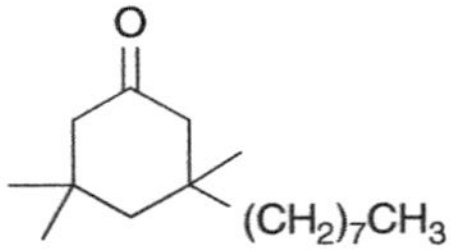 O $(CH_2)_7CH_3$	<3

(Continued)

Table 9.3 (Continued)

Entry	Halide	Cu(I) salt	Enone	Product[a]	% yield[b]
5	$Cl(CH_2)_7CH_3$	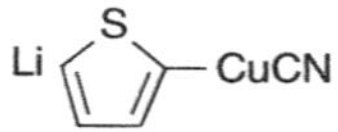 Li (2-thienyl)–CuCN + TMSCl & BF_3•OEt_2	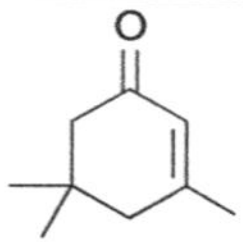	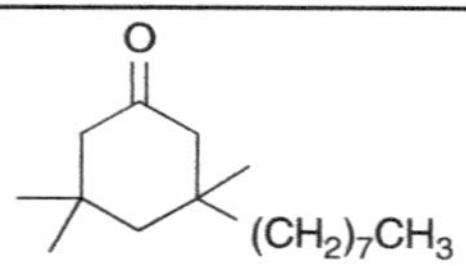 $(CH_2)_7CH_3$	84
6	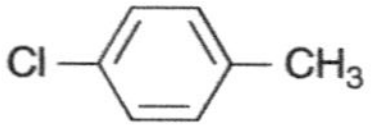 Cl–C_6H_4–CH_3	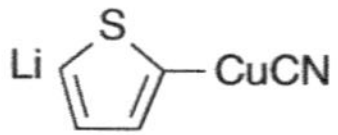 Li (2-thienyl)–CuCN	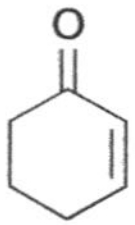	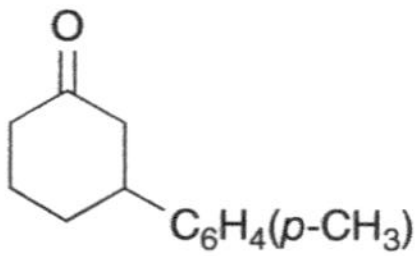 $C_6H_4(p\text{-}CH_3)$	68

[a] Most products were compared with authentic samples. The new substance, 3-(*p*-methylphenyl)cyclohexanone, has satisfactory IR, ^{1}H NMR, ^{13}C NMR, and high-resolution mass spectral data.
[b] Isolated yields.

a more reactive calcium cuprate species was produced, and the yield was greatly improved to 87% when lithium 2-thienylcyanocuprate [11] was used. This cuprate also underwent the conjugate addition with acyclic enones, for example, 2-hexene-4-one, to give 5-methyl-3-tridecanone in 47% yield. We have not attempted further optimization. Reaction of this calcium cuprate with a sterically hindered enone, for example, isophorone, produced <3% of the desired compound in 24 h. The isolated yield, however, increased to 84% when the additives BF_3 etherate and chlorotrimethylsilane [12] were used. In the aryl case, *p*-tolylcalcium cuprate also underwent this transformation with 2-cyclohexenone to give 3-(*p*-methylphenyl)cyclohexanone in reasonable yield (Table 9.3).

9.4 Preparation and Reactions of Calcium Metallocycles

Although magnesium complexes of 1,3-dienes prepared from magnesium and 1,3-dienes have received considerable attention in organometallic syntheses [13], the corresponding calcium complexes of 1,3-dienes have not yet been reported. We have observed that the calcium complexes can readily be prepared by the reaction of the highly reactive calcium with a wide variety of 1,3-dienes. The resulting bis-organocalcium reagents were found to readily undergo alkylation reactions with a variety of electrophiles in a high degree of regio- and stereospecific manner (Table 9.4).

The reactivity of the calcium metallocycles was higher than that of the corresponding magnesium analogues. In the 1,4-diphenyl-1,3-butadiene cases, the chemical yields were excellent and generally were higher than the corresponding magnesium analogues. For example, 1,4-diphenyl-1,3-butadiene/calcium complex reacted rapidly with 1,3-dibromopropane and 1,4-dibromobutane to form *trans*-1-phenyl-2-*trans*-β-styrenylcyclopentane and *trans*-1-phenyl-2-*trans*-β-styrenylcyclohexane in 91% and 53% isolated yield, respectively. The stereochemistry of these reactions was always stereospecific.

The observed regiochemistry was basically the same as that reported for magnesium. Reaction of (1,4-diphenyl-2-butene-1,4-diyl)calcium complexes with α,ω-alkylene dihalides usually gave 1,2-addition products, while 1,4-addition was always observed in reactions with dichlorosilane. In contrast to the magnesium complex, treatment of (1,4-diphenyl-2-butene-1,4-diyl)calcium complex with 1,2-dibromoethane yielded 7% of the 1,4-addition product *cis*-3,6-diphenylcyclohex-1-ene [14], along with 72% of the starting material. The yield of the 6-membered ring product was increased to 80%, and the amount of recovered starting material dropped to 8% when 1,2-dichloroethane was used. The higher reduction potential of 1,2-dichloroethane presumably eliminated most of the simple electron transfer pathway. Interestingly, reaction of this

Table 9.4 Reactions of 1,3-diene/calcium complex with organic dihalides.[a]

Entry	Diene	Li/Ar	CaX_2	Electrophile	Product[b]	% Yield[c]
1	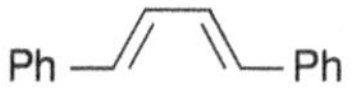 Ph–CH=CH–CH=CH–Ph	Li/Biph	CaI_2	$Br(CH_2)_3Br$	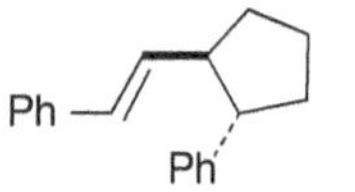 Ph, Ph	91
2	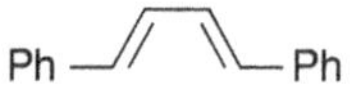 Ph–CH=CH–CH=CH–Ph	Li/Np	—	$Br(CH_2)_3Br$	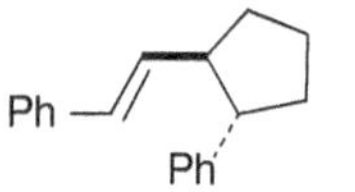 Ph, Ph	51[d]
3	 Ph–CH=CH–CH=CH–Ph	Li/—	CaI_2	$Br(CH_2)_3Br$	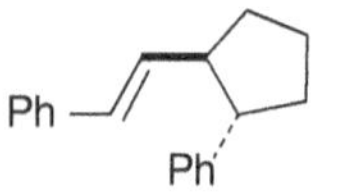 Ph, Ph	74[e]
4	 Ph–CH=CH–CH=CH–Ph	Li/Biph	CaI_2	$Br(CH_2)_4Br$	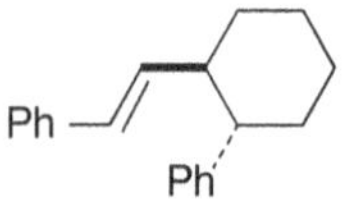 Ph, Ph	53
5	 Ph–CH=CH–CH=CH–Ph	Li/Biph	CaI_2	$Br(CH_2)_2Br$	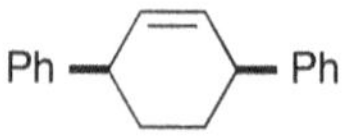 Ph, Ph	7[f]
6	 Ph–CH=CH–CH=CH–Ph	Li/Biph	CaI_2	$Cl(CH_2)_2Cl$	 Ph, Ph	80[g]
7	 Ph–CH=CH–CH=CH–Ph	Li/Biph	CaI_2	$ClCH_2Cl$	 Ph, Ph	47[h]
8	 Ph–CH=CH–CH=CH–Ph	Li/Biph	CaI_2	$(CH_3)_2SiCl_2$	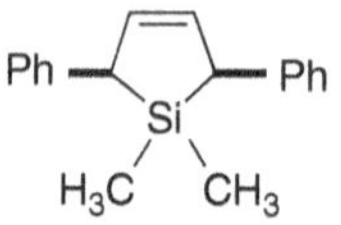 Ph, Ph, Si, H_3C, CH_3	—[i]
9	 2,3-dimethyl-1,3-butadiene	Li/Biph	CaI_2	$Cl(CH_2)_3Cl$	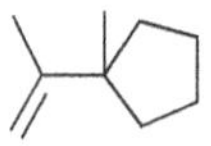	(98)[j]
10	 2,3-dimethyl-1,3-butadiene	Li/Biph	—	$Cl(CH_2)_3Cl$	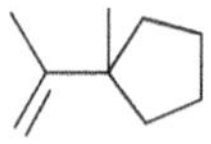	(25)
11	 2,3-dimethyl-1,3-butadiene	Li/Biph	CaI_2	$Cl(CH_2)_4Cl$	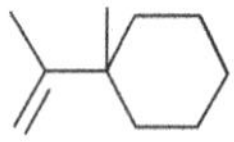	(36)[j]
12	 2,3-dimethyl-1,3-butadiene	Li/Biph	CaI_2	$Br(CH_2)_4Br$	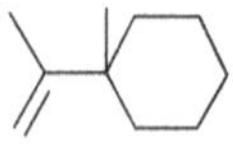	(54)[j]
13	 2,3-dimethyl-1,3-butadiene	Li/Biph	CaI_2	Ph_2SiCl_2	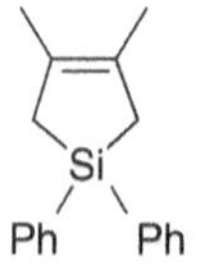 Si, Ph, Ph	(89)[k]

[a] The active calcium was prepared from 2.05 equiv of preformed lithium biphenylide and 1.0 equiv of CaI_2.

[b] The known products were compared with the authentic sample. All new substances have satisfactory spectroscopic data including IR, ^1H NMR, ^{13}C NMR, and high-resolution mass spectral data.

[c] Isolated yields. GC yields are given in parentheses.

[d] 31% starting material was recovered.

[e] No starting material was recovered.

[f] 72% starting material was recovered.

[g] 8% starting material was recovered.

[h] 43% starting material was recovered.

[i] Isolation was difficult because of the overlapping with biphenyl.

[j] Product was isolated by distillation.

[k] Product was isolated by reverse-phase thin-layer chromatography.

calcium complex with dichloromethane afforded only the 1,2-addition product *trans*-1-phenyl-2-*trans* β-styrenylcyclopropane [15] in 47% yield, along with 43% of 1,4-diphenyl-1,3-butadiene.

Reduction of 1,4-diphenyl-1,3-butadiene with 2.2 equiv of preformed lithium naphthalenide without the presence of Ca(II) salts, followed by the addition of 1,3-dibromopropane, also yielded the same cyclopentane derivative, but the yield was substantially lower than that obtained in the presence of calcium salts. Also of note is the fact that in the absence of calcium salts, over 30% of the starting material was recovered. It is possible that electron transfer from the butadiene dianions to the organic halides was facilitated in the absence of calcium salts, and these resulting radicals and/or anions did not efficiently add to the 1,3-diene. A similar result was also noted in the nonactivated diene system. The yield dramatically decreased from 94 to 25% in the similar experiments using 2,3-dimethyl-1,3-butadiene with 1,3-dichloropropane. In any event, the observed chemistry is dramatically different when the calcium salts are present. Direct reduction of the 1,3-dienes with lithium metal in the absence of electron carriers was also carried out. Reduction of 1,4-diphenyl-1,3-butadiene with 2.5 equiv of lithium metal in THF, followed by the sequential addition of 2.0 equiv of CaI_2 and 1,3-dibromopropane, yielded the same 5-membered ring product in 74% yield along with a small amount of unidentified high molecular weight material. Significantly, no starting material was found in the reaction workup. While the exact structures of the organometallic species involved have not yet been determined, the requirement and involvement of calcium ions are unequivocal. The crystal structure of the magnesium analogue of the 1,4-diphenyl-1,3-butadiene complex has been reported [16].

This chemistry can also be extended to 2,3-dimethyl-1,3-butadiene, which is a molecule that is much more difficult to reduce. The calcium complex was readily prepared by reaction of freshly distilled 2,3-dimethyl-1,3-butadiene with either the biphenylide complex or the calcium naphthalenide complex. Reaction of the resulting complex with 1,3-dichloropropane and 1,4-dichlorobutane gave the 5-membered ring product and 6-membered ring product in 94% and 36% yield, respectively. For the latter reaction, the yield was improved to 54% when 1,4-dibromobutane was used. The regiochemistry of the 2,3-dimethyl-1,3-butadiene/calcium complexes again paralleled that of the corresponding magnesium complexes. Similarly, treatment of (2,3-dimethyl-2-butene-1,4-diyl)calcium complex with dichlorodiphenylsilane yielded the 1,4-addition adduct in 89% yield.

In summary, a highly reactive form of calcium has been prepared by the lithium biphenylide or lithium naphthalenide reduction of calcium salts. This calcium will rapidly undergo an effective oxidative addition reaction with alkyl or aryl halides to yield the corresponding organocalcium compounds. These organocalcium compounds have been found to undergo Grignard-type additions to ketones. Addition of copper(I) salts leads to new complexes which are presumed to be the corresponding calcium cuprates. These calcium cuprates will undergo clean cross-coupling with acid chlorides to generate ketones as well as undergo conjugate 1,4-additions to α,β-unsaturated ketones. The activated calcium will react with 1,3-dienes to yield (2-butene-1,4-diyl) calcium complexes. These bis-organocalcium reagents can undergo dialkylation reactions with α,ω-alkylene dihalides and dichlorosilanes to form 3-, 5-, and 6-membered ring derivatives. Significantly, these reactions are stereospecific and are highly regioselective.

Typical Procedure for the Preparation of Active Calcium

Lithium (9.0 mmol) and biphenyl (9.8 mmol) in freshly distilled THF (20 ml) were stirred under argon until the lithium was completely consumed (ca. 2 h). To a well-suspended solution of CaI_2 or $CaBr_2$ in freshly distilled THF (20 ml), the preformed lithium biphenylide was transferred via a cannula at room temperature. The reaction mixture was stirred for 1 h at room temperature prior to use. (Note: Excess calcium salt was used in the oxidative addition reactions with organic halides. Details are described later in this section.)

Typical Grignard-Type Reaction

Activated calcium (3.07 mmol), prepared from lithium biphenylide (6.15 mmol) and excess CaI_2 (4.91 mmol) in THF (30 ml), was cooled to −78°C. The color turned green upon cooling. *p*-Chlorotoluene (324 mg, 2.56 mmol) was added via a disposable syringe at −78°C. The reaction mixture was allowed to warm up to −20°C and stirred at −20°C for 30 min. The reaction mixture was cooled back to −35°C, and excess cyclohexanone (510 mg, 5.20 mmol) was added via a disposable syringe at −35°C. The resulting mixture was gradually warmed to room temperature and was stirred at room temperature for 30 min. The reaction mixture was recooled to −35°C and neutral H_2O (20 ml) was added at −35°C. After being warmed to room temperature, the reaction mixture was filtered through a small pad of Celite and was washed with Et_2O (50 ml). The aqueous layer was extracted with Et_2O (3 × 30 ml), and the combined organic phases were washed with H_2O (15 ml) and dried over anhydrous $MgSO_4$. Removal of solvent and flash column chromatography on silica gel (100 g, 230–400 mesh), eluted sequentially with 20:1 hexanes/EtOAc, 15:1 hexanes/EtOAc, and 10:1 hexanes/EtOAc, afforded 1-(*p*-methylphenyl)cyclohexanol (417 mg, 86% yield) as white crystals: mp 53–55°C.

Typical Ketone Formation Reaction

The organocalcium reagent (2.72 mmol) was prepared from *p*-chlorotoluene (344 mg, 2.72 mmol) and highly reactive calcium (3.15 mmol) as described in the preceding text. CuCN·2LiBr (3.0 mmol) in THF (10 ml) was added via cannula at −35°C, and the reaction mixture was stirred at −35°C for 30 min. Benzoyl chloride (950 mg, 6.76 mmol) was added via a disposable syringe at −35°C, and the resulting mixture was gradually warmed to room temperature. Saturated aqueous NH_4Cl solution (20 ml) was added at room temperature. The reaction mixture was then filtered through a small pad of Celite and was washed with Et_2O (50 ml). The aqueous layer was extracted with Et_2O (2 × 30 ml), and the combined organic phases were washed with H_2O (3 × 15 ml) and dried over anhydrous $MgSO_4$. Removal of solvent and flash column chromatography in silica gel (100 g, 230–400 mesh, eluted sequentially with 20 : 1 hexanes/EtOAc, 15 : 1 hexanes/EtOAc, and 10 : 1 hexanes/EtOAc) yielded (4-methylphenyl)phenylmethanone (458 mg, 86% yield).

Typical Conjugate 1,4-Addition Reaction

The organocalcium reagent (2.66 mmol) was prepared from 1-chlorooctane (395 mg, 2.66 mmol) and highly reactive calcium (3.10 mmol) as described in the preceding text. Lithium 2-thienylcyanocuprate (0.25 M in THF, 14 ml, 3.50 mmol) was added via syringe at −50°C, and the reaction mixture was gradually warmed to −35°C over a 30 min period. The reaction mixture was cooled back to −50°C, and 2-cyclohexen-1-one (210 mg, 2.18 mmol) was added via a disposable syringe at −50°C. The resulting mixture was gradually warmed to room temperature. Saturated aqueous NH_4Cl solution (20 ml) was added at room temperature. The reaction mixture was then filtered through a small pad of Celite and was washed with Et_2O (50 ml). The aqueous layer was extracted with Et_2O (2 × 30 ml), and the combined organic phases were washed with H_2O (3 × 15 ml) and dried over anhydrous $MgSO_4$. Removal of solvent and flash column chromatography on silica gel (70 g, 230–400 mesh, eluted sequentially with 50 : 1 hexanes/EtOAc and 10 : 1 hexanes/EtOAc) gave 3-octylcyclohexanone (401 mg, 87% yield).

Typical Reaction of the Calcium Complex of 1,3-Diene

Highly reactive calcium (5.02 mmol) was prepared from CaI_2 (5.02 mmol) and lithium biphenylide (10.30 mmol) in THF (20 ml) as described in the preceding text. To this calcium solution was added *trans,trans*-1,4-diphenyl-1,3-butadiene (0.863 g, 4.18 mmol) in THF (10 ml) at room temperature. (An internal standard, *n*-dodecane, was added with starting material for the GC analyses in the cases of 2,3-dimethyl-1,3-butadiene.) After being stirred at room temperature for 30 min, the reaction mixture was cooled to −78°C, and excess 1,3-dibromopropane (1.020 g, 5.05 mmol) was added via a disposable syringe at −78°C. The reaction was monitored by GC (OV-17 column). (In the cases of 2,3-dimethyl-1,3-butadiene, GC yields were reported based on the analyses of reaction quenches by an OV-17 column.) The reaction mixture was gradually warmed to −60°C and stirred at −60°C for 1 h. Saturated NH_4Cl aqueous solution (20 ml) was then added to −40°C. The reaction mixture was filtered through a small pad of Celite and was washed with Et_2O (30 ml). The aqueous layer was extracted with Et_2O (2 × 30 ml). The combined organic phases were washed with H_2O and brine and dried over anhydrous $MgSO_4$. Removal of solvent and flash column chromatography on silica gel (200 g, 230–400 mesh, eluted sequentially with hexanes and 1% Et_2O/hexanes) afforded *trans*-1-phenyl-2-*trans*-β-styrenylcyclopentane [1] (940 mg, 91% yield).

9.5 Synthesis of Polyphenylcarbynes Using Highly Reactive Calcium, Barium, and Strontium: A Precursor for Diamond-like Carbon

Man-made diamond processes are both unwieldly and expensive processes. The popular commercial process involves a transformation of graphite to diamond at 2 500°F under 50 000 atm of pressure. Another commercial method is a process of chemical vapor deposition (CVD) which yields a diamond film with high thermal conductivity, optical transparency, hardness, inertness, and even high electrical mobilities (semiconductivity) [17, 18]. Both processes are slow, expensive, and difficult to control [19]. Recently, Bianconi and Visscher [20] reported the chemical synthesis of a polymer precursor—poly(phenylcarbyne) which could be transformed to diamond-like carbon by simple pyrolysis. The report immediately attracted the attention of the media and the public since "the discovery implies there is a chemical means of forming diamond without hydrogen or a plasma. The discovery also implies a chemistry-based layer-by-layer growth of diamond is feasible." However, the approach required a high-intensity ultrasonic immersion equipment and sodium/potassium alloy [20] which could limit the industrial applications; moreover, the yield of poly(phenylcarbyne) is only 25% by this approach. We reported a facile synthesis for poly(phenylcarbyne); the reaction can be completed under mild conditions, and the yield is almost double that of the previous method [21].

Highly reactive metals have been found to be a useful tool in the synthesis of special polymers, such as in chemical modification of halogenated polystyrene resins [22], and in the synthesis of highly regioregular and high electroconductive poly(alkylthiophenes) [23]. This experience encouraged us to investigate the synthesis of poly(phenylcarbyne) using these metals. The poly(phenylcarbyne) was synthesized by the following procedure [24]. Lithium (0.06 mol, ribbon from Aldrich) and biphenyl (0.066 mol) in a 100 ml flask were stirred in freshly distilled THF (40 ml) at room temperature under argon until the lithium was completely consumed (ca. 2 h). To a well-suspended solution of CaI_2 (0.03 mol, anhydrous from Cerac) in THF (40 ml) in a 250 ml flask, the preformed lithium biphenylide was transferred via cannula at room temperature. The solution of Rieke calcium was stirred for 1 h at room temperature (the Rieke calcium is dark green and is apparently homogeneous in THF). A solution of α,α,α-trichlorotoluene (0.019 mol, 99+% from Aldrich) in 10 ml of dry pentane was then added via a cannula at −78°C. The reaction mixture was then warmed to room temperature and refluxed for 4 h. The reaction mixture was then cooled to room temperature and worked up [25] by the reported procedure [20]. The resulting tan powder was dried at 100°C under vacuum for 24 h, giving 0.79 g (46%) of poly(phenylcarbyne) (Equation 9.1). Using $SrBr_2$ or BaI_2 instead of CaI_2, following the same procedure described in the preceding text, gave the same product with 42% yield.

$$2Li^+Ar^- + MX_2 \xrightarrow[\text{Room temp, 1 h}]{\text{THF, argon}} M(Ar)_2 + 2LiX$$

M = Ca, Sr, Ba; Ar = biphenyl; X = I, Br

$$PhCCl_3 + 1.5M(Ar)_2 \xrightarrow[\text{−78°C to reflux}]{\text{THF, argon}} \left(C(Ph) \right)_n$$

M = Ca, 46%; M = Sr, 42%; M = Ba, 42% (9.1)

The spectral analysis (FTIR, ^1H and ^{13}C NMR, UV, fluorescence) proved that the polymer obtained using this methodology was the same as the previously reported poly(phenylcarbyne) [20]. The poly(phenylcarbyne) is a random three-dimensional polymer network. Infrared spectra [26] showed only monosubstituted phenyl groups present (the strong absorbance at 698 cm^{-1} (δ($mono_1$)) and 756 cm^{-1} (δ($mono_2$)) for two out-of-plane vibrations of a monosubstituted benzene ring) [27]. No absorption bands of disubstituted benzene rings were found at 805 (for para-substituted) [27a, b], 790, and 870 cm^{-1} (for meta-substituted) [27c]. No absorption bands of aliphatic C=C were found at 1650 cm^{-1} [20, 28]. The ^1H NMR (500 MHz, $CDCl_3$) spectrum displayed a dominant aromatic proton resonance at δ 7.3 ppm. The very weak resonance around δ 3.5 ppm belonged to the terminal proton [29] of the polymer network. The ^{13}C NMR (125 MHz, $CDCl_3$) spectrum exhibited three resonances. One centered at 140 ppm was attributed to the ipso carbon of the phenyl ring, an intense resonance at 128 ppm was denoted to the other five carbons of the phenyl ring, and a very broad resonance centered at 50 ppm was denoted to the quaternary carbon of poly(phenylcarbyne) [20].

The UV-vis electronic spectrum exhibited an intense broad absorption which started at a wavelength $\lambda \leq 200$ nm and decreased to 460 nm (Figure 9.1). Corresponding to the electronic absorbance, an intensive, broad fluorescence peak with λ_{max} at 460 nm was found in the fluorescence spectrum of the polymer (Figure 9.1). Both UV-vis and fluorescence properties were consistent with the poly(phenylcarbyne) structure [20].

The elemental analysis of the polymer compared well with the empirical formula $(C_6H_5C)_n$ of poly(phenylcarbyne) [30]. The molecular weight was determined by gel permeation chromatography (GPC, relative to polystyrene standard, THF as solvent), $M_w = 4979$ and $M_n = 2791$, corresponding to a polydispersity index of 1.81. The poly(phenylcarbyne) is soluble in common organic solvents, and a brown transparent film is easily formed from the solutions.

The soluble and homogeneous forms [2] of highly reactive metals such as Rieke calcium, barium, or strontium are effective reagents for this type of polymerization. The polymerization failed when insoluble highly reactive metal powders such as Rieke magnesium [31] were used (Equation 9.2). The soluble biphenyl complexes M(biphenyl)$_2$ (M = Ca, Sr, or Ba) were sufficiently active to facilitate polymerization. In contrast only oxidative addition of Mg* into the carbon–chlorine bond occurred, and upon workup only α-chlorotoluene and α,α-dichlorotoluene were recovered quantitatively as products after 24 h of reflux. No coupling reaction or polymerization occurred.

1.5 equiv Rieke Mg / THF, Ar, room temp, 1 h

Reflux, 24 h → No coupling reaction

H⁺ → $CHCl_2$ + CH_2Cl ← H⁺

1 : 1 (9.2)

These results suggest a free-radical mechanism for the polymerization of α,α,α-trichlorotoluene mediated by the homogeneous forms of highly reactive metals (Scheme 9.1). The polymerization is assumed to proceed by the coupling

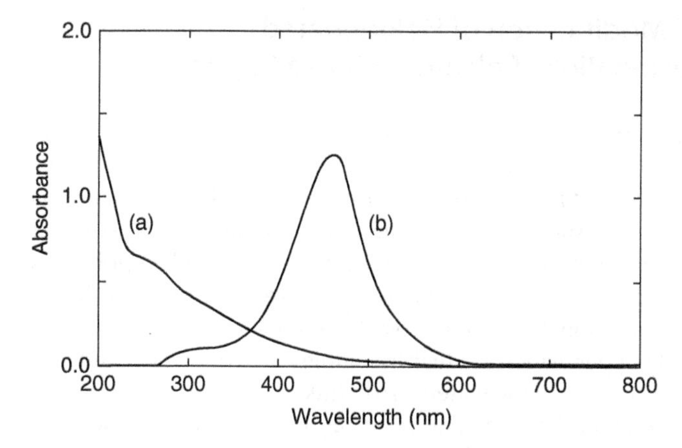

Figure 9.1 (a) UV-vis electronic spectrum (cyclohexane) and (b) fluorescence spectrum (cyclohexane, excitation wavelength = 300 nm) of poly(phenylcarbyne) obtained by this methodology.

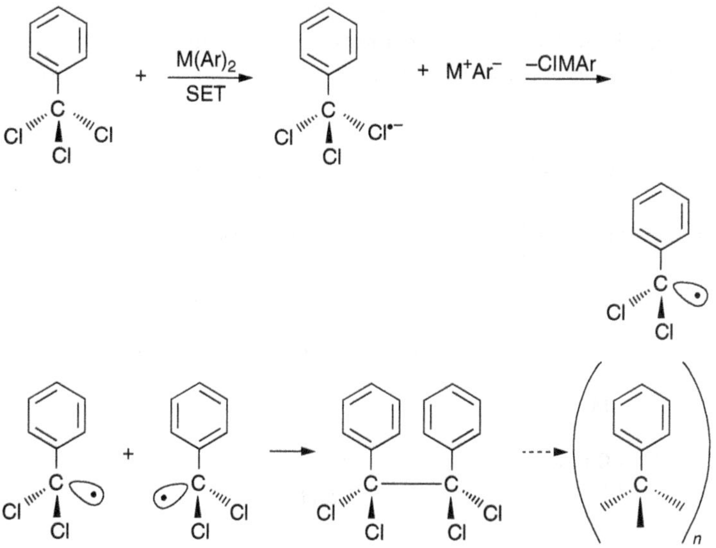

Scheme 9.1 Proposed reaction scheme. M = Ca, Sr, or Ba; Ar = biphenyl.

of free radicals which are generated by single-electron transfer (SET) between $M(biphenyl)_2$ and the carbon–chlorine bond [32], followed by exclusion of a chlorine anion. This novel approach to poly(phenylcarbyne) appears to be a general approach to several new materials.

9.6 Chemical Modification of Halogenated Polystyrenes Using Rieke Calcium or Rieke Copper

The chemical modification of cross-linked polymers has received considerable interest since the discovery of Merrifield resin and its use in peptide synthesis. Cross-linked polystyrene resins have been mainstays in solid-phase synthesis and organic synthesis. Few successful examples of functionalizing cross-linked polystyrene resins have been reported, as the insoluble cross-linked polymers are particularly resistant to reagents. Rieke metals have been found to be an effective tool for the functionalization of cross-linked polymers.

Rieke calcium and copper, which are highly soluble reactive metals, undergo direct oxidative addition to halogenated cross-linked polystyrene resins as shown in Schemes 9.2 and 9.3 [33]. The resulting insoluble polymeric aryl-calcium or aryl-copper reagents can be used to carry out a number of useful transformations on the polymers by reacting them with various electrophiles as shown in Table 9.5 and Table 9.6 [22].

Rieke calcium is exceptionally highly reactive. Thus, fluorinated, chlorinated, brominated, and chloromethylated cross-linked polystyrene resins are all successfully converted to the corresponding calcium reagents. It should be noted that few metals undergo oxidative addition to aryl fluorides. Thus, it is noteworthy that Ca* reacts with *p*-fluoropolystyrene at room temperature to

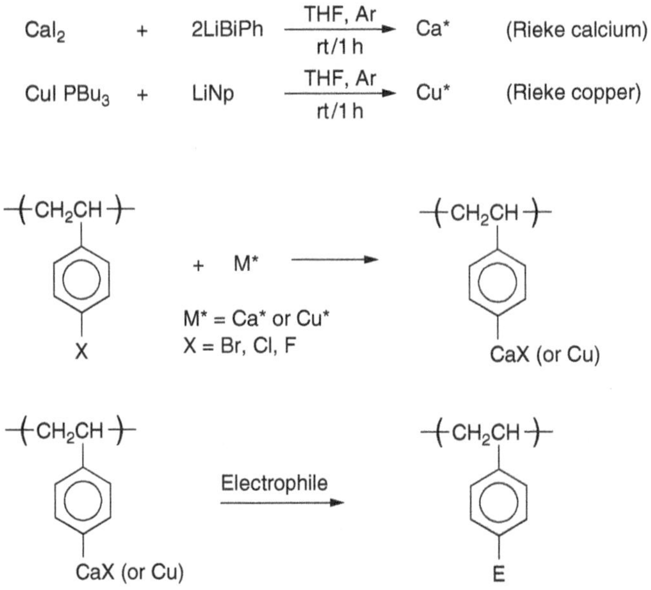

$$CaI_2 + 2LiBiPh \xrightarrow[rt/1\,h]{THF,\,Ar} Ca^* \quad \text{(Rieke calcium)}$$

$$CuI\,PBu_3 + LiNp \xrightarrow[rt/1\,h]{THF,\,Ar} Cu^* \quad \text{(Rieke copper)}$$

$$\text{-(CH}_2\text{CH)-}\,C_6H_4X + M^* \longrightarrow \text{-(CH}_2\text{CH)-}\,C_6H_4CaX \text{ (or Cu)}$$

$M^* = Ca^*$ or Cu^*
$X = Br, Cl, F$

$$\text{-(CH}_2\text{CH)-}\,C_6H_4CaX \text{ (or Cu)} \xrightarrow{\text{Electrophile}} \text{-(CH}_2\text{CH)-}\,C_6H_4E$$

Scheme 9.2 Modifications of halogenated polystryene resin using Rieke calcium or copper.

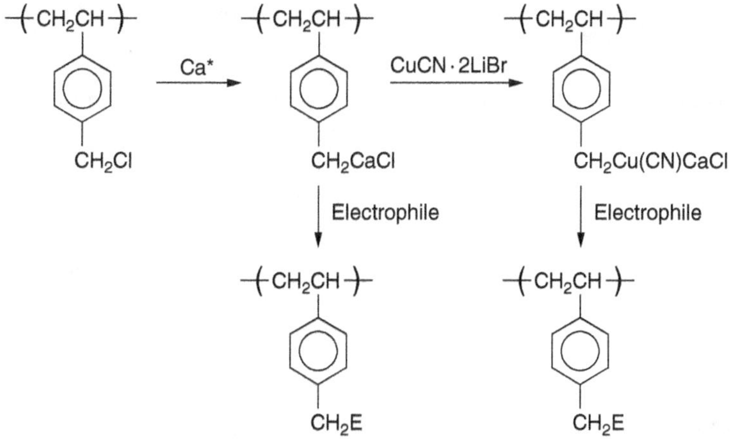

$-\!(CH_2CH)\!-$ with CH_2Cl → Ca* → CH_2CaCl → $CuCN \cdot 2LiBr$ → $CH_2Cu(CN)CaCl$

Electrophile → CH_2E ; Electrophile → CH_2E

Scheme 9.3 Reactions of organocalcium and calcium cuprate reagents prepared from chloromethylated polystyrene and Rieke calcium.

Table 9.5 Reactions of organocalcium reagents prepared from ρ-halopolystyrene.

Polystyrene polymer	Electrophile	Polystyrene product	Yield (%)
(P)-Br	CO_2	(P)-COOH	83
(P)-Br	$ClSiPh_3$	(P)-$SiPh_3$	72
(P)-Br	$ClPPh_2$	(P)-PPh_2	71
(P)-Br	C_6H_5CHO	(P)-CH(OH)C_6H_5	100
(P)-Cl	H_2O	(P)-H	100
(P)-Cl	CO_2	(P)-COOH	82
(P)-Cl	C_6H_5COCl	(P)-COC_6H_5	26
(P)-Cl	CH_3COCl	(P)-$COCH_3$	28
(P)-F	CO_2	(P)-COOH	78
(P)-F	$ClSiMe_3$	(P)-$SiMe_3$	26
(P)-F	CH_3COCl	(P)-$COCH_3$	59
(P)-F	C_6H_5CHO	(P)-CH(OH)C_6H_5	25

give the polymeric aryl-calcium reagent [34]. The polymeric calcium intermediates are converted to functionalized polystyrene resins by reaction with different electrophiles or converted to polymeric cuprate reagents, which then afford the ketone-containing polymers shown in Table 9.6.

Table 9.6 Reactions of calcium cuprate reagents from chloromethylated polystyrene and Rieke calcium with various electrophiles.

Polystyrene polymer	Electrophile	Polystyrene product	Yield (%)
(P)-CH_2Cl	C_6H_5COCl	(P)-$CH_2COC_6H_5$	100
(P)-CH_2Cl	CH_3COCl	(P)-CH_2COCH_3	100
(P)-CH_2Cl	$Br(CH_2)_3CN$	(P)-$(CH_2)_4CN$	37
(P)-CH_2Cl	$Br(CH_2)_3CO_2Et$	(P)-$(CH_2)_4CO_2Et$	80
(P)-CH_2Cl	C_6H_8O	(P)-$CH_2C_6H_9O$	100

References

1 Xiong, H.; Rieke, R.D. *J. Org. Chem.* 1989, **54**, 3247.
2 Wu, T.; Xiong, H.; Rieke, R.D. *J. Org. Chem.* 1990, **55**, 5045.
3 (a) Freeman, P.K.; Hutchinson, L.L. *J. Org. Chem.* 1983, **48**, 879. (b) Raston, C.L.; Salem, G. *J. Chem. Soc., Chem. Commun.* 1984, 1702. (c) Harvey, S.; Junk, P.C.; Raston, C.L.; Salem, C. *J. Org. Chem.* 1988, **53**, 3134. (d) Bogdanovic, B. *Acc. Chem. Res.* 1988, **21**, 261.
4 Lindsell, W.E. In *Comprehensive Organometallic Chemistry*; Wilkinson, G., Ed., Pergoman Press: Oxford, 1982; Vol. **1**, Chapter 4, pp 223–252.
5 (a) Kawabata, N.; Matsumura, A.; Yamashita, S. *Tetrahedron* 1973, **29**, 1069. (b) Kawabata, M.; Matsumura, A.; Yamashita, S. *J. Org. Chem.* 1973, **38**, 4268.
6 (a) Gowenlock, B.G.; Lindsell, W.E. *J. Organomet. Chem. Libr., Organomet. Chem. Rev.* 1977, **3**, 1. (b) Bogatskii, A.V.; Chumachenko, T.K.; Derkach-Kozkukhova, A.E.; Lyamtseva, L.N.; Suprinovich, E.S. *Zh. Obshch. Khim.* 1977, **47**, 2297.
7 Dubois, J.E.; Bauer, P.; Molle, G.; Daza, J. *C. R. Acad. Sci., Ser. III* 1977, **284**, 145.
8 Rieke, R.D.; Bales, S.E. *J. Am. Chem. Soc.* 1974, **96**, 1775.
9 Yu, S.H.; Ashby, E.C. *J. Org. Chem.* 1971, **36**, 2123 and references cited therein.
10 Knochel, P.; Yeh, M.C.P.; Berk, S.C.; Talbert, J. *J. Org. Chem.* 1988, **53**, 2390.
11 Lithium 2-thienylcyanocuprate was purchased from Aldrich Chemical Co. Lipshutz, B.H.; Koerner, M.; Parker, D.A. *Tetrahedron Lett.* 1987, **28**, 945.
12 (a) Corey, E.J.; Boaz, N.W. *Tetrahedron Lett.* 1985, **26**, 6019. (b) Alaxakis, A.; Berlan, J.; Besace, Y. *Tetrahedron Lett.* 1986, **27**, 1047.
13 (a) Erker, G.; Kruger, C.; Muller, G. *Adv. Organomet. Chem.* 1985, **24**, 1. (b) Yasuda, H.; Tasumi, K.; Nakamura, A. *Acc. Chem. Res.* 1985, **18**, 120. (c) Walther, D.; Pfuetrenreuter, C. *Naturwiss. Reih.* 1985, **34**, 789.
14 The stereochemistry of cis-3,6-diphenylcyclohex-1-ene was identified by converting the cyclohexene to 1,2-cyclohexanediol via the epoxide (see

Experimental Section for details). Treatment of cis-3,6-diphenylcyclohex-1-ene with *m*-chloroperbenzoic acid in the presence of K_2CO_3 in CH_2Cl_2 gave only a single product in 60% yield along with 20% of recovered starting material. The fully decoupled ^{13}C NMR spectrum gave only seven peaks which unambiguously proved that two phenyl groups were in cis geometry. Reaction of the epoxide with 6% $HClO_4$ in acetone yielded 1,4-diphenylcyclohexane-2,3-diol in 93% yield. The proton spin–spin coupling constants further verified that the two phenyl groups were cis.

15 Lomonosov, M.V. *Zh. Org. Khim.* 1977, **13**, 63.

16 The molecular formula in the solid state is $Mg(THF)_3$(s-cis-PhCh PhCH =CHCH=CHPh). Kai, Y.; Kanehisa, N.; Miki, K.; Kasai, N.; Mashima, K.; Yasuda, H.; Nakamura, A. *Chem. Lett.* 1982, 1277.

17 Plano, M.A.; Landstrass, M.I.; Pan, L.S.; Han, S.; Kania, D.R.; McWilliams, S.; Ager III, J.W. *Science* 1993, **260**, 1310.

18 Jin, S.; Graebner, J.E.; McCormack, M.; Tiefel, T.H.; Katz, A.; Dautremont-Smith, W.C. *Nature* 1993, **362**, 822.

19 Amato, I. *Science* 1993, **260**, 1424.

20 Visscher, G.T.; Nesting, D.C.; Badding, J.V.; Bianconi, P.A. *Science* 1993, **260**, 1496.

21 Rieke, R.D.; Chen, T. *Chem. Mater.* 1994, **6**, 576.

22 O'Brien, R.A.; Chen, T.-A.; Rieke, R.D. *J. Org. Chem.* 1992, **57**, 2667.

23 Chen, T.-A.; Rieke, R.D. *J. Am. Chem. Soc.* 1992, **114**, 10087.

24 All manipulations were carried out under an atmosphere of argon on a dual manifold/argon system. Lithium ribbon, biphenyl, and MX_2 (M = Ca, Sr, Ba; X = I, Br) were weighed out and charged into reaction flasks under argon in a Vacuum Atmosphere Co. dry box. Tetrahydrofuran was distilled from Na/K alloy under an argon atmosphere immediately before use.

25 Workup procedure: To the reaction mixture, distilled water (100 ml) was added. The mixture was then stirred for 30 min at room temperature. The organic layer was separated, filtered, and then concentrated to 30 ml in volume by vacuum. Addition of methanol (150 ml) to organic layer gave a tan precipitate, which was collected by filtration, washed by several portions of methanol (30 ml), and purified by reprecipitation twice from THF solution upon addition of ethanol. The biphenyl, which is soluble in ethanol and very soluble in methanol, and oligomer were washed out by the filtration and the purification earlier.

26 FTIR spectral data (neat polymer film cast from $CHCl_3$ solution onto NaCl disk): 3053(s), 3024(s), 2924(m), 1946(w), 1882(w), 1803(w), 1599(s), 1492(s), 1444(s), 1180(m), 1155(m), 1030(m), 910(w), 756(s), 698(s), and 667 cm^{-1}(w). The spectra show no absorption hand for —OH group.

27 (a)Yamamoto, T.; Hayashi, Y.; Yamamato, A. *Bull. Chem. Soc. Jpn.* 1978, **51**, 2091. (b)Kovacic, P.; Oziomek, J. *J. Org. Chem.* 1964, **29**, 100. (c)Kovacic, P.; Marchiona, V.J.; Koch, F.W.; Oziomek, J. *J. Org. Chem.* 1966, **31**, 2467.

28 Silverstein, R.M.; Bassler, G.C.; Morrill, T.C. *Spectrometric Identification of Organic Compounds*; John Wiley & Sons, Inc.: New York, 1981; p 110.

29 The chemical shift (δ) for methyl protons of toluene [$PhCH_3$] is 2.3 ppm and for methylene protons of bibenzyl ($Ph{-}CH_2CH_2{-}Ph$) is 2.9 ppm. In poly(phenylcarbyne), the terminal protons (>C(Ph)—CH(Ph)—C(Ph)<) are further deshielded since they bear more benzyl groups in the three-dimensional network of polymer. The chemical shift for these protons should be downfield to ~3.5 ppm.

30 Elemental anal. calcd for $(C_6H_5C)_n$: C, 94.36; H, 5.64. Found: C, 93.27; H, 5.79. No oxygen or chlorine was found, and no residue was detectable on combustion analysis of the sample.

31 Rieke, R.D.; Xiong, H. *J. Org. Chem.* 1991, **56**, 3109.

32 Walborsky, H.M.; Hamdouchi, C. *J. Org. Chem.* 1993, **58**, 1187.

33 O'Brien, R.A.; Chen, T.-A.; Rieke, R.D. *J. Inorg. Organomet. Polym.* 1992, **2**, 345.

34 Pittman Jr., C.U. *Polym. News* 1993, **18**, 79.

10

Barium

10.1 Preparation of Rieke Barium

As barium is in the same periodic row as calcium, it was anticipated that active barium could be prepared by essentially the same approach as used for preparing Rieke calcium. This turned out to be the case, and highly reactive Rieke barium was readily prepared essentially by the identical approach used for preparing Rieke calcium.

The general experimental procedure for preparing the Rieke barium is presented in the succeeding text. An oven-dried 20 ml Schlenk tube or 50 ml three-necked flask equipped with a Teflon-coated magnetic stirring bar is flushed with argon. Freshly cut lithium (15 mg, 2.2 mmol) and biphenyl (350 mg, 2.3 mmol) are put into the apparatus and covered with dry THF (10 ml), and the mixture is stirred for 2 h at 20–25°C (lithium was completely consumed). Into a separate oven-dried 50 ml Schlenk tube or 50 ml three-necked flask equipped with a Teflon-coated magnetic stirring bar is placed anhydrous BaI_2 (430 mg, 1.1 mmol) under an argon atmosphere; this is covered with dry THF (5 ml) and stirred for 5 min at room temperature. To the resulting yellowish solution of BaI_2 in THF is added at room temperature a solution of the lithium biphenylide through a stainless steel cannula under an argon stream. The reaction mixture is stirred for 30 min at room temperature, and the resulting dark brown suspension of reactive barium thus prepared is ready to use.

The resulting Rieke barium was found to undergo oxidative addition to a wide variety of organic halides.

Chemical Synthesis Using Highly Reactive Metals, First Edition. Reuben D. Rieke.

10.2 Oxidative Addition of Rieke Barium to Allylic Halides: Preparation of Stereochemically Homogeneous Allylic Barium Reagents

The resulting Rieke barium was found to undergo oxidative addition to a wide variety of organic halides. Yanagisawa and Yamamoto have carried out extensive studies on the preparation of stereochemically homogeneous allylic barium reagents [1, 2]. These reagents can be generated at −70°C or below and retain their stereochemical integrity even upon reaction with a wide variety of electrophiles (Scheme 10.1).

Corey and coworkers successfully applied this cross-coupling method to the synthesis of (3*S*)-2,3-oxidosqualene (Scheme 10.2; **5**) [3]. The utility of allylic barium reagents for nucleophilic substitution reactions was further demonstrated by the synthesis of cembrol A (**7**), in which ring closure of the epoxy compound occurred regioselectively (Scheme 10.3) [4].

Xiong and Corey also used Rieke barium in an enantioselective total synthesis of glabrescol, a chiral C_2-symmetric pentacyclic oxasqualenoid. The first step is shown in Scheme 10.4 [5].

In summary, the studies of allylic barium reagents have demonstrated the following main features: (i) allylic barium reagents are readily prepared by treatment of the corresponding allylic chlorides with Rieke barium; (ii) no stereoisomerization of the primary allylic barium compound is observed below −70°C; (iii) the barium reagent reacts selectively at the α-position with aldehydes, ketones, carbon dioxide, and allylic halides with complete retention of the stereochemistry of the starting halide; (iv) selective 1,4-addition reaction occurs with α,β-unsaturated ketones. The extraordinary α-selectivity and stereospecificity of the reactions of allylic barium reagents provide unprecedented routes to homoallylic alcohols, β,γ-unsaturated carboxylic acids, and 1,5-dienes and are broadly applicable in organic synthesis.

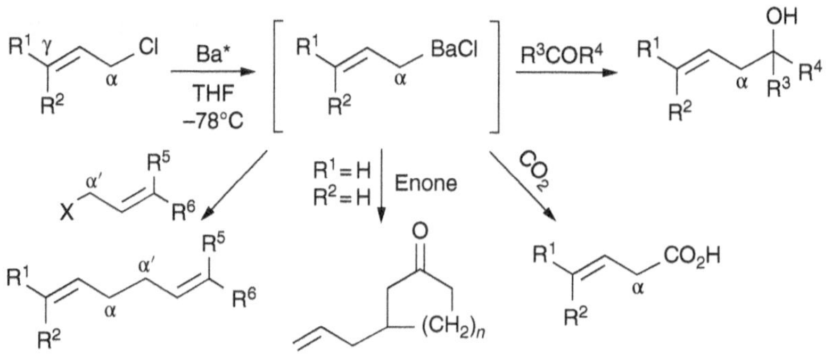

Scheme 10.1 Various allylation reactions using allylic barium reagents.

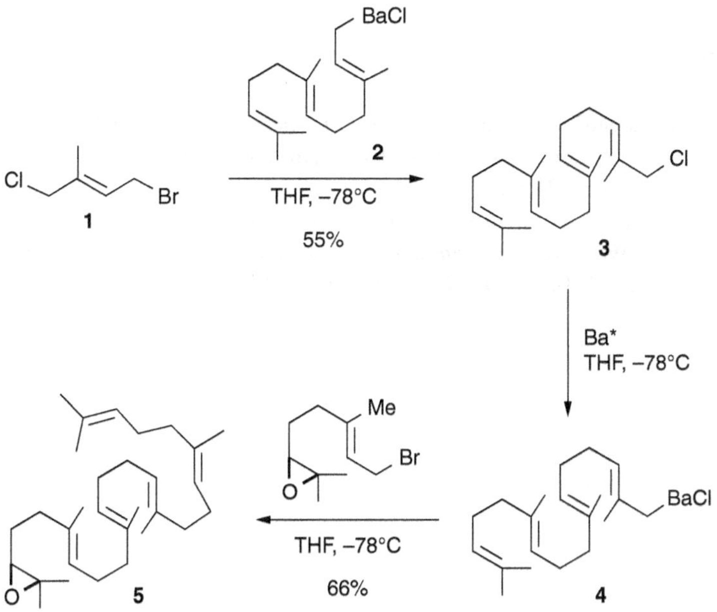

BaCl

2

THF, −78°C

55%

Cl

Br

1

Cl

3

Ba*
THF, −78°C

Me

Br

O

THF, −78°C

66%

BaCl

4

O

5

Scheme 10.2 Synthesis of (3*S*)-2,3-oxidosqualene.

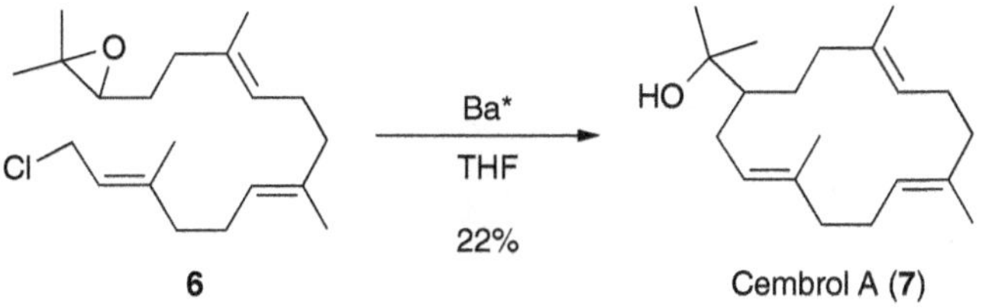

O

Cl

6

Ba*

THF

22%

HO

Cembrol A (**7**)

Scheme 10.3 Preparation of cembrol A.

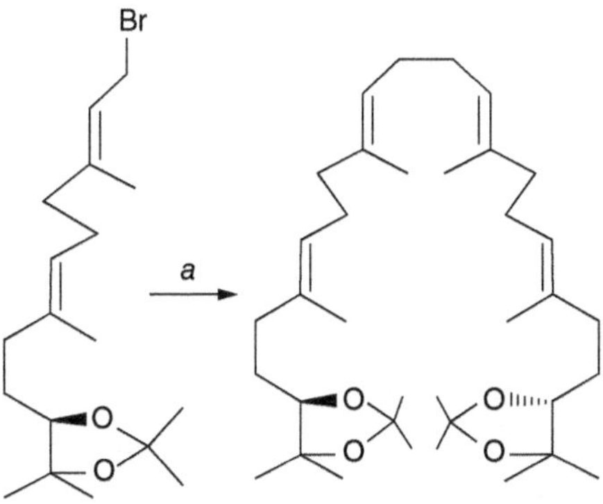

Br

a

O

O

O

O

O

O

[a]Conditions: (a) Rieke barium, THF, −78°C, 1 h (56%)

Scheme 10.4 Coupling of allylic barium reagent[a].

References

1 Yanagisawa, A.; Habaue, S.; Yamamoto, H. *J. Am. Chem. Soc.* 1991, **113**, 5893.
2 Yanagisawa, A.; Yamamoto, H. In *Active Metals*; Furstner, A., Ed.; VCH Publishers: New York, 1995, Chapter 2.
3 Corey, E. J.; Noe, M.C. *Tetrahedron Lett.* 1993, **34**, 5995.
4 Corey, E.J.; Shieh, W.-C. *Tetrahedron Lett.* 1992, **33**, 6435.
5 Xiong, Z.; Corey, E.J. *J. Am. Chem. Soc.* 2000, **122**, 9328.

11

Iron

11.1 Preparation of Highly Reactive Iron and Some Oxidative Addition Chemistry

Although iron has been known to man prior to 3000 BC, its use in chemistry had to wait until the twentieth century. Most of its early applications involved catalysis chemistry, primarily in the petrochemical industry. In 1983, we reported the preparation of an extremely reactive iron powder using the Rieke method [1–3].

Reduction of anhydrous iron(II) halides with 2.3 equiv of lithium or reduction of iron(III) halides with 3.3 equiv of lithium using naphthalene as an electron carrier in THF or DME yields a highly reactive iron powder. While the iron powders from iron(III) salts are slightly more reactive, the relative ease of handling the much less hygroscopic iron(II) salts makes them the preferred choice for preparing highly reactive iron powders. The iron powders show slow or little settling and the dry powders are pyrophoric.

The iron slurries show exceptional reactivity toward oxidative addition reactions with carbon halogen bonds. In fact, the reaction with C_6F_5I is so exothermic that the slurry has to be cooled to 0°C before the addition of C_6F_5I. The reaction of iron with C_6F_5Br is also quite exothermic; hence, even for this addition, the iron slurry is cooled to about 0°C. The organoiron compound formed in the prementioned reaction, solvated $Fe(C_6F_5)_2$, reacts with CO at room temperature and ambient pressure to yield $Fe(C_6F_5)_2(CO)_2(DME)_2$.

Reaction of allyl halides with the iron powders is rapid and exothermic and leads to near quantitative yields of the self-coupled product 1,5-hexadiene. Similarly, reaction of benzyl chloride with the iron powders at room temperature yields bibenzyl in 60–70% yields along with 20–25% of toluene. In contrast, reaction of aryl halides with the iron powders leads to reductive cleavage rather than self-coupling. Similarly, reaction of 1-bromoheptane with the iron powder in THF for 3 h at room temperature produced 35% heptane and no

Chemical Synthesis Using Highly Reactive Metals, First Edition. Reuben D. Rieke.

self-coupled product. Further reaction of this mixture for 1 h at reflux increased the yield to 70% heptane.

The iron slurries react readily with ethyl α-bromoacetate. The resulting organoiron species adds readily to aldehydes and ketones to produce β-hydroxy esters in excellent yields. Addition of a mixture of an aryl aldehyde and an allylic halide to the iron slurry produced good yields of the cross-coupled alcohol.

Finally, it would appear that these highly reactive iron powders will be of value as a general reducing agent. Reaction of nitrobenzene with 4 equiv of iron and 1 equiv of *n*-butanol in THF at room temperature was very exothermic. After 1 h at room temperature followed by reflux for 1 h, the reaction mixture gave 88% aniline upon workup.

While very little chemistry has been attempted using the Rieke iron, it is clear that it exhibits extremely high chemical reactivity toward a number of organic halides. It is anticipated that many new uses and reactions will be developed using this unusual form of iron.

Preparation of a Typical Iron Slurry

A 50 ml two-necked flask, equipped with a magnetic stirrer and a condenser topped with an argon inlet, is charged with freshly cut lithium (0.154 g, 0.0222 mol), 1.221 g (0.00963 mol) of $FeCl_2$ (quickly weighed in air), and naphthalene (0.187 g, 0.00146 mol) [2]. Glyme (18 ml) is syringed in and vigorous stirring started. The initially beige mixture slowly darkens and in about 3 or 4 h becomes black and greasy appearing. After some time, it reverts to a free-moving black slurry. The next day, a black slurry exists in a slightly green solution with no lithium present. The slurry is washed by allowing the slurry to settle, removing the solvent above the slurry by syringe and adding fresh glyme, and stirring the slurry and again letting it settle. Usually, after two such cycles, no green color is visible in the solution.

Very often the piece of lithium coats with the metal powder. In such cases, the argon flow is increased, and, with a curved spatula, the piece of lithium is rubbed against the side of the flask to expose a fresh surface of the lithium.

Preparation of $Fe(C_6F_5)_2(CO)_2(C_4H_{10}O_2)_2$

An iron slurry is prepared in the usual manner and then allowed to settle. The solvent above the slurry is removed (most of the iron is coated about the magnetic stirring bar), and the slurry is cooled to 0°C for the slow addition of 1 equiv of C_6F_5I per iron. After the addition, the flask is allowed to warm to rt and the brown mixture in the flask is stirred for 25 h even though the reaction appears to be complete in about 2 h. A slow stream of CO (large excess) is then bubbled through the thick brown solution for 11 h; no color change is apparent.

The volatiles are removed from the mixture under vacuum. The product is dissolved in 20 ml of CH_2Cl_2 and then anaerobically filtered. Slow solvent evaporation in an inert atmosphere results in large yellow crystals. The crystals are often covered with a brown tarry residue which is readily removed by washing the crystals with a small amount of 2:1 hexane/CH_2Cl_2 mixture. (Yield 60%; mp 103–114°C (sealed tube). Anal. Calcd for $C_{22}H_{20}F_{10}FeO_6$: C, 42.19; H, 3.22; F, 30.34. Found: C, 42.35; H, 3.30; F, 30.54.)

The analogous reaction carried out at about 55°C between an iron slurry and 1 equiv of C_6F_5Br, followed by bubbling a large excess of CO into the mixture, results in the same product and in comparable yield. Because of less tarry residues and hence easier workup, we find the reaction employing C_6F_5Br more attractive.

References

1 Kavaliunas, A.V.; Rieke, R.D. *J. Am. Chem. Soc.* 1980, **102**, 5944.

2 Kavaliunas, A.V.; Taylor, A.; Rieke, R.D. *Organometallics* 1983, **2**, 377.

3 Rieke, R.D.; Kavaliunas, A.V. *Organometallic Synthesis*; Elsevier Science Publishers B.V.: Amsterdam, 1988; Vol. **4**.

12

Palladium and Platinum

12.1 Preparation of Highly Reactive Palladium and Platinum and Some Oxidative Addition Chemistry

Three general approaches for the preparation of highly reactive palladium and platinum metal powders have been developed [1, 2]. As will be seen, each successive approach leads to more reactive metal powders.

Oxidative addition of RX to transition metals has been observed using the metal atom or metal vaporization approach of Skell [3, 4]. Klabunde has reported that nickel and palladium, when cocondensed with aryl halides, readily undergo oxidative insertion into the carbon–halogen bond [5–8]. Cocondensation of nickel or palladium with pentafluorobromobenzene and triethylphosphine gave good yields of the bromopentafluorobis(triethylphosph ine)metal complex. The corresponding solution reaction of common commercial nickel powders or palladium powders with aryl halides has not been observed due to the poor reactivity of these and most other transition metals toward oxidative addition.

Initially we tried the standard approach of reduction of NiI_2, $NiBr_2$, or $NiCl_2$ with potassium in refluxing THF. Finely divided black nickel powders were obtained; however, they showed rather limited reactivity toward oxidation insertion into carbon–halogen bonds. Similar results were found with palladium and platinum.

We have demonstrated with several of the main group elements that the reactivity of the resulting metal is highly dependent on such factors as the solvent, reducing agent, anion, or in the case of some metals the presence of additional alkali salts [2, 9, 10]. In the case of the transition metals, the presence of a triaryl- or trialkylphosphine during the reduction yields a highly reactive metal slurry.

Chemical Synthesis Using Highly Reactive Metals, First Edition. Reuben D. Rieke.

The reduction of $PdCl_2$ with potassium in the presence of triethylphosphine in THF yields a highly reactive black palladium slurry. Addition of pentafluorobromobenzene to the metal slurry produces a rapid reaction [11]. After 1 h the reaction was worked up, yielding 76% of the bromopentafluorophenylbis(triethylphosphine)palladium(II). Addition of iodobenzene to the palladium slurry at room temperature for 1 h produced the new complex *trans*-iodophenylbis(triethylphosphine)palladium(II) in 52% yield.

The reaction has been extended to unreactive halides. The exceptionally high reactivity of the slurries produced by this procedure is exemplified by the reaction of the palladium slurry with chlorobenzene to give the chlorophenylbis(triethylphosphine)palladium(II) in 54% yield based on the palladium halide used.

In a similar manner, highly reactive platinum slurries can be prepared by reduction of platinum halides in the presence of phosphines. The addition of pentafluorobromobenzene to the platinum slurry yielded the known *trans*-bromopentafluorophenyl-bis(triethylphosphine)platinum(II) in 40% yield, based on the platinum halide used. The reaction has been extended to other aryl halides.

In the case of activated palladium, reduction using lithium with naphthalene appears to give the most reactive powder. For example, such a powder reacts at room temperature with an excess of allyl chloride to afford (π-allyl $PdCl)_2$ in 47% yield. Palladium powder prepared by potassium reduction in refluxing glyme reacts with allyl chloride under the same conditions to yield <6% of the corresponding compound. Palladium powders produced by potassium reduction in refluxing ethers or via alkali metal–naphthalene reductions readily react with allyl bromide at room temperature to yield (π-allyl $PdBr)_2$ in >90% yield. The reaction between stoichiometric amounts of allyl iodide and palladium metal reduced via lithium–naphthalene is 60% complete in 2 min and essentially quantitative in 30 min. Fischer and Bürger [12] obtained (π-allyl $PdBr)_2$ from finely divided palladium and refluxing allyl bromide in 46% yield. Under similar conditions they reported that a reaction occurred with allyl iodide but no compound was isolated and with allyl chloride no reaction of the metal was observed after several days.

The lithium–naphthalene reduction has been successfully extended to other metals. Platinum metal so prepared reacts at room temperature with allyl iodide. Cobalt metal reacts with allyl bromide to yield 1,5-hexadiene without any evidence of an organocobalt compound. Iron powder reacts very vigorously with allyl bromide at room temperature to yield 1,5-hexadiene.

In this section, we describe in detail the preparation of metal slurries of palladium and platinum by alkali metal reduction of the compounds $[P(C_2H_5)_3]_2MX_2$ and the reaction of the resulting metal slurries with a variety of organic halides.

Preparation of Slurries

Potassium metal reduction of the palladium and platinum compounds $[P(C_2H_5)_3]_2MX_2$ is readily effected in refluxing ethers such as glyme or THF to yield finely divided and highly reactive metal slurries. Equally useful are mixtures of anhydrous metal salts and triethylphosphine; these react in a matter of minutes in ethers to yield the $[P(C_2H_5)_3]_2MX_2$ compounds. Reduction times with potassium vary considerably depending on the compound being reduced; for example, solutions of $[P(C_2H_5)_3]_2PtI_2$ are reduced in about 1 h, whereas some of the others take as long as 20 h. During the early stages of reduction, small amounts of the slurries syringed onto water inflame; the absence of such behavior is assumed to indicate complete reaction of potassium. Reductions involving lithium with naphthalene are easier to judge as complete, since the piece of lithium floats and is visible during reduction. Slurries resulting from $[P(C_2H_5)_3]_2MX_2$ are black and, in the case of palladium, exist in a yellow solution which undoubtedly contains the known tris- and tetrakis(triethylphosphine) metal compounds as well as perhaps other substances. In an attempt to determine the insoluble palladium in such a slurry, the powder was separated from the soluble material by centrifugation and careful anaerobic washing of the powder. Analysis of the powder showed that 52% of the starting palladium compound existed as a black insoluble solid. The removed yellow solution was allowed to react with bromobenzene and yielded 32% of *trans*-$[P(C_2H_5)_3]_2Pd(C_6H_5)Br$, based on the amount of *trans*-$[P(C_2H_5)_3]_2PdCl_2$ used. Thus the organometallic products obtained from these slurries are arising in part from the soluble material present and in the case of the nitrile derivatives likely arise entirely from them. However, in those reactions where the yields are 60–70% or greater, it is clear that a considerable portion of the product is arising by direct reaction with the palladium metal powder.

Reduction of the phosphine metal halide complexes can also be carried out with other alkali metals. For example, *trans*-$[P(C_2H_5)_3]_2PdCl_2$ was reduced with lithium using a small amount of naphthalene as an electron carrier. However, this procedure offers no particular advantage as the reduction times, reactivity of slurries, and final yields of oxidative insertion products are very similar to the reduction procedure using potassium. It must be pointed out, however, that for the reduction of transition metal halides in the absence of phosphines, the reduction using lithium and naphthalene is far superior to reductions using potassium with regard to reactivity of the resulting metal powders.

Palladium Compounds

Palladium slurries react with a variety of aryl halides and result in *trans*-$[P(C_2H_5)_3]_2Pd(R)X$ compounds. The trans geometry for all these compounds was established by NMR spectroscopy [14]. In general, increased temperatures during the reaction of the aryl halide with the slurry result in increased yields of the organopalladium compounds. However, in the case of iodobenzene, a reaction with a palladium slurry at elevated temperature resulted in no organopalladium compound but only biphenyl and *trans*-$[P(C_2H_5)_3]_2PdI_2$ as well as the ubiquitous palladium metal. This suggests that at elevated temperatures, *trans*-$[P(C_2H_5)_3]_2Pd(C_6H_5)I$ undergoes a reductive elimination reaction of the type observed for analogous nickel [15]. The palladium compounds are stable in the solid state and in solution, although exhibiting some photodecomposition which appears to be accelerated by impurities.

The product of the reaction between allyl bromide and a palladium slurry was not $[P(C_2H_5)_3]_2Pd(C_3H_5)Br$ but $P(C_2H_5)_3Pd(C_3H_5)Br$. The former compound has been obtained in a reaction of $Pd[P(C_2H_5)_3]_3$ with a twofold excess of allyl bromide [13]. $P(C_2H_5)_3Pd(C_3H_5)Br$ was obtained by allowing a palladium slurry to react with a fourfold excess of allyl bromide. A similar effect of excess allylic halide has been noted in a reaction between $Pd[P(C_6H_5)_3]_4$ and a large excess of 2-methallyl chloride which yielded $P(C_6H_5)_3Pd[CH_2C(CH_3)CH_2]Cl$ [16].

Reduction of PdI_2 with lithium in the presence of a catalytic amount of naphthalene and in the absence of triethylphosphine results in a palladium powder which reacts with C_6F_5I to give a deep brown solution of C_6F_5PdI. This compound in acetone immediately reacted with triethylphosphine and was isolated as *trans*-$[P(C_2H_5)_3]_2Pd(C_6H_5)I$ in 44% yield. The palladium slurry prepared in this manner does not react under these conditions with triethylphosphine; thus it can be ruled out that the product arises via a zerovalent triethylphosphine compound of palladium. From this same reaction mixture was isolated a 20% yield of *trans*-$[P(C_2H_5)_3]_2PdI_2$ as well as a 2% yield of $[P(C_2H_5)_3]_2Pd_2I_4$. The former compound was obtained as a by-product during a metal atom reaction of palladium with C_6F_5I followed by trapping the product with triethylphosphine [17]. The analogous compound, C_6F_5PdBr, has also been prepared electrochemically from the metal [18].

Platinum Compounds

Platinum slurries, obtained by potassium reduction of $[P(C_2H_5)_3]_2PtX_2$ in ethereal solvents, exhibit similar reactivities with aryl halides to those of palladium. The yields of products are, as in the case of palladium, varied with the highest being for the more reactive aryl halides. The compounds possess the trans geometry as determined by NMR spectroscopy [14] and in contrast to the palladium compounds exhibit no photosensitivity.

Summary

Highly reactive slurries of palladium and platinum have been prepared by alkali metal reduction of compounds $[P(C_2H_5)_3]_2MX_2$ in ethers using simple apparatus and procedures. These metals undergo oxidative insertion into carbon–halogen bonds of aryl halides and yield compounds of the type *trans*-$[P(C_2H_5)_3]_2M(R)X$.

Metal slurries prepared in the presence of triethylphosphine contain, besides the black insoluble metal powder, compounds of the metals in low oxidation states, most likely the known tri- and tetrakis phosphine metal compounds which are known to react with organic halides, and contribute to the yields of the products obtained here.

Highly reactive transition metal powders have been prepared in the absence of stabilizing ligands. The ability to prepare these metals with simple apparatus is of great importance not only to synthetic organic, organometallic, and inorganic chemistry but also to catalysis and surface chemistry.

Preparation of *trans*-$[P(C_2H_5)_3]_2Pd(C_6H_5)I$

A 50 ml three-necked flask equipped with a magnetic stirrer and a condenser topped with an argon inlet was charged with 0.959 g (0.00232 mol) of *trans*-$[P(C_2H_5)_3]_2PdCl_2$ and 0.181 g (0.00463 mol) of potassium (freshly cut and cleaned), and then 10 ml of freshly distilled THF was syringed in. The yellow solution was then refluxed with vigorous stirring for 22 h, after which time a black metal powder existed in a yellow solution. The reaction mixture was allowed to cool to room temperature, 0.26 ml (0.0023 mol) of C_6H_5I was syringed in, and the mixture was stirred for 1 h after which it was filtered, the solids were washed out with dichloromethane, and the solvent was removed from the filtrate under reduced pressure. The crude material was dissolved in hexane, treated with charcoal, and filtered. Slow evaporation of the solvent resulted in large yellow crystals of *trans*-$[P(C_2H_5)_3]_2Pd(C_6H_5)I$: 0.639 g (51%); mp 111–111.5°C. Very often the crystalline product is obtained somewhat oil covered and the recrystallization must be repeated.

Preparation of *trans*-$[P(C_2H_5)_3]_2Pd(C_6H_5)Br$

(i) A 50 ml three-necked flask, equipped with a magnetic stirrer and a condenser topped with an argon inlet, was charged with 0.840 g (0.00316 mol) of $PdBr_2$. An argon atmosphere was established in the apparatus, and 10 ml of freshly distilled glyme was syringed into the flask, followed by 0.94 ml (0.0064 mol) of $P(C_2H_5)_3$. Immediately the solution becomes yellow orange as *trans*-$[P(C_2H_5)_3]_2PdBr_2$ started forming, and after 10 min no $PdBr_2$ was visible in the stirred solution. A piece of potassium (0.247 g, 0.00632 mol) cut and cleaned under hexane was added to the reaction mixture, and stirring and heating were started. After 5 h at reflux the mixture contained a black slurry in a yellow solution. To it was added 0.66 ml (0.0064 mol) of C_6H_5Br, and reflux was maintained for an additional 14 h. The cooled mixture was filtered, solids were washed out with dichloromethane, and the solvent was removed from the filtrate under reduced pressure. The crude material was dissolved in hot hexane, treated with charcoal, and filtered. Slow evaporation of solvent resulted in large colorless crystals of *trans*-$[P(C_2H_5)_3]_2Pd(C_6H_5)Br$: 0.998 g (63%); mp 107–107.5°C (lit. [13, 17] 96–98°C, 103–104°C). (ii) The previously described reaction apparatus was charged with 1.104 g (0.00267 mol) of *trans*-$[P(C_2H_5)_3]_2PdCl_2$, 0.0376 g (0.00542 mol) of freshly cut and cleaned lithium, and 0.0324 g (0.000253 mol) of naphthalene. Glyme (10 ml) was syringed in and the mixture was stirred for 70 h at room temperature, after which time the usual black slurry in a yellow solution was obtained. A sample of 0.57 ml (0.0054 mol) of C_6H_5Br was syringed in, and the mixture was refluxed for 21 h. The usual workup yielded 0.612 g (46%) of *trans*-$[P(C_2H_5)_3]_2Pd(C_6H_5)Br$.

Preparation of *trans*-$[P(C_2H_5)_3]_2Pd(C_6H_5)CN$

A palladium slurry was prepared in the usual manner from 0.247 g (0.00632 mol) of potassium, 0.559 g (0.00315 mol) of $PdCl_2$, and 0.93 ml (0.0063 mol) of $P(C_2H_5)_3$ in 10 ml of glyme. It was then allowed to react at reflux for 37 h with 0.65 ml (0.0063 mol) of C_6H_5CN. The mixture was allowed to cool to room temperature and filtered under argon, and the solvent was removed under vacuum. Several recrystallizations from an 8 : 1 hexane–THF solvent mixture yielded 0.367 g (26%) of large colorless crystals of *trans*-$[P(C_2H_5)_3]_2Pd(C_6H_5)CN$: mp 112–116°C. This is somewhat different from the literature value of 85–90°C. IR (Nujol) 2125 cm^{-1} $\nu(C{\equiv}N)$, 1568 cm^{-1} $\nu(C{=}C)$ (lit. 2140, and 1580 cm^{-1}, respectively [13]). (Anal. Calcd for $C_{19}H_{35}NP_2Pd$: C, 51.18; H, 7.91; P, 13.89. Found: C, 51.30; H, 7.88; P, 14.10.)

Preparation of *trans*-$[P(C_2H_5)_3]_2Pt(C_6H_5)I$

A 50 ml three-necked flask equipped with a magnetic stirrer and a condenser topped with an argon inlet was charged with 1.406 g (0.00313 mol) of PtI_2. THF (8 ml) was syringed in followed by 0.93 ml (0.0063 mol) of $P(C_2H_5)_3$. A clean piece of potassium, 0.245 g (0.00627 mol), was added, and stirring and heating were started. After 20 min the reduction appeared to be complete; however, it was refluxed an additional hour and then allowed to cool. A black powder existed in a colorless solution. To it was added 0.35 ml (0.0031 mol) of C_6H_5I, and the mixture was stirred for 8 h at room temperature. It was then filtered, solids were washed out with dichloromethane, and the solvent was removed from the filtrate under reduced pressure. Repeated recrystallization from hexane, after treatment with charcoal, resulted in 1.219 g (66%) of very pale yellow crystals of *trans*-$[P(C_2H_5)_3]_2Pt(C_6H_5)I$: mp 120–121°C (lit. [19] 122°C).

Preparation of *trans*-$[P(C_2H_5)_3]_2Pt(C_6H_5)Br$

A platinum slurry was prepared from 0.647 g (0.00243 mol) of $PtCl_2$, 0.72 ml (0.0049 mol) of $P(C_2H_5)_3$, and 0.193 g (0.00494 mol) of potassium in 8 ml of glyme. To it was added 0.52 ml (0.0049 mol) of C_6H_5Br, and the mixture was refluxed for 25 h. The usual workup resulted in 0.818 g (57%) of large colorless crystals of *trans*-$[P(C_2H_5)_3]_2Pt(C_6H_5)Br$: mp 110–111°C (lit. [20] 111.0–111.3°C).

References

1 Rieke, R.D.; Wolf, W.J.; Kujundzic, N.; Kavaliunas, A.V. *J. Am. Chem. Soc.* 1977, **99**, 4159.
2 (a) Rieke, R.D.; Kavaliunas, A.V.; Rhyne, L.D.; Fraser, D.J.J. *J. Am. Chem. Soc.* 1979, **101**, 246. (b) Kavaliunas, A.V.; Rieke, R.D. *J. Org. Chem.* 1979, **44**, 3069.
3 Skell, P.S.; Wescott, L.D., Jr.; Goldstein, J.P.; Engel, R.R. *J. Am. Chem. Soc.* 1965, **87**, 2829.
4 Skell, P.S.; McGlinchey, M.J. *Angew. Chem.* 1975, **87**, 215.
5 (a) Klabunde, K.J. *Angew. Chem.*, 1975, **87**, 309. (b) Klabunde, K.J. *Angew. Chem. Int. Ed. Engl.*, 1975, **14**, 287.
6 Klabunde, K.J.; Efner, H.F. *J. Fluorine Chem.* 1974, **4**, 114.
7 (a) Klabunde, K.J.; Low, J.Y.F.; Efner, H.F. *J. Am. Chem. Soc.* 1974, **96**, 1984. (b) Klabunde, K.J.; Low, J.Y.F., *J. Am. Chem. Soc.*, 1974, **96**, 7674.
8 Klabunde, K.J.; Low, J.Y.F. *J. Organomet. Chem.* 1973, **51**, C33.
9 Rieke, R.D.; Bales, S.E. *J. Am. Chem. Soc.* 1974, **96**, 1775.
10 Rieke, R.D.; Ofele, K.; Fischer, E.O. *J. Organomet. Chem.* 1974, **76**, C19.
11 Typical quantities are $PdCl_2$ (1.22 g), Et_3P (1.63 g), and K (0.53 g) and were mixed with 25 ml of THF. The mixture was refluxed for 20 h yielding the black metal slurry. The slurry was cooled to −78°C and C_6F_5Br (1.7 g) was added. After 1 h, the reaction was warmed up to room temperature and worked up. Yields are based upon $PdCl_2$.
12 Fischer, E.O.; Bürger, G. *Z. Naturforsch. B* 1961, **16**, 702.
13 Schunn, R.A. *Inorg. Chem.* 1976, **15**, 208.
14 Jenkins, J.M.; Shaw, B.L. *Proc. Chem. Soc., Lond.* 1963, 279.
15 Parshall, G.W. *J. Am. Chem. Soc.* 1974, **96**, 2360.
16 Powell, J.; Shaw, B.L. *J. Chem. Soc. A* 1968, 774.
17 Klabunde, K.J.; Low, J.Y.F. *J. Am. Chem. Soc.* 1974, **96**, 7674.
18 Habeeb, J.J.; Tuck, D.G. *J. Organomet. Chem.* 1977, **139**, C17.
19 Booth, G.; Chatt, J. *J. Chem. Soc. A*. 1966, 634.
20 Coulson, D.R. *J. Am. Chem. Soc.* 1976, **98**, 3111.

13

Highly Reactive Uranium and Thorium

13.1 Two Methods for Preparation of Highly Reactive Uranium and Thorium: Use of a Novel Reducing Agent Naphthalene Dianion

Yellow glass containing uranium oxide dates back to the first century AD [1]. It was not recognized as uranium at that time. Klaproth recognized an unknown element in pitchblende in 1789. He was not able to isolate the pure metal but named it for the planet Uranus. The pure metal was finally isolated in 1841 by Eugene-Melchior Peligot, who reduced anhydrous UCl_4 with potassium. The radioactive nature of uranium was not understood until 1896, when Henri Becquerel detected its radioactivity. While the inorganic chemistry of uranium was extensively studied during the first half of the twentieth century, its use in organic chemistry had to wait until the second half of the twentieth century.

Highly dispersed and reactive metal powders have commanded a great deal of interest for their applications in catalytic and stoichiometric chemical syntheses, as well as their uses in materials science. Numerous methods exist for the preparation of these metal powders. Most of these methods for preparing metal powders involve the reduction of a metal salt or oxide by hydrogen (usually at high temperatures) [2–4] or by other chemical or electrochemical means [3, 5–10]. In addition to reductive techniques, pyrolysis [3, 11, 12] and metal atom vaporization [13–19] have been employed. An excellent review of the preparation, reactivity, and physical properties of unsupported metal particles has appeared [20].

Because of the commercial importance of uranium, a number of methods for generating finely divided chemically reactive uranium metal have been developed. Pyrophoric uranium metal powders have been prepared by thermal decomposition of uranium amalgam [21–23] or uranium hydride [24, 25]. Many methods have involved reduction of uranium oxides [26]. Other methods employed are melt electrolysis [26] and potassium reduction of $(\eta^5\text{-}C_6H_5)_4U$ [27].

Chemical Synthesis Using Highly Reactive Metals, First Edition. Reuben D. Rieke.

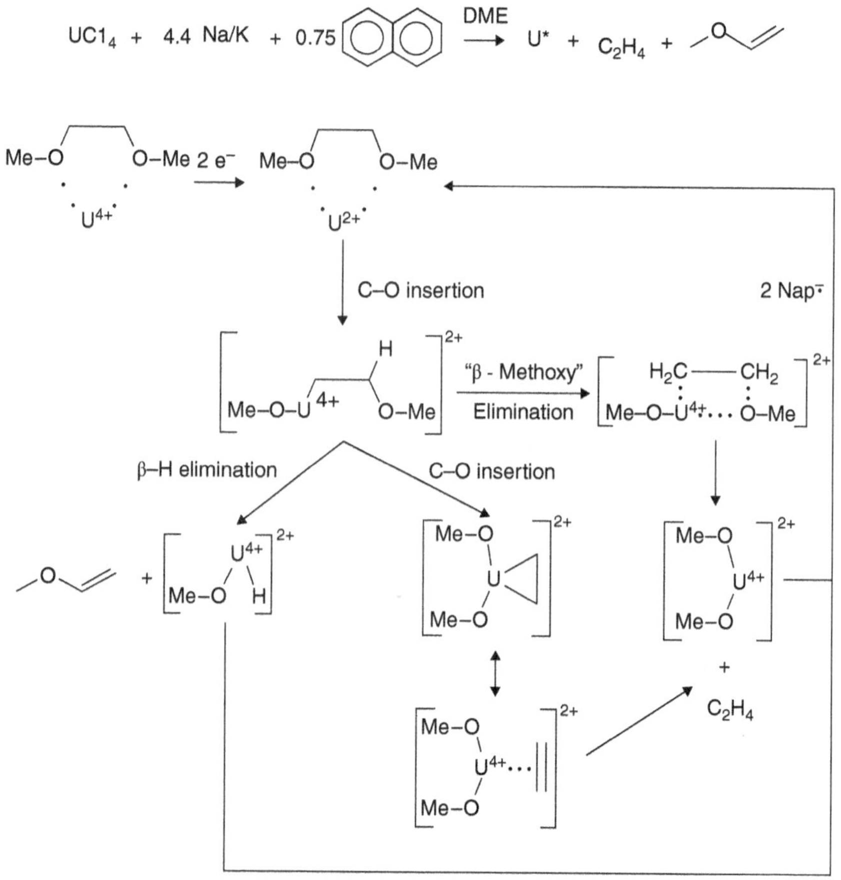

$$UCl_4 + 4.4\ Na/K + 0.75\ \text{(naphthalene)} \xrightarrow{DME} U^* + C_2H_4 + \text{(methyl vinyl ether)}$$

Scheme 13.1 Proposed reaction scheme.

Since 1972, we have published many reports describing convenient methods for the generation of highly reactive metal powders and their use in organic as well as organometallic synthesis [28–34]. Most of the active metals prepared by our group have been prepared in ethereal solvents [35]. Our initial report on the preparation of active uranium (**I**) employed 1,2-dimethoxyethane (DME) as a solvent (Scheme 13.1) [31]. We have since developed a method for preparing active uranium (**3**) in hydrocarbon solvents, which leads to a much cleaner and more controllable chemistry compared to **1** [34, 36].

The oxophilicity of titanium has been exploited in the well-known reductive carbonyl coupling reactions pioneered by McMurry [37–41] and others [42, 43] using low-valent titanium. This reaction has been shown to occur with many other early transition metals [44, 45]. The oxophilicity of the early transition metals, lanthanides, and actinides is well known. The standard enthalpies of formation for UO_2 (−259.3 kcal/mol) and ThO_2 (−293.1 kcal/mol) [46] are among the most negative known for any MO_2 metal oxide. There have been no investigations of the reductive ketone coupling reaction, however, with either lanthanide or actinide metals. There are many examples of coordinated carbonyl couplings in organometallic lanthanide [47–50] and actinide [51–58] η^2-acyl complexes. The complex $Cp_3U{=}CHPR_3$ has been shown to reductively couple bridging and terminal carbonyl ligands of a dinuclear iron complex [59]. In certain cases, a formal trimerization [49] or tetramerization of CO can even be effected [48, 52]. In none of these cases, however, has the organic fragment been removed intact from the metal center. One report has appeared where di-*n*-butyluranocene has been used in a reductive coupling reaction to give azobenzenes from nitrobenzenes [60]. In view of this coupling reaction of U(IV), the high oxophilicity of uranium, and the similarities between the chemistry of the actinides and group 4 transition metals, we have investigated the reactions of active uranium and thorium with oxygen-containing compounds (Scheme 13.1).

Results and Discussion

The reduction of UCl_4 in ethereal solvents by Li, Na, or K does not proceed to completion due to coating of the surface of the alkali metal [31, 61]. Active uranium can, however, be prepared by the reduction of UCl_4 using Na/K alloy in DME [31, 61] (**I**). We have shown this active uranium to be very reactive with allyl iodide and benzophenone [31]. Uranocene can be prepared directly in 56% yield by reaction of **I** with cyclooctatetraene (COT) in refluxing DME for 22 h. This is the same yield reported by Streitwieser in the sealed tube reaction of a pyrophoric form of uranium (prepared under vigorous conditions by repeatedly hydrogenating uranium metal and thermally decomposing the resulting hydride) with COT at 150°C [24].

Active uranium prepared by Na/K reduction in DME is also an extremely active polymerization catalyst and will polymerize 100 equiv of 1,3-butadiene at −4°C and atmospheric pressure in less than 1 h. This is much more reactive than the active uranium powder prepared by Chang et al. (via thermal decomposition of U/Hg) which polymerized 80% of 1,3-butadiene (1 atm) in 4 h at 70°C [62]. The resulting polybutadiene prepared by our method exhibited IR bands corresponding to *cis* (675 cm^{-1}), vinyl (910 cm^{-1}), and *trans* (970 cm^{-1}) morphologies [63].

When UCl_4 is reduced by Na/K alloy in tetrahydrofuran (THF) or DME, however, decomposition of the solvent is seen even at low temperatures. Few investigations of the reductive cleavage of DME have been reported [64–68]. When DME is used for the reduction of UCl_4, the major fragments from the solvent can be seen as gaseous components in the headspace above the reaction. The active uranium (**I**) reacts with DME to produce primarily ethylene and methyl vinyl ether [64] (GC/MS) (Scheme 13.1) [69]. Production of these compounds can be accounted for by the insertion of a low-valent uranium species into a carbon–oxygen bond of the DME. Complexes of this type have been identified for complexes of W and Ta [66, 67]. Insertion into the other C—O bond of the DME gives ethylene, whereas β-hydride elimination [70–76] gives methyl vinyl ether. If additional sodium naphthalide is added, further reduction of the solvent takes place, and more ethylene and methyl vinyl ether can be obtained. In this way, the uranium can be considered as mediating the reductive decomposition of DME into ethylene and methyl vinyl ether. Reactions of organic substrates with **I** in ethereal solvents give low yields and complicated product mixtures containing solvent fragments [77].

In order to eliminate the reaction of active uranium with the solvent, a preparation of active uranium in an inert solvent was sought. The use of hydrocarbon solvents eliminates the problem of reactivity with oxygen functionalities. We have observed that the most highly reactive metals are obtained from soluble reducing agents. A problem exists, however, in that there are very few strong reducing agents that are soluble in nonpolar solvents. Lithium naphthalide cannot be formed in hydrocarbon solvents.

Using the method of Fujita et al. [78], we were able to prepare a highly active form of cadmium in toluene [32]. This method works for uranium as well but is inconvenient as sonication is needed for each reaction. We have prepared and isolated the stable crystalline lithium naphthalide dianion derivative $[(TMEDA)Li]_2[Nap]$ **2** directly by sonicating a 1.6 M solution of TMEDA, Li, and naphthalene in toluene. When sonication is stopped after all of the Li has dissolved, the dianion crystallizes. This complex was prepared previously by deprotonation of 1,4-dihydronaphthalene [79] but has not been used in any synthetic chemistry. The 1,4-dihydronaphthalene that was employed is a fairly expensive and sensitive compound, precluding its widespread use in synthesis. Our procedure is much less expensive and amenable to preparative scale synthesis. We prepared 50–60 g quantities of this complex and found it to be indefinitely stable at room temperature when stored in an argon-filled dry box. Complex **2**, however, does decompose under nitrogen.

NMR characterization of **2** was difficult as the resonances were quite broad, presumably due to exchange between the dianion and small amounts of the paramagnetic radical anion. When the sample was cooled, these resonances sharpened until −47°C when the complex crystallized out of solution. Even at −47°C, the individual resonances were not resolved.

The UV/vis spectrum of complex **2** shows an absorption at 444 nm, which disappears in the presence of air. The metalation of naphthalene by potassium has been studied previously, and a visible absorption around 450 nm was attributed to the naphthalene dianion [80].

When complex **2** in benzene is quenched with water or saturated NH_4Cl, 1,4-dihydronaphthalene, 1,2-dihydronaphthalene, and naphthalene are produced

(3.4:1.0:2.4). The presence of dihydronaphthalenes, especially 1,4-dihydronaphthalene, shows that a naphthalene dianion is present. After 2 equiv of complex **2** was stirred with 1 equiv of UCl_4 in benzene for 1 h, followed by quenching with water or saturated aqueous NH_4Cl, only naphthalene is observed. This shows that **2** has been completely consumed in reducing the UCl_4 within 1 h at room temperature.

The potential reaction of the reducing agent **2** with hydrocarbon solvents like toluene and xylenes that contain benzylic hydrogens was investigated. When a solution of **2** in toluene (prepared *in situ* by sonication for 13 h) was quenched with methyl iodide or dimethyl sulfate, very small amounts (≤6%) of ethyl benzene or xylenes were observed. When a solution of **2** in benzene (prepared *in situ* by sonication for 13 h) was quenched with methyl iodide or dimethyl sulfate, negligible amounts (≤1.5%) of toluene were seen. Since the sonication conditions (42°C bath temperature, hot-spot peak temperature and pressure of about 3000 K and 300 atm, respectively, have been measured [81]; 13 h) are much more vigorous than those customarily used in the preparation of active uranium (room temperature, ≤1 h), it can be concluded that no reaction of complex **2** with the solvent is taking place during the reduction of UCl_4.

Unfortunately, the reactivity of active uranium prepared by this method **3** with acidic C—H bonds precludes its use with aliphatic ketones. Apparently, **3** reacts with the acidic α-protons of the carbonyl, generating a carbanion or equivalent which does aldol-type chemistry. Thus, acetophenone reacted with **3** to give primarily the aldol product 1,3-diphenyl-2-buten-1-one (dypnone), while cyclohexanone produced primarily 2-cyclohexylidenecyclohexanone. For this reason, only aromatic ketones will be considered in this study.

Reaction of active uranium with 2 equiv of benzophenone in refluxing xylenes gave the expected reductive coupling product tetraphenylethylene (TPE), as well as an unexpected product, 1,1,2,2-tetraphenylethane (TPA). Further reduction of tetraarylethylenes to tetraarylethanes has been observed previously in similar reductive coupling reactions of low-valent titanium [38, 82]. This hydrogenation reaction has been studied by Geise [83]. The source of the extra hydrogen in these titanium reactions has been implied to be from the reducing agent when $LiAlH_4$ is used or from the THF solvent when Mg or K are used as reducing agents [82]. This is not surprising, as $LiAlH_4$/transition-metal halide systems are well-known reducing agents for unsaturated organic compounds [84, 85]. In the uranium system, however, the hydrogen source is quite different. Reaction of active uranium with TPE in a refluxing mixture of xylenes/toluene (5:3) for 4 h showed no TPA. Even when active uranium was reacted with TPE under 1 atm of H_2 in a refluxing mixture of xylenes/toluene (5:3), no TPA was seen. Reaction of active uranium with TPA under similar conditions did not produce any TPE either. This demonstrates that in our system TPA was not formed by further reduction of TPE but must be formed during the coupling step at the metal surface. In addition, the

Table 13.1 Stoichiometry effect on reaction of benzophenone with U*.

$$U^* + 2Ph_2CO \xrightarrow[8h\ ref]{Xylenes} PH_2C{=}CPh_2 + Ph_2CHCHPh_2$$

Stoichiometry			% yield[a]	
UCl_4	"Li"[b]	Ph_2CO	$Ph_2C{=}CPh_2$	$Ph_2CHCHPh_2$
1	1	2	24	1
1	2	2	32	7
1	3	2	33	33
1	4	2	41	32
1	4	4	38	56
1	5	2	39	23

[a] GC yields.
[b] "Li" = $0.5[(TMEDA)Li]_2[Nap]$.

reaction of **3** (1 equiv) with benzophenone (2 equiv) in the presence of TPE (1 equiv) did not produce any appreciable additional amount of TPA. This shows that reduction of TPE by a possible hydrogen-transfer species, generated during the reaction of **3** with benzophenone, does not occur.

Active uranium reacts with diphenylmethanol (benzhydrol) to give exclusively TPA. A 60% yield of TPA was observed after 4 h in refluxing xylenes (2 : 1 benzhydrol/uranium). No reaction was seen at room temperature. No TPE was ever observed in this reaction. On the basis of these observations, it appears that the TPA formed in the coupling reaction of benzophenone arises from the initial reduction of benzophenone to a benzhydrol alkoxide followed by coupling of this alkoxide rather than by further reduction of the TPE [83, 86]. When 4,4′-dimethylbenzophenone is reacted with **3**, the intermediacy of 4,4′-di-methylbenzhydrol is observed [87].

No coupling of benzophenone was observed when benzophenone was reacted with UCl_4 (unreduced) or with $UCl_4 + 4TMEDA$ in refluxing xylenes. Clearly the coupling chemistry observed must depend on the presence of a low-valent uranium species.

By varying the stoichiometry in the reduction of UCl_4 by **2**, it can be seen that using four "Li" equivalents produces the greatest yield of coupled products (TPE + TPA) (Table 13.1).

When the UCl_4 is "over reduced" by using five "Li" equivalents, the yield of both TPE and TPA is diminished, primarily due to the formation of large amounts of diphenylmethane (DPM). This DPM presumably arises from the action of excess reducing agent on benzophenone in the presence of uranium.

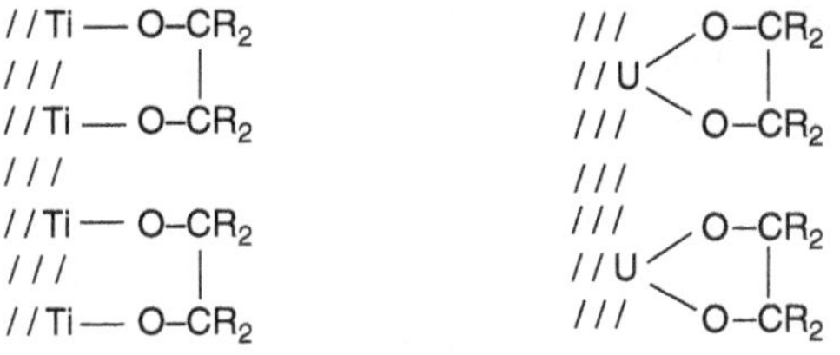

```
//Ti — O–CR2          ///   ,O–CR2
///        |          //U<     |
//Ti — O–CR2          ///   `O–CR2
///                   ///
//Ti — O–CR2          ///   ,O–CR2
///        |          //U<     |
//Ti — O–CR2          ///   `O–CR2
```

Figure 13.1 Surface metallopinacols on active titanium and uranium.

Table 13.2 Temperature effect on reaction of benzophenone with U*.

	% yield[a]		
Temp (°C)	**$Ph_2C{=}CPh_2$**	**$Ph_2CHCHPh_2$**	**Ph_2CH_2**
140	34	20	30
110	30	7	34
90	35	4	14
70	53	2	8
50	7	0	0

[a] GC yields.

For most of the reactions in this study, a stoichiometry of two ketones per uranium has been used (Table 13.1). This is in accord with the postulated stoichiometry of active uranium (U*) producing UO_2. When only a twofold excess of benzophenone is used, however, the combined yields of coupled products is 94%. This is in marked contrast to the titanium systems where each titanium is only capable of reacting with 1 equiv of ketone [82]. In the titanium systems, optimal yields of alkene are obtained with excess titanium, whereas we find that excess ketone produces the highest yields of alkene. This difference in M:O stoichiometry is suggestive of a different type of metallopinacol intermediate (Figure 13.1). Further evidence which suggests the presence of this type of intermediate is the fact that **3** reduces the α-diketone benzyl to diphenylacetylene, whereas active forms of titanium have been shown by two authors not to effect this transformation [40, 88].

The production of the hydrogenated products TPA and DPM is very temperature dependent. Large amounts of these compounds are formed in refluxing xylenes, whereas virtually no TPA or DPM is observed at 70°C (Table 13.2). This is in marked contrast to the titanium systems, where low temperatures in the formation of the low-valent titanium species have been shown to promote the hydrogenation of TPE [83]. TPE can be formed quantitatively by using **3** at 70°C with an excess of benzophenone.

In order to further probe the mechanistic course of this reaction, the possibility of a pinacolic intermediate was addressed. Metallopinacolic intermediates in reductive coupling reactions of ketones have been strongly implied by many groups on the basis of the observance of pinacols [38, 39, 82]. We have also observed the presence of pinacols in ketone coupling reactions that are quenched before completion. To investigate the possibility of a metallopinacolic intermediate further, the reaction of active uranium with benzopinacol was conducted. This reaction proceeded faster than the corresponding ketone reaction. In this reaction equal amounts of TPE and TPA were seen, whereas when benzophenone is reacted with **3**, more TPE than TPA is seen (TPE/TPA is about 4:3). The yield of TPE is similar in the two reactions with the major difference being that more TPA is produced from the benzopinacol reaction. Benzopinacol contains two acidic hydrogens that could account for the greater yield of TPA. When benzopinacol-d_2 was reacted with **3**, greater than or equal to 73% (MS) of the deuterium was incorporated into TPA. This clearly shows that the benzylic hydrogens in TPA can come from a uranium hydride. Hydridic organometallic uranium species are well known [53, 54, 56, 72–74, 76, 89–93].

The product distribution from the reaction of **3** with pinacols shows a similar temperature dependence to that of the ketone reactions, although the pinacol reactions proceed at a somewhat lower temperature. The reaction of **3** with benzopinacol at 50°C gives only TPE, with no TPA observed.

The source of the uranium hydride in the reaction of benzophenone with **3** is not as clear however. When the reaction of benzophenone with **3** was worked up with D_2O, no deuterium incorporation in TPA or DPM was seen, nor was any seen when toluene-d_8 was used as a solvent. All of the naphthalene is recovered, and no hydrogenated or substituted naphthalenes are observed. Elemental analysis of the solid remaining after reaction of active uranium with benzophenone (following removal of the solvent and Soxhlet extraction) showed substantial amounts of nitrogen (0.52 mol of N/U) contained in the solid $LiCl/UO_x$ matrix. This suggests that TMEDA (presumably coordinated to uranium) is reacting with the low-valent uranium generating a uranium hydride. This uranium hydride could be incorporated into the reaction products TPA and DPM, just as the uranium deuteride resulting from the reaction with benzopinacol-d_2 was incorporated.

The temperature dependence of the reaction of **3** with TMEDA could account for the fact that the hydrogenated products TPA and DPM are only seen at high temperatures (vide supra). It appears that the hydrogenated products TPA and DPM arise when the benzophenone is reacted at high temperatures with **3** containing uranium hydrides. These hydrides can be formed either from substrates containing acidic hydrogens or by thermal reaction of the low-valent uranium species with coordinated TMEDA.

The inherent chemistry described for the reaction of **3** with benzophenone and benzopinacol is not altered by substituting the phenyl rings of the ketone

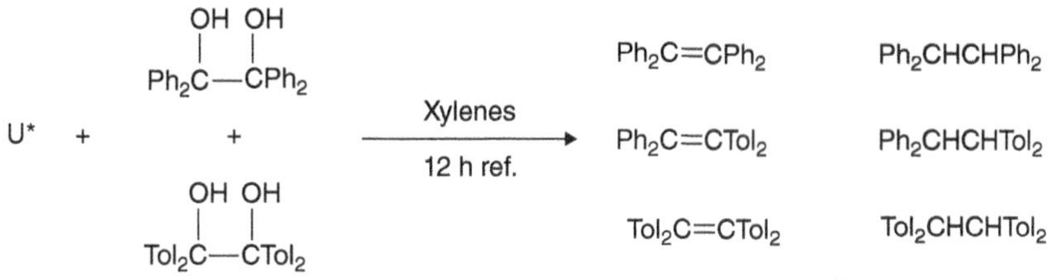

$$U^{*} + Ph_2C(OH)\text{—}C(OH)Ph_2 + Tol_2C(OH)\text{—}C(OH)Tol_2 \xrightarrow[\text{12 h ref.}]{\text{Xylenes}} \begin{array}{ll} Ph_2C{=}CPh_2 & Ph_2CHCHPh_2 \\ Ph_2C{=}CTol_2 & Ph_2CHCHTol_2 \\ Tol_2C{=}CTol_2 & Tol_2CHCHTol_2 \end{array}$$

Figure 13.2 Product of mixed pinacol reaction with U*.

or pinacol with methyl groups. Thus, the reaction of the ketone 4,4′-dimethylbenzophenone (Tol_2CO) or the pinacol 1,1,2,2-tetrakis(4-methylphenyl) ethane-1,2-diol (tetramethylbenzopinacol) **4** with active uranium gives tetrakis(4-methylphenyl)ethylene (TTE) and 1,1,2,2-tetrakis(4-methylphenyl) ethane (TTA). Reaction of a 1 : 1 mixture of benzophenone and Tol_2CO with **3** gives a statistical distribution of the six expected products (Figure 13.2).

In order to further examine the role of the pinacolic intermediate, a crossover experiment was conducted. In the reaction of a 1 : 1 mixture of **4** and benzopinacol with **3** in refluxing xylenes, a statistical distribution of all six coupled products was seen. The product distribution was identical with that seen from a 1 : 1 mixture of benzophenone and Tol_2CO.

When the reaction was conducted at 50°C, however, the results were very different. At this temperature no hydrogenated products (TPA, 1,1′-bis(4-methylphenyl)-2,2′-diphenylethane, or TTA) were seen. Initially only TPE and TTE were formed. After 2 h at 50°C, the formation of 1,1′-bis(4-methylphenyl)-2,2′-diphenylethylene (DPDTE) began to be observed. By 18 h at 50°C, the rate of formation of TPE and TTE had decreased, and the formation of the mixed ethylene was predominant (Figure 13.3).

This surprising result shows that C—C bond cleavage of the metallopinacol is extremely facile. Exchange of surface-bound metallopinacols prior to deoxygenation appears to be a likely possibility (vide infra) (Figure 13.4). These results can be rationalized by surface activation of the active uranium particles. In the low-temperature reaction, the product formed initially is that of the unexchanged metallopinacol. As the reaction proceeds, activation of the metal particle continues, which exposes more active sites on the surface. As the number of active sites on the surface of the metal particle grows, the probability of finding adjacent metallopinacols increases. In this way, the production of the exchanged metallopinacol deoxygenation product DPDTE is dependent on surface activation of the active uranium metal particle. This accounts for the initial formation of the unexchanged products TPE and TTE until the surface is activated enough for the formation of DPDTE to be observed. At 144°C, the reaction of the active uranium with pinacols is so fast that no induction period is observed before the formation of the exchanged deoxygenation product.

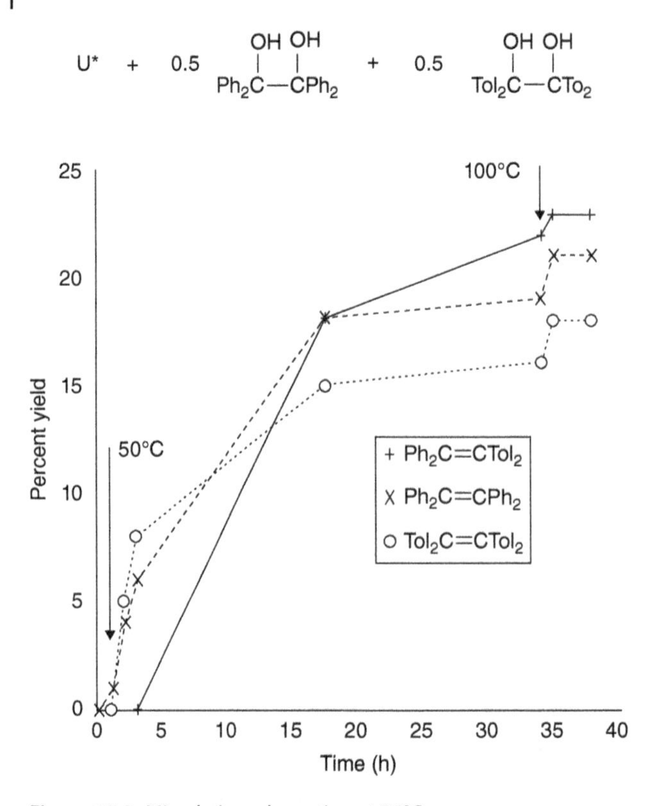

Figure 13.3 Mixed pinacol reaction at 50°C.

Figure 13.4 Pinacol exchange on the surface of U.

Further evidence for surface activation of active uranium by alcohols was seen in the reaction of active uranium with benzophenone in the presence of methanol. When the methanol was added along with the benzophenone to freshly prepared active uranium, TPE production could be observed even at room temperature (3 h, 13%). After 70 h at 50°C, 38% TPE was seen. In the absence of methanol, only negligible amounts of TPE can be formed even at 50°C.

At 50°C, arylpinacols are deoxygenated to give tetraarylethylenes, whereas at this temperature negligible coupling of aryl ketones is found (vide supra). When a mixture of an aryl ketone and an arylpinacol (2 : 1) is reacted with active uranium, the initial product formed was found to be the one resulting from coupling of the aryl *ketone*. Once again the pinacol appears to be acting initially to activate the surface of the metal particle.

Active thorium has been prepared in a manner analogous to that of active uranium. Active thorium reacts with DME, as does active uranium **I**. When benzophenone is reacted with active thorium in refluxing DME, the product mixture contains TPE and TPA in low yield (<10%), as well as many other coupled products containing solvent fragments. The reactivity of active thorium with aromatic ketones in hydrocarbon solvents is similar to that of active uranium, although the yields tend to be lower (3 days in refluxing xylenes; 11% TPE, 15% TPA) and the product distributions are more complex. These preliminary reactions with thorium have shown **3** to be much better for the reductive coupling reactions in this study.

In addition to aromatic solvents, we have seen reactivity of active uranium prepared in 1-decene and TMEDA. The yields of TPE and TPA resulting from preparation and reaction of active uranium with benzophenone in 1-decene are only slightly lower than those in aromatic solvents, presumably due to solubility considerations. There was little difference in the proportion of TPA to TPE. It appears that decene does not serve as either a hydride source (allylic hydrogens) or a hydride sink (via hydrogenation of the double bond). The preparation and reaction of active uranium with benzophenone in TMEDA gave yields of TPE comparable to other aromatic solvents at that temperature (≈20%). Negligible amounts of TPA were seen, however. The presence of a large excess of a basic solvent could serve to reduce the amount of metal hydrides present.

Conclusion

The hydrocarbon soluble naphthalene dianion equivalent $[(TMEDA)Li]_2[Nap]$ can be prepared in large amounts and used as a convenient reducing agent. This reducing agent allows active uranium and thorium to be prepared in inert hydrocarbon solvents. These active metals are extremely reactive with oxygen-containing compounds and have been shown to react with ethers, ketones, and alcohols. Active actinides also show a great propensity for formation of metal hydrides. Very weakly acidic C—H bonds in TMEDA can be activated by using these metals, with the resulting hydrides being transferred to organic functionalities. Aromatic ketones can be reductively coupled and aromatic pinacols are deoxygenated, presumably by way of a surface metallopinacolic intermediate to give tetraarylethylenes or tetraarylethanes in high yield. This reductive coupling reaction is effective without the need for a large excess of active metal, as in the titanium systems. Contrary to the active titanium systems of McMurry, active uranium and thorium are able to react with two ketones per metal atom. This implies having a mononuclear surface actinide metallopinacol, as opposed to binuclear titanium metallopinacols.

Preparation of Active Uranium in DME (1)

The following describes a typical preparation of active uranium in DME. In the dry box, UCl_4 (0.3575 g, 0.9412 mmol), Na (0.0269 g, 1.170 mmol), K (0.1177 g, 3.010 mmol), and naphthalene (0.0900 g, 0.7022 mmol) were placed in a two-necked 50 ml flask equipped with a Teflon-clad stir bar, condenser, vacuum adapter, and septum. The apparatus was removed from the dry box and connected to the argon line. Sodium–potassium alloy (NaK) was formed by heating the alkali metals with a heat gun until the K melted. The reaction flask was immersed in a cooling bath that was maintained at −60°C. Freshly distilled DME (5.5 ml) was added and stirring started. After 2–4 days of stirring at −65°C (when no more alloy could be seen), cooling was stopped, and the active uranium was ready for use after being allowed to slowly warm to room temperature.

Reaction of (1) with 1,3-Butadiene

1,3-Butadiene (8.2 ml, 94 mmol) was condensed into a Schlenk flask attached to an argon line and maintained with dry ice/2-propanol. The butadiene was transferred via cannula into a slurry of **1** (0.94 mmol of U) over 30–45 min by removing the dry ice/2-propanol cooling bath. The reaction flask was cooled by using a NaCl/ice bath, and the butadiene (bp −4°C) was kept refluxing by circulating −30°C acetone through the condenser until the butadiene was consumed. The reaction was monitored by periodically removing headspace samples with a gastight syringe and analyzing by GC. The slurry became very thick, with stirring barely possible after 1 h. The reaction was allowed to proceed 18 h. The uranium was removed by washing with 3 N HCl and the polymer analyzed by IR.

Preparation of $[(TMEDA)Li]_2[Nap]$ (2)

The following describes a typical preparation of **2**. In the dry box, lithium (5.3144 g, 0.76587 mol), cut into many small pieces, and naphthalene (49.0802 g, 0.382914 mol) were loaded into a 500 ml three-necked flask equipped with a septum, vacuum adapter, and fritted filtration tube. This flask was removed from the dry box, immersed in the ultrasonic cleaner, and attached to the manifold. Freshly distilled TMEDA (115 ml, 0.762 mol) and freshly distilled toluene (200 ml) were added to a flask by syringe. The reaction proceeded for 4 days with sonication until all of the lithium had reacted. Upon cessation of sonication, **2** crystallized. The flask was cooled to −50°C, and freshly distilled pentane (100 ml) was added. The solution was filtered through the sintered glass frit, and the crystals of **2** obtained were washed with cold pentane (2 × 275 ml). The majority of the pentane was removed under a flow of argon, and the crystals of **2** were transferred to another fritted tube in the dry box. This tube was attached to the vacuum line, and argon passed through the crystals, from bottom to top, overnight. Yields of **2** crystals obtained were typically around 50% (50 g). A considerable amount of **2** remained in the supernatant solution, and no attempt was made to isolate more crystals from this. The crystals of **2** were stored in the argon-filled dry box and were found to be stable indefinitely. Visible absorption at 444 nm of a less than 0.7 mM solution ($\varepsilon > 4000$) confirms the presence of a naphthalene dianion [80].

Preparation of Active Uranium in Hydrocarbon Solvents (3)

The following describes a typical preparation of active uranium in hydrocarbon solvents. In the dry box, UCl_4 (0.5020 g, 1.322 mmol) and **2** (1.0143 g, 2.7086 mmol) were placed in a two-necked 50 ml flask equipped with a Teflon-clad stir bar, vacuum adapter, and septum. On the vacuum line, freshly distilled solvent (20 ml) was added and stirring started. After 1 h of stirring at room temperature, the active uranium was ready for use.

Reaction of (3) with Ketones

To a slurry of **3** (0.506 g, 1.331 mmol) was added, via cannula, a solution of benzophenone (0.48560 g, 2.6649 mmol) and GC internal standard *n*-pentadecane (0.37 ml, 1.3 mmol) in 2 ml of freshly distilled xylenes. To the flask that contained the ketone and pentadecane was added an additional 2 ml of xylenes, and this solution was added to **3**. The reaction was stirred at room temperature for 1 h, followed by refluxing overnight. The reaction was quenched with 3 N HCl and filtered, and the products were extracted with $CHCl_3$. The coupled products could be obtained by addition of hexanes to a $CHCl_3$ solution. Separation was accomplished on silica gel by using hexanes/$CHCl_3$ elution.

Reaction of (3) with Pinacols

A solution of benzopinacol (0.48637 g, 1.3270 mmol) and GC internal standard *n*-pentadecane (0.37 ml, 1.3 mmol) in 10 ml of freshly distilled xylenes was warmed to dissolve (≈90°C). This solution was added to a slurry of **3** (0.507 g, 1.3 mmol) at room temperature via a cannula. The reaction was stirred at room temperature for 1 h, followed by refluxing overnight. The reaction was worked up in the same manner as the ketone reactions.

Preparation of $(p\text{-}CH_3C_6H_4)_2COHCOH(p\text{-}CH_3C_6H_4)_2$ [94]

MgI_2 (1.6016 g, 5.7586 mmol), Mg powder (0.3811 g, 15.68 mmol, 70–80 mesh), and 4,4′-dimethylbenzophenone (1.9535 g, 9.2900 mmol) were loaded into a 50 ml two-necked flask which was equipped with a Teflon-clad stir bar, septum, and condenser with vacuum adapter, in the dry box. Freshly distilled toluene (10 ml) and ether (5 ml) were added on the vacuum line, and stirring was started. The solution became thick and greenish with a large amount of a white precipitate. Heating was started immediately, causing the white precipitate to dissolve and the solution to look somewhat purplish. After the solution was refluxed overnight, H_2O (2 ml) was added. The Mg powder was dissolved in 3 N HCl and extracted with ether and toluene. The pinacol $(p\text{-}CH_3C_6H_4)_2COHCOH(p\text{-}CH_3C_6H_4)_2$ (1.73478 g, 4.10528 mmol, 88% yield based on 4,4′-dimethylbenzophenone) crystallized as a white solid upon addition of hexanes to the concentrated ether/toluene extract.

References

1 Kahn, B.E.; Rieke, R.D. *Organometallics* 1988, 7, 463.
2 Hofer, L.J.E.; Peebles, W.C. *J. Am. Chem. Soc.* 1947, **69**, 893.
3 Brauer, G. *Handbook of Preparative Inorganic Chemistry*; 2nd ed.; Academic: New York, 1965; Vol. **2**.
4 Anderson, J.R. *Structure of Metallic Catalysts*; Academic: New York, 1975.
5 Whaley, T.P. *Inorg. Synth.* 1957, **5**, 195.
6 Whaley, T.P. *Inorg. Synth.* 1960, **6**, 47.
7 Smith, T.W.; Smith, S.D.; Badesha, S.S. *J. Am. Chem. Soc.* 1984, **106**, 7247.
8 Whaley, T.P. In *Handling and Uses of Alkali Metals*; Advances in Chemistry 19; American Chemical Society: Washington, DC, 1957; p 129.
9 Chu, L.; Friel, J.V. *J. Am. Chem. Soc.* 1955, **77**, 5838.
10 Scott, N.D.; Walker, J.F. U.S. Patent 2177 412, 1939; *Chem. Abstr.* **1940**, *34*, 974.
11 Gilliland, W.L.; Blanchard, A.A. *Inorg. Synth.* 1946, **2**, 234.
12 (a) Rodier, G.; Moreau, C. *Pap. Sect. Inorg. Chem. 16th Int. Congr. Pure Appl. Chem.* 1957; (b) *Chem. Abstr.* 1960, **54**, 20448c.
13 Davis, S.C.; Severson, S.J.; Klabunde, K.J. *J. Am. Chem. Soc.* 1981, **103**, 3024.
14 Scott, B.A.; Plecenik, R.M.; Cargill, G.S. III; McGuire, T.R.; Herd, S.R. *Inorg. Chem.* 1980, **19**, 1252.
15 Klabunde, K.J.; Murdock, T.O. *J. Org. Chem.* 1979, **44**, 3901.
16 Timms, P.L. In *Cyrochemistry*; Moskovits, M., Ozin, G.A., Eds.; John Wiley & Sons, Inc.: New York, 1976.
17 Skell, P.S.; McGlinchey, M.J. *Angew. Chem. Int. Ed. Engl.* 1975, **14**, 195.
18 Timms, P.L. *Angew. Chem. Int. Ed. Engl.* 1975, **14**, 273.

19 Klabunde, K.J. *Acc. Chem. Res.* 1975, **8**, 393.
20 Davis, S.C.; Klabunde, K.J. *Chem. Rev.* 1982, **82**, 153.
21 Chang, C.T. British Patent 1 502579, 1978; *Chem. Abstr.* **1978**, *89*, 154476j.
22 Chang, C.C.; Sung-Yu, N.K.; Hseu, C.S.; Chang, C.T. *Inorg. Chem.* 1979, **18**(3), 885.
23 Wang, L.C.; Lee, H.C.; Lee, T.S.; Lai, W. C.; Chang, C.T. *J. Inorg. Nucl. Chem.* 1978, **40**, 507.
24 Starks, D.F.; Streitwieser, A.J., Jr. *J. Am. Chem. Soc.* 1973, **95**(10), 3423.
25 Seaborg, G.T.; Katz, J. J. *The Actinide Elements*; McGraw-Hill: New York, 1954; pp 133–138.
26 *Gmelins Handbuch der Anorganischen Chemie*; Springer-Verlag: Berlin, 1936; Vol. **55**, pp 40–43.
27 Kanellakopulos, B.; Fischer, E.O.; Dornberger, E.; Baumgartner, F. *J. Organomet. Chem.* 1970, **24**, 507.
28 Rieke, R.D.; Hudnall P.M. *J. Am. Chem. Soc.* 1972, **94**, 7178.
29 Rieke, R.D. *Top. Curr. Chem.* 1975, **59**, 1.
30 Rieke, R.D. *Acc. Chem. Res.* 1977, **10**, 301.
31 Rieke, R.D.; Rhyne, L.D. *J. Org. Chem.* 1979, **44**, 3445.
32 Burkhardt, E.R.; Rieke, R.D. *J. Org. Chem.* 1985, **50**, 416.
33 Rockfort, G.L.; Rieke, R.D. *Inorg. Chem.* 1986, **25**, 348.
34 Rieke, R.D.; Burns, T.P.; Wehmeyer, R.M.; Kahn, B.E. In *High-Energy Processes in Organometallic Chemistry*; Suslick, K.S., Ed.; ACS Symposium Series 333; American Chemical Society: Washington, DC, 1987; pp 223–245.
35 The early work on active metals was done by using a solvent which refluxed at a temperature above the melting point of the alkali metal used (i.e., K in DME). Active Al, In, and Cr were prepared in hydrocarbon solvents by only this method. We prefer lower-temperature reductions and soluble reducing agents.
36 Kahn, B.E.; Rieke, R.D. Presented at the 192nd National Meeting of the American Chemical Society, Anaheim, CA, September 1986.
37 McMurry, J.E. *J. Am. Chem. Soc.* 1974, **96**(14), 4708.
38 McMurry, J.E. *Acc. Chem. Res.* 1974, **7**(9), 281.
39 McMurry, J.E. *J. Org. Chem.* 1977, **42**(15), 2655.
40 McMurry, J.E.; Fleming, M.P.; Kees, K.L.; Krepski, L.R. *J. Org. Chem.* 1978, **43**(17), 3255.
41 McMurry, J.E. *Acc. Chem. Res.* 1983, **16**, 405.
42 Tyrlik, S.; Wolochowicz, I. *Bull. Soc. Chim. Fr.* 1973, 2147.
43 Mukaiyama, T.; Sato, T.; Hanns, J. *Chem. Lett.* 1973, 1041.
44 Fujiwara, Y.; Ishikawa, R.; Akiyama, F.; Teranishi, S. *J. Org. Chem.* 1978, **43**(12), 2477.
45 Dams, R.; Malinowski, M.; Geise, H.J. *Bull. Soc. Chim. Belg.* 1982, **91**(2), 149.
46 Barin, I.; Knacke, O. *Thermochemical Properties of Inorganic Substances*; Springer-Verlag: New York, 1973.

47 Jeske, G.; Lauke, H.; Mauermann, H.; Swepston, P.N.; Schumann, H.; Marks, T.J. *J. Am. Chem. Soc.* 1985, **107**(26), 8091.
48 Evans, W.J.; Wayds, A.L.; Hunter, W.E.; Atwood, J.L. *J. Chem. Soc., Chem. Commun.* 1981, 706.
49 Evans, W.J.; Grate, J.W.; Hughes, L.A.; Zhang, H.; Atwood, J.L. *J. Am. Chem. Soc.* 1985, **107**(12), 3728.
50 Namy, J.L.; Souppe, J.; Kagan, H.B. *Tetrahedron Lett.* 1983, **24**(8), 765.
51 Manriquez, J.M.; Fagan, P.F.; Marks, T.J.; Day, C.S.; Day, V.W. *J. Am. Chem. Soc.* 1978, **100**(22), 7112.
52 Fagan, P.J.; Manriquez, J.M.; Marks, T.J.; Day, V.W.; Vollmer, S.H.; Day, C.S. *J. Am. Chem. Soc.* 1980, **102**, 5393.
53 Fagan, P.J.; Moloy, K.G.; Marks, T.J. *J. Am. Chem. Soc.* 1981, **103**(23), 6959.
54 Katahira, D.A.; Moloy, K.G.; Marks, T.J. *Organometallics* 1982, **1**(12), 1723.
55 Moloy, K.G.; Marks, T.J.; Day, V.W. *J. Am. Chem. Soc.* 1983, **105**(17), 5696.
56 Moloy, K.G.; Marks, T.J. *J. Am. Chem. Soc.* 1984, **106**(23), 7051.
57 Moloy, K.G.; Fagan, P.J.; Manriquez, J.M.; Marks, T.J. *J. Am. Chem. Soc.* 1986, **108**(1), 56.
58 Tatsumi, K.; Nakamura, A.; Hofmannn, P.; Hoffmann, R.; Moloy, K.G.; Marks, T.J. *J. Am. Chem. Soc.* 1986, **108**(15), 4467.
59 Cramer, R.E.; Higa, K.T.; Pruskin, S.L.; Gilje, J.W. *J. Am. Chem. Soc.* 1983, **105**, 6749.
60 Grant, C.; Streitwieser, A.J., Jr. *J. Am. Chem. Soc.* 1978, **100**, 2433.
61 Rhyne, L.D. Ph.D. thesis, The University of North Carolina, Chapel Hill, NC, 1980.
62 Wen, T.C.; Chang, C.C.; Chuang, Y.D.; Chiu, J.P.; Chang, C.T. *J. Am. Chem. Soc.* 1981, **103**(15), 4576.
63 Richards, D.H. *Chem. Soc. Rev.* 1977, **6**(2), 235.
64 Vandenberg, E.J. *J. Polym. Sci., Polym. Chem. Ed.* 1972, **10**(10), 2887.
65 Britnell, D.; Drew, M.G.B.; Fowles, G.W.A.; Rice, D.A. *J. Chem. Soc., Chem. Commun.* 1972, 462.
66 Britnell, D.; Drew, M.G.B.; Fowles, G.W.A.; Rice, D.A. *Inorg. Nucl. Chem. Lett.* 1973, **9**(4), 415.
67 Fowles, G.W.A.; Rice, D.A.; Shanton, K.J. *J. Chem. Soc., Dalton Trans.* 1978, 1658.
68 Daroda, R.J.; Blackborow, J.R.; Wilkinson, G. *J. Chem. Soc., Chem. Commun.* 1980, 1098.
69 A reviewer has suggested that following the first C—O insertion, a second reduction step could occur generating another U(II). Direct β-H elimination or C—O insertion followed by loss of ethylene from this U(II) species would generate another U(II) species. We feel that this step is unlikely, rather, the U(II) species generated from the second reduction would likely insert into another DME molecule. The U(IV) species so generated could subsequently produce ethylene and methyl vinyl ether by the pathways indicated.

70 Kalina, D.G.; Marks, T.J.; Wachter, W.A. *J. Am. Chem. Soc.* 1977, **99**(11), 3877.
71 Maata, E.A.; Marks, T.J. *J. Am. Chem. Soc.* 1981, **103**, 3576.
72 Fagan, P.J.; Manriquez, J.M.; Maatta, E.A.; Seyam, A.M.; Marks, T.J. *J. Am. Chem. Soc.* 1981, **103**(22), 6650.
73 Bruno, J.W.; Kalina, D.G.; Mintz, E.A.; Marks, T.J. *J. Am. Chem. Soc.* 1982, **104**(7), 1860.
74 Bruno, J.W.; Marks, T.J.; Morss, L.R. *J. Am. Chem. Soc.* 1983, **105**(23), 6824.
75 Jeske, G.; Lauke, H.; Mauermann, H.; Schumann, H.; Marks, T.J. *J. Am. Chem. Soc.* 1985, **107**(26), 8111.
76 Bruno, J.W.; Stecher, H.A.; Morss, L.R.; Sonnenberger, D.C.; Marks, T.J. *J. Am. Chem. Soc.* 1986, **108**(23), 7275.
77 It has been previously reported by Rieke et al. that active uranium prepared in DME by a similar method reacts cleanly with allyl iodide and benzophenone [31]. This active uranium, however, was prepared at room temperature under conditions that were not as strictly anaerobic as those in the present report. It is likely that these conditions may have resulted in passivation of the active uranium powder and possible reduction of the DME decomposition reaction. Most importantly, the products of DME decomposition are primarily gaseous. The gaseous products were not analyzed in the previous communication [31].
78 Fujita, T.; Watanaba, S.; Suga, K.; Sugahara, K.; Tsuchimoto, K. *Chem. & Ind.* 1983, **4**, 167.
79 Brooks, J.J.; Rhine, W.; Stucky, G.D. *J. Am. Chem. Soc.* 1972, **94**, 7346.
80 Huynh Ba Gia; Jerome, R.; Teyssie, P. *J. Organomet. Chem.* 1980, **190**, 107.
81 Sehgal, C.; Steer, R.P.; Sutherland, R.G.; Verrall, R.E. *J. Chem. Phys.* 1979, **70**, 2242.
82 Dams, R.; Malinowski, M.; Westdorp, I.; Geise, H.Y. *J. Org. Chem.* 1982, **47**(2), 248.
83 Dams, R.; Malinowski, M.; Geise, H.J. *Bull. Soc. Chim. Belg.* 1982, **91**(4), 311.
84 Chum, P.W.; Wilson, S.E. *Tetrahedron Lett.* 1976, **17**, 15.
85 Ashby, E.C.; Linn, J.J. *J. Org. Chem.* 1978, **43**(13), 2567.
86 Chen, T.C.; Chan, T.H.; Shaver, A.J. *Organomet. Chem.* 1984, **268**, C1.
87 The GC conditions employed permitted separation of 4,4′-dimethybenzophenone from 4,4′-dimethylbenzhydrol, but benzophenone and benzhydrol coeluted.
88 Bauer, D.P.; Macomber, R.S. *J. Org. Chem.* 1976, **41**(15), 2640.
89 Manriquez, J.M.; Fagan, P.J.; Marks, T.J. *J. Am. Chem. Soc.* 1978, **100**(12), 3939.
90 Duttera, M.R.; Fagan, P.J.; Marks, T.J.; Day, V.W. *J. Am. Chem. Soc.* 1982, **104**(3), 865.
91 Duttera, M.R.; Day, V.W.; Marks, T.J. *J. Am. Chem. Soc.* 1984, **106**(10), 2907.
92 He, M.Y.; Xiong, G.; Toscano, P.J.; Burwell, J.L., Jr.; Marks, T.J. *J. Am. Chem. Soc.* 1985, **107**(3), 641.
93 Toscano, P.J.; Marks, T.J. *J. Am. Chem. Soc.* 1985, **107**(3), 653.
94 Gomberg, M.; Bachmann, W.E. *J. Am. Chem. Soc.* 1927, **49**, 236.

14

Aluminum

14.1 Preparation of Highly Reactive Aluminum and Reaction with Aryl Halides

Alum was used by the Greeks and Romans in medicines and in dying processes. In 1807, Davey proposed the name alumine for the as yet undiscovered metal in alum. Finally, in 1825 Hans Christian Oersted isolated an impure sample of aluminum by heating $AlCl_3$ with potassium amalgam and boiling off the mercury at the end of the reduction.

Little effort was expended to examine the reaction of aluminum with organic substituents until 1940, when Grosse and Mavity [1] found that $(C_6H_5)_3Al_2I_3$ could be prepared from aluminum and iodobenzene but similar reactions with chloro- or bromobenzene failed to react. Recently Wittenberg [2] attempted to increase the reactivity of aluminum metal by grinding with $AlCl_3$, $AlBr_3$, and NaCl. Other attempts to activate the metal have involved aluminum–magnesium alloys [3]. We reported a new procedure for the preparation of aluminum metal in a highly reactive state [4]. We reported a convenient and economical synthesis of phenylaluminum halides in high yield by the reaction of the activated aluminum and aryl halides at moderate temperatures and in organic solvents.

The highly active aluminum metal powder has been prepared by reducing anhydrous aluminum halides in organic solvents, such as THF, xylene, and triethylamine, under a nitrogen atmosphere. The most convenient combinations appear to be $AlCl_3$-K-THF and $AlCl_3$-K-xylene or $AlCl_3$-Na-xylene. The reactivity of the aluminum metal generated by this method has been investigated by the following experiments. Activated aluminum was allowed to react, respectively, with iodobenzene and bromobenzene in xylene under reflux with the ratio Al^*/C_6H_5X of 2 : 1; the reaction reached 100% after 5 min in both cases. In contrast, Grosse [1] prepared $(C_6H_5)_3Al_2I_3$ from aluminum shavings and neat iodobenzene at 100°C for 44 h and only obtained an 84% yield. Wittenberg's activated aluminum [2] reacted with bromobenzene in

Chemical Synthesis Using Highly Reactive Metals, First Edition. Reuben D. Rieke.

Table 14.1 Reactions of aryl halides with activated aluminum.

Expt. no.	ϕX	Activated Al from $AlCl_3$ + M	Condition to generate Al* time, h (temp, °C)	Al*/ϕX	Reaction temp., °C	Solvent	Reaction time	Yield (%)[a]
1	I	K	2 (66)	2	66	THF	30 min	86
2	I	K	2 (89)	2	89	Et_3N	60 min	76
3	I	K	3 (25)	2	110	Xylene	5 min	100
4	Br	K	10.5 (25) 1 (140)	2	140	Xylene	5 min	100
5	Br	K	3 (25)	2	110	Xylene	20 min	92
6	Br	Na	3 (25)	2	110	Xylene	20 min	86
7	Cl	K	1 (25)	4	140	Xylene	2 h	64

[a] Analyses were carried out with HP model 5750 research gas chromatograph equipped with 6 ft × 1/8 in. stainless steel column packed with 10% UCon-W 98.

xylene at 140°C for 5 h gave a 79% yield. Thus the reactivity of Rieke aluminum metal appears to be vastly superior to that of any of the methods described in the preceding text and allows reactions of this type to be carried out in solution at much lower temperatures:

$$AlCl_3 + 3M \rightarrow Al^*(\text{activated}) + 3MCl(M = K, Na)$$
$$3C_6H_5X + 2Al^* \rightarrow (C_6H_5)_2\,AlX + C_6H_5AlX_2$$

The reactions of chloro-, bromo-, and iodobenzene with activated aluminum in organic solvents have been studied, and the results are summarized in Table 14.1.

In conclusion, a highly reactive form of aluminum has been prepared. However, very little chemistry has been explored. It is apparent that many new and exciting reactions remain to be discovered with this new metal.

Experimental

The following procedure for carrying out these reactions is representative of the general technique. $AlCl_3$ (Baker 99.5%, 22 mmol) in 10% excess was placed in a 200 ml round-bottomed flask with a side arm equipped with a septum cap in a dry box, and then freshly distilled solvent (75 ml) was added followed by adding freshly cut potassium (2.34 g, 60 mmol) or Na (1.38 g, 60 mmol) into the flask under nitrogen. The mixture was heated to melt the alkali metal and then cooled down to room temperature and stirred either with a magnetic stirrer or high-speed stirrer (up to 20 000 rpm). The black, finely divided aluminum metal was generated within 1–3 h, and then 10 mmol of aryl halide and an internal standard was added. Aliquots were periodically quenched in 10% HCl solution, and the organic layer was analyzed by GLC. The yield was determined by measuring the appearance of benzene and disappearance of aryl halide.

References

1 Grosse, A.V.; Mavity, J.M. *J. Org. Chem.* 1940, **5**, 106.
2 Wittenberg, D. *Justus Liebigs Ann. Chem.* 1962, **654**, 23.
3 Mardykin, V.P.; Gaponik, P.N. *Izv. Vyssh. Uchebn. Zaved., Khim. Khim. Tekhnol.* 1969, **12**(7), 963. *Chem. Abstr.* **1970**, 31888m.
4 Rieke, R.D.; Chao, L.-C. *Synth. React. Inorg. Met.-Org. Chem.* 1974, **4**, 101.

15 Cobalt

15.1 Two Methods for Preparing Rieke Cobalt: Reaction with CO and Also Fischer–Tropsch Chemistry

Cobalt was discovered by George Brandt in 1739. Little work on its reaction with organic substrates appeared until well into the twentieth century. Our work on oxidative addition chemistry first appeared in 1980. Cobalt represents an interesting contrast to the many activated metal powders generated by reduction of metal salts. As will be seen, the cobalt powders are highly reactive with regard to several different types of reactions. However, in contrast to the vast majority of metals studied to date, they show limited reactivity toward oxidative addition with carbon–halogen bonds.

Two general approaches have been used to prepare the cobalt powders. The first method [1–3] used 2.3 equiv of lithium along with naphthalene as an electron carrier in DME to reduce anhydrous cobalt chloride to a dark gray powder, **1**. Use of cobalt bromides or iodides gave a somewhat less reactive form of **1**. Slurries of **1** were very reactive toward strongly electrophilic aryl halides, such as C_6F_5X (X = Br, I), yielding the solvated species $Co(C_6F_5)_2$ and CoX_2. More reactive cobalt, **2**, was prepared by dissolving lithium in DME containing excess naphthalene [4]. Excess naphthalene was used to insure rapid and complete dissolution of all the lithium. Addition of cobalt chloride or iodide to cold glyme solutions of lithium naphthalide resulted in rapid formation of black–gray cobalt metal, **2**. Addition of dry cobalt halides or their suspensions in DME gave a product with similar properties and reactivity.

In contrast to **1**, which rapidly settled to give a clear solution, **2** remained suspended in an opaque black solution. Centrifugation of a suspension of **2** gave a black solution and sediment. Washing the sediment with fresh DME produced a black solution even after several repetitions. Washing the sediment with less polar solvents, such as diethyl ether or light alkanes, gave clear or slightly cloudy colorless solutions. Switching back to DME regenerated the

Chemical Synthesis Using Highly Reactive Metals, First Edition. Reuben D. Rieke.

black solutions. This suggests that an intimate mixture of cobalt microparticles and lithium chloride exists. As fresh glyme is added, part of the lithium chloride matrix is dissolved, liberating more cobalt particles which color the suspension.

When dry, **2** was pyrophoric and almost completely nonferromagnetic, that is, very little was attracted to a strong bar magnet. This was in marked contrast to **1** or commercial samples of 325 mesh cobalt powders, which were not pyrophoric and were strongly attracted to a magnet held in their vicinity. Qualitatively, the magnetic properties of **2** were suggestive of superparamagnetism. Debye–Scherrer photographs from samples of **2** showed no, or at best weak and diffuse, lines which could not be assigned to any common modification of metallic cobalt. This suggested that the sizes of the cobalt crystallites were less than approximately 30Å. Surface analyses of the cobalt powders indicate that they are composed mainly of metallic cobalt along with the alkali salt as well as considerable carbonaceous matter. Some cobalt oxides or possibly hydroxides were also observed; however, it is highly likely that the majority of the oxide accumulation occurred on sample preparation.

The cobalt powders demonstrated high reactivity with CO. Only a very limited number of transition metals have been demonstrated to react directly with carbon monoxide (under mild conditions) to give reasonable quantities of metal carbonyl complexes. Nickel and iron are two prime examples. It is interesting to note that the first such report for chromium involved finely divided chromium prepared by the reduction of $CrCl_3{\cdot}3THF$ with potassium in benzene [5]. Similarly the cobalt powders could be used to prepare $Co_2(CO)_8$ in good yields under mild conditions. A slurry of **2** in hexanes reacted with CO at 1000–1400 psi at 80–110°C to give cobalt carbonyl in yields up to 79% [3]. It is interesting to note that when CO was added to dry **2** in a Parr bomb, it catalyzed the disproportionation of CO into CO_2 and carbon along with the evolution of considerable heat [4]. Cobalt is considered to bind CO in a nondissociative manner at room temperature; however, at higher temperatures the adsorption becomes dissociative [6]. Adsorption of CO on polycrystalline cobalt films is an exothermic process and is dependent on coverage and the type of binding site [7]. It would appear that due to the high surface area of the active cobalt powders along with the substantial CO pressures (initial pressure of 1130 psi), enough heat was released to initiate the disproportionation reaction.

Cobalt has been used as a Fischer–Tropsch catalyst in a variety of forms [8]. Thus it was not surprising to see that both active forms of cobalt powders were moderate Fischer–Tropsch catalysts. Reacting synthesis gas with **2** in batch reactor conditions at elevated pressure and temperatures generated methane as the primary product. The life spans of the catalyst and to a lesser extent the products were affected by whether a support was used or how the cobalt was deposited on the support. Catalytic activity was not especially high and amounted to 4–7 mol of methane/mol of cobalt.

The highly oxophilic nature of the cobalt powder was readily demonstrated by its reaction with nitrobenzene at room temperature. Reductive coupling was quickly carried out by **2** to give azo and azoxy derivatives. Nitrobenzene reacted with **2** to give azobenzene in yields up to 37%. In some cases small amounts of azoxybenzene were also formed. With 1,4-diiodonitrobenzene, **2** reacted to give low yields of 4,4′-diiodoazoxybenzene and 4,4′-diiodoazobenzene.

In marked contrast to the majority of activated metals prepared by the reduction process, cobalt showed limited reactivity toward oxidative addition with carbon–halogen bonds. Iodopentafluorobenzene reacted with **2** to give the solvated oxidative addition products CoI_2 and $Co(C_6F_5)_2$ or $Co(C_6F_5)I$. The compound $Co(C_6F_5)_2 2PEt_3$ was isolated in 54% yield by addition of triethylphosphine to the solvated materials. This compound was also prepared in comparable yield from **1** by a similar process.

From the reaction of **2** with iodobenzene at reflux, a low yield of biphenyl was obtained, while much of the aryl halide remained unchanged. Similarly, **1** showed little reactivity toward iodobenzene.

Reactions of **2** with alkyl halides were generally more successful for C—C bond formation. For example, bibenzyl was formed in good yield from the reaction of **2** with benzyl bromide. Dichlorodiphenylmethane and **1** reacted to give tetraphenylethylene in 63% yield. Similarly, diiodomethane reacted with **1** to give ethylene.

Preparation of Cobalt Powder (1)

The following describes a typical preparation of cobalt powder **1**. In a dry box, lithium (0.2054 g, 29.6 mmol), naphthalene (0.4344 g, 3.4 mmol), and cobalt chloride (1.7683 g, 13.6 mmol) were charged into a flask equipped with a Teflon-clad stir bar and a vacuum adapter. On the vacuum line, 25 ml of freshly distilled glyme was added to the flask, and the mixture was stirred for 14 h at room temperature. If desired, most of the naphthalene could then be removed from the mixture by allowing the dark gray product to settle out, decanting the supernatant, and repeating the process with fresh solvent.

Preparation of Cobalt Powder (2)

The following describes a typical preparation of **2**. In the dry box, lithium (0.3342 g, 48.2 mmol) and naphthalene (8.0600 g, 62.9 mmol) were placed in a flask equipped with a Teflon-clad stir bar and vacuum adapter. On the vacuum line, 70 ml of freshly distilled glyme was added, and the mixture stirred vigorously until all of the lithium had dissolved. The resulting black–green solution was cooled to −78°C, and a slurry of cobalt chloride (3.0673 g, 23.6 mmol) in 30 ml of glyme was added. This mixture was stirred for 0.5 h at −78°C and an additional 3 h at room temperature. The product obtained settled out of the opaque black supernatant very slowly. If desired, the powder may be separated by centrifugation. Washing the powder with fresh glyme always gave black supernatants, whereas washing with diethyl ether or hexanes gave slightly cloudy supernatants. Typical elemental analyses were as follows: Anal. Found: C, 7.09; H, 0.89; Cl, 36.93; Co, 40.30; Li, 9.42.

Preparation of $Co_2(CO)_8$ from Activated Cobalt

Method 1

In a separate preparation, 0.441 g (63.54 mmol) of lithium, 6.881 g (53.69 mmol) of naphthalene, and 4.018 g (30.95 mmol) of cobalt chloride were placed in a 100 ml flask, and 50 ml of diglyme was added to the mixture at −35°C. After stirring overnight at that temperature, there was little evidence of reaction. It was necessary to stir the mixture at −20°C for an additional 72 h. When the gray–black slurry was then transferred to a centrifuge tube as previously described (vide supra), a scrap of unreacted lithium (0.060 g, 8.62 mmol) was recovered from the reaction flask.

The cobalt powder was washed by centrifugation with six 30 ml portions of hexanes and loaded into the stirred bomb with 150 ml of hexanes. The bomb was pressurized with carbon monoxide to 95 atm and allowed to stir overnight. In this experiment, the pressure drop with time was not monitored as closely as in the following experiment. After 8 h the theoretical pressure drop was observed. Heating to 100°C overnight and cooling to room temperature failed to cause an appreciable change in the pressure of the remaining carbon monoxide. The bomb was then vented, and the contents were filtered to give a clear brown solution from which 3.856 g of $Co_2(CO)_8$ was isolated by concentration of the solution in vacuo. The yield, based on cobalt chloride, was therefore 73%.

Method 2

In a 100 ml round-bottom flask equipped with an argon inlet and a Teflon-clad stir bar were placed 0.691 g (108.0 mmol) of lithium and 14.085 g (109.9 mmol) of naphthalene. Attached to this flask via one of its side arms was a bent glass tube terminating in a 50 ml flask containing 6.754 g (52.64 mmol) of cobalt chloride. Glyme (70ml) was added to the lithium–naphthalene mixture, and the dark green solution allowed to stir overnight at −22°C to ensure complete dissolution of all the lithium. The solution was then chilled to −50°C, and the cobalt chloride added by tipping the apparatus and rotating the sidearm tube.

The solution thickened immediately and had to be warmed to −15°C to allow stirring. After stirring overnight at −15°C, the gray–black mixture was warmed to room temperature where it became sufficiently mobile to be transferred via an 18 gauge cannula to a prepurged centrifuge tube capped with a rubber septum. The solids were then separated from an opaque brown solution by centrifugation at 1200–1800 rpm. The solids were washed by centrifugation with three 25 ml portions of fresh glyme and six 25 ml portions of hexanes before being transferred to the stirred Parr bomb with 200 ml of hexanes. The hexane washing was found to be necessary to remove the ethereal solvent, which was found to be detrimental to the high-yield preparation of $Co_2(CO)_8$. The washing also serves to remove most of the naphthalene, simplifying the isolation of the pure metal carbonyl. The bomb was pressurized to 95 atm with carbon monoxide and allowed to stir (400 rpm) at room temperature (ca. 18°C), while the pressure drop versus time was monitored. After 2 h the pressure had dropped to approximately 90% of the theoretical amount and had ceased to drop as precipitously as in the early stages of the reaction. The reaction mixture was heated to 100°C, and a rapid decrease in carbon monoxide pressure was observed over a period of about 14 h, after which time the pressure continued to drop but at a much slower rate. The drop was monitored for 7 days and found to be linearly decreasing with time, suggesting slow leakage around the stirrer shaft. The reaction mixture was now cooled and vented, and the soluble components isolated from a gray residue by filtration. The solution was concentrated to a small volume in vacuo, and 7.093 g of reddish brown crystals were isolated from the mother liquor by filtration. These crystals were identified as $Co_2(CO)_8$ (79%) by comparison of their infrared solution spectrum to that reported in the literature: IR (literature values shown in parentheses) 2076 s (2075), 2065 wsh (2064) 2048 versus (2047), 2034 msh (2035), 2028 s (2028), 1869 m (1867), and 1860 m (1858) cm^{-1}. A small amount of $Co_4(CO)_{12}$ is apparently present in this sample as revealed by an extra peak at 2060 cm^{-1}. Other bands expected for this species are either too weak to observe in this sample or are obscured by those of $Co_2(CO)_8$.

In separate experiments using similar quantities of the same starting materials and the same procedures, a 17% yield of $Co_2(CO)_8$ was obtained after 1.7 h, 95 atm, and 20°C and a 47% yield of $Co_2(CO)_8$ after 16 h at 95 atm and 107°C.

Reaction of Co with Synthesis Gas

Method 1

A sample of **2** was prepared by adding $CoCl_2$ (3.0673 g, 23.6 mmol; 30 ml of glyme) to a solution of lithium naphthalide (0.3342 g, 48.2 mmol of lithium; 8.0600 g, 62.9 mmol of naphthalene; 70 ml of glyme) at −78°C. From the well-stirred metal slurry was removed a 15 ml sample (containing ca. 5.1 mmol of Co), and this was transferred to a 125 ml bomb equipped with a borosilicate glass liner and a Teflon-clad magnetic stir bar. The gas lines connected to the bomb were flushed with synthesis gas (3 : 1 H_2—CO), and the bomb was filled to 790 psi with the gas. At 295 K, the pressure decreased rapidly to 758 psi and remained constant for the succeeding 106 min. The bomb was now heated in a sand bath while the pressure and temperature were monitored. At about 477 K, the pressure began to decline smoothly, ultimately stabilizing at 460 psi at

483 K after 1215 min. The apparatus was cooled to room temperature, and the headspace gases and liquid phase analyzed by GC, giving the results shown in Table 15.1. Gas-phase products were identified by comparison of their retention time with authentic samples. Liquid-phase products were identified by retention time comparison and GC–MS. The recovered liquid phase was pale blue–green with most of the metal collected in a sticky mass.

Method 2

A 15 ml sample of the slurry described in Method 1 was treated with 1060 psi of synthesis gas (2.4 : 1 H_2—CO) at 296 K. The pressure decreased to 985 psi within 0.5 h and remained constant over the succeeding 0.5 h. The reactor was heated to about 470 K and kept at that temperature for 427 min. A sample of gas for GC and GC–MS analyses was removed at this time, giving the results listed in Table 15.1. GC–MS indicated that small amounts of formaldehyde, acetaldehyde, propanal, and acetic acid were also formed.

Hydrolysis of Active Cobalt

A sample of **2** was prepared from 0.3727 g (53.7 mmol) of lithium, 8.9290 g (69.7 mmol) of naphthalene, and 3.3438 g (25.8 mmol) of cobalt chloride. The dark gray powder was separated by centrifugation, washed with four 40 ml portions of distilled diethyl ether, and dried overnight in vacuo at room temperature to give 3.3350 g of loose free-flowing solid. Hydrolysis of about 1 g samples of this material with concentrated hydrochloric acid produced 8.5 mmol of gas/g of material. If it is assumed that the hydrolyzed samples were representative of the whole, this would be 17% more than the expected amount of gas if Co was the only reducing agent present; that is, $2H^+ + Co \rightarrow H_2 + Co^{2+}$. Hydrolysis of commercial 325 mesh cobalt was found to give 2–5% less gas than that calculated from this equation. GC analysis of the hydrocarbon portion of the evolved gas from one of the hydrolysis experiments showed C_1—C_3 hydrocarbons, which were identified by retention time and GC–MS (see Table 15.2, run 1). The intensely blue acid solutions were clear or only slightly turbid when decomposition was complete.

In another experiment, 1.0108 g of cobalt powder from the same preparation was treated with 825 psi of H_2 for 12 min. By this time the pressure decreased to 810 psi. The H_2 was vented and 250 psi of CO and 850 psi of H_2 were added sequentially. Within 3 min the pressure had decreased, stabilizing at 1060 psi. The bomb was heated to 413 K for 7 days. A plot of pressure versus temperature during the heating and cooling period, and monitoring of the pressure at 413 K, indicated that little or no gas-consuming reactions had occurred. GC analysis of the bomb headspace gases showed no evidence for the formation of significant amounts of hydrocarbons. The recovered cobalt (0.960 g) was gummy, and the interior of the bomb was coated with sublimed naphthalene.

Table 15.1 Production from the reaction of synthesis gas over cobalt powders.

Method	Catalyst	Pressure, psi	H_2:CO	Analyzed phase	% CH_4	% C_2H_4	% C_2H_6	% C_3H_6	% C_3H_8	% CH_3OH	% C_4[a]	% CH_3COOCH_3
1	2	790	3:1	Gas	88		5		5		2	
1	2	790	3:1	Liquid	1		3		10	3	17	66
2	2	1060	24:1	Gas	51	3	14	12	11	tr[b]	9	[c]
3	Co/Al_2O_3	900	3.1:1	Gas	56		3		6	tr[a]	35	
4	$2/Al_2O_3$	830	3.1:1	Liquid						82[d]		

[a] C_4 hydrocarbons.
[b] tr = trace.
[c] Product obscured by internal standard.
[d] Balance ethanol.

Table 15.2 Weight percent composition of the hydrocarbons formed in the hydrolysis of cobalt powders.[a]

	Run 1[b]	Run 2[c]	Run 3[b]	Run 4[d]
CH_4	50	24	50	46
C_2H_4	10	9	6	5
C_2H_6	19	8	10	9
C_3H_6	10	12	8	4
C_3H_8	10	12	5	7
C_4H_8		8	3	3
C_4H_{10}		8	3	6
C_5H_{10}		13		
C_5H_{12}		7	15	21

[a] In some cases, totals do not add to 100 due to a rounding off.
[b] **2**.
[c] **2** after treatment with H_2–CO, 1060 psi, 140°C.
[d] **2** after CO, 1130 psi, 87°C.

The recovered material was washed with pentane, leaving a free-flowing dark gray solid that was very strongly ferromagnetic.

Hydrolysis of a sample of the gummy material with concentrated hydrochloric acid gave a variety of hydrocarbons (see Table 15.2, run 2).

Reaction of Dry Cobalt Powders with CO

A sample of **2** (0.7654 g, 110.0 mmol of lithium; 14.8517 g, 115.9 mmol of naphthalene; 6.8920 g, 53.1 mmol of cobalt chloride; 150 ml of glyme) was prepared, and the powder washed with four 50 ml portions of hexanes and diethyl ether. From these washings were obtained 13.98 g (ca. 94%) of the naphthalene and 7.294 g of a gray–black cobalt powder. The powder was not ferromagnetic; for example, none was attracted to a strong bar magnet, and only a few particles were oriented by the magnet: Anal. Found: C, 7.09; H, 0.89; Cl, 36.93; Co, 40.30; Li, 9.42.

A 3.6065 g sample of this powder was treated with CO (1130 psi) at room temperature. The pressure dropped immediately, reaching 700 psi in 5 min with a large exotherm that made the bomb intolerably hot to the touch. Visual examination of the sample after the bomb was vented showed no obvious change in the powder. The sample was then repressurized to 1000 psi at room temperature. After 10 min, the pressure stabilized at 955 psi with no obvious exotherm. The bomb was placed in an 87°C oil bath for 2 h and then cooled and

Table 15.3 Calculated lattice spacings found for active cobalt powders.

Sample	Lattice spacing, Å							
2	5.50	3.44	2.78	2.30	2.05	1.82	1.71	1.54
Carburized **2**	5.55	3.43	2.81	2.30				
Annealed **2**	2.05	1.26	1.07					

vented. The powder was washed with hexanes to remove a small amount of $Co_2(CO)_8$ and naphthalene and then dried in vacuo to a constant mass of 4.4249 g. The $Co_2(CO)_8$ was oxidized by allowing the solution to stand in air, and the solid residues were collected on ashless filter paper and ignited to constant mass (0.00968 g). With the oxide residues weighed as Co_3CO_4, it was calculated that less than 0.5% of the cobalt in the powder prior to treatment with CO was converted to $Co_2(CO)_8$. Anal. found for the pentane-washed powder after CO treatment: C, 33.32; H, 1.10; Cl, 30.90; Co, 25.51; Li, 5.93.

Hydrolysis with concentrated hydrochloric acid, of samples before and after treatment with CO, gave 9.0 and 8.0 mmol of gas/g, respectively. The hydrocarbons are shown in Table 15.2, runs 3 and 4.

From hydrolysis of the carburized samples was obtained a finely divided black material that represented 29.89% of the sample mass. Ignition of this material resulted in a 95–99% decrease in mass and was accomplished with the formation of a small quantity of a refractory material.

In similar experiments CO_2 was shown to be present in the gases vented from the reactor by precipitation of $CaCO_3$ from a saturated $Ca(OH)_2$ solution through which the gases were dispersed.

Debye–Scherrer photographs of cobalt from this preparation before CO treatment, after CO treatment, and after annealing at 300°C for 14 h under argon were obtained. Analysis of these films was inconclusive; the calculated interplanar distances being in poor agreement with known metallic oxidic or carbidic phases (see Table 15.3).

Reaction of Cobalt with Aromatic Nitro Compounds

Nitrobenzene

A slurry of **2** (0.5861 g, 84.4 mmol of lithium; 14.0901 g, 109.9 mmol of naphthalene; 4.7055 g, 36.7 mmol of $CoCl_2$) was prepared, and the supernatant liquid removed by centrifugation. The metal powder was returned to the reaction flask via cannula with use of 100 ml of fresh glyme. Nitrobenzene

(1.44 g, 11.7 mmol) was added at once to the well-stirred room temperature slurry, resulting in a surprisingly vigorous exotherm. After about 2 min a 2 ml sample of the mixture was quenched with five drops of water. Analysis of the quenched reaction mixture by GC and thin-layer chromatography revealed that no nitrobenzene remained and that azobenzene and other highly colored products had formed. The reaction mixture was washed with five 40 ml portions of glyme, which were combined and extracted with diethyl ether. The diethyl ether was washed with 3 M HCl. The diethyl ether layer was extracted once again with water, dried over $MgSO_4$, and evaporated to give 3.28 g of a brown oil. The oil was placed on a silica gel column, from which 0.39 g (37%) of azobenzene was eluted with hexane–chloroform. Recrystallization from ethanol gave pure azobenzene, identified by comparison of its infrared spectrum and mixed melting point with authentic samples. Other materials removed from the column included naphthalene and several highly colored oily components which were not identified.

The aqueous extracts were combined, made basic (pH 9), and reextracted with diethyl ether to give a small amount (0.25 g) of brown oil. TLC analysis showed that a multicomponent mixture had resulted. This was not investigated further.

In another experiment, performed similarly but with 0.52 mol of $PhNO_2$/mol of Co, a 21% yield of azobenzene was obtained with numerous unidentified products. At reflux **2** also reacted with $PhNO_2$ (1 mol of $PhNO_2$/mol of Co) to give azobenzene (17%) or at room temperature to give azobenzene (18%) and azoxybenzene (9%), identified by their infrared spectra. Commercial 325 mesh cobalt powder was treated with a few drops of concentrated HCl in glyme until the strong blue color of Co(II) was observed, indicating that the surface oxide coating had been breached. The cleaned metal was washed with several portions of fresh glyme and treated with a nitrobenzene. No reaction was observed at room temperature or at reflux. It is apparent that much research is yet to be done to fully explore the range of chemistry that these highly reactive cobalt metal powders are capable of carrying out.

1-Iodo-4-Nitrobenzene

A slurry of **2** (0.2292 g, 33.0 mmol of lithium; 6.3502 g, 49.5 mmol of naphthalene; 1.830 g, 14.1 mmol of $CoCl_2$) at −78°C was treated with a room temperature solution of 1-iodo-4-nitrobenzene (2.5853 g, 11.5 mmol, freeze–thaw degassed three times) in 20 ml of glyme. After 1 h the mixture was brought to reflux for 2 h. TLC examination of the products showed that no 1-iodo-4-nitrobenzene remained and that 4,4′-dinitrobiphenyl was not a materially significant product. The reaction mixture was divided into two equal portions, one of which was worked up from chloroform–water. From the crude product were isolated 4,4′-diiodoazobenzene (0.1315 g, 10%) and 4,4′-diiodoazoxybenzene (0.1585 g, 12%) by column chromatography, and the compounds were characterized by infrared spectroscopy, mixed melting point (azo compound only), and high-resolution mass spectroscopy. Other compounds were recovered from the column as impure, highly colored oils, which were not characterized.

1,2-Dinitrobenzene

Caution! 1,2-Dinitrobenzene is a high explosive that is known to detonate when subjected to shock or temperature above its melting point [9]. To reduce the potential hazards in this experiment, less reactive **1** was used. To a slurry of **1** (0.2054 g, 29.6 mmol of lithium; 0.4344 g, 3.4 mmol of naphthalene; 1.7683 g, 13.6 mmol of cobalt chloride) was added a solution of 0.4740 g (2.8 mmol) of 1,2-dinitrobenzene in 40 ml of glyme at 0°C over a period of about 3 h. The mixture was warmed to room temperature and stirred for 4 h. TLC analysis showed only naphthalene, 1,2-dinitrobenzene, and an unknown material with a strong blue fluorescence (also observed in the reaction of **2** with $PhNO_2$). The solvent was removed, and the metal slurry washed several times with fresh portions of glyme. Water-diethyl ether extractions of the combined portions of glyme eventually yielded 0.3476 g of crystals. These were shown by TLC to be a mixture of naphthalene and 1,2-dinitrobenzene.

Reaction of Cobalt with C_6F_5I: Preparation of $(C_6F_5)_2Co{\cdot}2PEt_3$

A slurry of **2** (0.2860 g, 41.2 mmol of lithium; 6.8679 g, 53.6 mmol of naphthalene; 2.2900 g, 17.8 mmol of cobalt chloride) was prepared, and the product centrifuged, washed with two 30 ml portions of diethyl ether, and slurried back into the reaction flask with 30 ml of glyme. Pentafluorophenyl iodide (5.38 g, 18 mmol) was added to the metal slurry at 0°C. After 1 h, the blue–green mixture was warmed to room temperature for 14 h. The mixture was cooled to 0°C, and 4.70 g (40 mmol) of triethylphosphine was added. It was stirred for 5 h at room temperature and then worked up as previously described [1, 2] to give 3.05 g (54%) of crystalline $(C_6F_5)_2Co{\cdot}2PEt_3$, whose physical and spectroscopic properties are consistent with those previously reported.

Reaction with Benzyl Bromide

A slurry of **2** (0.2812 g, 40.5 mmol of lithium; 6.7343 g, 52.5 mmol of naphthalene; 2.2540 g, 17.6 mmol of cobalt chloride) was prepared, and the product separated and washed with three 30 ml portions of diethyl ether. The metal powder in 30 ml of fresh glyme was treated at room temperature with 5.46 g (32 mmol) of benzyl bromide, causing a brisk exotherm after a brief induction period. After 14 h the products were quantitatively determined by GC as bibenzyl (63%) and toluene (6%) with use of a known quantity of *n*-nonane as an internal standard and application of response factor corrections.

Reaction with Phenyl Halides: Phenyl Iodide

Method 1

A slurry of **2** (0.2472 g, 35.6 mmol of lithium; 5.9647 g, 46.5 mmol of naphthalene; 1.9765 g, 15.4 mmol of cobalt chloride) was prepared, and the product washed once with glyme. It was then reacted in 25 ml of glyme with 4.5955 g (22.5 mmol) of phenyl iodide. Quenches were taken periodically by withdrawing 1 ml samples and treating the samples with two drops of 1 M HCl. The samples were then quantitatively analyzed by GC with use of *n*-dodecane as an internal standard and application of response factor corrections. After 1 min, 59% of the phenyl iodide remained. After 20 h (last 3 h at reflux), 61% remained. No biphenyl was observed until the reaction mixture was refluxed, after which an 11% yield of that compound was found.

Method 2

A slurry of **1** (0.3166 g, 45.6 mmol of lithium; 0.5889 g, 4.6 mmol of naphthalene; 2.5429 g, 19.8 mmol of cobalt chloride) was prepared, and the powder washed once with 25 ml glyme. It was then reacted in 30 ml of glyme with 6.3406 g (31.1 mmol) of phenyl iodide. Quenches were taken as described in Method 1 (vide supra) and showed that 64% of the phenyl iodide remained unchanged after 1 min at room temperature. After 26 h (3 h at reflux), 60% of the phenyl iodide remained. No biphenyl was observed at any time.

Preparation of Tetraphenylethylene

Method 1

A slurry of **1** (0.3156 g, 50.7 mmol of lithium; 1.3021 g, 10.2 mmol of naphthalene; 7.7919 g, 24.9 mmol of cobalt chloride) was prepared, and the powder washed with 160 ml of glyme in two portions. The slurry in 20 ml of glyme was reacted with 3.70 g (15 mmol) of dichlorodiphenylmethane, causing a brief mild exotherm. After 20 h, GC analysis revealed that only a trace of Ph_2CCl_2 remained in the green solution. The organic products were extracted from 3 M HCl with chloroform, washed with several portions of water, and dried over magnesium sulfate. After concentration in vacuo, recrystallization of the residues from benzene–ethanol gave 1.58 g (63%) of pale yellow crystalline tetraphenylethylene, identified by its mixed melting point with an authentic sample and high-resolution mass spectroscopy.

Method 2

A slurry of **1** (0.1853 g, 26.7 mmol of lithium; 0.2254 g, 1.8 mmol of naphthalene; 4.002 g, 12.8 mmol of cobalt iodide) was chilled to −78°C, and 1.05 g (13 mmol) of cyclohexene, and 1.48 g (6.2 mmol) of dichlorodiphenylmethane were added. After it was stirred for 48 h at room temperature, the mixture was refluxed for 3 h and worked up as described in Method 1 to give 0.53 g (51%) of tetraphenylethylene.

Method 3

A slurry of **1** (0.2154 g, 31.0 mmol of lithium; 0.1840 g, 1.4 mmol of naphthalene; and 4.5599 g, 14.6 mmol of cobalt iodide; 45 ml of THF) was prepared. The metal powder was isolated by filtration under argon and dried in vacuo overnight. The cobalt powder (3.678 g) was slurried in 20 ml of distilled hexanes and treated with 2.72 g (11 mmol) of dichlorodiphenylmethane at room temperature, causing the formation of a green solution. After the mixture was stirred for 48 h, the products were worked up as previously described to give 0.80 g (42%) of tetraphenylethylene.

Reaction of Cobalt with Diiodomethane

A slurry of **1** (1.9500 g, 6.2 mmol of cobalt iodide; 0.5009 g, 12.8 mmol of potassium; 0.7878 g, 6.2 mmol of naphthalene; 20 ml of THF) was treated with 5.2331 g (19.5 mmol) of diiodomethane and 0.0714 g of *n*-tridecane (GC standard). After 14 h ethylene was identified by mass spectroscopy in the gases above the reaction mixture. Analysis of the reaction mixture by GC showed that 57% of the diiodomethane remained unchanged.

References

1 Kavaliunas, A.V.; Rieke, R.D. *J. Am. Chem. Soc.* 1980, **102**, 5944.
2 Kavaliunas, A.V.; Taylor, A.; Rieke, R.D. *Organometallics* 1983, **2**, 377.
3 Rochfort, G.L.; Rieke, R.D. *Organometallics* 1984, **23**, 787.
4 Rochfort, G.L.; Rieke, R.D. *Inorg. Chem.* 1986, **25**, 348.
5 Rieke, R.D.; Ofele, K.; Fischer, E.O. *J. Organomet. Chem.* 1974, **76**, C19.
6 Bell, A. *Catal. Rev. Sci. Eng.* 1981, **23**, 203.
7 Toyoshima, I.; Somorjai, G.A. *Catal. Rev. Sci. Eng.* 1979, **19**, 105.
8 Young, R.S. *Cobalt: Its Chemistry, Metallurgy and Uses.* ACS Monograph 149; Reinhold Publishing Corp.: New York, 1960.
9 Sax, N.I. *Dangerous Properties of Industrial Materials*, 4th Ed.; Van Nestrand Reinhold: New York, 1975.

16

Chromium

16.1 Preparation of Highly Reactive Chromium Metal and Its Reaction with CO to Yield $Cr(CO)_6$

Chromium was discovered by Louis Nicolas Vauquelin in France in 1797. There was little use of chromium in organic chemistry until the mid-twentieth century. Much of this work was initiated by Prof. E.O. Fischer in Munich, Germany. His discovery of the preparation of arene ($Cr(CO_3)$) complexes lead to him sharing the Nobel Prize in Chemistry in 1973 with Prof. Geoffrey Wilkinson of the United Kingdom.

The formation of metal carbonyls by the direct reaction of the metal with CO at elevated temperatures and pressures is a well-established procedure for a wide variety of metals. However, for certain metals this procedure fails due to the low reactivity of the metal. Chromium is such a metal, and all attempts to react the metal with CO have failed to yield any $Cr(CO)_6$. We would like to report the preparation of chromium powder by the Rieke process and the direct reaction of the metal with CO to yield $Cr(CO)_6$ [1].

The standard procedures for the preparation of $Cr(CO)_6$ involve the chemical reduction of chromium(II) or chromium(III) compounds in an autoclave under CO pressure [2]. It also has been reported that the electrochemical reduction of chromium compounds under CO pressure yields $Cr(CO)_6$ [3]. We have found that chromium metal prepared by the reduction of $CrCl_3$, or better $CrCl_3{\cdot}3THF$, with potassium in either THF or benzene is highly active and will react with CO (280 atm) at elevated temperatures to yield $Cr(CO)_6$. It has been observed that the generation of active magnesium by the reduction of $MgCl_2$ in the presence of KI leads to a much more reactive magnesium [4]. We found in this work that reduction of the chromium salts in the presence of KI produces higher yields of $Cr(CO)_6$ (35%). The optimum ratio of KI to $CrCl_3{\cdot}3THF$ was 1:1.

Chemical Synthesis Using Highly Reactive Metals, First Edition. Reuben D. Rieke.

Reduction in the presence of KBr gave similar yields; however, reduction in the presence of KCl gave yields of only 30%.

Much improved yields were obtained if the reduction of $CrCl_3{\cdot}3THF$ was carried out in benzene rather than THF. In these cases, the benzene was stripped off after the reduction was complete, and then THF was added to the black powder. The resulting slurry was then added to the autoclave, and reaction with CO gave yields of $Cr(CO)_6$ of 50% or better. Reduction of $CrCl_3{\cdot}3THF$ in benzene in the absence of KI resulted in yields of $Cr(CO)_6$ of only 30%. Also use of KBr or KCl in the benzene reductions resulted in lower yields of $Cr(CO)_6$.

In order to determine if the black material added to the autoclave was in fact chromium metal or a mixture of Cr(II) salts and potassium, the following experiments were carried out. The black material obtained from the reduction of $CrCl_3{\cdot}3THF$ with potassium in benzene was repeatedly extracted with THF to remove and to determine soluble chromium(II) salts. Also the amount of unreacted potassium was determined by repeated extraction of the black material with alcohol and water and subsequent acidic titration of the extracts. From these two determinations, it was established that a minimum of 58% of the $Cr(CO)_6$ formed in the autoclave reaction originated from chromium(0). In a separate experiment, the black material was extracted repeatedly with THF to remove all the soluble chromium salts and was then treated with CO under the same conditions as in the previous reactions. The yield of $Cr(CO)_6$ was 30% based on the material placed in the autoclave. Thus, it is clear that Cr(0) is produced in the reduction, and this Cr(0) will react with CO to yield $Cr(CO)_6$. It is difficult to determine the exact amount of $Cr(CO)_6$ which originated from Cr(0) in the black powder due to its instability. Thus, excessive manipulations of the black powder even under argon or very long reduction times in the generation of the black powders seem to deactivate the Cr(0).

We had recently reported on the preparation of highly reactive magnesium. We found that the reduction of chromium salts with the highly reactive magnesium in an autoclave leads to high yields of $Cr(CO)_6$. The use of magnesium generated in the presence of KI to reduce $CrCl_3{\cdot}3\ THF$ in the autoclave resulted in yields in excess of 83% of $Cr(CO)_6$.

In conclusion, a highly reactive form of chromium metal has been prepared by the Rieke reduction approach. This highly reactive chromium will react with CO to produce $Cr(CO)_6$ in moderate yields. At this point, no additional chemistry of these highly reactive metal powders has been attempted. However, it is likely that much remains to be discovered.

This work was carried out with Prof. E.O. Fischer and Prof. Karl Öfele at the Technical University in Munich, Germany, in the fall of 1973. This chapter is dedicated to the life and many chemical contributions of Prof. E.O. Fischer.

Experimental

Preparation of $Cr(CO)_6$ from Chromium Powder

The following is the general procedure used in the preparation of chromium powder. A dry three-necked 100 ml round-bottomed flask equipped with a reflux condenser and a glass magnetic stirrer was filled with argon. To this flask were added K (1.60 g, 0.041 mol) freshly cut under hexane, KI (2.28 g, 0.0137 mol), $CrCl_3 \cdot 3THF$ (5.14 g, 0.0137 mol), and purified benzene (50 ml) which had been stored under argon. The mixture was heated to 80°C for a total of 2 h. Initially, the $CrCl_3 \cdot 3THF$ dissolved in the benzene, giving a deep purple solution. However, before the potassium melted, a heavy pink–purple precipitate came out of solution. Within a few minutes after the potassium melted, the solution began to turn brown–black and become less viscous. After heating for a total of 2 h, the benzene was stripped off at room temperature. THF (25 ml) was then added to the brown–black powder, and the resulting slurry was placed in autoclave filled with argon. The autoclave was filled with high-purity CO (280 atm) and heated at 220°C for a total of 16–20 h. The resulting black slurry was filtered, and the black solid was washed with THF (5 × 5 ml). The combined filtrates were diluted with water, causing the $Cr(CO)_6$ to precipitate out. The product was collected by centrifugation and filtration, yielding 1.384 g (51% yield) based on material placed in the autoclave. The $Cr(CO)_6$ obtained had the same melting point and IR spectrum as an authentic sample. Appreciable longer reaction times, higher reaction temperatures, and the use of low-purity CO led to decreased yields of $Cr(CO)_6$.

Preparation of $Cr(CO)_6$ by Reduction with Activated Magnesium

Preparation of activated magnesium by reduction of $MgCl_2$ (2.04 g, 0.0214 mol) with potassium (2.50 g, 0.0384 mol) in the presence of KI (3.55 g, 0.0214 mol) in THF (50 ml) was carried out as previously described [4]. The black slurry was placed in an autoclave filled with argon. $CrCl_3 \cdot 3THF$ (3.37 g, 0.009 mol) was then added, and finally the autoclave was filled with high-purity CO (280 atm). The autoclave was heated for 16–20 h at 220°C. The workup was the same as described in the preceding text, yielding 1.56 g of $Cr(CO)_6$ (83% yield).

Extraction of Chromium Powder

Potassium (1.51 g, 0.0386 mol), $CrCl_3 \cdot 3THF$ (4.83 g, 0.0129 mol), and KI (2.14 g, 0.0129 mol) were heated in benzene (50 ml) at 90°C for 2 h. The black material was filtered from the brown benzene solution, and the brown–black residue washed with benzene (3 × 10 ml) and then with THF (6 × 10 ml) until the THF was colorless. The combined benzene and THF extracts were evaporated, and the remaining brown crystals were dissolved in water, and the chromium content was determined as Cr_2O_3, yielding 256 mg of Cr. Thus 38.1% of Cr remained only partially reduced in the form of chromium(II) salts. The black material was then extracted with *n*-butanol (5 ml), methanol (5 ml), and finally water (200 ml). The water wash was carried out until the water was neutral. Titration of the alkaline solution required 87.4 ml of 0.1 N HCl. This corresponds to 341 mg of K or 22.7% of the original amount.

References

1 Rieke, R.D.; Ofele, K.; Fischer, E.O. *J. Organomet. Chem.* 1974, **76**, C19.
2 (a) Natta, G.; Ercoli, R.; Calderazzo, F.; Rabizzoni, A. *J. Am. Chem. Soc.* 1957, **79**, 3611. (b) Fischer, E.O.; Hafner, W.; Ofele, K. *Chem. Ber.* 1959, **92**, 3050. (c) Nesmeyanov, A.N.; Anisimov, K.N.; Volkov, V.L.; Fridenberg, A.E.; Mikheev, E.P.; Medvedeva, A.V.; *Zh. Neorg. Khim.* 1959, **4**, 1827. (d) Podall, H.E.; Dunn, J.H.; Shapiro, H. *J. Am. Chem. Soc.* 1960, **82**, 1325. (e) Podall, H.E.; Prestridge, H.B.; Shapiro, H. *J. Am. Chem. Soc.* 1961, **83**, 2057. (f) Shapiro, H.; Podall H.E. *J. Inorg. Nucl. Chem.* 1962, **24**, 925.
3 Guainazzi, M.; Silvestri, G.; Gambino, S.; Filardo, G. *J. Chem. Soc. Dalton Trans.* 1972, 927.
4 Rieke, R.D.; Bales, S.E. *J. Chem. Soc., Chem. Commun.* 1973, 879.

Index

Chemical Synthesis Using Highly Reactive Metals, First Edition. Reuben D. Rieke.